Modeling of Unit Operations in the Food Industry

Unit Operations and Processing Equipment in the Food Industry

Modeling of Unit Operations in the Food Industry

Edited by

Seid Mahdi Jafari

Department of Food Materials and Process Design Engineering, Faculty of Food Science and Technology, Gorgan University of Agricultural Sciences and Natural Resources, Gorgan, Iran

Nantawan Therdthai

Department of Product Development, Faculty of Agro-Industry, Kasetsart University, Bangkok, Thailand

ELSEVIER

WP
WOODHEAD PUBLISHING
An imprint of Elsevier

Woodhead Publishing is an imprint of Elsevier
200 River's Edge Drive, Suite 430, Medford, MA 02155, United States
125 London Wall, London EC2Y 5AS, United Kingdom

ISBN: 978-0-12-822180-8

For information on all Woodhead Publishing publications visit our website at https://www.elsevier.com/books-and-journals

Publisher: Mara Conner
Acquisitions Editor: Vaishali Singh
Editorial Project Manager: Hilary Carr
Production Project Manager: Vasantharani Raja
Cover Designer: Raman Kumar

Typeset by TNQ Tech

Dedication

To our families, students, and colleagues, for their enduring support, inspiration, and collaboration throughout this journey.

Contents

Contributors

Reza Abedi-Firoozjah Student Research Committee, Kermanshah University of Medical Sciences, Kermanshah, Iran; Department of Food Science and Technology, School of Nutrition Sciences and Food Technology, Kermanshah University of Medical Sciences, Kermanshah, Iran

Kannan Aravamudan Department of Chemical Engineering, Indian Institute of Technology Madras, Chennai, India

P. Barciela Instituto de Agroecoloxía e Alimentación (IAA), Universidade de Vigo, Nutrition and Food Group (NuFoG), Campus Auga, Ourense, Spain; Nutrition and Food Group (NuFoG), Galicia Sur Health Research Institute (IIS Galicia Sur), Vigo, Spain

Samira Beikzadeh Department of Nutrition and Food Sciences, Maragheh University of Medical Sciences, Maragheh, Iran

B. Calvo Centro de Investigación Biomédica en Red en Bioingeniería, Biomateriales y Nanomedicina (CIBER-BBN), Madrid, Spain

Osvaldo H. Campanella Department of Food Science and Technology, the Ohio State University, Columbus, OH, United States

M. Carpena Instituto de Agroecoloxía e Alimentación (IAA), Universidade de Vigo, Nutrition and Food Group (NuFoG), Campus Auga, Ourense, Spain; Nutrition and Food Group (NuFoG), Galicia Sur Health Research Institute (IIS Galicia Sur), Vigo, Spain

F. Chamorro Instituto de Agroecoloxía e Alimentación (IAA), Universidade de Vigo, Nutrition and Food Group (NuFoG), Campus Auga, Ourense, Spain; Nutrition and Food Group (NuFoG), Galicia Sur Health Research Institute (IIS Galicia Sur), Vigo, Spain

Christian O. Díaz-Ovalle Tecnológico Nacional de México/I.T. Roque, Departamento de Ingenierías, Celaya, Guanajuato, México

P. Donn Instituto de Agroecoloxía e Alimentación (IAA), Universidade de Vigo, Nutrition and Food Group (NuFoG), Campus Auga, Ourense, Spain; Nutrition and Food Group (NuFoG), Galicia Sur Health Research Institute (IIS Galicia Sur), Vigo, Spain

Ozlem Caglar Duvarci Department of Chemical Engineering, Facultly of Engineering, Izmir Institute of Technology, Izmir, Türkiye

J. Echave Instituto de Agroecoloxía e Alimentación (IAA), Universidade de Vigo, Nutrition and Food Group (NuFoG), Campus Auga, Ourense, Spain; CIMO, LA SusTEC, Instituto Politécnico de Bragança, Campus de Santa Apolonia, Bragança, Portugal; Nutrition and Food Group (NuFoG), Galicia Sur Health Research Institute (IIS Galicia Sur), Vigo, Spain

Farinaz Esmi Student Research Committee, Tabriz University of Medical Sciences, Tabriz, Iran; Department of Food Science and Technology, Faculty of Nutrition and Food Sciences, Tabriz University of Medical Sciences, Tabriz, Iran

M. Teresa Fernández-Felipe Instituto de Agroquímica y Tecnología de Alimentos (IATA-CSIC), Paterna, Valencia, Spain

A. Ferrer-Mairal Instituto Agroalimentario de Aragon IA2, Universidad de Zaragoza-CITA, Zaragoza, Spain

Sandro M. Goñi Centro de Investigación y Desarrollo en Ciencia y Tecnología de los Alimentos (CIDCA), CCT CONICET La Plata, UNLP, CIC, La Plata, Argentina; Facultad de Ingeniería, UNLP, La Plata, Argentina

J. Grasa Centro de Investigación Biomédica en Red en Bioingeniería, Biomateriales y Nanomedicina (CIBER-BBN), Madrid, Spain

P. Guíu Instituto Agroalimentario de Aragon IA2, Universidad de Zaragoza-CITA, Zaragoza, Spain

E. Hernández-Alhambra Aragón Institute of Engineering Research (i3A), Universidad de Zaragoza, Zaragoza, Spain

Seid Mahdi Jafari Department of Food Materials and Process Design Engineering, Faculty of Food Science and Technology, Gorgan University of Agricultural Sciences and Natural Resources, Gorgan, Iran

Kimon Andreas G. Karatzas Department of Food & Nutritional Sciences, University of Reading, Reading, United Kingdom

M.A. Martínez Centro de Investigación Biomédica en Red en Bioingeniería, Biomateriales y Nanomedicina (CIBER-BBN), Madrid, Spain

Poliana Mendes de Souza African Sustainable Agriculture Research Institute (ASARI), University Mohammed VI Polytechnic (UM6P), Laâyoune, Morocco

Kyuya Nakagawa Department of Chemical Engineering, Faculty of Engineering, Kyushu University, Fukuoka, Japan

Laura Otero Institute of Food Science, Technology and Nutrition (ICTAN-CSIC). MALTA Consolider Team. Madrid, Spain

A.G. Pereira Instituto de Agroecoloxía e Alimentación (IAA), Universidade de Vigo, Nutrition and Food Group (NuFoG), Campus Auga, Ourense, Spain; Nutrition and Food Group (NuFoG), Galicia Sur Health Research Institute (IIS Galicia Sur), Vigo, Spain

A. Perez-Vazquez Instituto de Agroecoloxía e Alimentación (IAA), Universidade de Vigo, Nutrition and Food Group (NuFoG), Campus Auga, Ourense, Spain; Nutrition and Food Group (NuFoG), Galicia Sur Health Research Institute (IIS Galicia Sur), Vigo, Spain

Arícia Possas Department of Food Science and Technology, UIC Zoonosis y Enfermedades Emergentes (ENZOEM), CeiA3, Universidad de Córdoba, Córdoba, Spain

M.A. Prieto Instituto de Agroecoloxía e Alimentación (IAA), Universidade de Vigo, Nutrition and Food Group (NuFoG), Campus Auga, Ourense, Spain; Nutrition and Food Group (NuFoG), Galicia Sur Health Research Institute (IIS Galicia Sur), Vigo, Spain

Emmanuel Purlis Universidad de Buenos Aires, Facultad de Ciencias Exactas y Naturales, Departamento de Industrias, Buenos Aires, Argentina; CONICET – Universidad de Buenos Aires, Instituto de Tecnología de Alimentos y Procesos Químicos (ITAPROQ), Buenos Aires, Argentina

Dolores Rodrigo Instituto de Agroquímica y Tecnología de Alimentos (IATA-CSIC), Paterna, Valencia, Spain

M.L. Salvador Instituto Agroalimentario de Aragon IA2, Universidad de Zaragoza-CITA, Zaragoza, Spain

S. Seyyedi-Mansour Instituto de Agroecoloxía e Alimentación (IAA), Universidade de Vigo, Nutrition and Food Group (NuFoG), Campus Auga, Ourense, Spain; Nutrition and Food Group (NuFoG), Galicia Sur Health Research Institute (IIS Galicia Sur), Vigo, Spain

Claudia Siemer ELEA Technology GmbH, Quakenbrueck, Germany

Aurora Silva Instituto de Agroecoloxía e Alimentación (IAA), Universidade de Vigo, Nutrition and Food Group (NuFoG), Campus Auga, Ourense, Spain; REQUIMTE/LAQV, Instituto Superior de Engenharia do Porto, Instituto Politécnico do Porto, Porto, Portugal

Ozgur Tarhan Department of Food Engineering, Faculty of Engineering and Life Sciences, Uşak University, Uşak, Türkiye

Arisa Thamsuaidee ELEA Technology GmbH, Quakenbrueck, Germany; Laboratory of Food Chemistry, Department of Chemistry, National and Kapodistrian University of Athens, Athens, Greece

Nantawan Therdthai Department of Product Development, Faculty of Agro-Industry, Kasetsart University, Bangkok, Thailand

Theocharia Tsagkaropoulou Department of Food & Nutritional Sciences, University of Reading, Reading, United Kingdom

Hananeh Yazdanbakhsh Department of Food Science and Technology, School of Nutrition Sciences and Food Technology, Kermanshah University of Medical Sciences, Kermanshah, Iran

Mohammad Yousefi Food and Beverages Safety Research Center, Urmia University of Medical Sciences, Urmia, Iran

Preface

Food processing operations are fundamental to the transformation of raw materials into safe, high-quality, and value-added products. In response to increasing demands for efficiency, consistency, and innovation within the food industry, modeling has emerged as a critical tool for understanding and optimizing these processes.

Modeling of Unit Operations in the Food Industry, part of the *Unit Operations and Processing Equipment in the Food Industry* series, provides a comprehensive overview of the modeling approaches applied to a broad range of food processing operations. This book presents the essential principles and methodologies for describing, analyzing, and improving process performance across diverse food systems.

The scope of this volume includes the modeling of heat transfer and thermal processing, nonthermal technologies, and mass transfer operations. It also addresses key unit operations, such as drying, frying, baking, sterilization, evaporation, and extraction, as well as the emerging technologies, including high-pressure processing and radiofrequency applications. Throughout the book, emphasis is placed on the integration of experimental data, theoretical analysis, and computational methods to support robust and practical modeling approaches.

This volume is intended for undergraduate and postgraduate students, researchers, and professionals in food engineering and related disciplines. It serves both as a foundational resource for those developing an understanding of modeling in food processing and as a reference for practitioners seeking to enhance product quality, safety, and process efficiency.

The contributing authors are recognized experts in their respective areas, and each chapter reflects the current developments in modeling techniques and their application to food processing systems. The material has been organized to ensure clarity, consistency, and technical rigor.

It is hoped that this book will support the advancement of knowledge in food engineering and encourage the effective application of modeling tools in both academic research and industrial practice.

Seid Mahdi Jafari
Nantawan Therdthai

Introduction to mathematic modeling in food processes

1

Nantawan Therdthai[1] *and Seid Mahdi Jafari*[2]
[1]Department of Product Development, Faculty of Agro-Industry, Kasetsart University, Bangkok, Thailand; [2]Department of Food Materials and Process Design Engineering, Faculty of Food Science and Technology, Gorgan University of Agricultural Sciences and Natural Resources, Gorgan, Iran

1.1 Introduction to food processes

Food processes have evolved continuously over many decades. Initially, food was being processed primarily to preserve its quality and extend its shelf life. Over time, production expanded to an industrial scale for commercialization. As this occurred, food safety became a critical concern, necessitating attention to process control to avoid errors. Today, food processes also aim to develop new products that meet customer needs and address pain points associated with food consumption. The trend toward healthier foods presents a challenge, as processors strive to preserve and enhance bioactive compounds while maintaining functional properties. Simultaneously, there is a growing demand for ready-to-cook and ready-to-eat options to accommodate busy lifestyles. These objectives drive innovation in food process technologies, enhancing both product quality and process efficiency.

Food processes can be broadly classified into two categories: thermal and non-thermal processing. Thermal processing methods include drying, frying, baking, grilling, pasteurization, sterilization, ultra-high temperature (UHT) processing, chilling, and freezing (Fellows, 2009). Non-thermal processing methods encompass ozone treatment, cold plasma, sonication, pulsed electric fields (PEF), high-pressure processing (HPP), membrane technology, fermentation, and irradiation (Cullen et al., 2012).

Additionally, food processes can be categorized based on operational principles into three main groups: (i) processes related to mass transfer, (ii) processes related to heat transfer and thermal effects, and (iii) non-thermal processes. Each process has specific parameters that play critical roles on its operation. Understanding the relationships between these parameters and process responses is essential for effective operation, and this can be achieved through modeling. This book focuses on mathematical modeling to simulate key processes, aiding in process optimization, design, and control.

1.1.1 Mass transfer processes

Mass transfer processes, including membrane separation, extraction, evaporation, and drying, are designed to enhance mass flow rates and improve overall process

Modeling of Unit Operations in the Food Industry. https://doi.org/10.1016/B978-0-12-822180-8.00011-0

efficiency. In membrane separation, various operating parameters—such as pressure, feed flow rate, and temperature—play crucial roles in optimizing flux rates. Increasing the transmembrane pressure enhances the driving force for permeation (Saddiqi et al., 2024), while optimizing the feed flow rate can create turbulence at the membrane surface, reducing concentration of polarization. Additionally, applying heat to the membrane separation process can be beneficial, as higher temperatures generally increase diffusion rates. However, excessively high temperatures may damage the membrane or the product. Membrane configuration parameters also require careful design. To enhance mass flow dynamics, it is important to increase both the surface area and the wettability of the membrane. While larger membrane pore sizes can increase mass flow rates, they may reduce the membrane's selectivity, potentially compromising mass separation effectiveness. A significant concern in membrane separation is fouling, which can restrict the overall mass transfer rate. Fouling can be minimized by increasing crossflow velocity (Khalid et al., 2024) and using spacers or baffles to enhance turbulence, thereby reducing boundary layer thickness. Additionally, it is important to manage feed properties, such as viscosity and concentration of polarization, to avoid fouling and facilitate effective mass transfer.

For extraction, There are several technologies, including solvent extraction, cold pressing, steam distillation, supercritical fluid extraction, microwave-assisted extraction, and ultrasound-assisted extraction. Selecting the appropriate technology for the specific material is crucial. In solvent extraction, the choice of solvent and the ratio of solvent to material are vital for achieving optimal yield and obtaining active compounds in the extract. The choice of solvent significantly impacts extraction efficiency. Other important parameters that must be optimized to enhance extraction efficiency include extraction time, agitation speed, particle size, and extraction temperature. Longer extraction times and higher temperatures can increase yield but may also degrade sensitive compounds (Khotchai et al., 2024). Reducing the particle size of the material increases the surface area available for extraction, thus improving efficiency. However, excessively fine particles may lead to filtration challenges. Agitation during extraction enhances contact between the solvent and the solid matrix, improving mass transfer and extraction efficiency. In the case of supercritical fluid extraction, adjusting the pressure can enhance extraction efficiency by altering the solvent's density and solubility properties.

In evaporation and drying processes, water migrates from the inner layers of a product to the outer layers to reduce moisture content and water activity, thereby extending shelf life. Key process parameters include temperature, pressure, airflow rate, and surface area. Higher temperatures and increased airflow rates can enhance the mass transfer rate; however, it is crucial to avoid temperatures that could degrade product quality. Lowering the pressure (or creating a vacuum) reduces the boiling point of the liquid, allowing for evaporation at lower temperatures and minimizing thermal degradation. To maximize efficiency, equipment can be designed to increase surface area, such as with thin film evaporators (Azri et al., 2024). Additionally, implementing energy recirculation systems can improve energy efficiency and reduce operational costs.

Mass transfer processes are vital in food processing for enhancing product quality and efficiency, encompassing techniques such as membrane separation, extraction, evaporation, and drying. Collectively, these processes drive innovations that improve food quality, safety, and operational efficiency.

1.1.2 Heat transfer and thermal process

In heat transfer processes such as blanching, pasteurization, and sterilization, key parameters to enhance heat transfer include temperature gradient, heat transfer surface area, pressure, and fluid flow rate. A greater temperature difference between the heating medium and the product increases the driving force for heat transfer, while maximizing surface area through equipment design—such as using fins, tubes, or plates—can improve transfer rates. Increasing the flow rate of the heating medium reduces boundary layer thickness and promotes turbulent flow, enhancing mixing and heat transfer efficiency compared to laminar flow. Higher operating pressures raise the boiling point and improve heat transfer during sterilization. Proper equipment configuration is essential, including adequate insulation to minimize heat loss and mechanical agitation to ensure uniform temperature distribution. Employing counterflow or crossflow heat exchangers is often more effective than parallel flow designs, and using corrugated plates or spiral configurations can enhance turbulence (Haddadvand et al., 2024). Selecting a heating medium with high thermal conductivity is also crucial for optimizing thermal processes.

Regarding dry heating in baking, roasting, and frying, a combination of conduction, convection, and radiation is employed to cook food while developing desirable flavors and colors. To enhance heat transfer rates, it is essential to optimize operating parameters such as oven or fryer temperature and surface area. Configuring equipment effectively can improve process efficiency; for instance, installing convective fans helps circulate hot air for uniform heat distribution (Therdthai et al., 2004), while leaving space around food in the fryer or oven allows for better airflow. Regularly stirring or turning food during frying or roasting further ensures even heat distribution and enhances browning reactions. Additionally, food properties such as increased surface area, low surface moisture, and high thermal conductivity are preferred to accelerate heating processes.

Similarly, in processes aimed at removing heat from foods, such as cooling and freezing, several factors can significantly enhance the efficiency of heat removal. Increasing the temperature gradient between the food and the cooling medium is crucial, as a larger difference drives heat transfer more effectively. Enhancing airflow and maximizing contact surface area between the food and the cooling medium also contribute to improved cooling and freezing rates. For optimal performance, using a medium with higher thermal conductivity, such as brine, is preferred over air, which has lower thermal conductivity and may slow down the cooling process. Additionally, well-insulated cooling and freezing equipment is essential to maintain low temperatures, reducing energy consumption and improving overall energy efficiency. Proper insulation minimizes heat gain from the environment, further supporting rapid cooling or freezing. Employing techniques such as forced-air circulation or immersion in cooling mediums can also facilitate uniform heat removal, ensuring that food products reach desired temperatures quickly and effectively.

To improve heat transfer rates in food production, novel processes such as Ohmic heating and microwave heating have been increasingly utilized. In microwave heating, several parameters—including microwave power, frequency, and the dielectric properties of the materials—are crucial for controlling process efficiency. The dielectric properties of foods can vary based on their composition and temperature; for instance, foods with high moisture and salt contents tend to have higher dielectric properties (Therdthai and Visalrakkij, 2012), allowing them to absorb microwave energy more effectively. In constrat, foods in frozen form have significantly lower dielectric constant and dielectric loss (Bai-Ngew et al., 2011)), restrict them to absorb and covert the microwave energy into heat. To enhance uniformity within the microwave chamber, using a turntable tray in the batch operation can help distribute microwave energy evenly, minimizing hot and cold spots.

For Ohmic heating, the configuration of the electrodes—such as their shape, size, and spacing—has a significant impact on heat distribution. Optimizing the design to maximize surface contact with the food can improve the heat transfer rate. In continuous operation systems, adjusting the flow rate of the food is essential to ensure uniform heating and minimize the residence time required to reach desired temperatures (Rivera et al., 2024). By refining these parameters, both microwave and ohmic heating can lead to more efficient and uniform cooking processes, enhancing overall food quality and safety.

Heat transfer processes are critical in food processing. Effective heat transfer relies on optimizing key parameters, including temperature gradients, heat transfer surface area, pressure, and fluid flow rates. Overall, understanding and managing these heat transfer processes are essential for ensuring food quality, safety, and energy efficiency in production.

1.1.3 Non-thermal process

The use of non-thermal processes, particularly HPP and PEF processing, to inhibit microbial growth has gained popularity as a means to prevent quality degradation caused by heat (Hogan et al., 2005). In HPP, it is crucial to properly design and control parameters such as pressure (typically ranging from 300 to 600 MPa), holding time, and mass load. Monitoring the resultant temperature is essential to ensure that it remains low. Foods with high acidity (low pH) (Gouvea et al., 2020) and high salt content tend to be more effectively processed for microbial inhibition.

For PEF processing, key parameters include electric field strength, pulse duration, frequency of the electric pulses, exposure time to PEF, and the flow rate of the food through the PEF chamber. Generally, increasing electric field strength can enhance cell permeabilization, but excessively high levels may lead to negative effects (Ohshima et al., 2016). It is important to minimize exposure time and resultant temperature while still achieving adequate microbial inactivation. Short pulses may be effective for some applications, while longer pulses may be necessary for others, depending on the type of food. Additionally, the optimal frequency is specific to the food matrix. Adjusting the pH and electrical conductivity of the food can improve cell permeabilization and the overall effectiveness of PEF. Furthermore, the design of key equipment components, such as electrodes and chambers, should be optimized to maximize electric field distribution and ensure uniform treatment of the food.

Every non-thermal process involves particular parameters that are vital to its functioning. It is crucial to comprehend how these parameters relate to process outcomes for efficient operation, and this understanding can be gained through modeling.

1.2 Mathematic modeling, its steps and performance

1.2.1 Classification of mathematic modeling

Mathematical models can be classified based on state variables, the type of state variable, the structure of evolution equations, and the characteristics of parameters (Bellomo and Preziosi, 1995). Models are primarily categorized into dynamic models, where state variables depend on time, and static models, where state variables are independent of time. Additionally, models can be further divided into continuous models, which depend on distance (space), and discrete models, which are independent of distance.

In terms of equation structure, models can take various forms. When a state equation is expressed algebraically, it is called an algebraic model. If the equation is presented as ordinary differential equations (ODE), the model is classified as ODE; similarly, if it uses partial differential equations (PDE), it is known as PDE. Discrete models typically correspond to ODEs, while continuous models align with PDEs. Furthermore, all models can be classified as linear or non-linear, with non-linear behavior commonly observed in physical systems, while linear behavior occurs under specific conditions.

Models can also be distinguished based on their parameters. Deterministic models contain fixed parameters, while stochastic models include random parameters, reflecting real-world situations where uncertainty is prevalent. Mathematical models can take various forms, such as differential equation models, statistical models, data fitting models, probabilistic models, and game-theoretic models.

Additionally, models can be categorized into three main types: (i) physics-based models, which rely on physical laws; (ii) observation-based models, which use empirical data; and (iii) hybrid models, which combine elements of both approaches. This classification aids in selecting the appropriate modeling approach for specific applications and understanding the complexities of the systems being studied (Bellomo and Preziosi, 1995). Physics-based models are grounded in theoretical knowledge and established scientific principles. In the context of food processing, these models are constructed using fundamental concepts such as heat and mass transfer, fluid flow, mass and energy balances, and rheological properties. This foundation allows for a systematic and comprehensive understanding of the relationships among process parameters and responses, leading to what is often referred to as a "white box model."

This term signifies that the internal workings of the model are transparent and can be easily interpreted.

One of the significant advantages of physics-based models is their ability to provide detailed insights into the mechanisms governing food processes. By using well-defined equations and theoretical frameworks, these models can predict how changes in one parameter will affect others, thus facilitating process optimization. However, when addressing complex problems—especially those involving multiple interacting variables—computational time may increase significantly. The complexity of a problem is influenced by factors such as the number of parameters involved, the specific boundary conditions applied, and the assumptions made during model formulation.

Examples of physics-based models include deterministic models, computational fluid dynamics (CFD) models, and kinetic models. Deterministic models rely on fixed parameters and predictable outcomes, while CFD models simulate fluid flow and heat transfer in complex systems. Kinetic models focus on the rates of reactions and changes in concentration over time. Each of these modeling approaches leverages fundamental scientific principles to provide valuable insights into the behavior of food processes, aiding in the development of more efficient processing techniques and better control strategies. Overall, while physics-based modeling can offer a robust framework for understanding food processes, it is essential to recognize the trade-offs involved, particularly regarding computational demands. Nevertheless, the clarity and precision of these models make them indispensable tools in both research and industrial applications (Therdthai, 2021).

In contrast to physics-based models, observation-based models rely heavily on empirical data regarding both the quantity and quality of relevant parameters. The design of experiments is crucial for effectively collecting data, as this data serves as the foundation for establishing relationships among process parameters and responses. A larger dataset can enhance the reliability of these established relationships, leading to more accurate model predictions. However, the quality of the data is equally important; minimizing errors during data collection is essential to ensure that the relationships reflect actual phenomena rather than artifacts of measurement inaccuracies.

Observation-based models are good at handling complex systems with multiple interacting parameters. They are particularly robust and do not require extensive knowledge or background in food processes, making them accessible for various applications. However, one drawback is that the predicted outcomes can sometimes be difficult to interpret clearly, as the models may not provide insights into the underlying mechanisms driving the observed relationships.

Examples of observation-based models include data fitting models, statistical models, artificial neural networks, and fuzzy logic models. These are collectively referred to as "empirical models," as they are derived from observed data rather than theoretical principles. A well-known empirical model used in food processing, especially for drying simulations, is the Page model, established in 1949 (Simpson et al., 2017). This model has been widely adopted due to its effectiveness in predicting drying behavior based on empirical data, showcasing the value of observation-based approaches in practical applications. While observation-based modeling provides valuable tools for analyzing complex food processes, it is essential to carefully consider data collection methods and the interpretation of results to ensure that the models yield meaningful insights.

Hybrid models combine elements of both physics-based and observation-based approaches, leveraging the strengths of each to enhance overall model performance. By integrating the systematically explainable outputs of physics-based models with the robustness and flexibility of observation-based models, hybrid models offer a comprehensive framework for understanding complex systems.

The advantage of a physics-based model lies in its ability to provide clear insights into the underlying mechanisms of a process, allowing for detailed predictions based on established scientific principles. Meanwhile, observation-based models are good at handling real-process complexities and variability, as they rely on empirical data that captures the nuances of actual processes. By merging these two approaches, hybrid models can achieve a more accurate and reliable representation of systems, benefiting from both theoretical foundations and practical data.

This combined approach not only improves model performance but also enhances the interpretability of results. Hybrid models can be particularly useful in food processing, where factors such as heat transfer, mass transfer, and product characteristics must all be considered. By utilizing data to validate and refine the theoretical aspects of the model, practitioners can develop more effective processing strategies and optimize outcomes. Overall, hybrid modeling represents a powerful tool for addressing the challenges of complex systems, enabling a deeper understanding and more accurate predictions in various applications, particularly in the food industry.

1.2.2 Modeling steps

In the case of a discrete model, the first step involves selecting a state variable (or dependent variable) that accurately represents the physical state of the process system. Once the state variable is chosen, all relevant parameters must be clearly defined. The model is then constructed by implementing initial conditions, constraints, and boundary conditions that govern the system's behavior. After the model is generated, problem-solving techniques are applied to obtain a solution. Before proceeding to practical applications, the model's performance is analyzed to ensure its accuracy and reliability (Bellomo and Preziosi, 1995). If the model passes validation, it can be employed for prediction, simulation, and optimization of food processes (Fig. 1.1).

Similarly, for continuous models, the process begins with the selection of a state variable that represents the physical state of the processing system. All parameters must also be clearly defined, and it is essential to determine the interactions between the process system and its environment. Key concepts such as equilibrium points, conservation of mass, energy, and momentum related to the state variable must be thoroughly defined. A set of mathematical equations is then generated to describe the time and spatial evolution of the state variable, establishing the mathematical model. Following this, the model is analyzed and validated to ensure that its performance is adequate for simulating actual processes effectively.

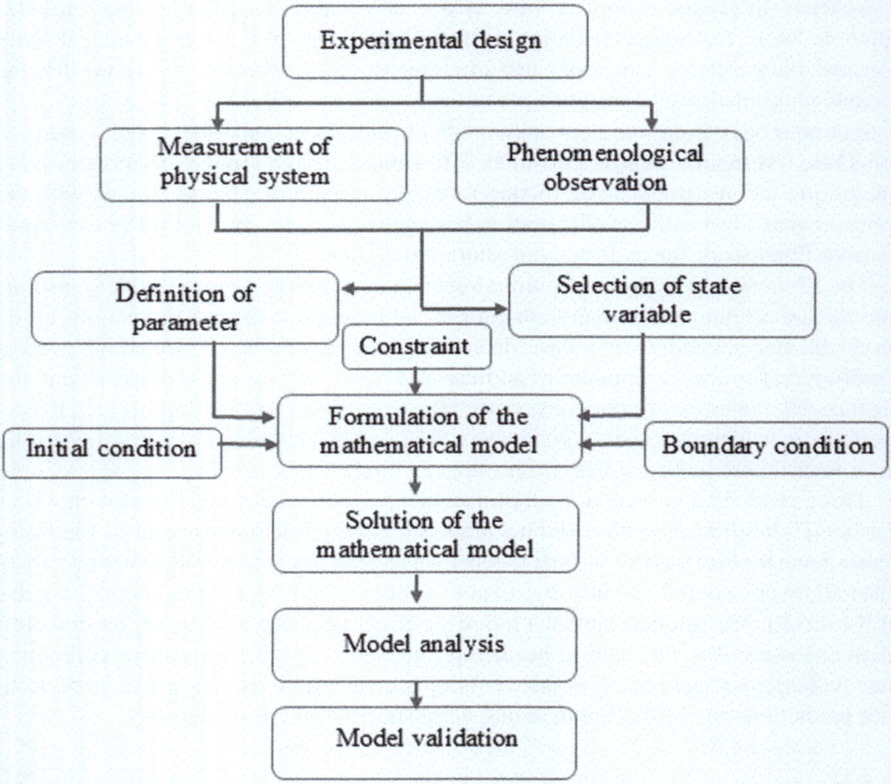

Figure 1.1 Modeling steps of a discrete model.

Through these structured steps, both discrete and continuous models can provide valuable insights into the dynamics of food processing systems, facilitating improvements in efficiency and product quality. Mathematical modeling is grounded in fundamental physical and chemical laws, primarily the conservation laws of mass, energy, and momentum. A mathematical model can consist of a single equation or a group of equations, including those related to continuity, energy, and motion.

Continuity Equation: The continuity equation is based on the principle of mass conservation, stating that the mass flowing into a system must equal the mass flowing out of the system in a steady state. Even though dynamic changes occur throughout the processing time, the mass entering the system should ultimately equal the mass exiting it. This can be represented mathematically as Eq. (1.1).

$$M_{in} - M_{out} = \frac{dM}{dt} \tag{1.1}$$

where $M = Mass$

$t = \text{Time}$

In food processes involving chemical reactions, components within the system can change to form new products or may disappear as reactants, leading to a situation where mass is not conserved. Therefore, the continuity equation for a specific component must account for these changes. This can be expressed mathematically as Eq. (1.2).

$$Mole_{in} - Mole_{out} + Mole_{reaction} = \frac{dMole}{dt} \tag{1.2}$$

For the reaction (Eq. 1.3) of 'a' that is changed into 'b', the continuity equation of component 'a' can be presented as Eq. (1.4).

$$a \xrightarrow{k} b \tag{1.3}$$

$$F_{a0}C_{a0} - F_{ai}C_{ai} - VkC_{ai} = \frac{dVC_{ai}}{dt} \tag{1.4}$$

where $F_{a0}C_{a0}$ = Flow of component 'a' into the system per time

$F_{ai}C_{ai}$ = Flow of component 'a' out of the system per time
VkC_{ai} = Rate of the use component 'a' to generate component 'b' from the reaction

Similarly, the continuity equation of component 'b' can be presented as Eq. (1.5).

$$F_{b0}C_{b0} - F_{bi}C_{bi} + VkC_{ai} = \frac{dVC_{bi}}{dt} \tag{1.5}$$

where $F_{b0}C_{b0}$ = Flow of component 'b' into the system per time

$F_{bi}C_{bi}$ = Flow of component 'b' out of the system per time

According to the first law of thermodynamics, energy (E), which includes internal, kinetic, and potential energy, can be conserved within a system. Heat (H) can be transferred to the system through conduction, convection, and radiation. Additionally, work (W) can be performed on or by the system. The rate of energy change within the system can be expressed mathematically as Eq. (1.6). This equation illustrates how energy is conserved by accounting for the heat entering and leaving the system, as well as any work done.

$$E_{in} - E_{out} + H - W = \frac{dE}{dt} \tag{1.6}$$

According to Newton's second law of motion, force is defined as mass times acceleration. The total forces acting in the j direction are equal to the rate of change in momentum in that same direction, reflecting the principle of conservation of momentum. In three-dimensional space, there are three axes: X, Y, and Z. Therefore, the balance of forces in all three directions must be considered. This can be expressed mathematically as Eq. (1.7).

$$F_x = \frac{dp_x}{dt}, F_y = \frac{dp_y}{dt}, F_z = \frac{dp_z}{dt} \tag{1.7}$$

where F = Forces in the respective directions

p = Momentum

Arrhenius equation: During food processing and storage, both chemical and enzymatic reactions can occur, leading to changes in the properties and qualities of food. For instance, during baking at high temperatures, caramelization occurs, resulting in the development of a brown crust and desirable flavors in baked goods. Prolonging the baking time can enhance the degree of browning. Similarly, dried bananas darken over time due to the Maillard reaction. The rate of this darkening can be slowed by removing oxygen from the packaging, a process known as vacuum packaging. This reduction in oxygen exposure is important because both oxygen and high temperatures accelerate reaction rates. Generally, the relationship between temperature and reaction rate is exponential, as described by the Arrhenius equation (Eq. 1.8). When the reaction rate is significantly influenced by increases in temperature, the activation energy is relatively large, indicating a strong temperature dependence of the reaction.

$$k = k_0 e^{-Ea/RT} \tag{1.8}$$

where k = Specific reaction rate

k_0 = Constant (pre-exponential factor)
Ea = Activated energy
R = Gas constant
T = Absolute temperature

The relationship between absolute temperature and reaction rate can be exponential when the reaction is limited by the chemical rate. However, if the reaction is limited by mass transfer, the relationship may not follow an exponential trend. The overall reaction rate is defined as the change in the concentration of each component (in moles per volume) divided by the stoichiometric coefficient of that component. For reactant components, the stoichiometric coefficients are represented as negative values, while for product components, they are positive. In the case of a first-order reaction where component A reacts to form component B, the overall reaction rate can be expressed by Eqs. (1.9)-(1.10). The specific reaction rate (k) is expressed as '1/s'.

$$A \xrightarrow{k} B \tag{1.9}$$

$$R = kC_A \tag{1.10}$$

In the case of a second-order reaction where component A reacts to form component B, the overall reaction rate can be expressed as Eq. (1.11). The specific reaction rate (k) is expressed as 'm^3/s·mole A'. This indicates that the rate of reaction depends on the square of the concentration of component A.

$$R = kC_A^2 \tag{1.11}$$

In the case of a first-order reaction where component A reacts with component B to form component C, the overall reaction rate can be expressed by Eqs. (1.12)-(1.13). In these equations,

k represents the specific reaction rate, which is expressed in units of m^3/s·mole B. This indicates that the rate of reaction is dependent on the concentrations of both components

$$A + B \xrightarrow{k} C \tag{1.12}$$

$$R = kC_A C_B \tag{1.13}$$

Thus the overall reaction rate can be generally presented as Eq. (1.14).

$$R = k_T C_A^a C_B^b \tag{1.14}$$

where R = Overall reaction rate.

C_A = Concentration of component A

C_B = Concentration of component B

a, b = Constant.

If the constant a is equal to 1, the reaction will follow first-order kinetics with respect to component A. Conversely, if a is equal to 2, the reaction will be classified as second-order. Additionally, the constants a and b can take on fractional values, allowing for more complex reaction dynamics.

After solving, the solution of equations provides the time and space evolution of the state variables or dependent variables.

1.2.3 Modeling performance

Many parameters are involved in food process operations. To accurately simulate the phenomena occurring during these processes, multiple parameters are incorporated into a model, which can increase its complexity. However, the number of added parameters must equal the number of equations, resulting in a degree of freedom equal to zero. Once the equations are generated, they can be solved using various techniques alongside specified boundary and initial conditions.

The performance of the model's predictions is validated to assess its accuracy and precision. This validation ensures that the established model can be reliably used for predictions in actual processes. Valid models can then be utilized for process optimization, design, and control. Key indicators of model performance (Therdthai, 2021) include the following:

Fitness of the model: To evaluate the model's fitness, the correlation between independent and dependent parameters in the equations is assessed. A higher correlation indicates better fitness. Generally, a fitness value greater than 0.7 is considered acceptable, with values closer to 1.0 being preferred.

Determination of error: The difference between the actual value and the predicted value is defined as the error. The amount of error can be quantified using the sum of squared errors (SSE), as expressed in Eq. (1.15):

$$SSE = \sum_{i=1}^{n} \left(Y_{pi} - Y_i\right)^2 \tag{1.15}$$

where Y_{pi} = Predicted value.

Y_i = Actual value

n = A number of data points.

When using the SSE to assess model performance, it is crucial to specify the number of data points included in the analysis. This information helps to justify the validity of the model, as a larger dataset can provide more reliable insights into the model's predictive accuracy. A model evaluated on a limited number of data points may yield misleading results, so understanding the sample size is essential for robust validation.

In some cases, the mean squared error (MSE) is preferred over SSE. The MSE represents the average of the squared differences between actual and predicted values, making it a useful metric for evaluating model performance in terms of mean values. This approach provides a normalized measure of error, which can be particularly advantageous when comparing models or assessing performance across different data sets. The mean squared error can be expressed as Eq. (1.16). By using MSE, practitioners can capture the overall tendency of the model to predict values accurately while also accounting for variability in the data.

$$MSE = \frac{1}{n} \sum_{i=1}^{n} \left(Y_{pi} - Y_i\right)^2 \tag{1.16}$$

Ultimately, whether using SSE or MSE, it is vital to consider the context and characteristics of the data to ensure an appropriate evaluation of model performance. These metrics provide valuable insights into how well a model can represent real-world phenomena, guiding improvements and refinements in food process modeling.

Additionally, the root mean square error (RMSE) is another important metric used to assess model performance. RMSE provides a direct quantification of the variation from the actual data range, making it particularly intuitive for interpretation. It is calculated by taking the square root of the mean squared error, as expressed in Eq. (1.17).

$$RMSE = \left[\frac{1}{n}\sum_{i=1}^{n}\left(Y_{pi} - Y_i\right)^2\right]^{1/2} \tag{1.17}$$

RMSE is particularly useful because it is in the same units as the original data, allowing for straightforward comparison between predicted and actual values. This characteristic makes RMSE an effective tool for understanding how well a model fits the data and for assessing its predictive accuracy. By incorporating RMSE along with other metrics such as SSE and MSE, researchers can gain a comprehensive understanding of a model's performance, ensuring more informed decisions in food process modeling and optimization.

In qualitative analysis, a scatter plot of errors against predicted values is commonly employed to determine the pattern of errors. This visual representation is essential for assessing the reliability of a model. If the errors appear random and are evenly distributed around zero, it indicates that the model is likely reliable and accurately predicts the outcomes. Conversely, if a discernible pattern emerges in the errors, it suggests that the model may require revision to achieve a more random distribution of errors.

An example of scatter plots depicting model errors is shown in Fig. 1.2. The scatter plots in Fig. 1.2a and c demonstrate poor model performance, characterized by distinct patterns in the errors. In contrast, the scatter plots in Fig. 1.2b, d, e, and f exhibit better performance, with errors that are smaller and appear more random, indicating a more accurate fit to the data.

Interestingly, Fig. 1.2g does not display large errors overall; however, it reveals a systematic pattern where underprediction occurs at values below 0.5, while overprediction is evident at values above 0.5 (Bai-Ngew et al., 2011). This highlights the importance of careful interpretation of scatter plots, as they can reveal specific areas where the model fails to perform adequately.

For a comprehensive evaluation of model performance, scatter plots should be used in conjunction with quantitative indicators such as MSE. By combining qualitative assessments from scatter plots with quantitative metrics, researchers can gain a more holistic understanding of a model's strengths and weaknesses, ultimately leading to more informed decisions in the modeling process.

1.3 Application of mathematical modeling in food processes

1.3.1 Mathematical modeling for process optimization

Food process optimization has become increasingly important in the food industry, aimed at improving and refining the methods, techniques, and conditions used in

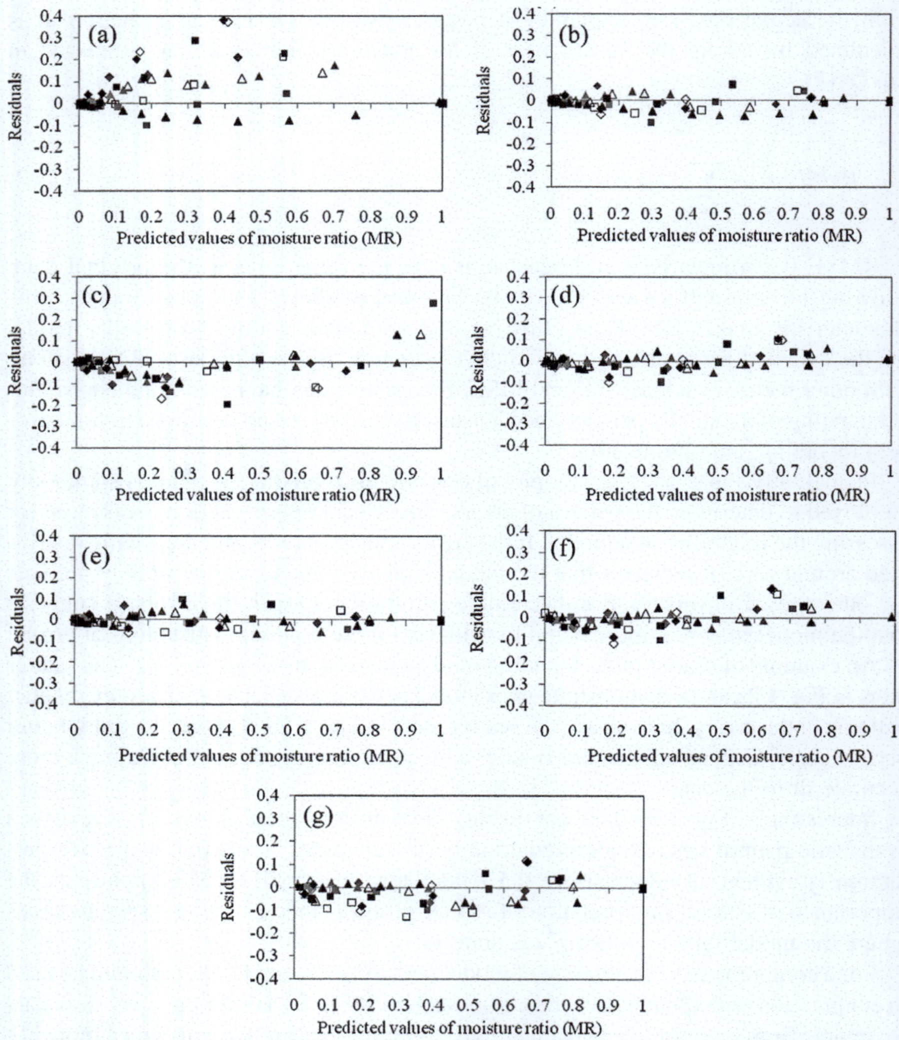

Figure 1.2 Scatter plots of thin layer models: (a) Lewis, (b) Page, (c) Henderson and Pabis, (d) Mod. Henderson and Pabis, (e) Logarithmic, (f) Two-term model and (g) Wang and Singh. ◊MV 7.23, □MV 5.49, Δ MV 3.88 represented drying condition. White, gray, and black represented pretreatment methods, including control, chilling, and freezing, respectively. (Reprinted from Bai-Ngew, S., Therdthai, N. and Dhamvithee, P. Journal of Food Engineering, 104, 119, 2011. With permission from Elsevier.).

food processes to achieve optimal performance. Key aspects of process optimization include efficiency, quality, safety, and cost management.

Efficiency in food processes involves maximizing processing speed, minimizing losses, and achieving the highest yield. By optimizing these factors, manufacturers can improve productivity and reduce waste. Regarding quality, the conditions of the process must be fine-tuned to ensure that the final product meets the desired physical, chemical, microbiological, nutritional qualities, and sensory standards as required by consumers. Safety is also paramount; processes must be designed to ensure safe operations and produce food that is safe for consumption. Finally, cost optimization is essential for maintaining competitiveness in the marketplace. This includes not only reducing production costs but also minimizing waste, particularly in alignment with green industry practices (Therdthai, 2021).

To effectively implement process optimization, models are utilized to guide the adoption of new technologies, refine existing processes, and continuously monitor and adjust operations based on performance data. Before optimization can take place, a systematic approach known as Design of Experiments (DOE) must be conducted to establish the relationships between process parameters and outputs, such as food quality, safety, yield, and efficiency. Various modeling techniques will be discussed in the subsequent chapters of this book.

Once relationships are established, optimization tools such as Linear Programming, Non-linear Programming, and Response Surface Methodology (RSM) can be employed to identify the best processing conditions based on specific optimization objectives.

Linear Programming is a mathematical approach used to find optimal solutions for processes, such as maximizing profit or minimizing cost, under the assumption of linear relationships. In cases where the relationships are non-linear, Non-linear Programming methods are appropriate. These methods can optimize processes while considering constraints related to factors like color, texture, and nutritional content. Both Linear and Non-linear programming share core components, including an objective function, constraints (limitations or requirements of the problem), and decision variables.

For Linear Programming, the Simplex method is commonly used; this iterative approach navigates along the edges of the feasible region defined by the constraints. Alternatively, the Interior–point method approaches the optimal solution from within the feasible region. For Non-linear programming, various solving methods are available, such as Gradient Descent, Newton's Method, Sequential Quadratic Programming (SQP), and Genetic Algorithms. Gradient Descent iteratively finds local minima using the gradient of the objective function, while Newton's Method employs second-order derivative information to locate optima. SQP solves a series of quadratic approximations, and Genetic Algorithms utilize heuristic search inspired by natural evolution, making them effective for complex non-linear problems (Adjei et al., 2024).

RSM is another widely used tool for food process optimization, praised for its efficiency and insightful approach to exploring complex relationships with fewer experiments compared to traditional methods. RSM is versatile and applicable across various fields, including manufacturing, biotechnology, and agriculture. It allows for the exploration of relationships between process parameters and one or more response variables. By fitting a model to experimental data, RSM aids in optimizing process conditions (Van Tai et al., 2024).

Common experimental designs used in RSM include Central Composite Design (CCD) and Box-Behnken Design. For CCD, the design incorporates factorial points, axial (star) points, and center points, providing a comprehensive view of the response surface. The Box-Behnken Design employs a rotational, spherical, or cubic arrangement of experimental points, emphasizing interactions between factors. Once experiments are conducted to gather data, a second-order polynomial (quadratic model) is typically used for modeling. The analysis of the data is often displayed through contour plots or surface plots, which illustrate the effects of parameters and their interactions. The fitted model is then utilized for process optimization, employing gradient-based optimization or numerical methods as solving techniques.

To verify the accuracy and robustness of the model and the optimization results, an additional set of data should be collected. Alternatively, cross-validation techniques can be employed. Various software programs, such as Design-Expert, Minitab, and JMP, facilitate the application of RSM.

Examples of food process optimization and their benefits can be summarized in Table 1.1, highlighting the practical implications of these modeling techniques in the food processes.

1.3.2 Mathematical modeling for process control

Mathematical models play a vital role in monitoring and controlling process performance in food processing, where precise control is essential for achieving desired outcomes such as safety, quality, and efficiency. The primary objectives of applying modeling in process control include understanding process dynamics, predicting future behavior, designing control strategies, and optimizing processes.

By developing models that accurately represent the processes involved, food manufacturers can gain insights into how different variables interact and influence overall performance. Then, the models can help forecast how processes will behave under various conditions based on current inputs. This predictive capability is crucial for planning and decision-making in the future. In addition, effective modeling allows for the development of sophisticated control strategies and algorithms that can regulate processes to maintain desired outputs, ensuring consistency in product quality. Lastly, by analyzing models, manufacturers can identify the best processing conditions that maximize efficiency, yield, and production rates while minimizing costs.

Key process control strategies that utilize models include Model Predictive Control (MPC), Proportional-Integral-Derivative (PID) Control, and Adaptive Control. MPC uses a dynamic model to predict future process behavior and optimize control actions over a defined prediction horizon. It excels in handling constraints associated with multiple inputs and outputs, making it highly effective in complex food processing scenarios (Schutz et al., 2024). PID control is a widely used strategy where models aid in tuning and setting optimal parameters. This method is effective for maintaining desired outputs in processes with relatively stable dynamics. Adaptive Control is particularly beneficial for systems that exhibit time-varying or uncertain dynamics. Adaptive control adjusts parameters in real-time based on feedback from the established models, allowing for responsive adjustments to maintain optimal performance.

Table 1.1 Food process optimization.

Process	Parameters	Model	Optimization	References
3D-printing	• LF-NMR data, gelatinization temperature, • Starch content, gelatinization time	BP-ANN modeling with levenberg–marquardt training algorithm	The BP-ANN model can predict the rheological properties (G′ and G″) of corn starch gels based on low-field nuclear magnetic resonance (LF-NMR) data. The optimal conditions for 3D printing occur at a gelatinization temperature of 75–85°C, with starch content in the range of 15%–20%	Jiao et al. (2024)
Extrusion	• Temperature • Screw speed	Regression modeling and genetic algorithms	The response surface regression model employs two process parameters to predict various process responses, including pressure, motor torque, and specific mechanical energy. It also assesses product responses such as radial expansion ratio, density, water absorption index, and water-soluble index.	Kowalski et al. (2018)
Foam-mat drying	• Egg albumin • Maltodextrin content • Drying temperature	Regression model and response surface methodology (RSM)	The concentrations of albumin and maltodextrin, along with the drying temperature, are optimized to maximize total phenolic content and drying rate.	Van Tai et al. (2024)

Continued

Table 1.1 Food process optimization.—cont'd

Process	Parameters	Model	Optimization	References
Spray freezing	• Initial concentration • Contact angle • Overall temperature transition • Heat transfer coefficients • Droplet sizes	The multi-stage solidification model for freezing a droplet filled with a binary mixture (analytical model)	A multi-stage, hybrid analytical-numerical model for droplet solidification is developed, incorporating a gradient-based optimization algorithm. It simulates five stages in binary mixtures—supercooling, nucleation, recalescence, equilibrium freezing, and subcooling—effectively capturing dynamic variations in temperature and solute concentration during the phase change process.	Xu et al. (2024)
Desalination/ Membrane technology	• Salt concentration • Feed flow rate • Condenser • Evaporator inlet temperatures	Improvized Gray wolf optimizer aided artificial neural network predictive model (IGWO-ANN)	The predictive IGWO-ANN model effectively simulates and analyzes permeate flux in desalination plants, maximizing performance while saving energy, time, and costs. By identifying optimal operating conditions, it improves efficiency and addresses challenges like fouling and scaling, contributing to sustainable water management solutions.	Mahadeva et al. (2024)

Software tools that support predictive process control include MATLAB®/Simulink, GAMS (General Algebraic Modeling System), and ASPEN Plus/HYSYS. These tools enable the simulation and optimization of processes, facilitating better decision-making.

For advanced adaptive controllers, Artificial Neural Networks (ANNs) represent a powerful technique. ANNs, inspired by the structure and function of the human brain, are particularly effective in managing complex, non-linear systems where traditional control methods may struggle. They consist of interconnected nodes (neurons) organized in layers: an input layer, hidden layers, and an output layer. ANNs learn the relationships among parameters through data training, adjusting weights and biases to minimize errors between predicted and actual outputs (Zhou and Therdthai, 2010).

Common training algorithms for ANNs include Backpropagation and Gradient Descent, which help fine-tune the model based on training data. In the context of process control, ANNs are used to determine control actions based on input signals and desired outputs, providing adaptive control capabilities that enhance overall system performance. ANN controllers are gaining popularity due to their ability to handle non-linearities and adapt to changes in system dynamics over time (Liu et al., 2024).

Software programs that assist with ANN design and integration include MATLAB®/Simulink, TensorFlow/PyTorch, and LabVIEW. However, some challenges remain. Key concerns include data requirements, computational Intensity, and overfitting risk. ANNs require significant amounts of training data to achieve accurate and reliable performance. Training and running neural networks can be computationally intensive, necessitating powerful hardware. However, overtraining can give adverse effect. ANN models may become overly specialized to the training data, reducing their ability to generalize to new, unseen data.

Recently, ANN controllers have found applications in robotics, particularly for controlling robotic arms and autonomous systems where precise and adaptive control is necessary. They are also utilized in imaging technology for real-time quality monitoring in food products, allowing controllers to be trained to recognize and reject poor-quality items based on visual characteristics.

Similarly, Fuzzy Logic Controllers (FLCs) are designed to handle uncertainty and imprecise information, mimicking human decision-making processes. Unlike classical binary logic, fuzzy logic accommodates degrees of truth, making FLCs suitable for complex, non-linear systems where traditional mathematical models may be challenging to establish (Schutz et al., 2024). In the food industry, FLCs are applied in automation systems for managing intricate manufacturing processes, enabling efficient control in situations where precise modeling is difficult. FLCs have also been successfully implemented in robotics for tasks such as navigation and manipulation within food processing and service environments. Software tools for FLC design include MATLAB®/Simulink with the Fuzzy Logic Toolbox, FuzzyTech, and Fuzzy Logic Designer.

While both ANN controllers and FLCs provide modern modeling capabilities and mimic human-like decision-making, they differ in several key aspects. Firstly, the complexity of the System, ANN controllers require careful selection of network architecture, training algorithms, and tuning of hyperparameters. The need for large datasets for effective learning can make this approach computationally intensive. In contrast, FLCs utilize a rule-based design with a set of rules and membership functions, making them simpler and more intuitive for individuals with domain expertise. Addition. Another aspect is adaptability and learning. ANN controllers can adapt to changing conditions over time if retrained with new data, while FLCs operate based on fixed rules, requiring manual adjustments to accommodate new situations. FLCs do not necessarily require extensive data for design; they rely on linguistic rules and membership functions, making them less computationally intensive once the fuzzy rules are established.

Overall, ANN controllers are well-suited for highly complex, non-linear systems where data-driven insights are available, offering high accuracy and precision. FLCs, on the other hand, are ideal for systems characterized by uncertainty, leveraging human expertise encoded in heuristic rules to maintain robustness even with simplified models. Generally, FLCs are less computationally intensive compared to ANN controllers, particularly once the fuzzy rules are established.

The application of mathematical models in designing process controllers for food processes offers significant potential for enhancing efficiency, quality, and cost-effectiveness. This is summarized in Table 1.2, which reviews some model applications in food processing systems.

1.3.3 Mathematical modeling for shelf-life estimation

The shelf-life of a product refers to the period during which it remains usable, safe, and effective under specified storage conditions. Accurate shelf-life estimation is crucial for ensuring food quality and safety, as it directly impacts consumer health and satisfaction, as well as economic viability for manufacturers. To estimate shelf-life effectively, it is essential to understand the mechanisms of chemical, biological, and physical changes that affect product stability over time. By selecting appropriate models and accurately determining relevant parameters, food scientists can predict shelf-life, ensuring that products maintain their quality and safety.

Several mathematical models assist in estimating shelf-life based on various factors, including temperature, humidity, and the nature of the product itself. Common models used for this purpose include the Arrhenius model, kinetic models, predictive microbiology models, and empirical models. Each of these models offers unique insights into the processes that govern shelf-life, and their application can significantly enhance our understanding of food preservation.

Arrhenius Model: It is widely used to estimate the effect of temperature on the rate of chemical reactions, a key determinant of shelf-life. Based on the Arrhenius equation, this model allows for the calculation of the reaction rate constant (k) at a specific temperature (Ti). The core idea is that as temperature increases, the rate of chemical reactions also increases, leading to faster deterioration of food quality. By calculating k at different temperatures, one can predict the shelf-life of a product at various storage conditions (Muniandy et al., 2024). Specifically, the shelf-life can be derived as inversely proportional to the k value, defined as the time at which 10% of the product's quality remains. This model provides a straightforward approach to

Table 1.2 Modeling for food process control.

Process	Controller	Advantage	References
Fermentation	Fuzzy control	The Takagi–Sugeno (TK) fuzzy controller approach is used to optimize temperature, pressure, pH, and moisture content during the fermentation of gari, a staple food in west Africa made from cassava. This approach minimizes energy consumption and enhances the production of high-quality fermented products.	Odetunji and Kehinde (2005)
Anaerobic digestion process	Fuzzy-dual PID controller for centralized control	The fuzzy-PID controller effectively manages temperature disturbances in the digester, improving control performance by reducing peak overshoot, rise time, and settling time for digester temperature control.	Anand et al. (2021)
Drying	Artificial neural network controller	The intelligent controller is designed to regulate microwave vacuum drying based on desired aromatic profiles. A PEN-3 electronic nose is integrated into the dryer to monitor changes in aroma and send data to a data acquisition system for real-time control of the drying process. Input parameters consist of real-time signals from the electronic nose at various stages of drying, while the output is the predicted vacuum level, which can be used for real-time vacuum control.	Liu et al. (2024)

Continued

Table 1.2 Modeling for food process control.—cont'd

Process	Controller	Advantage	References
Microwave heating	Back propagation neural network (BPNN) with adaptive particle swarm optimization (APSO) controller	The BPNN-APSO controller continuously adjusts the neural network's weights and biases to regulate the microwave input power, minimizing the error between the achieved and preset temperature. This control design enhances the heating uniformity of the microwave oven and improves the precision of controlled parameters.	Yao et al. (2024)
Raw material grading	Artificial neural network controller	Hyperspectral imaging technology provides extensive spatial information and collects spectral data, which are valuable for artificial neural network (ANN) controllers to detect and reject peaches with chilling injury.	Pan et al. (2016)

understanding how temperature fluctuations can impact the longevity of a product, making it an invaluable tool for manufacturers who seek to optimize storage conditions.

Kinetic Model: This model offers another layer of insight into food preservation by describing the deterioration of food quality over time. These models can be classified based on the rate of degradation. If the rate of degradation remains constant, the zero-order kinetic model is appropriate (Chen et al., 2024). However, when the rate varies, particularly in response to external factors, the n-order kinetic model—especially the first-order kinetic model—should be employed. First-order kinetic models are particularly useful in food science, as they can effectively account for the diminishing rate of degradation as the product approaches its shelf-life limit. These models provide a mathematical framework for predicting how various storage conditions can affect the stability of food products, allowing manufacturers to make informed decisions about packaging and storage (Hertog et al., 2016).

Predictive Microbiology Models: These models are essential for understanding the effects of microbial growth on shelf-life. These models utilize complex algorithms to consider multiple factors affecting microbial proliferation during storage, including temperature, pH, and humidity. For example, the Weibull distribution is commonly used to predict the time until degradation occurs, while the Baranyi model effectively addresses microbial growth by encompassing both lag and exponential growth phases. The Gompertz model is another widely used predictive microbiology model that describes microbial growth through a logistic curve. This model is particularly effective for estimating how long it will take for microbial counts to reach unacceptable levels, which is vital for determining shelf-life. By utilizing these predictive microbiology models, manufacturers can better manage food safety and quality, ensuring that products remain safe for consumption throughout their intended shelf-life.

Empirical Models: They offer a complementary approach to shelf-life estimation, relying on experimental data rather than theoretical principles. These models are particularly useful when detailed mechanistic models are unavailable, as they depend on data fitting and statistical analysis to estimate shelf-life. Empirical models can be adapted based on real-world observations, making them flexible and practical for various food products. These models are often employed in situations where the underlying mechanisms are complex or poorly understood. For instance, empirical models can capture the effects of multiple variables on shelf-life without requiring a detailed understanding of the interactions among those variables. This makes them especially valuable for new products or formulations where traditional models may not yet apply.

Therefore, mathematical modeling plays a crucial role in shelf-life estimation for food products. The Arrhenius model, kinetic models, predictive microbiology models, and empirical models each offer unique advantages and insights into the factors affecting product stability. By leveraging these models, food scientists and industry can make informed decisions about storage conditions, packaging, and product formulation, ultimately leading to enhanced quality and safety.

Understanding shelf-life not only helps manufacturers comply with regulatory requirements but also fosters consumer trust by ensuring that products meet quality standards. As food science continues to evolve, the integration of advanced mathematical modeling techniques will remain pivotal in optimizing food preservation strategies, thereby reducing waste and improving overall food security. The example of application of these models in shelf-life estimation for food processes is provided in Table 1.3, highlighting their significance in food science and technology.

1.3.4 Mathematical modeling for equipment design

Mathematical modeling is a crucial tool in equipment design, enabling engineers to optimize designs, predict performance, and reduce costs. By integrating theoretical knowledge with practical application, mathematical models facilitate well-informed design decisions that can significantly enhance efficiency and effectiveness in various industries, including food processing.

To begin the modeling process, the representation of equipment should be simplified, focusing on key components and their interactions. This initial simplification helps identify the primary parameters and physical phenomena that will influence performance. Once the critical components are established, the appropriate types of mathematical models must be selected. Two common models for equipment design are Finite Element Analysis (FEA) and Computational Fluid Dynamics (CFD).

Table 1.3 Modeling for shelf-life estimation.

Product	Model	Shelf-life indicator	References
Rice bran	Zero-order model, fractional conversion model, and back propagation neural network model with levenberg–Marquardt training	Kinetic and BP-ANN models are used to simulate free fatty acid (FFA) generation. Input parameters, including lipase activity, water activity, packaging materials, and storage temperature, help predict changes in FFA levels. By understanding the inhibitory effects of packaging materials and storage conditions on FFA generation, solutions for extending shelf life can be identified. In comparison, the BP-ANN model outperforms the kinetic model, offering more accurate predictions and greater versatility.	Chen et al. (2024)
Cottage cheese	Logistic, gompertz, and Barayni models	The Barayni model is superior to the logistic and gompertz models for predicting microbial growth (including Cocci, lactobacillus, Coliform, and total plate count) during the storage of indian cottage cheese. The shelf life is estimated based on microbial growth at different storage temperatures.	Chaturvedi et al. (2023)
Orange juice	Data fitting model with the linear and non-linear relationship.	The data fitting model with a non-linear relationship, such as polynomial and log-logistic models, outperforms linear models in simulating the microbial kinetics and electronic nose (E-nose) profiles of orange juice (with and without thermosonication treatment) during storage. Moderate thermosonication treatment can extend the shelf life of orange juice.	Wahia et al. (2020)

UHT milk	A statistical model with partial least Squares discriminant analysis (PLS-DA)	A statistical model using PLS-DA is employed to predict variations in sensory quality and E-nose profiles of UHT milk during storage. This model effectively discriminates between shelf life and brand of UHT milk samples based on flavouromics data during storage.	Xi et al. (2024)
Kiwifruit (raw material)	A kinetic model	The kinetic model uses input data on the initial firmness of the kiwifruit, the amount of intact structural cell wall compounds, and the initial activity of enzymes that break down these compounds to predict changes in firmness during storage at 0°C for 24 weeks. The model can also predict the ripening stage and remaining shelf life of the kiwifruit.	Hertog et al. (2016)

Finite Element Analysis (FEA) is a computational technique used to obtain approximate solutions to complex engineering problems, particularly those involving stress, strain, and deformation in materials and structures. The fundamental principle of FEA is to divide a complex structure into smaller, simpler parts known as elements. These elements can be one-dimensional, two-dimensional, or three-dimensional, each characterized by its material properties and defined behavior based on physical laws.

The entire structure is represented by a mesh formed by these elements. FEA solves governing equations of physics, typically based on mechanics principles, and is derived from differential equations that describe material behavior. Each element's equations are assembled into a global system that represents the entire structure. To ensure accuracy, FEA predictions are often compared with experimental data or analytical solutions, and sensitivity analyses are conducted to assess how changes in design parameters affect equipment performance (Aslam et al., 2024).

Although widely used, FEA has limitations. It may not capture all physical phenomena accurately, particularly in nonlinear or dynamic scenarios. The model's accuracy relies on the quality of the mesh and the precision of the material properties used in simulations.

CFD is another powerful modeling approach extensively used to simulate fluid behavior in various applications, from simple flows to complex interactions with structures. CFD excels in analyzing intricate flow scenarios and provides insights into flow behavior that may be challenging to measure experimentally, often reducing the need for extensive physical testing.

The CFD model employs numerical analysis and algorithms to solve and analyze problems related to fluid flows based on the principles of fluid dynamics, governed by the Navier–Stokes equations. These equations describe how fluids behave under various forces and include fundamental components such as the Continuity Equation (ensuring mass conservation), Momentum Equation (derived from Newton's second law), and Energy Equation (governing energy conservation relevant in thermal flows) (Zhou and Therdthai, 2019).

Similar to FEA, the fluid domain in CFD is divided into a finite number of small control volumes or elements through a meshing process. The mesh can be structured (regular grid) or unstructured (irregular shapes), depending on the geometry's complexity. Once the mesh is established, material properties, boundary conditions, and initial conditions are specified. The Finite Volume Method (FVM) is commonly employed to solve the governing equations, particularly for complex, nonlinear problems.

CFD offers various visualization tools that generate contour plots, vector plots, and streamlines, allowing engineers to analyze flow behavior, including velocity fields, pressure distributions, temperature gradients, and flow patterns (Therdthai et al., 2004). Like FEA, CFD predictions are validated against experimental data or analytical solutions, and sensitivity analyses are conducted to evaluate how changes in design parameters impact equipment performance. However, CFD simulations often require significant computational resources, particularly for high-fidelity analyses.

Beyond FEA and CFD, other models can be employed in equipment design, such as Multibody Dynamics (MBD) models and Reliability Engineering models. MBD models simulate the motion and interaction of interconnected rigid or flexible bodies, which is particularly useful for analyzing systems such as robotic arms, automotive suspensions, and machinery. These models assess kinematics, dynamics, and forces involved in mechanical systems. Reliability Engineering models are used to evaluate the reliability and failure rates of equipment under various conditions, crucial for designing safety-critical systems. Furthermore, optimization algorithms, such as genetic algorithms and simulated annealing, can help find the best design parameters or configurations for equipment.

In practice, these modeling approaches are often used in combination, depending on the design problem's complexity and specific requirements. Integrating multiple methodologies can provide a comprehensive understanding of performance, efficiency, and reliability in equipment design. For instance, FEA and CFD might be combined to assess both structural integrity and fluid behavior simultaneously, offering a holistic view of how equipment will function under real-world conditions.

Therefore, mathematical modeling plays a pivotal role in the design and optimization of equipment for food processes. By employing techniques such as FEA and CFD, engineers can make informed design decisions that enhance efficiency, safety, and performance. The integration of various modeling approaches allows for a nuanced understanding of complex systems, ensuring that food processing equipment meets the demands of the industry. Some examples of application of these models in food process equipment design are listed in Table 1.4, illustrating their significance in advancing food technology and ensuring quality and safety in food production.

1.4 Conclusion

Food processing plays a crucial role in transforming and stabilizing foods, ensuring safety for consumption, extending shelf life, enhancing nutritional value, and fostering innovation. These operations can be conducted using thermal or non-thermal processes, making it essential to carefully select appropriate methods to achieve desired food quality. Additionally, process parameters must be well-defined and controlled to optimize food production performance.

However, the number of process parameters can be substantial, and their interactions may range from simple to complex. Understanding the impact of these parameters on process operations, food quality, and overall efficiency is vital for designing optimal processing conditions for each application.

Mathematical modeling is an effective tool for establishing relationships between process parameters and their corresponding performance outcomes. These relationships can be linear or non-linear, with complexity varying based on specific processing scenarios. By understanding how process parameters influence outputs, it becomes possible to optimize processing conditions to maximize efficiency, including yield and production rate. Model simulations, coupled with real-time data acquisition, enable online optimization and control of process parameters to achieve optimal operational performance. Moreover, processing equipment can be designed to ensure uniform flow and temperature distribution based on model simulations.

Table 1.4 Modeling for food processing equipment design.

Equipment/Process	Model	Design parameter	References
Fluidized bed dryer and twin screw granulator	Population balance model (PBM)	The model-based design of a twin-screw granulator and a fluid bed dryer aims to identify the key unit operations in the continuous industrial production of powders and tablets. Equipment properties, powder characteristics, and operational conditions (such as liquid feed rate in the granulator, air temperature, and drying time) can be optimized to achieve the desired granule size distribution, porosity, moisture content, and tablet strength.	Wang et al. (2020)
Solar dryer	Computational fluid dynamics (CFDs) model	The configuration of the indirect solar dryer, which includes a solar collector unit, drying chamber, exhaust chimney, solar-powered fan, and dryer outlet, is designed to optimize airflow and temperature distribution within the drying chamber. Additionally, a CFD model is used to minimize the condensation of water during the drying process.	Demissie et al. (2019)
Ohmic sterilization system	A CFD model with hexahedral elements using a finite element-based numerical analysis	A continuous ohmic heating sterilization system, featuring heating and cooling sections, is simulated using a CFD model. The simulation describes the temperature distribution within the continuous ohmic chamber, the temperature field in the 90 degrees pipeline connecting the chamber to the cooling zone, and the temperature distribution of the counterflow heat exchanger in the cooling section. The configuration of the ohmic equipment is then designed based on the CFD simulation results.	Rivera et al. (2024)

Baking oven	A 3-D CFD model	A continuous baking oven is simulated using a 3-D CFD model. The oven configuration, which includes the combustion duct, convective fan, and conveyor movement, is designed to achieve optimal baking temperature profiles in each baking zone.	Therdthai et al. (2004)
Freeze dryer	A 3-D CFD model	A 3-D CFD model is used to simulate a closed chamber with stationary shelves. This simulation aids in optimizing the geometry of the freeze-dryer chamber, resulting in a potential reduction of freezing time by 50%.	Piechnik et al. (2024)

Models in food processing can be physics-based, observation-based, or a combination of both. Their accuracy and reliability are crucial for effectively analyzing and simulating food processing performance. Selecting appropriate models, collecting accurate input data, specifying initial and boundary conditions, and using suitable solving methods are challenging but essential steps for improving model accuracy and precision.

References

Adjei, B.A., Sebil, C., Otoo, D., Ackora-Prah, J., 2024. A quadratically constrained mixed-integer non-linear programming model for multiple sink distributions. Heliyon 10 (19), e38528. https://doi.org/10.1016/j.heliyon.2024.e38528.

Anand, K., Mittal, A.P., Kumar, B., 2021. Modelling and simulation of dual heating of substrate with centralized temperature control for anaerobic digestion process. J. Clean. Prod. 325, 129235. https://doi.org/10.1016/j.jclepro.2021.129235.

Aslam, R., Alam, M.S., Vishwakarma, R.K., Maqsood, S., 2024. Ohmic heating assisted vacuum evaporation chamber: finite element analysis and evaluation of parameters for processing of amla juice. Comput. Electron. Agric. 224, 109249. https://doi.org/10.1016/j.compag.2024.109249.

Azri, N.A., Mantripragada, H., Patel, R., Kowall, C., Cormack, G., Proust, N., Enick, R., Veser, G., 2024. Process intensification for production of dispersants via integration of reaction and separation in a horizontal thin film evaporator. Chem. Eng. J. 489, 151541. https://doi.org/10.1016/j.cej.2024.151541.

Bai-Ngew, S., Therdthai, N., Dhamvithee, P., 2011. Characterization of microwave vacuum-dried durian chips. J. Food Eng. 104 (1), 114–122. https://doi.org/10.1016/j.jfoodeng.2010.12.003.

Bellomo and Preziosi, 1995. Modelling Mathematical Methods and Scientific Computation. CRC Press Inc., Florida, USA.

Chaturvedi, K., Basu, S., Singha, S., Das, K., 2023. Predictive microbial growth modelling for an effective shelf-life extension strategy of Chhana (Indian cottage cheese). Food Control 149, 109697. https://doi.org/10.1016/j.foodcont.2023.109697.

Chen, C., Wang, X., Jing, P., Jiao, S., 2024. Kinetic analysis and back propagation neural network model for shelf-life estimation of stabilized rice bran. J. Food Eng. 380, 112168. https://doi.org/10.1016/j.jfoodeng.2024.112168.

Cullen, P.J., Tiwari, B.K., Valdramidis, V.P., 2012. Novel Thermal and Non-Thermal Technologies for Fluid Foods. Academic Press, p. 526. https://doi.org/10.1016/C2009-0-61145-4.

Demissie, P., Hayelom, M., Kassaye, A., Hailesilassie, A., Gebrehiwot, M., Vanierschot, M., 2019. Design, development, and CFD modeling of indirect solar food dryer. Energy Proc. 158, 1128–1134. https://doi.org/10.1016/j.egypro.2019.01.278.

Fellows, P.J., 2009. Food Processing Technology: Principles and Practice, third ed. Woodhead Publishing. 913pp.

Gouvea, F.S., Padilla-Zakour, O.I., Worobo, R.W., Xavier, B.M., Walter, E.H.M., Rosenthal, A., 2020. Effect of high-pressure processing on bacterial inactivation in açaí juices with varying pH and soluble solids content. Innov. Food Sci. Emerg. Technol. 66, 102490. https://doi.org/10.1016/j.ifset.2020.102490.

Haddadvand, R., Sohrabi, N., Yan, T.X., Nadalinia, F., Karouei, S.H.H., 2024. Numerical study on the influence of flow direction and fluid type on heat transfer and pressure drop in a two-tube spiral heat exchanger with innovative conical turbulators. Case Stud. Therm. Eng. 61, 104933. https://doi.org/10.1016/j.csite.2024.104933.

Hertog, M.L.A.T.M., Jeffery, P.B., Gwanpua, S.G., Lallu, N., East, A., 2016. A mechanistic model to describe the effects of time, temperature, and exogenous ethylene levels on softening of kiwifruit. Postharvest Biol. Technol. 121, 143–150. https://doi.org/10.1016/j.postharvbio.2016.08.002.

Hogan, E., Kelly, A.L., Sun, D.-W., 2005. Part 1: High-pressure processing of foods: An overview. In: Sun, D.-W. (Ed.), Emerging Technologies for Food Processing. Academic Press, pp. 3–32. https://doi.org/10.1016/B978-012676757-5/50003-7.

Jiao, X., Ren, G., Law, C.L., Li, L., Cao, W., Luo, Z., Pan, L., Duan, X., Chen, J., Liu, W., 2024. Novel strategy for optimizing of corn starch-based ink food 3D printing process: Printability prediction based on BP-ANN model. Int. J. Biol. Macromol. 276 (Part 2), 133921.

Khalid, O., Gidstedt, S., Lipnizki, F., Rudolph-Schöpping, G., 2024. Direct Membrane Filtration (DMF) of municipal wastewater – A study on the prevention and remediation of fouling. J. Water Process Eng. 67, 106235. https://doi.org/10.1016/j.jwpe.2024.106235.

Khotchai, W., Therdthai, N., Ritthiruangdej, P., 2024. Effect of microwave-assisted extraction on quality and taste profiles of crude extracts from split gill mushroom. J. Agric. Food Res. (Submitted).

Kowalski, R.J., Li, C., Ganjyal, G.M., 2018. Optimizing twin-screw food extrusion processing through regression modeling and genetic algorithms. J. Food Eng. 234, 50–56.

Liu, P., Song, F., Zheng, Y., Li, Z., Song, C., Li, J., Jin, G., 2024. Intelligent control of microwave vacuum drying based on online aroma monitoring. J. Food Eng. 380, 112148. https://doi.org/10.1016/j.jfoodeng.2024.112148.

Mahadeva, R., Kumar, M., Diwan, A., Manik, G., Dixit, S., Das, G., Gupta, V., Sharma, A., 2024. Improvised grey wolf optimizer assisted artificial neural network (IGWO-ANN) predictive models to accurately predict the permeate flux of desalination plants. Heliyon 10 (13). https://doi.org/10.1016/j.heliyon.2024.e34132.

Muniandy, A., Benyathiar, P., Ozadali, F., Mishra, D.K., 2024. Development of predictive model for the novel ultra-accelerated shelf-life test (UASLT) for shelf-life of packaged beverage. LWT 210, 116686. https://doi.org/10.1016/j.lwt.2024.116686.

Odetunji, O.A., Kehinde, O.O., 2005. Computer simulation of fuzzy control system for gari fermentation plant. J. Food Eng. 68 (2), 197–207. https://doi.org/10.1016/j.jfoodeng.2004.05.032.

Ohshima, T., Tanino, T., Kameda, T., Harashima, H., 2016. Engineering of operation condition in milk pasteurization with PEF treatment. Food Control 68, 297–302. https://doi.org/10.1016/j.foodcont.2016.03.047.

Pan, L., Zhang, Q., Zhang, W., Sun, Y., Hu, P., Tu, K., 2016. Detection of cold injury in peaches by hyperspectral reflectance imaging and artificial neural network. Food Chem. 192, 134–141. https://doi.org/10.1016/j.foodchem.2015.06.106.

Piechnik, E., Smolka, J., Palacz, M., Tolstorebrov, I., Eikevik, T.M., Stebel, M., Haida, A.J.N., Ciesielska, A., Bodys, J., 2024. Experimentally validated CFD-tool for a freezing simulation in a small-scale freeze-dryer. J. Food Eng. 367, 111888. https://doi.org/10.1016/j.jfoodeng.2023.111888.

Rivera, J., Gratz, M., Jaeger, H., Schottroff, F., 2024. Characterization and optimization of continuous ohmic thermal sterilization based on the development of a predictive computational toolbox. Innov. Food Sci. Emerg. Technol. 96, 103792. https://doi.org/10.1016/j.ifset.2024.103792.

Saddiqi, H.A., Javed, Z., Ali, Q.M., Ullah, A., Ahmad, I., 2024. Modelling and predicting lift force and trans-membrane pressure using linear, KNN, ANN and response surface models during the separation of oil drops from produced water. J. Water Process Eng. 66, 106014. https://doi.org/10.1016/j.jwpe.2024.106014.

Schutz, D.R., Mercaldi, H.V., Peñaloza, E.A.G., Silva, L.J.R., Oliveira, V.A., Cruvinel, P.E., 2024. Advanced embedded generalized predictive controller based on fuzzy gain scheduling for agricultural sprayers with dead zone nonlinearities. J. Process Control 135, 103164. https://doi.org/10.1016/j.jprocont.2024.103164.

Simpson, R., Ramírez, Nuñez, H., Jaques, A., Almonacid, S., 2017. Understanding the success of Page's model and related empirical equations in fitting experimental data of diffusion phenomena in food matrices. Trends Food Sci. Technol. 62, 194–201. https://doi.org/10.1016/j.tifs.2017.01.003.

Therdthai, N., 2021. Chapter 15: Modeling and optimization of food processes. In: Jafari, S.M. (Ed.), Engineering Principles of Unit Operations in Food Processing. Volume 1: Unit Operations and Processing Equipment in the Food Industry. Woodhead Publishing (an imprint of Elsevier). https://doi.org/10.1016/B978-0-12-818473-8.00004-9.

Therdthai, N., Visalrakkij, T., 2012. Effect of osmotic dehydration on dielectric properties, microwave vacuum drying kinetics and quality of mangosteen. Int. J. Food Sci. Technol. 47, 2606–2612. https://doi.org/10.1111/j.1365-2621.2012.03141.x.

Therdthai, N., Zhou, W., Adamczak, T., 2004. Three-dimensional CFD modelling and simulation of the temperature profiles and airflow patterns during a continuous industrial baking process. J. Food Eng. 65 (4), 599–608. https://doi.org/10.1016/j.jfoodeng.2004.02.026.

Van Tai, N., Van Hao, H., Ngoc Han, T.T., Ngoc Giau, T., Minh Thuy, N., Van Thanh, N., 2024. Effect of foaming conditions and drying temperatures on total polyphenol content and drying rate of foam-mat dried banana powder: Modeling and optimization study. J. Agric. Food Res. 18, 101352.

Wahia, H., Zhou, C., Mustapha, A.T., Amanor-Atiemoh, R., Mo, L., Fakayode, O.A., Ma, H., 2020. Storage effects on the quality quartet of orange juice submitted to moderate thermosensation: Predictive modeling and odor fingerprinting approach. Ultrason. Sonochem. 64, 104982. https://doi.org/10.1016/j.ultsonch.2020.104982.

Wang, L.G., Omar, C., Litster, J., Slade, D., Li, J., Salman, A., Bellinghausen, S., Barrasso, D., Mitchell, N., 2022. Model driven design for integrated twin screw granulator and fluid bed dryer via flowsheet modelling. Int. J. Pharm. 628, 122186. https://doi.org/10.1016/j.ijpharm.2022.122186.

Xi, Y., Yang, Y., Chi, X., Wang, W., Sun, B., Ai, N., 2024. Characterization of the flavor profile of UHT milk during shelf-life via volatile metabolomics fingerprinting combined with chemometrics. Food Res. Int. 191, 114705. https://doi.org/10.1016/j.foodres.2024.114705.

Xu, M., Akhtar, S., Mohit, M., Sasmito, A.P., 2024. Development of a multi-stage model for freezing of a suspended binary solution droplet. Int. J. Heat Mass Tran. 221, 125115.

Yao, Q., Zheng, X., Wang, R., Liang, W., Liu, T., Chu, W., 2024. Control of thermal uniformity in microwave heating process by BPNN and adaptive particle swarm optimization. Heliyon, e37971. https://doi.org/10.1016/j.heliyon.2024.e37971.

Zhou, W., Therdthai, N., 2010. Chapter 35: Artificial neural networks in food processing. In: Farid, M.M. (Ed.), Mathematical Modeling of Food Processing. CRC Press, Taylor and Francis, LLC, 999 pp.

Zhou, W., Therdthai, N., 2019. Chapter 9: Three-dimensional CFD modeling of continuous industrial baking process. In: Sun, D.-W. (Ed.), Computational Fluid Dynamics in Food Processing. 2nd edition. CRC Press, Taylor and Francis, LLC, 614 pp.

Modeling of thermal sterilization of non-newtonian fluids in a batch retort

2

Kannan Aravamudan
Department of Chemical Engineering, Indian Institute of Technology Madras, Chennai, India

2.1 Introduction

Canned food items are convenient options in today's fast-paced world. To have a long shelf life, the canned food should be sterilized and be safe from microorganisms such as *Clostridium botulinum*, *Clostridium thermosaccolyaticum*, *Bacillus* spp., etc (Jimenez et al., 2024). The spores of the microorganisms are heat-resistant, and the bacteria produce toxins that are fatal to humans even at low concentrations (Calderón-Alvarado et al., 2016; Padmavati and Anandharamakrishnan, 2013). Thermal sterilization is a well-established and popular technique despite relatively newer nonthermal processing developments, which apply pulsed electric field, high hydrostatic pressure, and ultraviolet rays (Farid and Abdul Ghani, 2004). Thermal sterilization is reputed to be reliable and effective, especially for ensuring shelf-stable products (Park and Yoon, 2018). In thermal sterilization, food may be canned first and then sterilized. Alternatively, in aseptic processing, the food is sterilized first and then packaged.

Stationary, batch retorts are used with steam as sterilizing medium to achieve sterilization. When in-package pasteurization of liquids in bottles and flexible pouches is used, vigorous agitation may not be appropriate (Datta and Teixeira, 1988). Agitated retorts may not be economical when low volumes of liquid foods are to be processed. When canned food is sterilized, it should be ensured that the coldest spot of the food material within the can receives adequate thermal treatment so that most of the microorganisms are killed. Hence, when deciding the sterilization duration, the location and temperature of the coldest spot must be estimated accurately.

Sterilization of foods such as canned tuna, thick syrups, purees and concentrates are typically described by pure conduction (Kumar et al., 1990). If conduction is assumed to be the sole heat transfer mechanism, the thermal sterilization problem with simple boundary conditions becomes amenable to analytical solution. Using the analytical solution applicable in the food can domain, the temperature may then be predicted with time and location quite easily. The coldest spot will be at the geometric center of the can and will not migrate from there with time.

Examples of liquid foods sterilized in cans include beer, evaporated milk, thin soups and broth, fruit and vegetable juices and vegetable soups (Datta and Teixeira,

Modeling of Unit Operations in the Food Industry. https://doi.org/10.1016/B978-0-12-822180-8.00015-8

1988; Kumar et al., 1990). When such liquid foods are sterilized, the heat transfer gets relatively more complicated than conduction owing to the manifestation of natural convection. This arises due to the unstable density gradients between the cold and hot regions of the can, leading to circulation and rapid heating of the fluid. Another complicating matter is the non-Newtonian nature of the food in which the viscosity is not only a function of temperature but also a function of the rate of shear, with both varying as a function of time. The modeling of the natural convection in non-Newtonian fluids must be done rigorously by accounting for the coupled heat and momentum transfer equations. The accurate numerical solutions of the governing equations will not only enable the prediction of the evolution of temperatures and velocities with time in the domain, but also help track the location of the slowest heating zone (SHZ). The SHZ refers to a core region in the food can that takes the longest time relative to the other regions to reach the final sterilization temperature and hence represents the limiting step of the overall sterilization process (Varma and Kannan, 2005; 2006). Hence, care should be exercised to ensure that this zone receives adequate sterilization.

Due to natural convection engendered flow, the SHZ is not fixed at one location (Abdul Ghani et al., 2003). Physical sensors, such as thermocouples if present in the food container, may interfere with the velocity profiles and consequently influence the temperature profiles (Abdul Ghani et al., 2003). Further, the SHZ may not be tracked accurately through temperature sensors. The temperature probes positioned at fixed locations inside the can cannot track the migration of the SHZ to different locations within the can. Even with solid food sterilization, the presence of a wireless sensor and its location affected the temperature evolution and sterility values at the cold spot, and the disturbance was more pronounced for smaller cans (Marra and Romano, 2003).

A few studies have carried out experimental studies to confirm the heating trends in food sterilization. Siriwattanayotin et al. (2006) studied thermal sterilization of sucrose syrup in food cans. They measured temperatures at two different axial locations during the heating process. Kannan and Sandaka (2008) reported transient midpoint temperature variation with time using a thermocouple during the thermal sterilization of 0.85% w/w carboxy methyl cellulose (CMC) solution. However, experimental measurements are static and applicable to select locations, while the SHZ movements are dynamic.

Due to the inherent difficulties with experimental tracking of heating of the food product, mathematical modeling approach has been extensively applied to predict the temperature patterns in the thermal sterilization process (Datta and Teixeira, 1988; Farid and Abdul Ghani, 2004; K. Datta and A. A. Teixeira, 1987; Naveh et al., 1983; Nicolaï et al., 1998). Mathematical modeling approach has been used to analyze heat transfer in industrial ovens (Mirade et al., 2004), and ventilated packaging systems (Zou et al., 2006; Khurana and Karwe, 2009). CFD based simulations and postprocessing of results have provided insight into many applications involving heating/cooling of food products, for example, drying, refrigeration and sterilization (Norton and Sun, 2006). Many CFD studies have investigated in detail the role played by convection in the thermal sterilization processes carried out in a still retort, for example, Abdul Ghani et al. (1999, 2002); Datta and Teixeira (1988); Kannan and Sandaka (2008); Varma and Kannan (2005, 2006). From CFD simulations (Abdul Ghani et al., 1999), demonstrated that bacterial deactivation was considerably influenced by natural convection induced flow patterns as well as temperatures. The highest concentrations of bacteria were found to occur at locations where velocity and temperature were minimum. Denys et al. (2004) studied the transient temperature patterns in egg-shaped bodies filled with CMC solution. They concluded from their CFD simulations that significant differences in the transient temperature profiles would result if natural convection effects were ignored and only conduction was considered. Augusto and Cristianini (2010) deduced from CFD simulations that only when the viscosity of the liquid food item increased to about 10^5–10^6 that of water, the food could be assumed to behave like a solid, and heat transfer may be assumed to occur by conduction. In the studies of Siriwattanayotin et al. (2006); Kannan and Sandaka (2008) mentioned in the previous paragraph, the experimental data conformed with the convective heating trends predicted from CFD simulations.

If the time for thermal sterilization for liquid foods is based on the conduction model predictions, to avoid time-consuming and complex numerical simulations, it is likely that the longer sterilization times suggested may result in overprocessing of food. The sterilization times are shorter for liquid foods as convective heating will be faster than conduction heating. Martins et al. (2008) observed that food engineers faced challenges to monitor, diagnose, and control the quality and safety of food products. The consumers' emphasis on sensory and nutritional characteristics in addition to safety and extended shelf life must also be borne in mind (Norton and Sun, 2006), and hence overprocessing must be avoided. The enhancement offered by convective heat transfer over conductive heat transfer may be quantified. Further, most studies had mainly evaluated heating by only conduction and its augmentation by convection (Kumar et al., 1990; Kumar and Bhattacharya 1991; Varma and Kannan 2006). Early literature did not focus sufficiently on estimation of heat transfer fluxes and development of a correlation for the heat transfer coefficient, or equivalently Nusselt number for heating a food can with isothermal walls. This gap was addressed by Kannan and Sandaka (2008). It is also necessary to estimate the effect of augmentation by convective heating on the lethality of the sterilization process over conductive heating. To address this gap (Koribilli et al., 2011), estimated enhancement factors for lethality using CFD studies.

Based on the above, the scope of this chapter is to detail the CFD work in the literature pertaining to natural convection phenomena occurring in food cans. The heat transfer regime change to natural convection after pure conduction is discussed, in addition to the movement of the slowest heating zone. The relative contribution of heat transfer rates from different positions on the can surface to the overall heat transfer rate is found. Correlations for the Nusselt number in the convection-dominated regime are developed. Conduction and convection model predictions are compared and the enhancement in heating rate as well as lethality is determined. Insight into the heat transfer rates, fluxes, and regime change from conduction mode to convection mode is obtained. Innovative modifications to the can geometry and orientation for reducing sterilization times are described. Further, optimization studies in batch thermal sterilization are discussed. Finally, we conclude by discussing the prospects in the field of thermal sterilization in a batch retort.

2.2 Materials and methods

2.2.1 Modeling of a typical nonNewtonian liquid food

For the analysis of thermal sterilization, Carboxy Methyl Cellulose (CMC) at 0.85% w/w is chosen as a typical nonNewtonian fluid system. The system may be used to represent tomato puree, carrot puree, green bean puree, apple sauce, apricot, and banana purees, which are canned and thermally sterilized (Steffe et al., 1986). The pseudoplastic nature of this system may be characterized as shown in Eq. (2.1) (Kumar and Bhattacharya, 1991).

$$\mu = 0.002232 \exp\left(\frac{30740}{RT}\right)\dot{\gamma}^{n-1} \tag{2.1}$$

Here n is taken to be 0.57. It may be seen that the viscosity η depends on the rate of shear ($\dot{\gamma}$). The viscosity is expressed in Pa.s, T in K and activation energy in J/mol. This model has been used by several researchers in literature, for instance by Abdul Ghani et al. (1999), Koribilli et al. (2011), Kumar and Bhattacharya (1991), Varma and Kannan (2006).

2.2.2 Analytical solutions to transient heat conduction in a cylinder

The overall equation for transient heat conduction in a finite cylinder of radius R and length L may be expressed as follows (Eq. 2.2).

$$\rho_s C_{p_s} \frac{\partial T}{\partial t} = k_s \frac{\partial^2 T}{\partial x^2} + \frac{k_s}{r}\frac{\partial}{\partial r}\left(r\frac{\partial T}{\partial r}\right) \tag{2.2}$$

Here, ρ_s, C_{p_s} and k_s are the density, specific heat, and thermal conductivity of the solid medium, respectively. The equation includes the accumulation of thermal energy on the LHS, while the RHS is the sum of heat transfer rates due to axial and radial conduction, respectively. For the thermal sterilization problem, it is assumed that the boundaries of the cylinder are all maintained at a constant temperature T_f. Hence, the boundary, axial, and initial conditions may be simply stated as follows.

At

a) $r = R,\ 0 \le x \le L,\ T = T_f$
b) $x = 0,\ 0 \le r \le R,\ T = T_f$
c) $x = L,\ 0 \le r \le R,\ T = T_f$
d) $r = 0,\ 0 \le x \le L,\ \frac{\partial T}{\partial r} = 0$

e) $t = 0, 0 \leq r \leq R, 0 \leq x \leq L, T = T_i$

In the analysis carried out in this chapter, the surrounding fluid temperature T_f is specified at 121°C, which is the temperature of condensing steam used as the sterilization medium. The initial temperature T_i was specified to be 40°C. The temperature is made non-dimensional as shown below (Eq. 2.3).

$$\theta = \left(\frac{T(r,t) - T_f}{T_i - T_f} \right) \tag{2.3}$$

The cylindrical food can is characterized by finite dimensions along the axial as well as radial directions. In contrast, we have a classical heat conduction analysis of an infinite cylinder where radius (R) is finite, but the length is infinite. Similarly, we also have the conduction heat transfer analysis for the one-dimensional infinite plate of finite width (L) but infinite length. From these two idealizations, it is possible to construct a finite cylinder whose radius R is the same as that of the infinite cylinder and height L is the same as the width of the infinite plate. The superposition involved is illustrated in Fig. 2.1 below.

For these two idealized geometries, that is, infinitely long cylinder and infinitely long plate, analytical solutions are available to the respective transient heat conduction equations. It may be proved elegantly by the principle of superposition (Holman and Bhattacharyya, 2011) that the solution to the transient heat conduction problem in a cylinder of finite length L and radius R may be simply obtained from the product of two dimensionless temperatures as shown below (Eq. 2.4).

$$\theta_{FC}\,(x, r, t) = \theta_{IP}(x, t) \times \theta_{IC}(r, t) \tag{2.4}$$

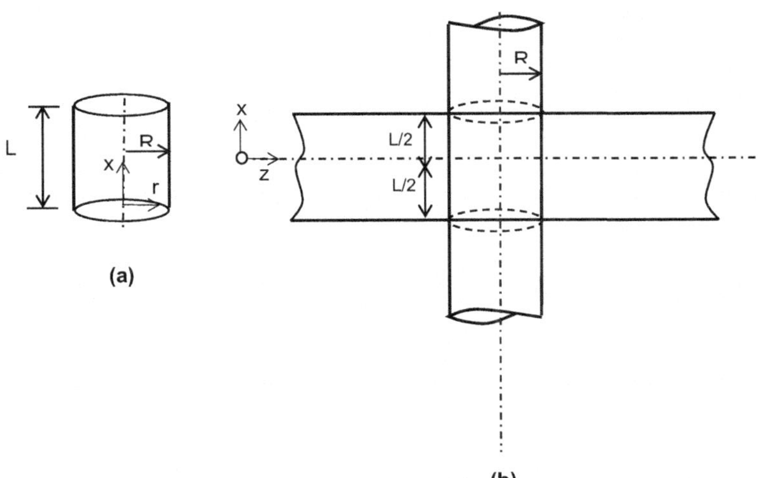

Figure 2.1 Representation of (a) the finite cylinder in terms of the infinitely long cylinder of radius R and (b) infinitely long plate of width L

Here, $\theta_{IP}(x,t)$ is the dimensionless temperature in an infinitely long plate of width L. Next, $\theta_{IC}(r,t)$ is the dimensionless temperature in an infinitely long cylinder with radius R. $\theta_{FC}(x,r,t)$ is the temperature profile in a finite cylinder of radius R and height L. We next insert the solution for the infinite plate (θ_{IP}) and infinite cylinder (θ_{IC}) in the general solution θ_{FC}. Without going into the details, the analytical solutions for θ_{IP} and θ_{IC} are given as follows. Derivations of the analytical solutions may be seen in standard texts such as (Kakaç et al., 2018).

The derivation of $\theta_{IP}(x,t)$ for the infinitely long plate is discussed first. The precondition to use the analytical solution is the identification of roots M_n (n = 1, 2, ..., ∞) to the following transcendental Eq. (2.5).

$$\mathrm{Cos}(M_n) = 0 \tag{2.5}$$

The solution to the above equation for a given n value is simply

$$M_n = \frac{(2n-1)}{L}\pi \tag{2.6}$$

Usually, the first few roots (say around 10 at the most) suffice to get the accurate analytical solution. We may now substitute these roots given by Eq. (2.6) to obtain the following series solution (Eq.2.7).

$$\theta_{IP}(x,t) = \frac{4}{\pi}\sum_{n=1}^{\infty}\frac{(-1)^{n+1}}{(2n-1)}e^{-\left\{\left(\frac{2n-1}{L}\right)^2\right\}\pi^2\alpha t}\cos\left\{\left(\frac{2n-1}{L}\right)\pi x\right\} \tag{2.7}$$

The derivation of $\theta_{IC}(x,t)$ for the infinite cylinder is discussed next. Like before, the precondition to use the analytical solution is the identification of roots M_n (n = 1, 2, ..., ∞) to the following transcendental Eq. (2.8).

$$J_0(M_n) = 0 \tag{2.8}$$

Here, J_0 is the Bessel function of orders 0. Usually, the first few roots (say around 10 at the most) suffice to get the accurate analytical solution. We may now substitute these roots in the following series solution Eq. (2.9).

$$\theta_{IC}(r,t) = 2\sum_{n=1}^{\infty}\frac{1}{M_n}\left[\frac{J_1(M_n)}{J_0^2(M_n)+J_1^2(M_n)}\right]e^{-\left(M_n^2\frac{\alpha t}{R^2}\right)}J_0\left(M_n\frac{r}{R}\right) \tag{2.9}$$

Here, J_1 is the Bessel function of order 1. Using Eq. (2.8), the above Eq. (2.9) may be simplified to give Eq. (2.10).

$$\theta_{IC}(r,t) = 2\sum_{n=1}^{\infty}\frac{1}{M_n}\left\{\frac{J_0\left(M_n\frac{r}{R}\right)}{J_1(M_n)}\right\}e^{-\left(M_n^2\frac{\alpha t}{R^2}\right)} \tag{2.10}$$

From Eqs. (2.4), (2.7) and (2.10), the dimensionless temperature for the finite cylinder may be obtained as given in Eq. (2.11).

$$\theta_{FC}(x,r,t) = \frac{8}{\pi}\left[\sum_{n=1}^{\infty}\frac{(-1)^{n+1}}{(2n-1)}e^{-\left\{\left(\frac{2n-1}{L}\right)^2\right\}(\pi^2 \alpha t)}\cos\left\{\left(\frac{2n-1}{L}\right)\pi x\right\}\right]$$
$$\left[\sum_{n=1}^{\infty}\frac{1}{M_n}\left[\frac{J_0\left(M_n\frac{r}{R}\right)}{J_1(M_n)}\right]e^{-\left(M_n^2\frac{\alpha t}{R^2}\right)}\right] \tag{2.11}$$

2.2.3 Modeling of natural convection heat transfer in a cylindrical can

CFD simulations involving natural convection differ from pure conduction as the former must account for fluid flow owing to thermally induced density gradients inside the food can. Natural convection simulations are also more involved than forced convection simulations. As the fluid density is a function of temperature, the flow equations are coupled with the thermal energy equation in natural convection. In forced convection, the flow equations may be solved first to establish the velocity field in the domain and then, using these, the thermal energy equation may be solved.

The assumptions made to simulate the natural convection-driven thermal sterilization process in a food can are (Abdul Ghani et al., 1999) given as follows:

a) Average fluid properties are used for reference density (ρ_{ref}) of 950 kg/m^3, specific heat (C_p) of 4100 J/kg.°C, thermal conductivity (k) of 0.7 W/m.K (for the nominal case) and coefficient of volume expansion (β) of 0.0002°C^{-1}. Food medium thermal conductivities of 0.35 and 1.4 W m^{-1} K^{-1} were also considered for parametric analysis
b) Boussinesq equation may be used as the buoyancy source term in the momentum equation to relate the effect of temperature on the fluid density
c) No slip boundary conditions are assumed between the fluid and the inner walls of the food can
d) The temperature instantaneously reaches 121°C at the walls of the can due to heating provided by the condensing medium steam, which has a high heat transfer coefficient.
e) The food can material has very little thermal resistance, and the temperature of the fluid contacting the wall is negligibly different from that of the wall temperature
f) The temperature varies only in the axial and radial directions. Variation in the azimuthal direction is absent due to the axis-symmetry of the geometry under consideration
g) Viscous heat dissipation may be neglected as the convection velocities are very low in the domain

The governing transport equations for the convective heat transfer associated with thermal food sterilization are given below (Eqs. 2.12–2.14).

i. The continuity equation

$$\frac{\partial \rho}{\partial t} + \nabla.(\rho \boldsymbol{V}) = 0 \tag{2.12}$$

Here, $\boldsymbol{V}$ represents the velocity vector.

ii. The momentum equation

$$\left(\frac{\partial \rho \mathbf{V}}{\partial \mathrm{t}} + \nabla.(\rho \mathbf{V} \otimes \mathbf{V})\right) = \nabla.(-\mathrm{p}\boldsymbol{\delta} + \eta(\nabla \mathbf{V} + (\nabla \mathbf{V})^{\mathrm{t}})) + \mathrm{S_M} \tag{2.13}$$

iii. The energy equation

$$\left(\frac{\partial \rho \mathrm{h_{total}}}{\partial \mathrm{t}} - \frac{\mathrm{D}p}{\mathrm{Dt}} + \nabla.(\rho \mathbf{V} \mathrm{h_{total}})\right) = \nabla.(\mathrm{k}\nabla \mathrm{T}) \tag{2.14}$$

Here, $\mathrm{h_{total}}$ is the specific total enthalpy. In Eq. (2.13) the momentum source term S_M is expressed in terms of the Boussinesq approximation as shown below

$$\mathrm{S_M} = -\rho_{\mathrm{ref}}\beta(\mathrm{T} - \mathrm{T_{ref}})\mathrm{g} \tag{2.15}$$

When the buoyancy term is applied, the pressure term in the momentum equation (Eq. 2.13) excludes the hydrostatic gradient (Varma and Kannan, 2006).

2.2.4 *Food can geometry*

First, the geometry of the 'Regular' cylindrical food can was specified at 111.11 mm height and 80.96 mm diameter. It was oriented in the vertical direction. As mentioned earlier, for further parametric analysis, the thermal conductivity of the fluid varied between 0.35 and 1.4 W/m.K. Two additional cans termed 'Medium' (L = 124 mm and D = 68.26 mm) and 'Long' (L = 150 mm and D = 65.00 mm) were also defined. Here, L refers to the total height of the food can. Hence, the aspect ratio (height to diameter ratio) was varied between 1.37 and 2.31. The 'Regular' and 'Medium' cans were of standard dimensions as specified by the Can Manufacturing Institute, Washington, while the 'Long' can was included to complete parametric study.

2.2.5 *Initial and boundary conditions*

For the convective heat transfer problem, the temperature conditions stipulated initially and at the walls of the food can are identical to those specified for the pure conduction model (section 2.2b). In addition, the velocities for the fluid initially, and at the boundaries are specified as described below.

At

a) $r = R, 0 \le x \le L, T = T_f, V_r = V_x = 0$
b) $x = 0, 0 \le r \le R, T = T_f, V_r = V_x = 0$
c) $x = L, 0 \le r \le R, T = T_f, V_r = V_x = 0$
d) $r = 0, 0 \le z \le L \frac{\partial T}{\partial r} = 0, V_r = 0, \frac{\partial V_x}{\partial r} = 0$
e) At $t = 0, 0 \le r \le R, 0 \le z \le L, T = T_i, V_r = V_z = 0$

As in the conduction case described above, the surrounding fluid temperature, T_f is specified at 121°C. The initial temperature was specified to be 40°C. The transport equations (Eqs. 2.12 – 2.14) along with the initial and boundary conditions as listed above were solved using CFX v5.7 licensed from CFX Ansys Ltd. For viscous liquids such as CMC and carrot orange soup, the dimensionless Grashoff number is very low and is in the range of 10^{-2} to 10^{-1} based on the maximum temperature difference and viscosity values. In contrast, water has a Grashoff number ranging from 10^9–10^{10}. Due to the low Grashoff number encountered for the working fluid CMC, the flow regime is laminar (Kannan and Sandaka, 2008).

2.2.6 Computational mesh details

The meshing of the cylindrical domain considered plays an important role in the accuracy of the temperatures and velocities obtained. A mesh option was used with fine grids near the walls where the gradients in temperatures and velocities are high. The mesh sizes from the wall toward the core grew in a geometric progression so that sudden jumps in grid sizes were avoided. The grid sizes near the wall were progressively reduced until mesh independence was achieved. Further details on the meshing may be obtained from Kannan and Sandaka (2008).

2.2.7 Solution methodology and validation

The first-order backward Euler method was used to solve the transient transport equations. The root -mean square of the residuals was computed and minimized to less than 10^{-4}. The numerical scheme chosen viz. high-resolution scheme, maintained second-order accuracy (Varma and Kannan, 2006). The computer processing time taken usually consumes about 10 h for simulating the heat transfer in the different geometries. The procedures and CFD calculations in the present study were validated by matching the midpoint temperatures as well as temperature contours was reported (Kumar and Bhattacharya, 1991).

2.3 Results and discussion

2.3.1 Estimation of heat transfer coefficient

Predictive modeling is invaluable in thermal sterilization to estimate the time required for the desired rise in temperature inside the food can. Previous CFD studies had focused mainly on finding temperature contours, velocity profiles, and SHZ migration at different time steps during the sterilization process. To complement these findings, heat transfer flux analysis for developing a Nusselt number (Nu) correlation to describe the natural convection problem inside a food can with isothermal surfaces becomes necessary. In literature, correlations for Nu were developed at steady state for idealized geometries such as flow past single infinitely long heated walls, parallel walls at different temperatures, inclined plates, rectangular cavities, and concentric cylinders (Farid and Abdul Ghani, 2004). In these correlations, the temperatures are different on two different surfaces.

For heated cylinders and spheres, natural convection effects in the fluid surrounding the body were analyzed for developing Nu correlations (Bergman et al., 2017). In the present application, the food medium experiencing natural convection is confined within the can. A further distinguishing feature is the transient nature of heat transfer in the food can as opposed to the usually steady state heat transfer in the idealized geometries mentioned above.

Farid and Abdul Ghani (2004) had observed that a heat transfer coefficient correlation is not available for liquids contained in vertical and horizontal cylindrical cans. They adopted the form of the analytical solution developed for heat conduction in an infinite cylinder (Eq. 2.10). The volume-to-area ratio as the characteristic dimension was used instead of cylinder radius, to account for heat transfer from all sides of the can. An effective thermal diffusivity was defined and incorporated in the expression of the Nusselt number based on Catton (1978). The purpose of this was to account for convection in rectangular cavities of width and height equal to the radius and length of the food can. They used this analogy for the vertical can based on the similarity between the convection rolls encountered in the cavity and in the food can. After performing several simulations with food items of varying viscosities and food can geometries, a correlation for SHZ was developed by the authors. This was used to predict the sterilization times for liquid foods in food cans.

(Kannan and Sandaka, 2008) estimated the wall heat transfer flux at different locations in the food can from postprocessing of the CFD simulation results. The heat transfer flux at the wall may be defined as follows (Eq. 2.16).

$$q_w = k \left. \frac{\partial T}{\partial r} \right|_{r=R} \tag{2.16}$$

Here, the negative sign is absent in the Fourier's law equation, because the fluid temperature increases with increasing radial distance from the axis. Based on this heat flux, the heat transfer coefficient may be defined using a suitably defined bulk reference temperature. The definition of the heat transfer coefficient is given below (Eq. 2.17).

$$q_w = h\left(T_w - T_{ref}\right) \tag{2.17}$$

Here T_w and T_{ref} are the wall and reference temperatures respectively. The fine print involving the definition of reference temperature should be clearly stated as different stipulations of reference temperatures lead to different values of the estimated heat transfer coefficients based on a specified heat flux. The overall heat transfer coefficient for the food can heating may be defined in terms of the volume-averaged temperature $< T_V >$ in the domain. The volume-averaged temperature is the temperature that would be obtained if the entire contents of the can were to be mixed instantaneously. The volume-averaged temperature is defined as shown below in Eq. (2.18). In this equation L refers to the total height of the food can.

$$< T_V > = \frac{\int_0^L \int_0^{2\pi} \int_0^R T r dr d\theta dz}{\int_0^L \int_0^{2\pi} \int_0^R r dr d\theta dz} \tag{2.18}$$

Using $< T_V >$ as the reference temperature, we get the following definition (Eq. 2.19) for the overall volume averaged heat transfer coefficient $\left(h^V_{avg}\right)$ in the food can.

$$q_w = h^V_{avg}(T_w - < T_V >) \tag{2.19}$$

2.3.2 Estimation of heat transfer rates from the surfaces

For each can geometry, the heat transfer rates at the top, curved wall, and bottom surfaces were determined at various time instances (Kannan and Sandaka, 2008). The heat transfer rate contributions from these surfaces, as well as their sum, are depicted in Fig. 2.2 for the regular cylinder (Sandaka, 2006). The heat transfer rate to the cylindrical segment initially was around 100 W but decreased by two orders of magnitude as the end of the sterilization process approached due to equalization of temperature differences within the food can. Up to 80 s, there was almost an equal contribution from the top and bottom surfaces. It may be inferred that until this duration, conduction was the chief mode of heat transfer, and disruptions due to

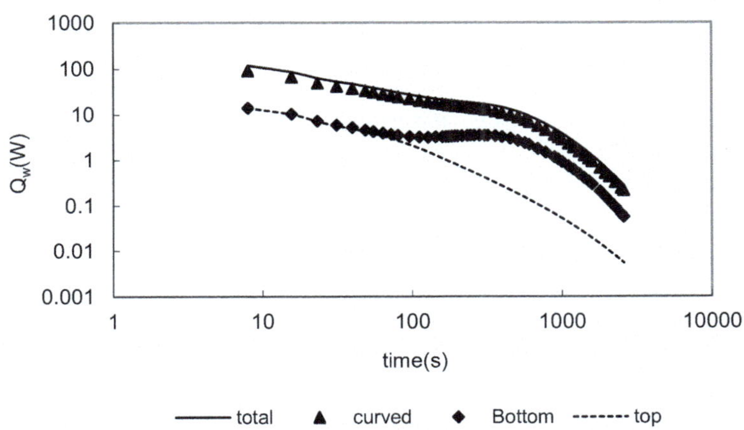

Figure 2.2 Comparison of heat transfer rates from the top, bottom and curved surfaces of the "regular" food can and their total contribution. Thermal conductivity of food medium = 0.7 W/m.°C.

convection currents were yet to set in. Subsequently, the heat transfer rates, however, decline substantially, especially from the top surface, indicating that the temperature driving force for heat transfer has declined as well. Variations may be observed between the main heat transfer surfaces that is, the bottom surface and the curved sidewall, with the latter dominating. This leads to the idea that any modifications to increase the convective heat transfer rates may be implemented at the curved sidewalls rather than at the top and bottom surfaces of the food can.

2.3.3 Rayleigh number analysis

From the heat transfer coefficients based on volume averaged temperatures and the length of the can as the characteristic dimension, the Nusselt number was defined. Consistent with convection literature, the Nusselt numbers were plotted as a function of Rayleigh number as defined below (Eq. 2.20). Here L refers to the total height of the food can.

$$< \mathrm{Ra} >= \frac{g\beta\rho \mathrm{L}^3 < \Delta \mathrm{T} >}{< \mu > \alpha} \tag{2.20}$$

As volume averaged temperature difference and volume averaged viscosity were used in Eq. (2.20), the Rayliegh number is also defined as volume averaged. When Nu was plotted versus Ra, a peculiar trend involving branching of Nusselt numbers was encountered. For a given Ra, it was possible to find two Nusselt numbers near the branching point, which goes against the increasing trend of Nu with Ra in conventional heat transfer correlations. However, the transient nature of the heat transfer should be borne in mind when analyzing this peculiar Nusselt number variation. A few notable trends could be discerned from the Nu versus Ra profiles (Fig. 2.3). For a given aspect ratio S, the Nusselt number curve is found at lower Ra at higher k values. For a given k, the Nu curve is found at higher Ra at higher S values.

This indicates that the heat penetration rate into the can will be slower at lower k and higher aspect ratios. The interesting trends may be explained based on the variation in volume averaged viscosity, and temperature with time.

Kannan and Sandaka (2008) carried out a detailed analysis for the variations in volume averaged viscosity $<\mu>$ and temperature driving force $<\Delta T>$ with time. They found that the ratio $\frac{<\Delta T>}{<\mu>}$ for a food material in a food can of specified aspect ratio and thermal conductivity passed through a maximum with time during the heating process. Even though both parameters decreased with time, the decrease in $<\mu>$ was faster than $<\Delta T>$ at lower times, leading to a maximum. Subsequently, $<\Delta T>$ tended to zero while the $<\mu>$ reached a constant value, and hence the ratio declined. The $\left(\frac{<\Delta T>}{<\mu>}\right)_{max}$ could be correlated with k and S as given below (Eq. 2.21).

$$\left(\frac{< \Delta \mathrm{T} >}{< \mu >}\right)_{max} = 8.22(\mathrm{kS})^{0.144} \tag{2.21}$$

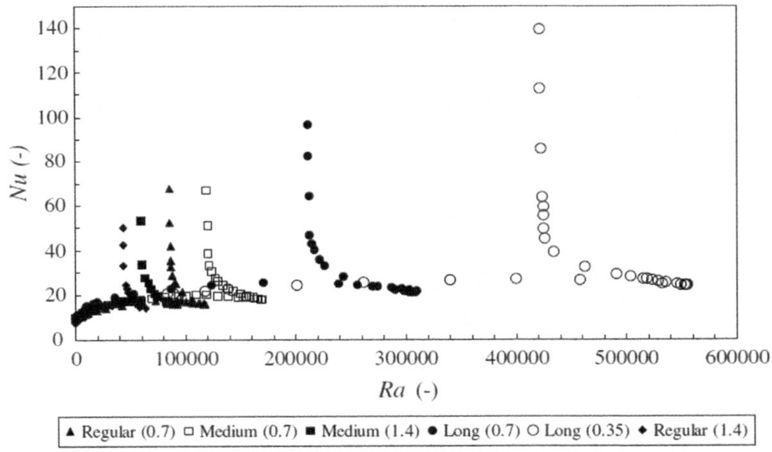

Figure 2.3 Variation of Nusselt number with Rayleigh number at different aspect ratios. Food can aspect ratios and food medium thermal conductivities in W/m.°C are shown in parentheses.
Reprinted from Journal of Food Engineering 88, Kannan, A., Sandaka, P.Ch.G., Heat transfer analysis of canned food sterilization in a still retort, 213–228, 2008, with permission from Elsevier.

Using Eq. (2.20), a maximum volume averaged Rayleigh number may be defined as shown below (Eq. 2.22).

$$< \mathrm{Ra}_{\mathrm{max}} > = \frac{g\beta\rho L^3 \left(\dfrac{< \Delta T >}{< \mu >}\right)_{max}}{\alpha} \tag{2.22}$$

Then, the actual <Ra> values defined by Eq. (2.20), were scaled by $<\mathrm{Ra}_{\mathrm{max}}>$ given above (Eq. 2.22) to obtain the volume-averaged reduced Rayleigh number $<\mathrm{Ra}_r>$. The Fourier number at which <Ra> passed through a maximum (Fig. 2.4) was termed the critical Fourier number Fo_c. The numerical value of Fo_c was 0.028. The actual Fo values were also divided by this Fo_c to obtain the reduced Fo_r. These scaling steps helped to express the $<\mathrm{Ra}_r>$ versus Fo_r variation in a unified form as given by Eq. (2.23).

$$\mathrm{Ra}_r = \frac{1.18 \exp\left(1.72\mathrm{Fo}_r\right)}{1 + 0.75 \exp\left(2.02\mathrm{Fo}_r\right)} \tag{2.23}$$

The unified form of the correlation is shown in Fig. 2.4.

This correlation is applicable for the aspect ratios in the range 1.37–2.31 and thermal conductivities in the range 0.35–1.4 W/m.K. From this figure, it is evident that there are the same $<\mathrm{Ra}_r>$ values at two Fo_r values, one below and the other above

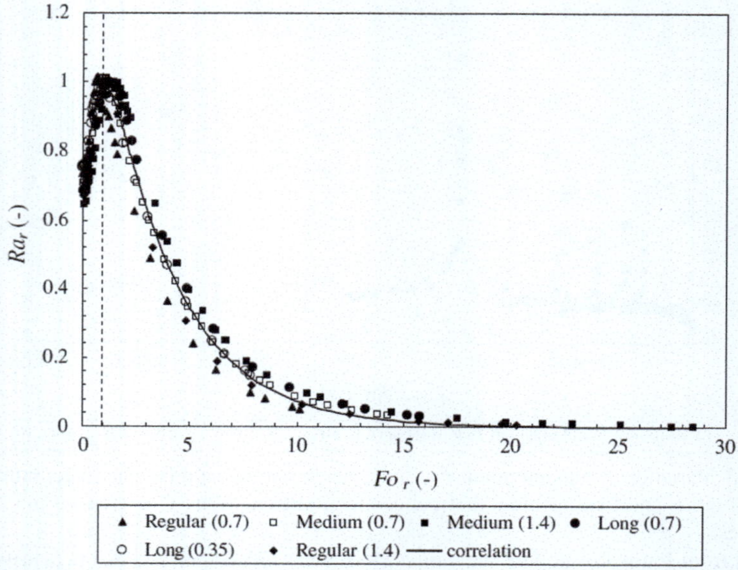

Figure 2.4 Variation of reduced Rayleigh number with reduced Fourier number for food cans of different aspect ratios and thermal conductivities. The vertical dashed line from one on the abscissa indicates the critical reduced Fourier number.
Reprinted from Journal of Food Engineering 88, Kannan, A., Sandaka, P.Ch.G., Heat transfer analysis of canned food sterilization in a still retort, 213–228, 2008, with permission from Elsevier.

Fo_C. The heat transfer regime at subcritical Fourier number corresponds to that of conduction, while those at supercritical Fourier number are characteristic of convection. The conduction regime is characterized by uniform kernels of temperatures, while in the convection regime, the convection currents distort the temperature profiles. The SHZ is in the geometric center at subcritical Fourier numbers, while it is near the bottom in the supercritical contour plot. The circulating fluid in the latter case helps the warmer liquid to penetrate the middle regions of the food can, thereby pushing the SHZ toward the bottom. Hence, the critical Fourier number plays an important role and represents the regime transition. As indicated above, the critical Fourier number took the value of 0.028. The justification for this value is discussed in the form of scaling analysis in the next section.

2.3.4 Scaling analysis for critical Fourier number

The order of magnitude analysis for the critical Fourier number was based on Bejan (2009). The time at which the onset of convection occurs was estimated by Eq. (2.24).

$$t_c \approx \left(\frac{\nu L}{g\beta\Delta T\alpha}\right)^{0.5} \tag{2.24}$$

Details of this derivation are given in Koribilli et al. (2011). Here ν is the kinematic viscosity of the food medium. Expressing in terms of dimensionless time, Eq. (2.24) may be given in terms of the critical Fourier number($\alpha t_c / R^2$) to yield Eq. (2.25).

$$Fo_c = \left(\frac{L}{R}\right)^2 (Ra)^{-\frac{1}{2}} \tag{2.25}$$

The next step will be to identify the orders of magnitudes of the different variables in Eqs. (2.20) and (2.25). Both geometric parameters and fluid properties must be considered to have an order of magnitude estimate for Fo_c. When convection sets in, it is reasonable to assume the temperature difference to be the difference between the wall and initial fluid temperatures which are 121 and 40°C respectively. Hence $< \Delta T >$ may be set an order of magnitude of 10^2 °C. The kinematic viscosity values in the domain at incipient convection was around 0.002–0.004 m^2/s and hence the order of magnitude was set at 10^{-3} m^2/s. The height of the can L varied between 0.111 and 0.15 m and hence its order of magnitude is 10^{-1}. The order of magnitudes of the remaining parameters are summarized in Table 2.1 (Koribilli et al., 2011).

Using the order of magnitudes presented in Table 2.1 in Eq. (2.25), the value of Fo_c is 0.032, which is in fair agreement to the value of 0.028 found by Kannan and Sandaka (2008).

2.3.5 Assumption regarding viscosity of the fluid medium

In sections 2.3c and 2.3d, the Rayleigh numbers involved the volume-averaged viscosity. In the CFD simulations, the power law model given by Eq. (2.1) was used without any assumption on the rate of shear. For analysis purposes, the correlations involving Ra_r and Ra_{max} developed earlier apply for Newtonian fluids. Hence, Kannan and Sandaka (2008) as well as Koribilli et al. (2011) approximated the food medium as a Newtonian fluid for want of heat transfer correlations describing transient convective heat transfer of nonNewtonian fluids in food cans with isothermal walls. This approximation as a "Newtonian like fluid" was also carried out in earlier studies

Table 2.1 Actual and order of magnitude values for different variables.

Sl. No.	Parameter	Value	Order of magnitude
1	R	32.5–40 mm	10^{-2} m
2	g	9.8 m^2/s	10 m^2/s
3	β	0.0018°C^{-1}	10^{-3}°C^{-1}
4	ρ	950 kg/m^3	10^3 kg/m^3
5	α	9×10^{-8}–3.6×10^{-7} m^2/s	10^{-7} m^2/s

(Abdul Ghani et al., 1999; Farid and Abdul Ghani, 2004; Kumar and Bhattacharya, 1991) for the same system, and the shear rate was fixed at 0.01 s^{-1} for the entire sterilization process. The shear rates in the fluid were small due to low velocities engendered by natural convection. The use of a constant rate of shear led to the food medium assuming Newtonian behavior with its viscosity solely being a function of temperature. Koribilli et al. (2011) tested this approximation by using a volume-averaged viscosity defined in terms of volume-averaged temperature and volume-averaged shear rate as shown below in Eq. (2.26).

$$< \mu > = 0.002232 \exp\left(\frac{30740}{R < T >}\right)(< \dot{\gamma} >)^{n-1} \tag{2.26}$$

Using the results from rigorous CFD simulations, the volume-averaged viscosity data were correlated as a function of <T> only as per Eq. (2.26). The volume-averaged rate of shear $< \dot{\gamma} >$ was taken as a parameter and fitted to the data. The value of $< \dot{\gamma} >$ was estimated at 0.0118 $s^{-1,}$ which was in reasonable agreement with the previously assumed value of 0.01 s^{-1}.

2.3.6 Prediction of volume-averaged temperatures in subcritical regime ($Fo < Fo_c$)

Based on the discussions above, the next step is to predict the volume-averaged and slowest heating zone temperatures in the food can, in this conduction-dominated regime. Upon substituting Eq. (2.11) into the following dimensionless temperature expression (Eq. 2.27), the dimensionless volume-averaged temperature in the domain may be captured at desired time instants.

$$< \xi_V > = \frac{\int_0^L \int_0^{2\pi} \int_0^R \xi r dr d\theta dz}{\int_0^L \int_0^{2\pi} \int_0^R r dr d\theta dz} \tag{2.27}$$

The resulting integration results in the following analytical expression (Eq. 2.28) for the dimensionless volume averaged temperature.

$$< \xi_{FC} > = \left(\frac{32}{\pi^2}\right)\left[\sum_{n=1}^{\infty} \frac{1}{(2n-1)^2} e^{-\left\{\left(\frac{2n-1}{L}\right)^2 (\pi^2 \alpha t)\right\}}\right]\left[\sum_{n=1}^{\infty} \frac{1}{M_n^2} e^{-\left(M_n{}^2 \frac{\alpha t}{R^2}\right)}\right] \tag{2.28}$$

The dimensionless volume averaged temperature in the finite cylinder $< \xi_{FC} >$ may be converted to actual volume averaged temperature (<T>) using Eq. (2.3).

The slowest heating zone temperature may be captured in the conduction dominated regime by simply setting $r = 0$ and $x = 0$ in Eq. (2.11).

As seen above, the initial heating phase coming under the subcritical regime may be dealt analytically for both volume averaged temperature as well as the slowest heating zone temperature. Further, the CFD simulations may be carried out for conduction heat transfer, considering the food medium to be a fictitious solid but with the same density, thermal conductivity, and thermal diffusivity as that of the liquid food. From the simulation results, both the slowest heating zone temperature at the geometric center of the food can, and the volume-averaged temperature may be validated with the analytical solutions given above. These independent approaches will mutually validate the analytical as well as CFD simulations.

2.3.7 Correlation for Nusselt number in supercritical regime ($Fo > Fo_c$)

This regime is characterized by convection and analytical solutions developed for the conduction model no longer apply. To predict the volume-averaged temperature in the food can, the simple lumped system transient heat transfer approach may be adopted. Since $\langle T \rangle$ is a measure of the overall temperature in the food medium, it may be considered as a lumped parameter. The utility of predicting $\langle T \rangle$ lies in its ability to predict the overall lethality treatment received by the food can as a function of time.

To carry out the lumped system analysis, an appropriately defined heat transfer coefficient, or equivalently Nusselt number, is required. The total heat flux at the curved wall of the cylindrical food can was obtained from CFD simulations (Kannan and Sandaka, 2008). The total heat transfer flux into the entire food container may be well approximated by the heat flux from the curved wall (Fig. 2.2). The top wall contributed negligibly to heat flux at longer times as the temperature driving force declined. Though the heat flux at the bottom wall was high, its contribution to the total heat transfer rate was small, as the bottom surface area was only ~18% of the curved surface area. The driving force used in the heat flux calculations is based on $\langle T \rangle$ as the reference temperature (Eq. 2.19). From the total curved wall heat flux and the temperature driving force (T_W-$\langle T \rangle$), the heat transfer coefficient h_{avg}^{V} may be found at different time instants. The CFD-based Nusselt numbers may be calculated using Eq. (2.29).

$$Nu = \frac{h_{avg}^{V} L}{k} \tag{2.29}$$

To use the CFD-based Nusselt numbers in the lumped system approach, they must be consolidated in the form of a correlation. From natural convection literature, it may be anticipated that Nu is primarily governed by the Rayleigh number. To develop such a correlation, the Nusselt number trend as a function of Fo was observed and illustrated in Fig. 2.5.

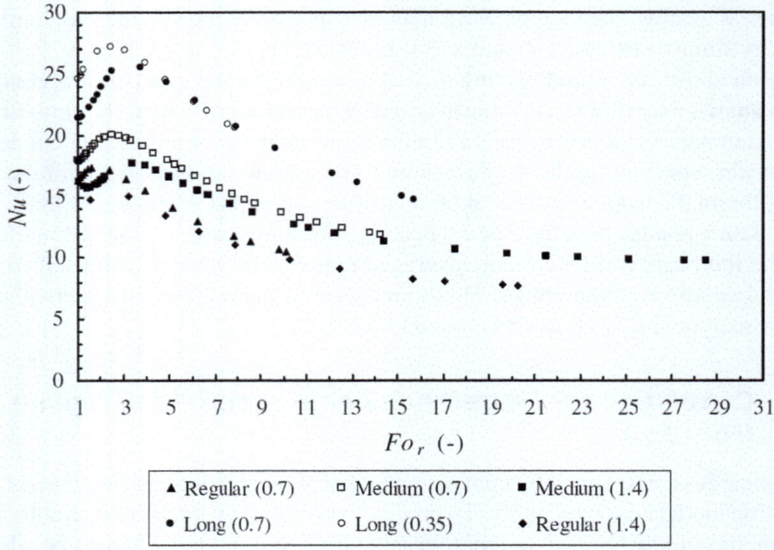

Figure 2.5 Variation of Nusselt number with reduced Fourier number in the supercritical regime.
Reprinted from Journal of Food Engineering 88, Kannan, A., Sandaka, P.Ch.G., Heat transfer analysis of canned food sterilization in a still retort, 213–228, 2008, with permission from Elsevier.

The correlation development is involved in the supercritical regime as conduction is not the sole mechanism of heat transfer. Natural convective flow would influence the temperature distribution in the food can and hence <T>. It is expected that the wall heat fluxes in the supercritical regime would be markedly different from those in the subcritical regime due to fluid flow. However, conduction may still contribute significantly at supercritical Fourier numbers not much higher than one. Hence, both conduction and convection effects must be suitably accommodated in the correlation for Nu in the supercritical regime.

Figure 2.5 reveals that Nusselt number passes through a maximum with Fo_r. This trend is distinct and may not be ignored (Kannan and Sandaka, 2008). From the trends in this figure, the aspect ratio is more influential than thermal conductivity. For long cylinders, the Nusselt numbers are not much affected despite a 2-fold variation in k. Thermal conductivity has a small inverse effect on Nu, especially before the occurrence of maximum (Fig. 2.5). Lower thermal conductivity and higher aspect ratio led to higher Nusselt number. The maximum in the Nusselt number was correlated to the aspect ratio and thermal conductivity as given by Eq. (2.30).

$$Nu_{max} = 12.84\frac{S^{0.69}}{k^{0.18}} \tag{2.30}$$

The actual Nusselt numbers were scaled by Nu_{max} to get reduced Nusselt numbers, which were then correlated with Fo_r and Ra_r. The reduced Rayleigh number expression was given by Eq. (2.23). The reduced Nusselt number was correlated as shown by Eq. (2.31).

$$Nu_r = 1.05Ra_r^{0.16} - 0.2\frac{Ra_r^{7.96}}{Fo_r^{0.5}} \tag{2.31}$$

The presence of a maximum in the Nusselt number (Fig. 2.5) indicates that there are two competing effects in the supercritical regime. Slightly above the critical Fourier number, the convective effects are tempered by conductive heat transfer, which may still be relevant. However, when convection starts dominating with increasing Fo_r beyond 1, the second term in Eq. (2.31) becomes increasingly small as Fo_r increases. At higher Fo_r the Ra_r decreases as well (Fig. 2.4), making the second term vanish quickly. At higher Fourier numbers, the Nusselt number simplifies to the conventional form, that is, a function of Rayleigh number as given by Eq. (2.32).

$$Nu_r = 1.05Ra_r^{0.16} \tag{2.32}$$

2.3.8 Estimation of volume averaged temperatures in the supercritical regime (Fo > Fo$_c$)

The lumped parameter transient heat transfer equation may be written as shown in Eq. (2.33).

$$\rho VC_p\frac{d<T>}{\partial t} = Nu_rNu_{max}\left(\frac{k}{L}\right)A(T_w - <T>) \tag{2.33}$$

In the above equation, V is the volume of the food can, and A is the heat transfer area. The Nusselt number obtained by multiplying Eq. (2.31) by Eq. (2.30) may be used in the above ordinary differential equation (ode) and numerically integrated. An ode rather than a partial differential equation (pde) arises as the volume-averaged temperature has been averaged over space and is only a function of time. In the supercritical regime, direct analytical integration is precluded as the Nu is a nonlinear function of time, as shown by Eq. (2.31). It may also be noted that the can aspect ratio and food medium thermal conductivity are incorporated through Nu_{max} (Eq. 2.30). The $<T>$ value estimated by the conduction model (Eq. 2.28) at Fo_r of unity is specified as the initial condition. The ODE45 routine in MATLAB® (Mathworks, Massachusetts) was used for numerical integration of the ode.

i) Comparison of CFD predictions of $<T>$ with alternate models

The volume averaged temperatures in both the subcritical and supercritical regimes may be obtained from CFD simulations. To cross validate these in the two regimes, the predictions from the conduction model (Eq. 2.28) and lumped parameter model

(Eq. 2.33) were respectively used. Typical results are presented in Fig. 2.6 where a good agreement between CFD and alternative models' predictions may be seen. Further, if conduction heat transfer were to be assumed throughout the thermal sterilization process, the temperature rise would be significantly slower than the actual case where convection effects are dominant. For conduction heat transfer, CFD simulations were carried out by solving the appropriate thermal energy equation in cylindrical coordinates (Eq. 2.34), assuming the food to be a solid medium but with otherwise the same values of C_p, ρ, and k. The resulting temperature distribution was volume averaged using (Eq. 2.18) to get volume averaged temperature <T>.

$$\rho C_p \frac{\partial T}{\partial r} = k\left[\frac{1}{r}\frac{\partial}{\partial r}\left(r\frac{T}{\partial r}\right) + \left(\frac{\partial^2 T}{\partial z^2}\right)\right] \tag{2.34}$$

These CFD conduction heat transfer results were validated with the analytical model predictions, as also shown in Fig. 2.6. Also, the clearcut transition from conduction to convection mode of heat transfer beyond the critical Fourier number is noteworthy. Since the purpose of thermal sterilization is to deliver adequate lethality of treatment to the food container contents, it is discussed next.

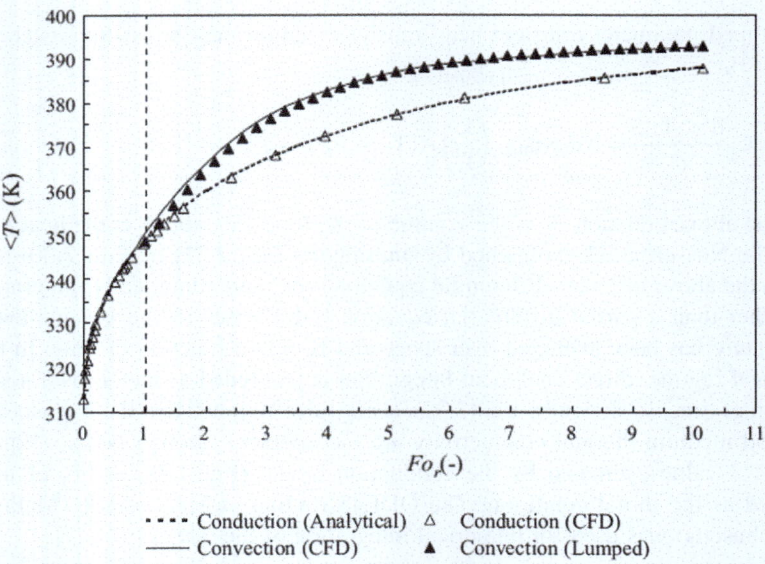

Figure 2.6 Comparison of volume-averaged temperatures predicted from conduction and convection-augmented modes of heat transfer for the regular food can.
Reproduced with permission from Springer Nature Food Bioprocess Technology 4, Koribilli, N., Aravamudan, K., Aditya Varadhan, M.U.S.V., Quantifying Enhancement in Heat Transfer Due to Natural Convection During Canned Food Thermal Sterilization in a Still Retort. 429–450, 2011.

2.3.10 Enhancement in the lethality of sterilization due to natural convection

The duration of the sterilization process was specified at 2589 s based on the SHZ temperature reaching the desired value. Based on the volume-averaged temperature, the total integrated lethality that was delivered during this period, was estimated by (Koribilli et al., 2011) using the following equation (Eq. 2.35)

$$F_{<T>} = \int_0^{2589} 10^{\frac{<T>-T_{ref}}{z}} dt \qquad (2.35)$$

The reference temperature used was 121°C. The value of z was fixed at 10°C. The z-value is a temperature change that is required to increase the rate of thermal destruction by a factor of 10 (Deák, 2014). The lethality calculations based on Eq. (2.35) were done after finding the <T> values in different ways, as listed below:

a) Using the simplest conduction model (Eq. 2.28) to predict the <T> for the entire duration of the thermal sterilization process, that is, both in the subcritical and supercritical regimes. Obviously, the model predictions of temperature rise will be conservative in the convection-dominated supercritical regime

b) Using the lumped parameter model (Eq. 2.33) in the supercritical regime, while the analytical solution to the heat conduction equation (Eq. 2.28) is used in the subcritical regime

c) Using the CFD simulation results carried out with the full-fledged model equations (Eqs. 2.12 – 2.14) for the entire duration of the sterilization process

The lethality of the treatment based on <T> is illustrated for the 'Regular' food can in Fig. 2.7.

From the above figure, there is a good match between the CFD model-based lethality and those predicted by the analytical conduction model in the subcritical regime and the lumped parameter transient heat transfer model in the supercritical regime. The CFD model-based lethality, and the analytical conduction model lethality agree in the subcritical region as convection had yet to set in. However, in the supercritical region, there is a significant enhancement in lethality due to the dominance of natural convective heating over conductive heating, showing a significant discrepancy between the two approaches. A small difference between the <T> values predicted by the convection and conduction model (Fig. 2.6) translates into a much larger discrepancy between the predicted lethality values (Fig. 2.7). This may be attributed to Eq. (2.35) where the temperatures are raised to the power of 10. For the 'Regular' food can with k value of 0.7 W/m.°C, the total lethality based on <T> at the end of the treatment was 659.9 s with convection and only 144.8 s with conduction, that is, an enhancement due to convection over pure conduction by a considerable 4.6. Koribilli et al. (2011) carried out calculations with different food cans (Regular, Medium, and Long) as well as different food medium thermal conductivities and reported enhancements based on <T> in the lethality. The lethality calculations based on <T> provide an estimate of the overall lethality received by the entire food can at different times.

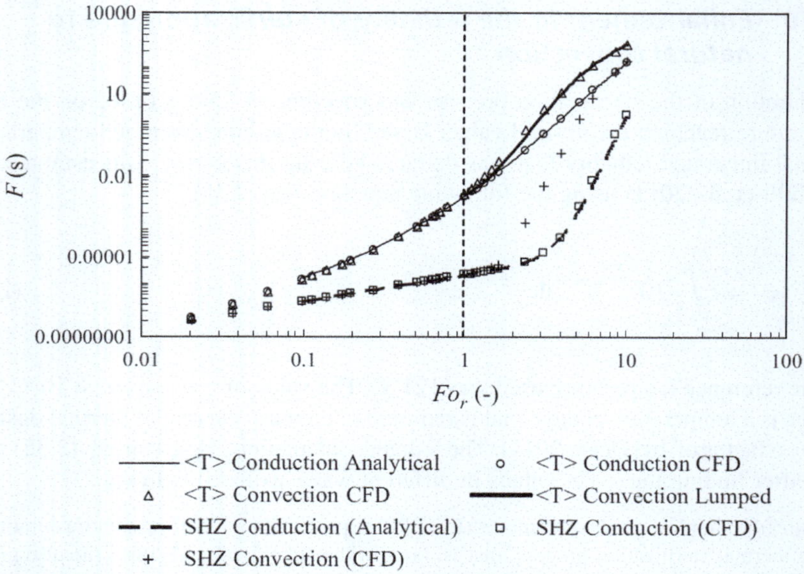

Figure 2.7 Comparison of lethality of thermal sterilization predicted by sole conduction and from natural convection enhanced heat transfer for the 'regular' cylinder with k value of 0.7 W/m.oC.
Reproduced with permission from Springer Nature Food Bioprocess Technology 4, Koribilli, N., Aravamudan, K., Aditya Varadhan, M.U.S.V., Quantifying Enhancement in Heat Transfer Due to Natural Convection During Canned Food Thermal Sterilization in a Still Retort. 429–450, 2011.

In addition to lethality based on volume-averaged temperatures, the total integrated lethality estimates may be based on the SHZ temperature as well. The latter may be found from CFD calculations in both the subcritical and supercritical regimes. The total integrated lethality based on the SHZ temperature may be defined according to Eq. (2.36).

$$F_{T_{SHZ}} = \int_0^{2589} 10^{\frac{T_{SHZ}-T_{ref}}{z}} dt \tag{2.36}$$

In this equation, T_{SHZ} is the slowest heating zone temperature. This gives an estimate of the minimum lethality received by the food can contents. The lethality trends based on SHZ are also shown in Fig. 2.7.

The SHZ temperature will always be lower than <T> except at t = 0. Hence, the lethality of the treatment received in the SHZ will be considerably lower than that based on the volume-averaged temperature (Fig. 2.7). In the conduction-dominated subcritical regime, the SHZ temperature may be set to the temperature at the geometric center of the food can and may be estimated from an analytical solution (Eq. 2.11). There is a good match between the lethality values based on the sole conduction heating from CFD and the analytical solution. However, these values, though matching the lethality values predicted by the full convection-based CFD model in the subcritical regime, are far lower in the supercritical regime. This indicates that the lethality based on the SHZ temperature is also considerably enhanced by natural convection over sole conduction in the food can. For the 'Regular' food can with a k value of 0.7 W/m.°C, the lethality based on SHZ temperature at the end of the treatment was 164.6 s with convection and only 1.75 s with conduction, that is, an enhancement due to convection over pure conduction was as much as 94.1. Koribilli et al. (2011). carried out calculations with different food cans (Regular, Medium, and Long) as well as different food medium thermal conductivities and reported enhancements in the lethality based on the SHZ temperature as well.

From the above discussion, it may be concluded that even though the temperature discrepancy between the true convection mode of heating and the approximate pure conduction mode of heating may not be much, it causes orders of magnitude discrepancy in the lethality of treatment, especially in the slowest heating zone. While the CFD simulations are expensive, they are imperative to track the accurate evolution of temperature and integrated lethality over the duration of the entire thermal sterilization process. Accurate estimation of the minimum and overall lethality received by the food can help the process designer to specify correct times for thermal sterilization so that the microorganisms are sufficiently killed without compromising on the taste, appearance, and nutritive value of the food product. To enhance the rate of thermal sterilization so that the can receives adequate heat treatment in a shorter time, various methods have been proposed in the literature. These are briefly summarized in the following sections.

2.3.11 Enhancement of sterilization rates through rotation of the food can

Abdul Ghani et al. (2003) studied the effect of rotating a can containing carrot orange soup using CFD simulations. The food can was oriented horizontally, and its speed of rotation (N) was 10 rpm. These authors assumed the highly viscous fluid to be shear rate independent and hence Newtonian. Laminar natural convective flow could be justified as the Grashoff number was low. Viscous heating was ignored as the fluid velocities were quite small in the food can. The food can rotation was accounted for in the boundary condition where the radial velocity was equated to the rotational speed by Eq. (2.37).

$$V_r = R\left(\frac{2\pi N}{60}\right) \tag{2.37}$$

The other boundary and initial conditions were the same as those specified for a stationary can as discussed in section 2.2e.

As a result of the rotation of the food can, the vertical velocities were 10–100 times higher than those obtained in a stationary food can after 1000 s of heating. The SHZ was found to split into two regions by the combined effects of natural and forced convection. The can's rotation led to the SHZ volume occupying less than 5% of the total volume at the completion of the heating, which took 3000 s. This was smaller by 10% and 20% when compared with the SHZ volumes occupied in the immobile can sterilized in a vertical and horizontal orientation, respectively.

2.3.12 Enhancement of sterilization rates through geometry changes

Varma and Kannan (2006) detailed the enhancement in food sterilization times without external agitations through modifications in the geometry and orientation of the food can. In addition to the commonly used cylindrical food can (CYL), they also proposed a conical food can having the same volume as that of the Regular cylindrical can. Both cone pointing upward (CPU) and cone pointing downward (CPD) configurations were considered by the authors. The conical can's heights were also fixed to be the same as that of the cylindrical can at 111.11 mm. CFD simulations were carried out using the same initial and boundary conditions as specified in section 2.2d. The food item considered was modeled as 0.85% w/w carboxy methyl cellulose (CMC).

The authors estimated the maximum Grashof numbers using the food can height as the characteristic length dimension, maximum temperature, and minimum viscosity in the domain. The Grashoff number was 1200 at the beginning of the heat treatment and reduced to three at the end after 2590 s. These low values of the Grashoff numbers during the entire course of the thermal sterilization justified the choice of laminar flow. In all three geometries, the temperature difference between the wall and the core region led to density differences, and the ensuing buoyancy forces led to circulatory flow patterns. The buoyancy forces initially increase due to widening differences in temperature between the wall and core region (Varma and Kannan, 2006). The circulatory flow engendered by natural convection was characterized by maximum velocities in the order of 10^{-4} m/s. The fluid moved upwards at the sides of the cylindrical food can and circulated toward the center. The temperature contours will be uniform kernels in the case of pure conduction. However, due to the circulatory flow, the temperature contours are distorted and the SHZ tends to migrate toward the bottom of the food can (Fig. 2.8).

Subsequently, due to the flow of the heated fluid to the core region, the temperature differences reduced. Hence, the velocities in the domain will initially increase and then approach zero when temperatures become nearly uniform in the domain. The velocities in the CPU were found to decline the fastest due to faster propagation of heat, thereby reducing the temperature gradients and buoyancy forces. The CPU configuration was found to sterilize the food contents faster than the regular food can, which in turn achieved faster sterilization than the CPD. These may be seen from the temperature contour plots for the three configurations given in Fig. 2.8. The cross-sectional area of the SHZ is maximum for CPD and minimum for.

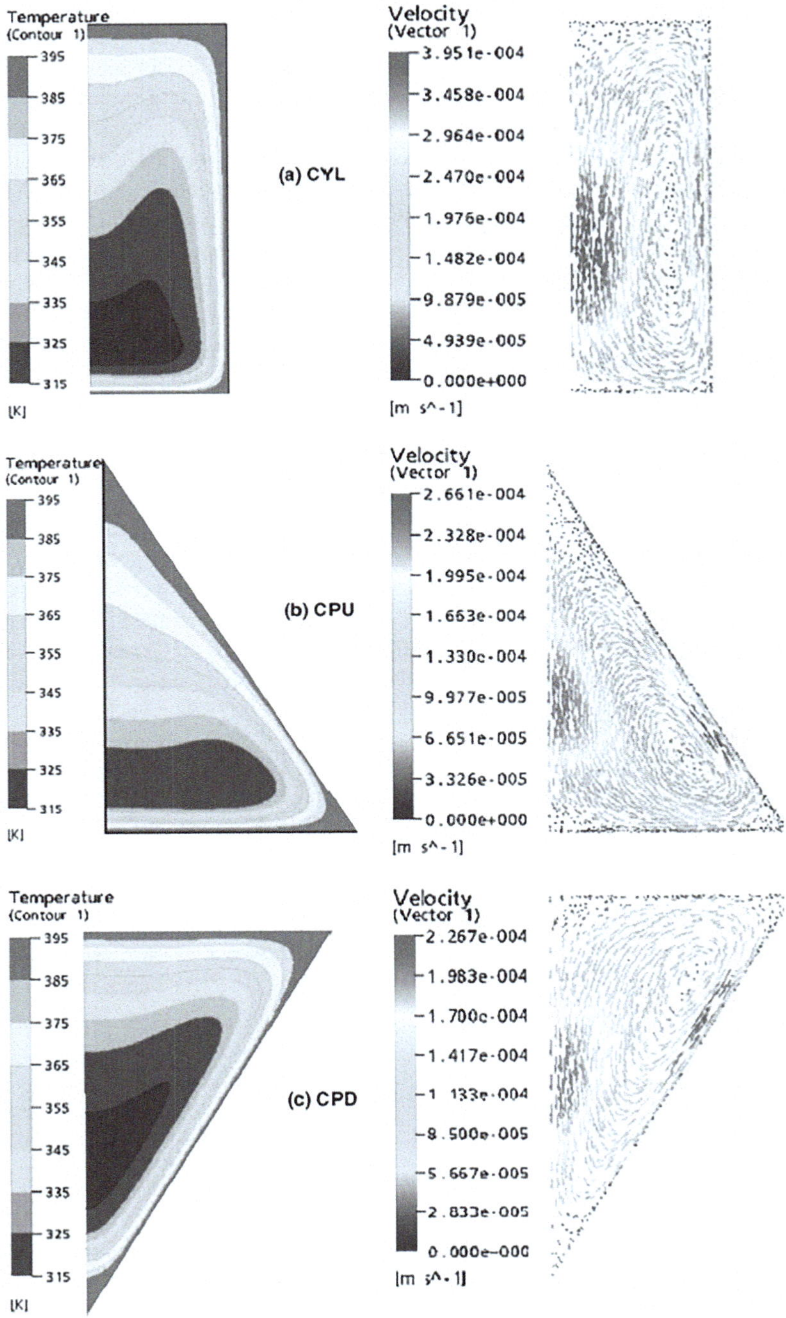

Figure 2.8 Plots of temperature and velocity contours for the three geometries at time of 240.44 s. one-half cross section is only shown due to axis-symmetry.
Reprinted from Journal of Food Engineering, 77, Varma, M.N., Kannan, A., CFD studies on natural convective heating of canned food in conical and cylindrical containers, 1024–1036, 2006, with permission from Elsevier.

CPU, while that for the Regular cylinder fell in between. Figure 2.9 traces the SHZ temperature with time. It may be seen that the SHZ temperature rises the fastest in CPU orientation, followed by the cylinder, and then the CPD orientation. For the CYL and CPU, the SHZ moved axially from the center in a downward direction and then moved toward the bottom right of the container. In the case of CPD, the SHZ moved down axially but did not move to the right. The SHZ settled at about 10% height from the bottom. Eventually, when the convection currents declined due to equalization of temperatures, the SHZ started migrating toward the center of the can as conduction heat transfer again became prominent.

Kannan and Sandaka (2008) had observed that in a cylindrical food can, the heat transfer through the curved surface dominated the heat transfer from the top and bottom surfaces. Hence, they opined that any geometric modifications to the food container may be done at the curved surface, where the heat transfer is much greater rather than from the top and bottom surfaces. In the CPU, heat transfer from the top region is obviated, while in the CPD, it is dominant, leading to poor performance. In the CPU, the region at the top is comparatively smaller than the other two geometries, and hence it heats the fastest. In the CPD arrangement (Varma and Kannan, 2006), observed that the bottom region heated the slowest as the convection was relatively less prominent. They considered three planes in each of the food geometries. These were located at 37.77, 55.55, and 74.43 mm from the base and were termed bottom, mid, and top planes, respectively. They found that the temperature rise at the midpoint of the plane (between the axis and the curved surface) may be ranked as follows.

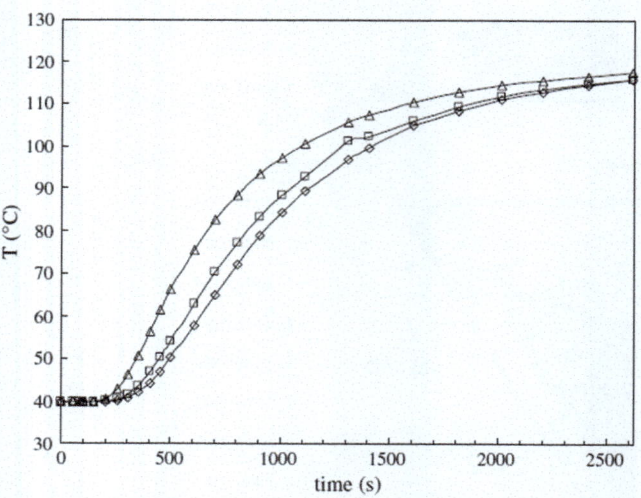

Figure 2.9 Variation of the SHZ temperature with time for CPU (Δ), cylinder (□) and CPD (◇). Reprinted from Journal of Food Engineering, 77, Varma, M.N., Kannan, A., CFD studies on natural convective heating of canned food in conical and cylindrical containers, 1024–1036, 2006, with permission from Elsevier.

CPU (top plane) > CPU (mid-plane) > CYL(top plane) > CPU (bottom plane) > CYL (mid-plane) > CPD (top plane) ≈ CYL (bottom plane) > CPD (mid-plane) > CPD (bottom plane).

The above trend clearly indicates that the CPD arrangement overall experiences slower heat transfer even when compared to the CYL geometry.

Varma and Kannan (2005) aimed to reduce the sterilization times by studying the effects of truncating the conical geometry of the food can and altering its orientation. Equal volume food cans were used irrespective of whether the geometry was cylindrical, full, or truncated cone. Using the nonNewtonian fluid CMC (0.85% w/w) as the model food medium, CFD simulations were carried out under laminar flow conditions. The full cone gave the shortest sterilization time when oriented vertically. The sterilization time was reduced by 17% when compared to the cylinder. The cylinder heated faster when oriented horizontally than vertically. The sterilization times for truncated cones were intermediate to those of the full cone and cylinder (Varma and Kannan, 2005). It was noteworthy that even a small inclination of the food cans may reduce the convection velocities and increase the sterilization time. Hence, care must be exercised to ensure that the food cans are aligned vertically, even without a small deviation in the food retort. When geometric modifications are made to the conventional cylindrical food can for esthetic reasons, the sterilization time-temperature history available for the cylinder may, however, not be applied to the new geometries, even though the volume and height are the same.

Augusto and Cristianini (2010) studied thermal sterilization in brick-shaped packages. Unlike conical and cylindrical geometries, these authors found that the orientation of the brick-shaped package did not alter the sterilization duration. The model food used was water and a viscous liquid that was 10 times more viscous than water, with the latter exemplifying concentrated milk, egg products, and concentrated juice.

Borah et al. (2020) studied thermal sterilization of processed food stuff in containers of circular and elliptical cross-section. They studied the effect of aspect ratio and orientation of the food can as well as the fluid rheological properties (shear rate thinning, Newtonian, and shear rate thickening) on the heating rates and lethality of treatment. A long food container, when positioned horizontally lead to circular flow patterns, which in turn result in better heat distribution. Hence, the required sterilization time was reduced by 16% when the long cylindrical can was oriented horizontally rather than vertically. Changing the can cross-section from circular to elliptical further reduced the sterilization time by 22%. The authors concluded that the geometry and orientation of the food can significantly affect the sterilization times, irrespective of the shear rate behavior of the nonNewtonian food medium.

In the above discussions, the importance of reducing sterilization time without compromising on the lethality of the treatment is brought out. Reducing sterilization times reduces energy consumption and batch time. Optimization of thermal sterilization processes is indeed challenging owing to the transient and distributed nature of the problem, often coupled with momentum and heat transfer and involving nonlinear quality constraints (García et al., 2006). Further studies involving optimization of operating conditions for food can sterilization are discussed next.

2.3.13 Optimization of thermal sterilization processes

Chen and Ramaswamy (2002) adopted a variable retort temperature (VRT) strategy to optimize the conduction heating of food. Artificial intelligence tools were adopted to develop predictive models to simulate conduction heating of foods. The responses were surface cook value, process time, and average quality retention. The surface cook value relates the loss of quality in the food product during a high-temperature treatment process to the equivalent cooking of the food at 100°C (Lund, 1977). Predictive models were developed using genetic algorithm (GA) and artificial neural networks (ANN) tools for estimating surface cooking value, processing time, and average quality retention. Artificial intelligence (AI) tools were better than conventional approaches to model development and optimization, especially when the problem was complex and nonlinear (Chen and Ramaswamy, 2002). Variable retort temperature profiles were established by using sine and exponential functions. This optimization strategy reduced process time by 20% and surface cook value by 7%–10% when compared to the best constant retort temperature process.

García et al. (2006) demonstrated an optimal retort temperature policy for maximizing the final nutrient retention in the pork puree food product. The constraints adopted were microbial lethality at the end point and the final temperature at the hottest point. Even with conduction heating, the authors claimed that the nonlinear and constrained nature of food processing models made the dynamic optimization approach daunting as well as time-consuming. The optimal control profiles may be very uneven and may involve singular arcs (García et al., 2006). The computationally demanding simulations revealed that some of the retort temperature profiles were peculiar with spikes, and different temperature profiles gave nearly the same final nutrient value. The authors warned that such behavior may have serious ramifications in real-time implementation. The authors smoothened the optimal temperature policy, improved the robustness, and increased computational speed by using successive reoptimizations and a host of other numerical refinements (García et al., 2006).

Simpson et al. (2020) provided assessments and outlook for thermal sterilization using variable retort temperatures. They observed that most optimization studies in the past had mostly focused on maximizing quality retention rather than reducing process time and hence energy consumption. Hence, the authors analyzed the VRT method through two different objectives. The first was to reduce operating time while maintaining product quality. The second was to enhance product quality by controlling energy consumption and operating time.

The transient heat conduction model was used with retort temperature as the control signal and food temperature inside the package as the state variable. The authors noted that since some of the constraints were nonconvex, only locally optimal solutions could be attained by the nonlinear programming solver. The results indicated that when the food product was more sensitive to temperature, the VRT strategy was beneficial. When the z-value (Eq. 2.35) was increased from 10 to 60°C, the improvement in food surface quality retention due to adopting a VRT strategy was not significant. The authors concluded that when food surface quality retention was the objective, the VRT did not lead to substantial improvements that could be detected by consumers. Hence, it could not be justified to replace the established CRT strategy. They alternately suggested focusing on reducing process time and energy consumption without compromising the food quality.

2.4 Conclusions

Thermal sterilization in a batch retort is still the popular mode of operation to ensure food safety and long shelf-life. Factors such as container geometry, its orientation and aspect ratio, food properties, especially viscosity and thermal diffusivity, and rheological behavior, that is, Newtonian or nonNewtonian have a strong influence on the thermal sterilization rates and food quality value. A systematic study is recommended to study the influence of all these factors using food systems of relevance to the food industry.

A comprehensive CFD study may be performed to account for the large number of variables mentioned above. CFD simulations require reliable estimates of the properties of the food system for their predictions to be realistic. Rheological characteristics may be measured for different food systems using suitable rheometers and incorporated into the mathematical model. CFD simulations for nonNewtonian liquid food may be corroborated by careful experiments using sensors that may not disturb the flow patterns unduly. CFD simulations with and without sensors may indicate locations where the flow patterns are not affected to carry out this experimental validation. Such studies will help to check if the CFD predicted heating trends are being followed at specified locations in the food can.

Simplistic conduction models may not hold for liquids and mixtures of liquids and solids. Previous literature has clearly shown that natural convection effects strongly influence thermal sterilization by making the heating faster. Hence, overprocessing of the food should be avoided to minimize loss of nutrient retention, color, appearance, and flavor. The required time for the SHZ to attain the minimum sterilization temperature must be carefully noted down through accurate simulations and used in planning the sterilization schedules. Numerical experiments using CFD simulations are very handy to get insight into the sterilization processes.

Due to rapid growth in computational power and availability of many CFD/Multiphysics software such as Fluent and COMSOL, solving the Navier-Stokes equations as well as thermal energy equations has become routine. However, solving partial differential equations accurately is considerably more complex and expensive to solve every iteration during optimization exercises. Therefore, a suitable black-box surrogate model using an ANN may be developed through proper training and validation. Such surrogate models may also be very helpful when studying the efficacy of variable retort temperature for predicting responses such as process time, energy consumption, and nutrient quality retention when optimizing for the most favorable temperature profile at the food's surface. Literature appears to be scarce on the adoption of VRT for natural convection-based sterilization processes. This gap needs to be filled. Also important is to ensure that the gain from adopting new strategies, such as geometry modifications, orientation changes like inclination, and variable retort temperatures produce a noticeable improvement in the desired food quality and operating objectives and is economically justifiable as well.

The thermal sterilization process in a simple batch retort has several exciting ramifications to heat transfer theory, which may still be explored. There is scope for developing suitable Nusselt number correlations for thermal food cans of different shapes, cross-section, orientation, especially for nonNewtonian food products. Most heat transfer correlations are for regular geometries involving external flows. Nusselt number correlations are useful in finding the volume-averaged temperature in the food can. However, the accurate prediction of the location and temperature of the slowest heating zone at different times through rigorous heat and momentum transfer analysis still needs to be addressed for various geometries.

When water-based systems are used, it should be ensured that laminar flow assumptions are still valid. In addition, as more food items involving solid particles in liquid gravy are getting packaged, their sterilization characteristics, such as sterilization duration, slowest heating zone temperature, and location, may be rigorously studied through CFD simulations.

References

Abdul Ghani, A.G., Farid, M.M., Chen, X.D., Richards, P., 1999. An investigation of deactivation of bacteria in a canned liquid food during sterilization using computational fluid dynamics (CFD). J. Food Eng. 42, 207–214. https://doi.org/10.1016/S0260-8774(99)00123-5.

Abdul Ghani, A.G., Farid, M.M., Chen, X.D., 2002. Numerical simulation of transient temperature and velocity profiles in a horizontal can during sterilization using computational fluid dynamics. J. Food Eng. 51, 77–83. https://doi.org/10.1016/S0260-8774(01)00039-5.

Abdul Ghani, A.G., Farid, M.M., Zarrouk, S.J., 2003. The effect of can rotation on sterilization of liquid food using computational fluid dynamics. J. Food Eng. 57, 9–16. https://doi.org/10.1016/S0260-8774(02)00215-7.

Augusto, P.E.D., Cristianini, M., 2010. Evaluation of geometric symmetry condition in numerical simulations of thermal process of packed liquid food by computational fluid dynamics (CFD). Int. J. Food Eng. 6. https://doi.org/10.2202/1556-3758.1695.

Bejan, A., 2009. Convection heat transfer, third ed. Wiley-India, New Delhi.

Bergman, T.L., Lavine, A.S., Incropera, F.P., DeWitt, D.P., 2017. Incropera's principles of heat and mass transfer, eighth ed., global edition. Wiley, Hoboken, NJ.

Borah, R., Gupta, S., Mishra, L., Chhabra, R.P., 2020. Heating of liquid foods in cans: Effects of can geometry, orientation, and food rheology. J. Food Process. Eng. 43, e13420. https://doi.org/10.1111/jfpe.13420.

Calderón-Alvarado, M., Alvarado-Orozco, J., Herrera-Hernández, E., Martínez-González, G., Miranda-López, R., Jiménez-Islas, H., 2016. Effect of two viscosity models on lethality estimation in sterilization of liquid canned foods. Food Sci. Technol. Int. 22, 496–515. https://doi.org/10.1177/1082013215627393.

Catton, I., 1978. Natural convection in enclosures. In: Proceeding of international heat transfer conference 6. presented at the international heat transfer conference 6, Begellhouse, Toronto, Canada, pp. 13–31. https://doi.org/10.1615/IHTC6.2350.

Chen, C.R., Ramaswamy, H.S., 2002. Modeling and optimization of variable retort temperature (VRT) thermal processing using coupled neural networks and genetic algorithms. J. Food Eng. 53, 209–220. https://doi.org/10.1016/S0260-8774(01)00159-5.

Datta, K., Teixeira, A.A., 1987. Numerical modeling of natural convection heating in canned liquid foods. Trans ASAE (Am. Soc. Agric. Eng.) 30, 1542–1551. https://doi.org/10.13031/2013.30600.

Datta, A.K., Teixeira, A.A., 1988. Numerically predicted transient temperature and velocity profiles during natural convection heating of canned liquid foods. J. Food Sci. 53, 191–195. https://doi.org/10.1111/j.1365-2621.1988.tb10206.x.

Deák, T., 2014. Food technologies: pasteurization. In: Encyclopedia of food safety. Elsevier, pp. 219–224. https://doi.org/10.1016/B978-0-12-378612-8.00257-2.

Denys, S., Pieters, J.G., Dewettinck, K., 2004. Computational fluid dynamics analysis of combined conductive and convective heat transfer in model eggs. J. Food Eng. 63, 281–290. https://doi.org/10.1016/j.jfoodeng.2003.06.002.

Farid, M., Abdul Ghani, A.G., 2004. A new computational technique for the estimation of sterilization time in canned food. Chem. Eng. Process. Process Intensif. 43, 523–531. https://doi.org/10.1016/j.cep.2003.08.007.

García, M.-S.G., Balsa-Canto, E., Alonso, A.A., Banga, J.R., 2006. Computing optimal operating policies for the food industry. J. Food Eng. 74, 13–23. https://doi.org/10.1016/j.jfoodeng.2005.02.011.

Holman, J.P., Bhattacharyya, S., 2011. In: Indian, S. (Ed.), Heat transfer in SI units, tenth ed. McGraw-Hill, New Dehli.

Jimenez, P.S., Bangar, S.P., Suffern, M., Whiteside, W.S., 2024. Understanding retort processing: A review. Food Sci. Nutr. 12, 1545–1563. https://doi.org/10.1002/fsn3.3912.

Kakaç, S., Yener, Y., Naveira-Cotta, C.P., 2018. Heat Conduction, fifth ed. Taylor and Francis, CRC Press, Boca Raton. https://doi.org/10.1201/b22157. [2018.

Kannan, A., Sandaka, P.C.G., 2008. Heat transfer analysis of canned food sterilization in a still retort. J. Food Eng. 88, 213–228. https://doi.org/10.1016/j.jfoodeng.2008.02.007.

Khurana, M., Karwe, M.V., 2009. Numerical prediction of temperature distribution and measurement of temperature in a high hydrostatic pressure food processor. Food Bioprocess Technol. 2, 279–290. https://doi.org/10.1007/s11947-008-0096-4.

Koribilli, N., Aravamudan, K., Aditya Varadhan, M.U.S.V., 2011. Quantifying enhancement in heat transfer due to natural convection during canned food thermal sterilization in a still retort. Food Bioprocess Technol. 4, 429–450. https://doi.org/10.1007/s11947-009-0232-9.

Kumar, A., Bhattacharya, M., 1991. Transient temperature and velocity profiles in a canned non-Newtonian liquid food during sterilization in a still-cook retort. Int. J. Heat Mass Tran. 34, 1083–1096. https://doi.org/10.1016/0017-9310(91)90018-A.

Kumar, A., Bhattacharya, M., Blaylock, J., 1990. Numerical simulation of natural convection heating of canned thick viscous liquid food products. J. Food Sci. 55, 1403–1411. https://doi.org/10.1111/j.1365-2621.1990.tb03946.x.

Lund, D.B., 1977. Design of thermal processes for minimizing nutrient retention. Food Technol. 71–78.

Marra, F., Romano, V., 2003. A mathematical model to study the influence of wireless temperature sensor during assessment of canned food sterilization. J. Food Eng. 59, 245–252. https://doi.org/10.1016/S0260-8774(02)00464-8.

Martins, R.C., Lopes, V.V., Vicente, A.A., Teixeira, J.A., 2008. Computational shelf-life dating: complex systems approaches to food quality and safety. Food Bioprocess Technol. 1, 207–222. https://doi.org/10.1007/s11947-008-0071-0.

Mirade, P.S., Daudin, J.D., Ducept, F., Trystram, G., Clément, J., 2004. Characterization and CFD modelling of air temperature and velocity profiles in an industrial biscuit baking tunnel oven. Food Res. Int. 37, 1031–1039. https://doi.org/10.1016/j.foodres.2004.07.001.

Naveh, D., Pflug, I.J., Kopelman, I.J., 1983. Simplified procedure for estimating the effect of a change in heating rate on sterilization value. J. Food Protect. 46, 16–18. https://doi.org/10.4315/0362-028X-46.1.16.

Nicolaï, B.M., Verboven, P., Scheerlinck, N., De Baerdemaeker, J., 1998. Numerical analysis of the propagation of random parameter fluctuations in time and space during thermal food processes. J. Food Eng. 38, 259–278. https://doi.org/10.1016/S0260-8774(98)00108-3.

Norton, T., Sun, D.-W., 2006. Computational fluid dynamics (CFD) – an effective and efficient design and analysis tool for the food industry: A review. Trends Food Sci. Technol. 17, 600–620. https://doi.org/10.1016/j.tifs.2006.05.004.

Padmavati, R., Anandharamakrishnan, C., 2013. Computational fluid dynamics modeling of the thermal processing of canned pineapple slices and titbits. Food Bioprocess Technol. 6, 882–895. https://doi.org/10.1007/s11947-012-0892-8.

Park, H.W., Yoon, W.B., 2018. Computational fluid dynamics (CFD) modelling and application for sterilization of foods: A Review. Processes 6, 62. https://doi.org/10.3390/pr6060062.

Sandaka, P.C., 2006. Heat transfer analysis in caned food sterilization. Indian Institute of Technology, Madras.

Simpson, R., Jiménez, D., Almonacid, S., Nuñez, H., Pinto, M., Ramírez, C., Vega-Castro, O., Fuentes, L., Angulo, A., 2020. Assessment and outlook of variable retort temperature profiles for the thermal processing of packaged foods: plant productivity, product quality, and energy consumption. J. Food Eng. 275, 109839. https://doi.org/10.1016/j.jfoodeng.2019.109839.

Siriwattanayotin, S., Yoovidhya, T., Meepadung, T., Ruenglertpanyakul, W., 2006. Simulation of sterilization of canned liquid food using sucrose degradation as an indicator. J. Food Eng. 73, 307–312. https://doi.org/10.1016/j.jfoodeng.2004.08.008.

Steffe, J.F., Mohamed, I.O., Ford, E.W., 1986. Rheological properties of fluid foods: data compilation. ASAE Publication.

Varma, M.N., Kannan, A., 2005. Enhanced food sterilization through inclination of the container walls and geometry modifications. Int. J. Heat Mass Tran. 48, 3753–3762. https://doi.org/10.1016/j.ijheatmasstransfer.2005.03.024.

Varma, M.N., Kannan, A., 2006. CFD studies on natural convective heating of canned food in conical and cylindrical containers. J. Food Eng. 77, 1024–1036. https://doi.org/10.1016/j.jfoodeng.2005.07.035.

Zou, Q., Opara, L.U., McKibbin, R., 2006. A CFD modeling system for airflow and heat transfer in ventilated packaging for fresh foods: I. Initial analysis and development of mathematical models. J. Food Eng. 77, 1037–1047. https://doi.org/10.1016/j.jfoodeng.2005.08.042.

Modeling of baking and roasting processes

3

Emmanuel Purlis [1, 2] *and Sandro M. Goñi* [3, 4]
[1]Universidad de Buenos Aires, Facultad de Ciencias Exactas y Naturales, Departamento de Industrias, Buenos Aires, Argentina; [2]CONICET – Universidad de Buenos Aires, Instituto de Tecnología de Alimentos y Procesos Químicos (ITAPROQ), Buenos Aires, Argentina; [3]Centro de Investigación y Desarrollo en Ciencia y Tecnología de los Alimentos (CIDCA), CCT CONICET La Plata, UNLP, CIC, La Plata, Argentina; [4]Facultad de Ingeniería, UNLP, La Plata, Argentina

3.1 Introduction

In general terms, baking and roasting can be considered as cooking unit operations or processes, since they are applied to transform a raw material into an edible or ready-to-eat food product. Baking refers to the transformation of dough (or batter) into products like breads, cakes, and cookies/biscuits, while roasting is associated with the cooking of meat cuts and meat products. In addition, there are other foods that are baked, for example, snacks, bread-like products like toasts, and roasted, for example, vegetables, plant-based meat analogs, seeds, grains, nuts. The common feature of these unit operations is the equipment, that is, baking and roasting are carried out in an oven at high temperature (usually above 150–160°C), traditionally using hot air as heating medium (indirect heating). As in other unit operations, there are numerous designs and configurations of ovens and heating strategies, aimed at improving the process efficiency, for example, use of radiant elements, volumetric heating, etc. But overall, baking and roasting can be defined as heat transfer (driven) unit operations, even if mass transfer plays a major role in these processes (especially in the products).

On the other hand, depending on the food material being processed, different physical and bio-chemical changes may take place within the product. For instance, (bread) dough is a soft/deformable solid-like (viscoelastic) foam or continuous macromolecular network composed by gluten proteins (and water), filled with starch granules, containing dispersed gas bubbles (Cuq et al., 2003), while meat may be modeled as a polymer gel composed by proteins and water (van der Sman, 2007a). However, both baking and roasting processes involve complex, simultaneous and coupled phenomena, concerning multiple components, phases and scales. Furthermore, they represent multi-objective problems, since there are several operational targets to achieve, for example, quality (doneness, texture, color), safety (microbial inactivation, harmful compounds), and processing (yield, energy efficiency) targets. In this context, trial-and-error or "cook-and-look" practices are not able to provide efficient and robust solutions, even if they represent a common way of working in many food

Modeling of Unit Operations in the Food Industry. https://doi.org/10.1016/B978-0-12-822180-8.00008-0

companies, especially at small- and medium-scale production. On the other hand, modeling can be applied to better understand (gain insights), design, control and optimize processes in order to find systematic solutions (Purlis, 2014). Moreover, the use of models can help in saving time, resources, and efforts involved in experimental trials and also in the training of operators and experts. For instance, it may be less expensive to develop and validate a robust model than performing numerous experiments to build a data-driven model or train an expert, or either find an "optimal" solution via experimental design and response surface methodology. Of course, the robustness and capability of a model will depend on its "nature", for example, assumptions, hypotheses, simplifications, etc., and its validation against experimental data. Finally, models are building blocks of digital twins, a state-of-the-art methodology related to digital transformation, which is under current development and implementation in food processing (Purlis, 2024a).

This chapter aims to provide an overview about different modeling approaches that have been applied to describe baking and roasting processes. Such approaches range from simple and phenomenological models to mechanistic or physics-based formulations including solid mechanics to describe large deformation during processing. The basic methodology to build such overview consisted in classifying or grouping modeling approaches according to their distinct features, capabilities and limitations, and providing some key examples and references. For instance, such classification could be used as a guide by practitioners to select the more convenient framework, depending on the specific objective of the model or task to accomplish. That is, models are tools, so different formulations can be used depending on the specific problem to be solved. Considering the differences in materials and physics of each unit operation, the chapter has been divided into two main parts, that is, baking and roasting. Each part begins with a (basic) description of transport phenomena and other relevant changes occurring during the corresponding process, in order to provide an adequate background to further discuss modeling approaches.

3.2 The baking process

3.2.1 Overview of the baking process

The baking process has many variations regarding both products and equipment/heating strategies. In order to provide an organized and useful overview of modeling approaches, it may be convenient to first distinguish some basic aspects of this unit operation. On the one hand, modeling of baking can be divided into product modeling and oven modeling:

- Product modeling refers to developing models to describe and predict changes occurring in the food material during processing, for example, temperature and water content profiles, porosity profile/change, volume variation (deformation), texture development, chemical reactions (e.g., browning), etc.
- Oven modeling means to model the surroundings of the product, for example, heat, mass and momentum transfer in the oven cavity, and also computation of energy demand of

the operation. For instance, typical equipment models involve CFD (computational fluid dynamics) studies, for example, Boulet et al. (2010), Chhanwal et al. (2010), Khatir et al. (2013).

As in other unit operations, heat and mass transfer coefficients are commonly used to represent the oven ambient behavior in product modeling via boundary conditions, considering than conjugate modeling can be a difficult task (and it is not always necessary). On the other hand, baking is applied mainly to three types of materials/products:

- Bread-like: semi-solid dough as initial state; no mold is required, although it is used in some cases like sandwich bread.
- Cake-like: liquid-like batter as initial state; a mold or container is required.
- Biscuits, cookies and crackers: semi-solid (thin) dough/mix; no mold is generally used (dough sheet or roll is cut prior to baking).

We will focus here on modeling changes occurring in the product, since oven modeling is more related to equipment design studies, and mostly concerning bread and cake, considering the complexity of underlying physics (e.g., temperature and water content profiles, large deformation) and also the relative number of published articles in the topic.

3.2.2 Transport phenomena and other changes during baking

Transport processes and other changes such as texture and color development depend on material composition and properties, and they are driven by operating conditions. Overall, bread and cake can be considered as hygroscopic, porous, soft/deformable materials, mainly made from biopolymers (flour) and water (cake-like products contain other ingredients, e.g., sugar, fat, egg), and leavening agents (yeast or baking powder). Operating conditions generally involve high temperature (above 160–180°C) and low relative humidity in the oven. Before giving a general description of transport phenomena occurring during baking, it would be helpful to present some typical experimental trends (based on bread baking). For instance, Fig. 3.1 shows representative temperature profiles for outer zones/surface and core (e.g., Purlis and Salvadori, 2009a; Sakin et al., 2007). At surface (crust), temperature increases continuously, tending to oven temperature over time. In the core zone (crumb), temperature does not change significantly at the beginning, but then rises quite rapidly to attain 100°C approximately, describing a typical sigmoid trend or S-curve. In a common baking process (e.g., industrial practice), inner zones or crumb (i.e., below the crust) do not surpass the observed plateau at ca. 100°C. Fig. 3.2 depicts representative profiles of water/moisture content during baking. Similar to temperature profiles, two different marked trends are observed: outer zones suffer dehydration from the beginning of the process, while moisture content in inner zones (crumb) does not change significantly, and can even increase during baking (Purlis and Salvadori, 2009a; Wagner et al., 2007). As a result, weight loss is continuous during the process, although a constant drying rate period is not observed: initial dough/batter is not saturated in liquid water to show a complete drying rate curve. On the other hand, during the first part of baking, the dough is deformed and overall volume increases. After dough/

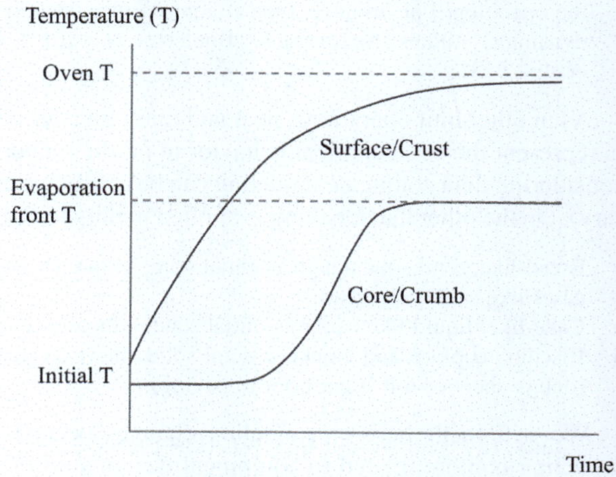

Figure 3.1 Representative temperature profiles for (bread) baking process. Temperature and time are shown in arbitrary units.

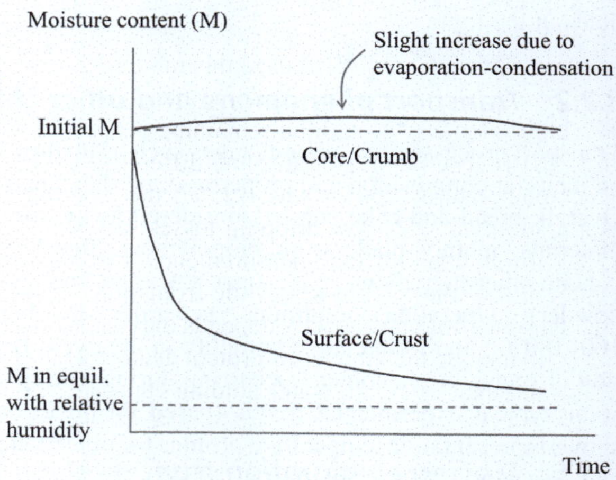

Figure 3.2 Representative moisture content profiles for (bread) baking process. Moisture content and time are shown in arbitrary units.

crumb transition, expansion stops, and toward the end of the process, some shrinkage can be observed in the presence of a hard, dry crust, for example, in French bread baking (Sommier et al., 2005).

To explain and describe these typical experimental trends, a general (basic) set of hypotheses is presented here, that can be used to further model the main phenomena observed during the baking process, mainly at macroscale. Please note that other mechanisms, assumptions and/or simplifications may be proposed depending on the specific model formulation, but this will be discussed later. That is, our aim here is to provide a general framework to present the modeling approaches afterward. So, considering the typical/conventional operating conditions of baking, when dough/batter is placed in an oven, the following mechanisms may take place (Fig. 3.3; Nicolas et al., 2014, 2016; Purlis, 2016; Purlis et al., 2021):

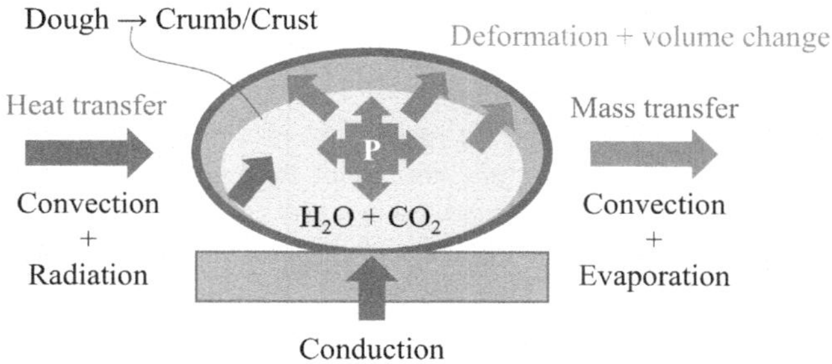

Figure 3.3 Schematic summary of phenomena occurring during (bread) baking. See text for detailed description.

- Heat is transferred from surroundings/oven ambient to product surface by convection (hot air), radiation (from hot walls and surfaces, and possibly from radiative heaters), and conduction (from supporting devices, e.g., tray, belt, mold, etc.). In addition, relative humidity of surrounding air is lower than water activity at product surface, establishing the driving force for dehydration. Usually, air is circulated within the oven (forced convection) to avoid stagnant zones and increase heating uniformity, and also to enhance heat and mass transfer at product surface.
- Therefore, product surface starts to increase its temperature (quite rapidly) and also to lose water by evaporation. These rapid changes at surface establish the respective gradients for heat and mass transport inside the product, since at core, temperature is lower and water content is higher.
- Subsequently, inside the product, heat is transferred from surface to core by conduction. In addition, the so-called evaporation-condensation mechanism has been proposed to explain the rapid heating of inner zones in a porous medium. When pores become interconnected (Fig. 3.3), convective contribution due to vapor movement can occur as well. Regarding evaporation, it mostly happens in a narrow zone or (moving) front underneath the surface, that is, the so-called evaporation front. This zone or front remains at 100°C approximately, corresponding to saturation temperature (at normal pressure), due to (high) latent heat of vaporization of water. At a given point, once water is evaporated and latent heat is not further consumed, the front can advance toward the core.
- This advance of the evaporation front is translated into the development of a dry outer zone or crust (similar to a phase change or Stefan problem). That is, dehydration occurs mostly in the evaporation zone/front and then water vapor is transported toward the surface and oven ambient. Underneath the evaporation front, a concentration gradient exists, although transport of liquid water may be not significant (evaporation occurs faster). On the other hand, part of water vapor can be transported toward the core, where temperature is lower, increasing moisture content via condensation.
- Simultaneously, in the first stages of baking, dough is deformed due to pressure generated by thermal expansion of gases (CO_2 and water vapor) inside closed, unconnected pores, resulting in volume increase or the so-called oven rise. When dough is transformed into crumb due to protein denaturation and starch gelatinization, which increases the resistance to deformation, and pores become interconnected and thus gases are allowed to flow reducing pressure and so driving force, deformation and volume expansion stop. Finally, in the presence of a dry, hard crust (e.g., French bread), some contraction or shrinkage can be observed toward the end of the process.

On the other hand, another typical change in products during baking is the development of color, especially at surface. Browning is due to heat-driven (nonenzymatic) chemical reactions, that is, the Maillard reaction and caramelization of sugars. In addition, these reactions are associated with generation of flavor and aroma compounds, but also with the production of harmful components such as acrylamide (Purlis, 2010).

Overall, despite presenting a summarized description of changes occurring in the products during the baking process, it is quite evident that it is a complex problem involving different, coupled and simultaneous transport phenomena, bio-chemical reactions, and phase transitions, and concerning multiple components, phases, and scales. All these changes determine the quality of products and operating aspects like energy demand.

3.3 Modeling of baking process

In view of the discussed complexity of the problem, different approaches or frameworks can be adopted to build models in order to design, predict, control, and optimize the baking process. Following, we provide an overview of such approaches, ranging from simple phenomenological models to comprehensive physics-based formulations. We propose a classification just as a guide to understand and discuss different approaches, their advantages and drawbacks, bearing in mind that all models are tools and thus have specific objectives and limitations, especially when dealing with complex problems like baking. For instance, we have divided modeling approaches into the following "categories":

- Simple phenomenological models.
- Diffusive phenomenological models.
- Multiphase porous media models.
- Multiscale models.

3.3.1 Simple phenomenological models

The general goal of this approach is to describe the main phenomena observed during the process, or to predict some processing parameters/outputs (e.g., baking time), by using simple equations or methodology, with a focus on practical situations. These models are based on phenomenological knowledge of the process and strongly depend on experimental data, that is, they are data-driven models with some physics-oriented formulation to add a sense of interpretability to parameters/coefficients and results, in order to avoid a full black-box modeling approach. For instance, these models represent a limiting case of reduced order models, where complexity in formulation is reduced but also the capability of explanation. In particular, we have worked on this modeling approach with the objective of filling the gap between complex models and actual industrial practice, and also for providing simple tools to researchers and practitioners with limited expertise in modeling and simulation (Purlis, 2019b, 2020, 2023).

In the first example, Purlis (2019b) proposed three simple methods to predict the critical or minimum baking time of bread, defined as the time required to achieve a complete dough/crumb transition, which ensures the (minimum) acceptability of the product. This transition has to be evaluated at the coldest point of the product, where temperature needs to reach a minimum value of 96°C. So, simple methods were focused on predicting the time to achieve this temperature at the core of the product. To develop such methods, a previously validated heat and mass transport model (Purlis and Salvadori, 2009a, 2009b) was used to generate "actual" or reference data, considering the traditional baking of French bread (e.g., baguette). For the first simple method, mass transfer was neglected and the boundary condition for heat transfer was simplified by using a prescribed temperature condition at 100°C after achieving this value by using (initially) a convective flux condition. In other words, this simplified heat transfer model assumes that evaporation front remains at surface instead of moving toward to the core. The second simple method included a further simplification: a constant (fictitious) value for thermal conductivity was used, based on fitting of temperature profile (vs. reference data). The third method involved a correlation for the critical baking time based on concepts of dimensional analysis, where the following dimensionless groups were considered: Fourier number or dimensionless (critical) time, dimensionless temperature (evaluated at the core), and Biot number. The parameters of the proposed correlation were obtained by regression. Overall, all three simple methods demonstrated a good accuracy to adjust/predict the critical baking time.

The second example is focused on the estimation of water loss during bread baking (Purlis, 2020), another important metric, since it is associated with energy demand and yield of the unit operation. Again, the same heat and mass transfer model was used as a reference method to generate data from a conventional bread baking process (Purlis and Salvadori, 2009a, 2009b). Two simple methods were proposed; the first one involved a lumped system, considering the classical approach of negligible internal resistance to mass transfer. So, it was assumed that the average water content of bread, M (kg/kg, dry basis), follows an ordinary differential equation:

$$V \frac{dM}{dt} = -A\, k_{\mathrm{m}}\, M \tag{3.1}$$

Where V and A are the volume (m^3) and surface area (m^2) of bread, respectively, t is time (s), and k_{m} is a mass transfer coefficient (m/s), which was correlated with operating conditions (oven temperature and heat transfer coefficient). The second method utilized the semiinfinite system hypothesis (or "short times" solution), assuming that bread baking is an "incomplete" moving boundary problem, given that the evaporation front does not reach the core. Similar to the first method, a mass transfer coefficient was obtained by regression. Overall, good prediction results were found for both methods, but the lumped solution is easier to implement.

The third and last example of simple modeling focuses on time-temperature profiles during baking, as they are essential to explain and characterize different changes related to quality, such as dough/crumb transition and surface browning (Purlis, 2023). Two equations were proposed to model temperature variation: for the core position, a modified Gompertz equation to capture the typical sigmoid trend; for the surface, an adapted Page model. In both cases, only two fitting parameters were included and a common spreadsheet was used for regression. The equations were evaluated by using a heat and mass transfer model (as in previous methods) and published experimental data. Good performance was found in all cases tested, including also reported data for frying and roasting (time-temperature profiles similar to those for baking).

In summary, these examples demonstrate that the complexity of the baking problem can be reduced significantly for some calculations, especially by applying phenomenological knowledge to propose pertinent simplifications. But, as previously noted, these models have limitations regarding capability of explanation and strongly depend on experimental data. Nevertheless, they can be a useful tool to solve some practical problems and an alternative to pure empirical or black-box methods.

3.3.2 Diffusive phenomenological models

This approach involves simplified transport models to mainly provide temperature and moisture content profiles, and derived quantities such as weight loss, by using macroscale equations. The term "diffusive phenomenological" obeys to the main simplification of these models: only diffusive mechanisms for heat and mass transfer are proposed, and thus effective or apparent transport properties/coefficients are included to describe other possible mechanisms. Such simplification requires using experimental data to obtain these effective coefficients by a fitting procedure. Besides, deformation can be coupled by using either experimental data or solid mechanics modeling. So, although these models may describe well temperature and moisture content variations, they are limited in terms of universality and prediction of other changes like porosity profile. On the other hand, they are able to provide valuable information for describing quality changes like surface browning and starch gelatinization, especially via kinetic modeling (e.g., Purlis, 2014).

To illustrate this approach, we revise here the formulation developed by Purlis and Salvadori (2009a, 2009b, 2010), where bread baking was considered as a moving boundary problem (or Stefan problem) with volume variation. In this sense, dough/bread was modeled as a domain that can contain three zones or regions (simultaneously): (i) crumb (wet inner zone), where temperature is always below 100°C and dehydration does not occur; (ii) crust (dry outer zone), where temperature increases above 100°C and dehydration happens; and (iii) evaporation front, between crumb and crust, where temperature remains at 100°C and phase change (water vaporization) occurs. In addition, the evaporation front moves (toward the core) according to phase change dynamics, determined by vaporization of water and related latent heat. The main assumptions of the model were the following:

- Conventional baking of bread is studied, where sample is considered as an infinite cylinder, so via axial symmetry assumption, a one-dimensional problem is obtained (i.e., transport in radial direction only).
- Dough/bread is a homogeneous and continuous material; the porous medium concept is incorporated by using effective or apparent thermophysical properties.

- Heat is transferred by conduction inside bread according to Fourier's law, where an effective thermal conductivity is used to account for the evaporation-condensation mechanism (which is actually coupled with mass transfer).
- Regarding mass transfer, it is supposed that only liquid diffusion in the crumb and only vapor diffusion in the crust take place, following the Fick's law.
- Water evaporates at 100°C (normal pressure).

Following these assumptions, the heat and mass balances can be written as, respectively:

$$\rho\, C_p\, \frac{\partial T}{\partial t} = \frac{1}{r}\frac{\partial}{\partial r}\left(r\, k\, \frac{\partial T}{\partial r}\right) \tag{3.2}$$

$$\frac{\partial W}{\partial t} = \frac{1}{r}\frac{\partial}{\partial r}\left(r\, D\, \frac{\partial W}{\partial r}\right) \tag{3.3}$$

Where ρ is density (kg/m^3), C_p is specific heat (J/(kg K)), k is thermal conductivity (W/(m K)), T is temperature (°C or K), W is moisture content (kg/kg, dry basis), D is water (liquid or vapor) diffusivity (m^2/s), r is the radial coordinate (m), and t is time (s). Besides, boundary conditions describing convective flux were used in both cases, plus radiation for heat transfer.

An essential aspect of the proposed moving boundary problem formulation is the definition of the thermophysical properties so phase change occurring during baking is incorporated into the model. For instance, this can be done by using different and appropriate properties for crumb and crust zones, according to specific features like porosity and moisture content, and a mathematical function to describe the "jump" in the evaporation front. In this way, the specific heat is computed by the following expression including the heat of water vaporization:

$$C_p = C_p^* + \lambda\, W\, \delta\left(T - T_{\mathrm{f}}, \Delta T\right) \tag{3.4}$$

Where C_p^* is specific heat of crumb/crust depending on water content and temperature, λ is the heat of water vaporization (J/kg), δ is a delta-type function describing the enthalpy jump (defined by the sum of two smoothed Heaviside functions with different sign), which depends on temperature T, temperature of phase change T_{f}, and a temperature difference (ΔT) over which phase change occurs (smooth change). So, the second term of the right side of Eq. (3.4) accounts for the enthalpy jump or phase change directly without using an auxiliary equation or balance, as in other formulations for moving boundary problems. For the case of thermal conductivity, density, and mass diffusivity, the following (general) expression was used:

$$k, \rho, D = \begin{cases} (k, \rho, D)_{\mathrm{crumb}} \text{ if } & T \le T_{\mathrm{f}} - \Delta T \\ \\ (k, \rho, D)_{\mathrm{crust}} \text{ if } & T > T_{\mathrm{f}} + \Delta T \end{cases} \tag{3.5}$$

Where the change in properties from crumb to crust is computed via a smoothed Heaviside function with continuous derivative (to improve convergence of numerical methods).

Furthermore, volume change can be coupled to this formulation. For instance, Purlis and Salvadori (2010) utilized a prescribed boundary velocity and a deformed mesh method to describe the oven rise, based on experimental data obtained during the baking process. Overall, this moving boundary problem formulation provides good results, especially regarding the description of typical features of the process, for example, sigmoid shape of temperature variation in the core, movement of evaporation front, and development of the crust. On the other hand, as previously mentioned, these so-called diffusive phenomenological models depend on effective transport properties and/or coefficients, mostly obtained by fitting model outputs with experimental profiles. In this regard, Cevoli et al. (2020) implemented an inverse method to estimate the effective moisture diffusivity of a cake-type product, based on a diffusive model similar to the one presented above. This example represents a systematic way of obtaining transport coefficients based on a proposed model and experimental data.

To summarize, this diffusion-based modeling approach is able to describe the main characteristics of the baking process by using a physics-guided framework with some simplifications (bearing in mind the comprehensive set of phenomena). Considering the state of the art, it is a useful approach that can provide detailed and practical information to compute quality and operational aspects, and it is still implementable without having a high-level of expertise in modeling and simulation. However, these models are not able to explain some important changes, mainly related to structure.

3.3.3 Multiphase porous media models

This approach refers to developing models that consider a comprehensive description of the material, phenomena, and changes occurring during the process (see Section 3.2.2), mainly at macroscale. In this sense, such models deal with different phases/species, as well as other mechanisms besides heat conduction and mass diffusion. This generates formulations with more than two balance equations (and corresponding boundary conditions), and additional transport coefficients, in comparison to the diffusive modeling approach (previous section). In some cases, these coefficients still depend on experimental data (effective or apparent values), giving that a theoretical description may be complex (e.g., involving multiscale modeling; see next section). Nonetheless, the premise is to develop physics-based or mechanistic models, avoiding the (extensive) use of empirical correlations. Another important feature is the modeling of deformation by using a solid mechanics framework. Similar to previous approach, other changes related to quality can be coupled to obtain a full description of the process, in order to gain insights and perform design, control and/or optimization.

Next, we provide a general description of this modeling approach, mainly based on the seminal work of Datta et al. These authors have applied this so-called multiphase porous media modeling framework to various heat and mass transport problems in food processing, and specially, they developed comprehensive models for baking of bread (Zhang et al., 2005; Zhang and Datta, 2006) and cake (Seranthian and Datta, 2023). Another important contribution to this approach has been made by Nicolas et al. for the case of bread baking (Nicolas et al., 2014, 2016, 2017). Other similar examples can be found in the literature, where some variations of these models were developed (e.g., El Helou et al., 2024; Ousegui et al., 2010; Yang et al., 2025). Although there is some difference in the final mathematical problem between the mentioned formulations (i.e., final set of equations), the fundamentals are essentially the same. So, the main hypotheses of this approach are the following:

- Three phases and four components are considered in the transport processes: solid (dough matrix, hygroscopic), liquid (water), and gas (binary mixture of water vapor and carbon dioxide).
- These mass species and phases are continuous in the (fictitious, macroscale) porous medium. Homogenization is applied so effective properties are used.
- Local thermal equilibrium is assumed, that is, there is only one temperature for all phases at any spatial location ("point").
- Pressure in the liquid water is the sum of gas pressure and capillary pressure.
- The solid matrix is a linear viscoelastic material; other rheological models could be used, for example, Yang et al. (2025) proposed a viscous model.
- The volume of the solid matrix remains constant (incompressible material).
- Deformation is driven by internal gas pressure.

Governing equations for heat and mass transfer are based on the conservation of energy and mass species, that is, liquid water, water vapor and CO_2 gas. The energy equation includes convective contribution of fluid phases, heat conduction and evaporation:

$$\rho_{\text{eff}}\, C_{p,\text{eff}}\, \frac{\partial T}{\partial t} + \sum_i C_{p,i}(n_i \cdot \nabla T) = \nabla \cdot \left(k_{\text{eff}} \nabla T\right) - \lambda\, \dot{I} \tag{3.6}$$

Density, specific heat and thermal conductivity are effective (subscript "eff") properties involving all phases (homogenization assumption); in the convective term, i accounts for each species (liquid water, water vapor, CO_2) and n is the convective flux of each species; $\dot{I}$ is the rate of evaporation (or condensation) of water.

Governing equations for mass transport of liquid water, gas and water vapor are the following, respectively:

$$\frac{\partial c_{\text{w}}}{\partial t} + \nabla \cdot n_{\text{w,G}} = -\dot{I} \tag{3.7}$$

$$\frac{\partial c_{\text{g}}}{\partial t} + \nabla \cdot n_{\text{g,G}} = \dot{I} + R_{CO_2} \tag{3.8}$$

$$\frac{\partial c_{\text{v}}}{\partial t} + \nabla \cdot n_{\text{v,G}} = \dot{I} \tag{3.9}$$

Where c_i is concentration (kg/m^3); in the gas phase (Eq. 3.8), it appears a CO_2 generation term due to action of leavening agents, that usually depends on temperature. The evaporation term represents a closure of the system of equations (Purlis, 2019a); Datta et al. have proposed the use of the so-called nonequilibrium approach to model explicitly this term, for example:

$$\dot{I} \propto K_{\text{evap}}\left(p_{\text{v,eq}} - p_{\text{v}}\right) \tag{3.10}$$

Where K_{evap} is a fitting/tuning parameter, $p_{\text{v,eq}}$ is the water vapor pressure in equilibrium (obtained from sorption isotherm) and p_{v} is the actual vapor pressure (to be solved, since it is related to vapor concentration via ideal gas law). Another formulation is possible by assuming equilibrium, that is, $p_{\text{v}} \approx p_{\text{v,eq}}$ (Purlis, 2019a). Back to Eqs. (3.7)–(3.9), it is worth noting that convective flux accounts for solid movement (subscript "G" indicates the ground frame):

$$n_{i,\text{G}} = n_{i,\text{s}} + c_i v_{\text{s}} \tag{3.11}$$

Where subscript "s" represents the flux with respect to the solid and v_s is the velocity of the solid due to deformation. For each species, the corresponding mass flux (relative to the solid) is:

$$n_{\text{w,s}} = \rho_{\text{w}}\, v_{\text{w,s}} - D_{\text{w,cap}} \nabla c_{\text{w}} \tag{3.12}$$

$$n_{\text{g,s}} = \rho_{\text{g}}\, v_{\text{g,s}} \tag{3.13}$$

$$n_{\text{v,s}} = \rho_{\text{v}}\, v_{\text{g,s}} - \frac{C^2}{\rho_{\text{g}}} M_{\text{v}}\, M_{CO_2}\, D_{\text{bin}} \nabla x_{\text{v}} \tag{3.14}$$

Where velocities of liquid (v_w) and gas (v_g) phases are calculated by using the Darcy's law and $D_{w,\text{cap}}$ is the capillary diffusivity. The complete formulation (and notation) can be found in the articles of reference. It is important to summarize the transport mechanisms considered for each species:

- Liquid water flux is due to convection driven by gas pressure (Darcy's law) and capillary pressure-driven flow (Eq. 3.12).
- Gas (mixture of water vapor and CO_2) flux is due to Darcy flow (Eq. 3.13).
- Water vapor flux is due to Darcy flow and binary molecular diffusion (Eq. 3.14).

So far, we (briefly) describe the heat and mass transport part of the model, now we present a summary of the model for large deformation, which is based on solid (continuum) mechanics concepts. In this regard, the equation for conservation of linear momentum (after some basic manipulation) reads:

$$\nabla \cdot \sigma_{\text{eff}} = \nabla p_{\text{g}} \tag{3.15}$$

Where σ_{eff} is the effective stress in the solid porous matrix, which variation is due to gas pressure gradient established during baking, that is, deformation is driven by internal gas pressure. In brief, the effective stress is related to the Jacobian J (equal to volume change ratio of the material, so it allows calculating porosity change), the deformation gradient tensor F (related to J via its determinant, det(F)), and the second Piola-Kirchhoff stress tensor (S). Besides, F is linked to strain tensor and thus to total material displacement and the solid velocity (v_s), which is used for solving the heat and mass transport equations. The deformation gradient can be formulated using a multiplicative split to account for elastic deformation and deformation due to moisture loss (e.g., shrinkage toward the end of baking). The governing equation for motion is complemented with a constitutive law for the material, which relates the stress to strain via rheological properties. For instance, Seranthian and Datta (2023) utilized the Kelvin-Voigt linear (viscoelastic) model, describing a parallel arrangement of a purely viscous damper of viscosity η, and a purely elastic spring of modulus of elasticity E. This constitutive equation is expressed as a first-order linear differential equation:

$$\sigma(t) = E\,\varepsilon + \eta\,\dot{\varepsilon} \tag{3.16}$$

Where ε and $\dot{\varepsilon}$ are the elastic strain and strain rate, respectively. In addition, the elastic strain is related to the relaxation time, which can be used to establish a temperature-dependent relationship to model rheological changes (stiffening) during the baking process.

Although we have not given a detail description of this modeling framework, we can say that it is a powerful approach, since it considers a comprehensive set of mechanisms and changes, and also allows different physics-based couplings, for example, the temperature dependence of rheological properties, the influence of moisture loss on deformation. From a modeling perspective, the most important aspect is the coupling between deformation (described by solid mechanics) and transport processes inside the product. Overall, we can mention two drawbacks or limitations: (1) this modeling framework involves a certain level of knowledge and expertise regarding physics and also simulation (even using commercial software); (2) the obtained models still depend on experimental data to evaluate effective transport coefficients. Nevertheless, without incorporating further complexities, it represents the most complete framework to model the baking process, in terms of capability of prediction and explanation or insights gaining.

3.3.4 Multiscale models

So far, we have discussed modeling at one single (spatial) scale, that is, the product scale or macroscale (or continuum). Even in the last modeling approach, where a "sufficient" detailed description of the product/process is considered, the porous medium is treated as a (idealized) continuum via homogenization and local equilibrium concepts. However, in the case of baking, a further objective would be to describe and predict explicitly the changes at pore scale or mesoscale and their relationships with macroscale properties (e.g., texture), considering the key role of the dispersed gas phase (bubbles or pores) in the product/process (van der Sman, 2022). For instance, such mesoscale properties and changes have been included via effective or apparent properties based on empirical data, in a partial and indirect manner. This (multiscale) problem can be tackled then via the multiscale modeling paradigm. In brief, multiscale models involve a hierarchy of interconnected sub-models that aim to describe the behavior of the material and different processes at multiple scales in a comprehensive way (Ho et al., 2013). The main advantage of this approach is the incorporation of physics occurring at smaller scales into macroscale models, so (a priori) fewer assumptions and empirical data are required for defining material properties, equilibrium relationships and driving forces (Datta et al., 2012).

Given the already discussed complexity of the baking process and the inherent challenge of developing multiscale models, examples of this framework are rather scarce in the literature. It is worth mentioning that some authors have worked on the development of pore scale or mesoscale models with focus on a gas cell or bubble as domain, that is, dough material enclosing a certain volume of gas. Although a proper multiscale model was not reported in those articles, they represent a building block (submodel) and a good starting point. For instance, de Cindio and Correra (1995) modeled the thermal expansion of a gas bubble ("microsystem") assuming linear viscoelasticity for the surrounding dough, while Fan et al. (1999) considered a viscous material in a similar system. Bikard et al. (2012) focused on the dough/crumb transition, also working on a gas cell model. These models aimed at describing macroscopic variables and changes from the behavior of the representative unit of the system (e.g., bubble, cell or pore) at mesoscale.

On the other hand, Lucas et al. (2015) actually proposed a multiscale (one-dimensional) formulation to describe the so-called evaporation-condensation-diffusion mechanism in bread baking. Briefly, this model assumes two phases over which different transport mechanisms occur, that is, dough and gas bubbles or pores. In general terms, the governing equations and fluxes for the dough phase (or fraction) are similar to those of the macroscale formulation previously presented, including modeling of deformation (viscous material in this case). For the volume fraction occupied by bubbles, the authors considered the transport of water vapor, CO_2 (initially solubilized in liquid water) and air, and proposed different mechanisms depending on the "state" of the pore network, that is, if pores are closed or open:

- Closed pores: during the first stages of baking, bubbles are closed and the evaporation-condensation-diffusion mechanism takes place. Water vapor is transferred by diffusion through the pore filled with stagnant CO_2, condensates on the cold side of bubble (idealized as 1D), then it diffuses as liquid water through the dough wall, until new vaporization occurs, and so forth. This process involves latent heat due to phase change, that is, evaporation-condensation.
- Open pores: when bubbles become interconnected during the last stages of baking, due to deformation and structural changes, gas transport is governed by convective flux (Darcy's law).

The shift between closed and open status of pores depends on two different temperature values (model parameters): the temperature of dough stiffening (viscosity change) and the temperature of pore opening (permeability change). These parameters were tuned to fit experimental profiles, but they are related to actual changes observed during baking, that is, they are consistent with physics. This new multiscale formulation predicts some phenomena previously reported during a typical baking operation, for example, an increase of water content at the core and end of global expansion or oven rise due to dough/crumb transition.

In summary, the multiscale modeling approach would provide the proper framework to develop comprehensive models by considering explicitly the physics at different relevant scales for the baking process (van der Sman, 2022). For instance, such a complete model would be a valuable tool for rational design of both recipe and process, avoiding common extensive trial-and-error practice. However, this formulation requires a sound understanding of physics (of the system), as well as modeling and simulation expertise to implement the formulation (more details about multiscale modeling aspects can be found in the cited articles).

3.3.5 Final remarks

To conclude the first part of this chapter, it is worth recalling a "general rule" of modeling: there is always a trade-off or compromise situation between complexity (formulation, implementation), dependence on experimental data, and level of universality or physics involved in formulation and outputs of the model (Fig. 3.4). This so-called rule especially applies for complex materials/processes like baking of bread and cake, which can be considered as a multiscale problem, where more than one approach is possible. The ultimate decision on modeling framework will depend on the specific

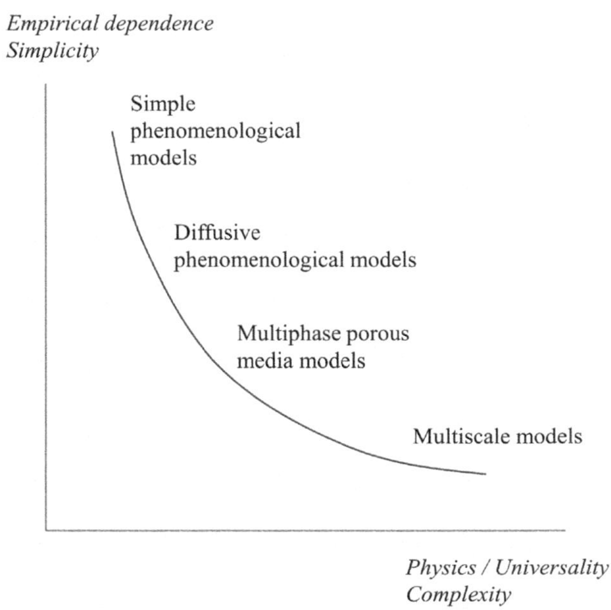

Figure 3.4 Conceptual trade-off curve for modeling approaches.

task or objective and the available resources and tools. Finally, regardless of the level of simplicity, it is always wise to have knowledge about the physics of the process in order to interpret and explain results in a consistent and logical way.

3.4 The roasting process

3.4.1 Overview of roasting process

The roasting process is an intense, hot dry-air cooking treatment, where the food is heated at elevated temperatures (generally higher than 150°C) in an oven. Roasting is applied to meats, vegetables, seeds and nuts. Also, pan roasting can be applied to seed and grains, that is, contact heating with hot surfaces (Sruthi et al., 2021). In the case of meat, roasting induces a series of phenomena and chemical reactions which confer unique and desirable features to food, related to taste, flavor and color; it also improves safety, palatability and digestibility, and extends shelf-life (Pathare and Roskilly, 2016). At the same time, harmful or toxic compounds can be formed and safety conditions (microbial inactivation) must be achieved.

3.4.2 Transport phenomena and other changes during roasting

In the following sections we focus on meat roasting, and finally, the roasting of other types of products will be briefly mentioned. In the meat roasting process diverse physical phenomena take place. First, the more relevant phenomena involve the heat and mass transfer during the process (Pathare and Roskilly, 2016), which determine different technological outputs like roasting time and process yield. Second, meat is a highly moist porous medium that can suffer large deformation (Datta, 2007), depending on the type of meat and operating conditions; considering such size variation help to a better prediction of technological outcomes and understanding of the process. Third, a moving evaporation front may be established, from the sample surface to inner regions, best known as a Stefan problem (Bonacina et al., 1973). In this chapter we will focus on the first aspect, that is, heat and moisture transfer, and we will discuss other phenomena briefly.

During roasting, heat is transferred from oven ambient to the meat surface by a combination of mechanisms: conduction and convection from the hot dry or partially moist oven air, radiation from the oven walls and/or electrical resistances inside the oven, and conduction from the tray where the meat is placed. Such combination results in an uneven heat flow over the meat sample surface; for modeling purposes, a uniform boundary condition is usually adopted. The heat flow inside the meat sample is mainly due to conduction, though heat transfer associated with the convective flow of moisture and fat is also present.

The most complex description of the physical phenomena is related to the moisture transfer, mainly due to material structure. For instance, when a meat sample is cooked in a high moisture or saturated environment like a water bath (immersion cooking), it is observed a significant moisture loss from the sample. Oillic et al. (2011) found weight losses greater than 40% during water bath cooking of *semimembranosus* beef muscle, while Bottani and Volpi (2009) cooked beef and turkey samples in industrial steam ovens and reported an absence of moisture vaporization, but still weight losses ranged from 15% to 18%. These observations demonstrate that only the moisture content does not determine the moisture flow inside the sample, since water moves to a region with higher water content. To understand such behavior, we should first consider the components and structure of meat: a porous medium made from a myofibrillar protein structure arranged as a collection of parallel fibers, held together by a network of connective tissue of collagen fibers (Bounton et al., 1975). In such structure, different water populations are present: (i) physically bound water, which is the intracellular water and wall water, and (ii) free water, which lies in the intercellular space (Khan et al., 2016). The intercellular water is held in such protein structures by capillary forces (Tornberg, 2005, 2013), so when such proteins are altered by external stimuli (e.g., temperature), the capillary forces and thus the water state change. In this regard, Yu et al. (2017) and Agafonkina et al. (2019) reported the denaturation temperature for main proteins: myofibrillar proteins (myosin, titin, actin, and others), around 40–70°C; stromal proteins (collagen, elastin, reticulin), peak at about 60–70°C; sarcoplasmic proteins (hemoglobin, myoglobin, and several enzymes), around 40–90°C.

In this way, the generally accepted explanation for moisture transport is based on the effect of temperature on the different meat components. Increasing temperature causes protein denaturation, reducing the water holding capacity of the proteins, and thus resulting in water de-binding, so less bound or free moisture appears in the sample. Studies using proton nuclear magnetic resonance T_2 relaxometry during heating confirmed the changes in different water populations (Bertram et al., 2006). At the same time, also denaturation induces a shortening or contraction of the protein network structure; such shrinkage creates a mechanical pressure/force that squeezes the released moisture toward lower pressure regions (Bell et al., 2006; Bengtsson et al., 1976; Dhall et al., 2012; Godsalve et al., 1977; Hung et al., 1978; Lepetit, 2008; Palka and Daun, 1999; Tornberg, 2005, 2013; van der Sman, 2007a, 2007b, 2013). Such lower-pressure regions correspond to surface, but also the core of the meat sample. For example, if a meat sample is punctured (fork or similar object) during roasting, a jet of liquid may escape from the sample when the puncture is released. Thorvaldsson and Skjöldebrand (1996) found that when beef roasting starts, moisture moves from the external regions toward the sample core, but later the flow is inverted, when core reached about 70°C.

When the squeezed moisture due to proteins denaturation arrives to the meat surface, it can be transferred to the ambient by evaporation or dripping (Fig. 3.5), that is, liquid passing across the surface (Burfoot and Self, 1989; Feyissa et al., 2013; Hung et al., 1978). The dripping is not only water, but also fats, proteins, minerals and different soluble compounds (the drip loss of fats can start once melting temperature is reached). The dripping loss can be observed when the heat transfer rate is low or moderate. As the liquid passes through the sample surface, the moisture may evaporate rapidly or slowly (Goñi and Salvadori, 2010). If low or moderate heat and mass transfer exists, this liquid may form droplets which may begin to flow over the sample surface toward lower regions (Fig. 3.5); then the drip losses can accumulate in the tray or fall to the tray if a netting tray is used (in any case, the liquid collected in the tray may evaporate during the process).

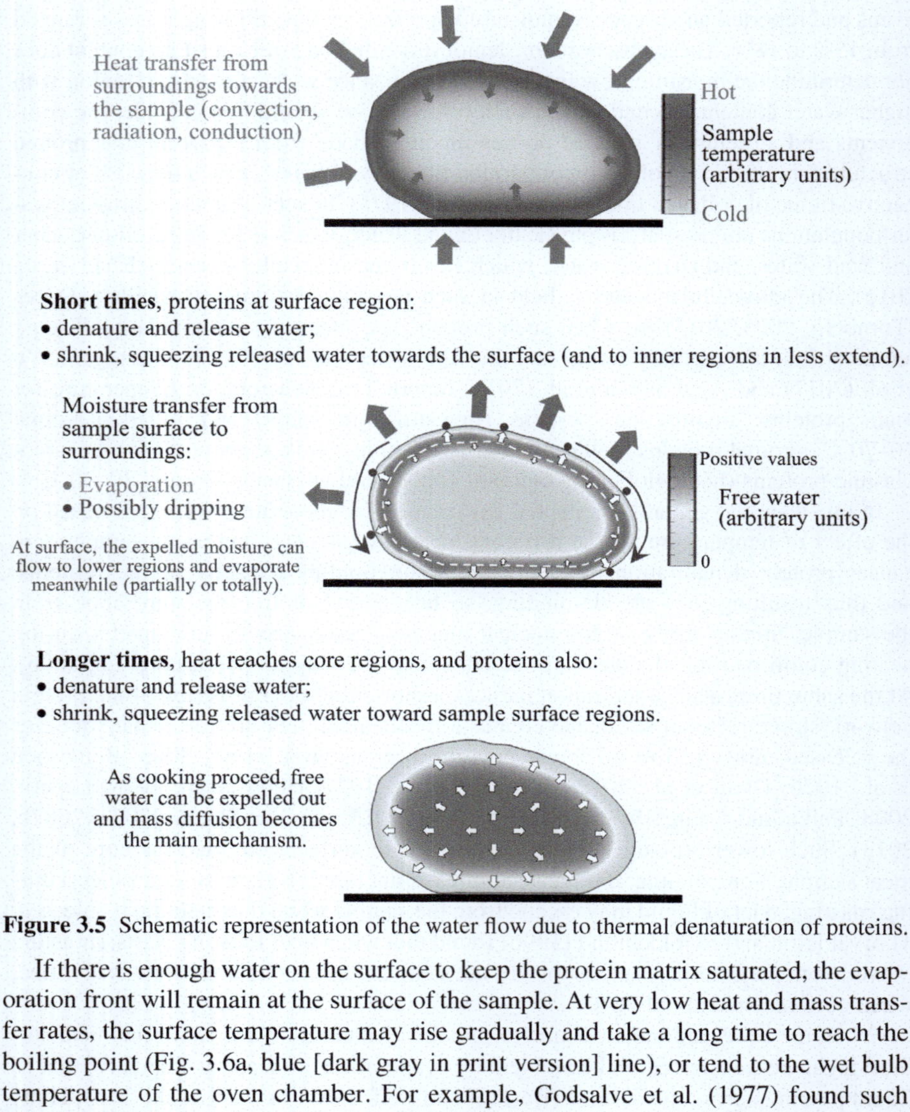

Figure 3.5 Schematic representation of the water flow due to thermal denaturation of proteins.

If there is enough water on the surface to keep the protein matrix saturated, the evaporation front will remain at the surface of the sample. At very low heat and mass transfer rates, the surface temperature may rise gradually and take a long time to reach the boiling point (Fig. 3.6a, blue [dark gray in print version] line), or tend to the wet bulb temperature of the oven chamber. For example, Godsalve et al. (1977) found such behavior during roasting of beef *semitendinosus* muscle at 121°C, while Bengtsson et al. (1976) found surface temperatures below 100°C during the roasting of beef samples at 175 and 225°C. As heat and mass transfer rates are increased, the surface temperature can reach the boiling point and remain constant because the internal water flow is enough to keep the surface sufficiently moist (Fig. 3.6a, red [light gray in print version] dashed line), for example, Goñi and Salvadori (2010) found this behavior for beef *semitendinosus* muscle. In these cases, the dripping can be observed clearly.

On the other hand, for very high heat transfer rates, high evaporation rate is achieved. If the internal moisture transfer rate is not sufficient to keep the surface

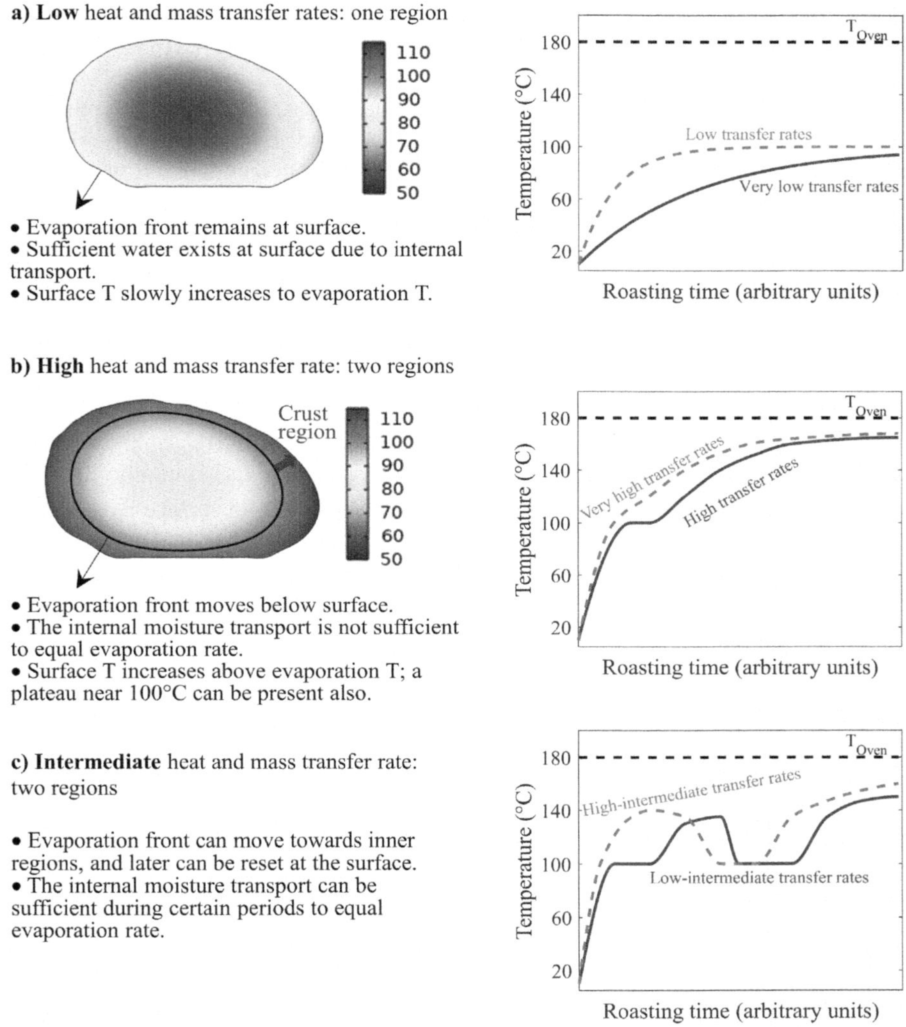

Figure 3.6 Schematic representation of surface temperature behavior during meat roasting at (a) low, (b) high, and (c) intermediate heat and mass transfer rates.

wet, the sample surface may pass directly through the boiling point without exhibiting a plateau (Fig. 3.6b, red [light gray in print version] dashed line), with the subsequent formation of an internal evaporation front. This front determines two regions: (i) a hot and dehydrated crust, with water content below water holding capacity, and (ii) the core, with water content above water holding capacity. For instance, Skjöldebrand and Hallström (1980) found such behavior during roasting of minced meat loaves. At high heat transfer rates, a plateau at the boiling point can be established for some time, while the internal moisture transfer rate keeps the surface wet. Later, as the surface moisture falls below the water holding capacity, the evaporation front is displaced below the sample surface, generating a crust (Fig. 3.6b, blue [dark gray in print version] line). In each of these conditions there is no dripping, so the total weight loss is equal to the evaporation loss.

When an intermediate heat and mass transfer rate is used, a combination of the previous behaviors can be observed. Initially, the surface temperature may show a plateau as long as the internal water transfer rate is sufficient to keep the surface wet. Later, if sufficient water is not available, the evaporation front may move inside the sample, leading to a gradual increase in surface temperature and the formation of a crust. As inner regions heat up and water is forced to move to the surface, a sudden drop in surface temperature can be observed due to the liquid entering these regions. The surface may again remain in a plateau while the internal pressure driven water transfer is active; eventually, the water flux decreases and the evaporation front begins to move inward again (Fig. 3.6c, blue [dark gray in print version] line). If a (slightly) higher heat transfer rate is established, such first plateau period may be absent (Fig. 3.6c, red [light gray in print version] dashed line). For instance, Goñi (2010) measured the surface temperature of six samples of bovine *semitendinosus* muscle roasted in a domestic oven; for one sample, a slight increase in surface temperature was observed (lower than 10°C), followed by a decrease after a while. Such behavior indicates the drying followed by a sudden surface rehydration (and cooling) due to liquid water coming from inside. Godsalve et al. (1977) measured surface temperature of beef *semitendinosus* muscle; for samples roasted (dry air) at 149, 177 and 204°C, the measured surface temperature follows the last general trend: it gradually increases above 100°C, then drops rapidly to close to 100°C and maintains this value for a certain time; finally, it begins to rise again (note that temperature decrease of more than 50°C in few minutes was measured under certain conditions). For some samples, the behavior described in Fig. 3.6b was also observed, with the surface temperature rising above 100°C and continuously rising to the oven temperature, that is, not sufficient water was available to maintain the evaporation front.

The general conditions for having one behavior or another are hard to define; it will depend on a number of factors related to the roasting conditions (oven temperature, air velocity, heating source - electric, gas -) and the meat sample (source, muscle, shape, size, etc.). In addition, one sample can exhibit several behaviors simultaneously. For example, the top surface can have a temperature close to 100°C, keeping the evaporation front on it; the lower surface regions can be under 100°C due to droplets flowing; corners and edges exposed to higher heat transfer rates can be dry enough to form a crust with a temperature above 100°C.

Regarding dripping losses, they can be easily confirmed by visual inspection, but are difficult to measure and only a few works reported such values. For instance, Bengtsson et al. (1976) determined both evaporative and total weight losses during roasting of beef samples: evaporative loss was present in the first stage, and when surface temperature reached 65–70°C, dripping loss began and represented near 50% of the total weight loss. Hung et al. (1978) measured the amount and size of drops per minute during roasting of beef samples from frozen state, and found two dripping periods, with the drip loss between 12% and 18% of total weight loss. Besides, the authors built a useful "flowchart of water migration in meat during cooking", summarizing different phenomena. Larson et al. (1992) reported drip losses between

21.9% and 25.6% and total weight losses between 29.2% and 32.4%, for cooking pork leg in oven bags at 82–163°C. Goñi and Salvadori (2010) found that dripping represents 53.2% of total weight loss during roasting of beef *semitendinosus* muscle. Schönfeldt and Strydom (2011) roasted 15 different beef cuts at 160°C under dry and moist conditions; the total weight loss ranged from 20.33% to 40.32%, while drip loss was between 4.76% and 31.62%.

Finally, it is worth noting that the high variability observed in reported drip loss data may obey to differences in the utilized experimental procedures, beyond a variation in roasting conditions and tested samples. Additionally, as pointed by Goñi and Salvadori (2010) and Kondjoyan et al. (2013), the drip moisture remains partially at the surface, so certain fraction is evaporated; therefore, when the drip loss is finally measured, a lower value is obtained (considering the total exerted liquid).

In summary, the roasting of meat involves complex, simultaneous and coupled transport phenomena and also structural changes (e.g., protein denaturation, large deformation), which depend on operating conditions and characteristics of food material. Experimental studies demonstrate the complex coupling between heat transfer, thermal denaturation of proteins (major component of meat matrix together with water), and mass transfer occurring during the process. In particular, the relationship between (heat-driven) mechanical deformation and transport of liquid water plays a key role and represents a challenge for modeling purposes.

3.5 Modeling of roasting process

Different approaches have been developed to describe meat roasting. Considering product modeling, a simple (and arbitrary) classification can be the following:

- Models with only heat transfer by conduction.
- Models with heat transfer by conduction, and empirical calculation of mass loss.
- Models with heat transfer by conduction and mass transfer by diffusion.
- Models with heat transfer by conduction and convection, and moisture transport by pressure driven convective flow and diffusion.
- General (multiphase) porous media approach.

In the next sections we will describe the main aspects of these approaches, and other models that do not fit in this classification. We do not attempt to cover all approaches, but the most common ones. The complexity in modeling has augmented along the years, because of the increase of computational power, the growth and spread of programming skills, and the emergence of more general-purpose, easy-to-use software for solving partial differential equations' systems, which have capabilities to solve strongly coupled problems.

3.5.1 Heat conduction models without mass transfer; evaporation at surface

A first approach is to consider only energy transfer inside the meat sample. General main assumptions are the following:

- Energy transfer is due to conduction (Fourier's law).
- Moisture evaporation takes place at surface, that is, it is assumed that during roasting sufficient water exist at the surface to keep the evaporation front on it. Internal mass transfer is neglected.
- Isotropic thermal properties.
- Uniform heat transfer coefficient over the sample surface.
- Uniform initial condition.
- No deformation is considered.

Under these assumptions, the energy balance read as:

$$\rho\, C_p \frac{\partial T}{\partial t} = \nabla \cdot (k \nabla T) \tag{3.17}$$

Where ρ is density (kg/m^3), C_p is heat capacity (J/(kg K)) and k is thermal conductivity (W/(m K)). Thermal properties can be constant or a function of the local temperature. Given the particular structure of the meat samples, the anisotropy of the tissue can be considered (Thorvaldsson and Skjöldebrand, 1996), in which case parallel and perpendicular models for thermal conductivity are available. As boundary condition for the heat balance, it is assumed that evaporation takes place at surface:

$$-n \cdot k \nabla T = h(T_s - T_{air}) + \varepsilon\sigma\left(T_s^4 - T_{oven}^4\right) + \lambda \dot{m}_{evap} \tag{3.18}$$

The left-hand side represents the conductive heat transfer inside the sample, where n is the outer normal surface vector. The right-hand side describes the external heat flow: the first term is the convective heat flow, the second term is the radiative heat transfer, and the last term is the energy associated with surface moisture evaporation. h is heat transfer coefficient (W/(m^2 K)), ε is meat emissivity, σ is Stefan–Boltzmann constant, 5.67×10^{-8} W/(m^2 K^4), λ is latent heat of water vaporization (2257 kJ/kg at 100°C), the subscript "air" refers to air temperature, "s" refers to meat surface temperature, and "oven" to the oven cavity walls or heating elements. The use of a single heat transfer coefficient, radiation or an effective heat transfer coefficient is discussed in Cepeda et al. (2013), as well as correlations for estimating such values. The local evaporation rate $\dot{m}_{evap}$ (kg/(m^2 s)) can be estimated as:

$$\dot{m}_{evap} = k_g\left(a_w P_{sat}(T_s) - \frac{RH}{100} P_{sat}(T_{air})\right) \tag{3.19}$$

Where k_g is the mass transfer coefficient (m/s), a_w is meat water activity at the surface temperature, RH is oven relative humidity (%), and P_{sat} is the saturation pressure of water (Pa) at the given temperature. Since the model does not predict a surface moisture content, the water activity can be a function of the surface temperature, $a_w = a_w(T_s)$, and also an average moisture content could be employed. When a tray is considered in the model, a different boundary condition may be set in the corresponding boundaries:

$$-n \cdot k\nabla T = h_{\text{sample-tray}}\left(T_s - T_{\text{tray}}\right) \tag{3.20}$$

The subscript "tray" refers to tray temperature. Table 3.1 summarizes some related articles which used this approach. The water evaporated from the sample can partially

Table 3.1 Models describing heat transfer (using Fourier's law) on meat products during roasting.

References	Product	Roasting conditions	Geometry and solution method	Additional details
Chang et al. (1998)	Turkey	162.8°C (without preheating)	2D irregular; finite element method	Experimental average moisture loss rate in boundary condition
Goñi et al. (2008)	Half beef *semitendinosus* muscle	172.8–223.5°C	3D irregular; finite element method	21.6% error prediction of weight loss
Obuz et al. (2002)	Beef *semitendinosus* muscle (0.13–0.19 m length, 0.08–0.1 m diameter)	163°C	Finite cylinder; finite difference method	22% error prediction of weight loss
Schillaci et al. (2022)	Beef burgers and vegetarian patties	Laser heat source at the boundary	3D finite cylinder; finite volume method	Mass loss from experimental data
Singh et al. (1984)	Beef *semimembranosus* muscles, 15 × 8 × 5.5 cm	175, 225°C	2D rectangular 8 × 5.5 cm; AID finite difference method	Model of accumulation of water vapor in the oven
Townsend et al. (1989)	Beef, 0.25 to 4 length over diameter ratios	177°C	Finite cylinder; finite difference method	Model with phase change (melting, boiling) incorporated

accumulate in the oven, which will modify the RH values and affect the driven force for evaporation; Singh et al. (1984) provide an approach to handle such variations.

The total evaporative rate ($\dot{M}_{\text{evap}}$, kg/s) can be estimated by integration of the local water vapor flow over the sample surface S:

$$\dot{M}_{\text{evap}} = \int_S \dot{m}_{\text{evap}}\, dS \tag{3.21}$$

The total evaporative loss (EL, kg) can be calculated by integrating $\dot{M}_{\text{evap}}$ over roasting time:

$$\text{EL} = \int_0^{t_{\text{final}}} \dot{M}_{\text{evap}}\, dt \tag{3.22}$$

Regarding prediction of weight loss, this simple formulation usually underestimates such values since it does not take into account drip losses (among other simplifications). As previously discussed, if enough water is available at the surface to maintain the evaporation front, it can be also lost by dripping. For example, Obuz et al. (2002) reported this underestimation for five out of six beef *semitendinosus* samples roasted at 163°C, with an average prediction error of 22%. Similarly, Goñi et al. (2008) found an underestimation of weight loss for (all) six beef *semitendinosus* samples roasted at different temperatures, with an average prediction error of 21.6%. Note that for very high heat transfer rates, where most of the expelled moisture would be evaporated from the sample surface, this model could predict weight loss more accurately. However, in these situations the evaporation front could move toward inner region, generating an outer crust zone. So, although weight loss prediction could be acceptable, prediction on cooking time may be affected due to proposed simplifications.

This modeling approach can be used to estimate the energy consumed by the sample during roasting, for example, two measures can be calculated, the sensible heat and the latent heat. The sensible heat is related to the temperature increase; the energy per time (J/s) can be estimated by integration of the local energy accumulation over the sample volume:

$$\dot{E}_{\text{sensible}} = \int_V \left(\rho\, C_p \frac{\partial T}{\partial t}\right) dV \tag{3.23}$$

The latent heat is associated with water evaporation (phase change); the energy per time (J/s) related to evaporation can be estimated by integrating the evaporation rate over the sample surface:

$$\dot{E}_{\text{latent}} = \int_S \left(\lambda\, \dot{m}_{\text{evap}}\right) dS \tag{3.24}$$

Then, the total energy consumption, E_{total} (J), can be obtained by calculating:

$$E_{\text{total}} = \int_0^{t_{\text{final}}} \left(\dot{E}_{\text{sensible}} + \dot{E}_{\text{latent}}\right) dt \tag{3.25}$$

For instance, Goñi and Salvadori (2014) found a good agreement (12% difference) between the previous equations and a rough macroscopic approximation:

$$E_{\text{total,macroscopic}} = m_0\left(C_{p,ave}\ \Delta T + \frac{\text{WL }(\%)}{100}\lambda\right) \tag{3.26}$$

Where m_0 is the sample raw mass (kg), $C_{p,\text{ave}}$ is an average specific heat of the sample, ΔT is an average temperature increase during roasting, and WL (%) is the percentage weight loss. All weight loss is assumed to be due to evaporation, but this is not necessarily true. For the temperature increase, usually initial temperature is assumed uniform, but final temperature may be different along the sample. Please note that energy associated to meat denaturation and fat melting could be considered as well, but it is not significant in comparison to water vaporization enthalpy and sensible heat.

3.5.2 Heat conduction models with empirical mass transfer

Following the previous approach, in order to incorporate an estimation of dripping losses and improve the model capability to predict weight losses, some empirical modifications have been proposed. For example, in several works the variation of local water and/or fat contents was described by a kinetic model:

$$\frac{dC_i}{dt} = -K_i\left(C_i - C_{i,\text{eq}}\right) \tag{3.27}$$

Where C is the component concentration, subscript i refers to water or fat, subscript "eq" refers to an equilibrium value, and K_i is a rate constant (which can depend on the local temperature and composition). The equilibrium values are defined as a function of the temperature and the initial composition of each component:

$$C_{i,\text{eq}} = f\left(T, C_{i,0}\right) \tag{3.28}$$

For water, the equilibrium value is often called water holding capacity. Both equations are coupled to the heat transfer model and solved together. If a local prediction for moisture and fat composition is available, often the thermal properties used in the heat transfer model are expressed as a function of composition and temperature. Table 3.2 mentions some selected articles that employed this approach, for different cooking methods.

When using Eq. (3.28) to calculate equilibrium values, care must be taken. As temperature increases, equilibrium values diminish; but if temperature diminishes,

Table 3.2 Models describing heat transfer (using Fourier's law) in meat products, and empirical mass transfer models, for different heating conditions.

References	Product	Roasting conditions	Geometry and solution method	Additional details
Dalvi-Isfahan (2023)	84 mm diameter, 12 mm thickness hamburger patty	Grilling at 150°C	Cylindrical shape; finite element method	Shrinkage, microorganism inactivation
Goñi and Salvadori (2010)	Half beef *semitendinosus* muscle	Roasting, 172.8–223.5°C	3D irregular; finite element method	Rate constant *K* as a function of local temperature
Goñi and Salvadori (2012)	Half beef *semitendinosus* muscle	Roasting, 150–230°C	3D irregular; finite element method	Multi-objective optimization
Kondjoyan et al. (2013)	Beef *semitendinosus* muscle	Water bath (50, 70, 90°C), roasting (dry, steam assisted, pure stem)	Bricks; finite element method	Rate constant *K* as a function of distance from the sample surface
Ou and Mittal (2006, 2007)	Hamburger patty, 88.5 mm diameter 9.5 mm height	Pan roasting	1D, axial direction; finite difference method	Water and fat transfer model. Double-sided, or simple-sided with flipping
Pan et al. (2000)	Hamburger patty, 120 mm diameter 10 mm height	Double contact heating, 177°C	1D; finite difference method	E. Coli O157:H7 destruction model coupled to heat transfer model
Sargolzaei et al. (2011)	Hamburger, 14 mm height 117 mm diameter	Roasting, 114, 152, 204°C	1D; finite difference method, 2D fluent	No evaporation is considered; moving melting front

equilibrium values do not recover (e.g., water holding capacity), although the equation will predict such recovery.

3.5.3 Heat conduction models with mass transfer by diffusion

As an improvement of previous simple formulations, now mass transfer is assumed to occur due to diffusion. The energy balance remains the same as before, for example, Eqs. (3.17)–(3.20), but a mass balance is proposed to describe moisture transport by using the Fick's law of diffusion:

$$\frac{\partial c}{\partial t} = \nabla \cdot (D \nabla c) \tag{3.29}$$

Where c is the moisture content (kg/m^3), and D is the moisture diffusion coefficient (m^2/s). Initial condition for moisture balance usually assumes a uniform value. At the surface, the mass transport rate by diffusion equals the convective moisture transfer rate to the oven cavity (similar as before):

$$-n \cdot D \nabla c = k_g \left(a_w P_{sat}(T_s) - \frac{RH}{100} P_{sat}(T_{air}) \right) \tag{3.30}$$

Where k_g is the mass transfer coefficient (m/s); now water activity can be set as a function of the surface moisture content and temperature: $a_w = a_w$ $(T_s,\ c_s)$. The evaporative mass loss can be obtained as described earlier (Eqs. 3.21–3.22). Less frequently, the water content at the surface is assumed to reach an equilibrium value (Dirichlet boundary condition), for example,: $c = c_{eq}$

Since a local value for moisture content is available now, the sample properties used in the model can be expressed as a function of composition and temperature. Table 3.3 shows some articles that utilized this approach. This kind of models can be useful to predict total weight loss in comminuted meat products, formulated with a variety of food ingredients that retain or absorb the water released from the meat proteins during cooking. As pointed out by Burfoot and Self (1989), unrealistic large diffusion coefficients may be required to predict a coherent total weight loss, since the actual mass transfer mechanism is not purely diffusive. That is, such (effective or apparent) diffusivity values try to capture the complex behavior originated from pressure driven mechanism, as it was previously discussed. In this sense, although a mass transfer mechanism is considered, given the complex transport phenomena of the process, this approach is still phenomenological and strongly depends on experimental data (note that a similar discussion was given for the baking process).

Finally, it is worth mentioning that the following version of the energy balance may be found in the literature:

$$\rho\, C_p \frac{\partial T}{\partial t} = \nabla \cdot (k \nabla T) + \lambda \frac{\partial c}{\partial t} \tag{3.31}$$

Table 3.3 Models for heat (using Fourier's law) and mass transfer (using Fick's law) of meat products during roasting.

References	Product	Roasting conditions	Geometry and solution method	Additional details
Ahmad et al. (2015)	6.7 cm diameter and 18 cm length sausages (lean meat, beef fat, spice mixture)	Low air temperature; boiling water	1D cylinder, finite difference method and method of lines	
Bisceglia et al. (2013)	Pork meat, cube and cylinder shapes (50 g)	175°C	Cubes and finite cylinders, finite element method	
Chen et al. (1999)	Cylindrical chicken patties (1 cm height, 6.2 cm diameter)	135–218°C	2D axisymmetric, finite element method	Transient oven air conditions; 1.2% error for cooking yield
Huang and Mittal (1995)	Meatballs of beef and spices mixture	140°C, natural and forced convection, boiling	1D spherical coordinates; finite difference method	Estimation of heat transfer coefficient and diffusivity
Isleroglu and Keymak-Ertekin (2016)	Beef *semitendinosus* muscle	Steam-assisted Hybrid oven, 180–240°C	Cylindrical shape; finite element method	Different boundary condition for air and pan contact
Niamnuy et al. (2008)	Shrimp	80–120°C (jet spouted bed)	3D irregular shape; finite element method	Deformation, solid mechanics
Rinaldi et al. (2011)	Mortadella bologna, cylindrical shape	80–100°C	105 × 260 mm, 240 × 780 mm, diameter × length; finite difference method	No significant differences between experimental and simulated weight loss

This expression should be avoided since it means that any internal moisture variation implies evaporation, which is not necessarily true (Datta, 2007). Besides, the discussed approach assumes that evaporation takes place at the sample surface only.

3.5.4 Heat and mass transfer models with convection

As it was previously highlighted, pressure driven flow is an essential mechanism of moisture migration inside the samples. So far, the revised modeling approaches do not include this mechanism. Instead, some shortcuts or empirical solutions are proposed to obtain estimations in agreement with experimental data. A mathematical model that explicitly describes the convective water flux inside meat during roasting was developed by van der Sman (2007a, 2007b). This model is based on the Flory-Rehner theory (soft matter approach) and uses the Darcy's law to describe pressure-driven water flow. The model agreed well with the generally accepted explanation of liquid water transport mechanism given by Godsalve et al. (1977), and it can predict both evaporative and dripping losses. Following, we present this formulation in a concise way.

First, the excess water content or water demand (WD) is used to describe the chemical potential, which determines the moisture flow:

$$\mathrm{WD} = c - \mathrm{WHC}(T) \tag{3.32}$$

Where c is water content (kg/m^3) and WHC is the water holding capacity at the local temperature of the meat sample. Initially, both values are equal and WD is zero, so no free water is available to flow. When temperature increases, denaturation of meat proteins starts and thus WHC diminishes, so water demand increases (free water appears in the sample). Besides, the shrinkage of the protein network (also due to denaturation) exerts pressure on the released water, namely swelling pressure (P_{sw}). Such variable is assumed to be proportional to the water demand:

$$P_{\mathrm{sw}} = E \times \mathrm{WD} = E(c - \mathrm{WHC}(T)) \tag{3.33}$$

Where E is a material property (Pa m^3/kg) that may depend on water content and temperature. In this model, the Flory-Rehner theory has been linearized to simplify the formulation; on other hand, it is assumed that WHC depends only on temperature (meaning that protein denaturation and water release are instantaneous, which is an approximation). The model states that the swelling pressure gradient is the driving force for the liquid water flux (kg/(m^2 s)) inside the product, which follows the Darcy's law:

$$\rho_{\mathrm{w}} u = -\frac{\kappa}{\nu} \nabla P_{\mathrm{sw}} \tag{3.34}$$

Where ρ_w is bulk density of water (kg/m^3), u is liquid velocity (m/s), ν is kinematic viscosity (m^2/s), and κ is the material permeability (m^2). For instance, the (unknown) properties can be lumped into a single (effective) coefficient γ (m^2/s), which can be determined from experimental data:

$$\gamma = \frac{E\,\kappa}{\nu} \tag{3.35}$$

The liquid water flux is then incorporated in the mass conservation equation to describe the variation of moisture content inside the product:

$$\frac{\partial c}{\partial t} + \nabla(\rho_{\mathrm{w}}\,u) = 0 \tag{3.36}$$

Some authors have proposed the inclusion of an additional diffusive term (Feyissa et al., 2009; Blikra et al., 2019):

$$\frac{\partial c}{\partial t} + \nabla(\rho_{\mathrm{w}}\,u) = \nabla\cdot(D\nabla c) \tag{3.37}$$

The energy balance considers the heat transfer by conduction and the convective contribution due to liquid water flux:

$$\rho\,C_p\frac{\partial T}{\partial t} + C_{p,\mathrm{w}}\,c\,u\nabla T = \nabla\cdot(k\nabla T) \tag{3.38}$$

Where $C_{p,\mathrm{w}}$ is heat capacity of liquid water (J/(kg K)). The boundary condition for the momentum equation (Darcy's law) establishes that the swelling pressure is zero at the surface, that is, the water content is equal to the water holding capacity:

$$P_{\mathrm{sw}} = 0 \rightarrow c = \mathrm{WHC}(T_{\mathrm{s}}) \tag{3.39}$$

Boundary conditions for the heat and mass balances given by Eqs. (3.18) and (3.30) can be used again, respectively. Furthermore, the liquid flow at the surface can be considered; additional terms should be added accordingly (Feyissa et al., 2009):

$$-n\cdot\left(k\nabla T + C_{p,\mathrm{w}}\,c\,u\,T\right) = h(T_{\mathrm{s}} - T_{\mathrm{air}}) + \varepsilon\sigma\left(T_{\mathrm{s}}^4 - T_{\mathrm{oven}}^4\right) + \lambda\dot{m}_{\mathrm{evap}} \tag{3.40}$$

$$-n\cdot(D\nabla c + c\,u) = k_{\mathrm{g}}\left(a_{\mathrm{w}}P_{\mathrm{sat}}(T_{\mathrm{s}}) - \frac{\mathrm{RH}}{100}P_{\mathrm{sat}}(T_{\mathrm{air}})\right) \tag{3.41}$$

Different assumptions can be proposed in this regard, since not all water reaching the surface evaporates necessarily. For instance, Blikra et al. (2019) included a factor in the boundary condition to account for the fraction of energy used for evaporation. Table 3.4 summarizes selected articles that uses a similar approach.

The weight loss due to evaporation can be calculated as in the previous models (Eqs. 3.21–3.22). On the other hand, the local drip loss rate, $\dot{m}_{\text{drip}}$ (kg/(m^2 s)), can be calculated at the surface as:

$$\dot{m}_{\text{drip}} = -\frac{\kappa}{\nu}\nabla P_{\text{sw}} \tag{3.42}$$

Then, the total dripping flow rate ($\dot{M}_{\text{drip}}$, kg/s) can be estimated by surface integration of the local rate:

$$\dot{M}_{\text{drip}} = \int_S \dot{m}_{\text{drip}}\, dS \tag{3.43}$$

And the total drip loss (DL, kg) can be computed by integrating $\dot{M}_{\text{drip}}$ over roasting time:

$$\text{DL} = \int_0^{t_{\text{final}}} \dot{M}_{\text{drip}}\, dt \tag{3.44}$$

Finally, the total weight loss WL (kg) is calculated as the sum of the weight losses due to both evaporation and dripping:

$$\text{WL}(t) = \text{EL}(t) + \text{DL}(t) \tag{3.45}$$

For instance, Goñi (2010) used this modeling approach to simulate the roasting of *semitendinosus* muscle in a domestic oven. A constant value was used for $\gamma = 2.05 \times 10^{-8}$ m^2/s, and the remaining properties as indicated in Goñi and Salvadori (2010). The maximum absolute difference between the experimental and simulated temperature profiles was 6.12%; the average error in weight losses due to evaporation was 28%, for dripping losses was 18.7%, and for total weight loss was 13.5%. From experimental measurements, it was obtained that dripping losses represent 53.2% of the total losses, while from simulations it was 52.7%. It was assumed that the water holding capacity depends only on temperature. A possible improvement can be to include the kinetics of protein denaturation (Ishiwatari et al., 2013; Yu et al., 2014) and their effect on water holding capacity.

3.5.5 Moving evaporation front models

In the models mentioned in the previous sections the evaporation front remains at the surface. But when high heat and mass transfer rates are used, and large samples are

Table 3.4 Models accounting for pressure driven water flow.

References	Product	Roasting conditions	Geometry and solution method	Additional details
Blikra et al. (2019)	Cod loin, brick 24 × 33 × 13 mm	Near 150°C	3D brick; finite element method	Factor *f* at the boundary condition to control heat transfer and experimental factor of evaporation
Chapwanya and Misra (2015)	1.5–4.0 cm thick beef steak	Double-sided pan cooking	1D; finite difference method	Osmotic pressure from flory-huggins theory
Feyissa et al. (2009)	Lean meat, 20 mm radius, 54 mm length	175°C	Cylindrical geometry; finite element method	With shrinkage
Feyissa et al. (2013)	Pork meat (*Longissimus dorsi*)	175°C, dry air	Bricks; finite element method	Factor *f* at the boundary condition to control heat transfer
Fukuoka et al. (2007)	Cylindrical beef fillet, 20 mm radius, 54 mm length	Pan cooking	2D; finite element method	Convection neglected in energy balance
Goñi (2010)	Half *semitendinosus* beef muscle	172.8–223.5°C	3D irregular; finite element method	13.5% prediction error of weight loss
Moya et al. (2021b)	Beef sample, 3D brick	Double-sided pan cooking	Brick; finite element method	With shrinkage, solid mechanics
Nelson et al. (2020)	Rectangular 5.5 × 8.0 cm beef steak	225°C	Rectangle; finite difference method	Fully nonlinear equations; anisotropic properties; with shrinkage
Rabeler and Feyissa (2018)	Chicken breast (skinless and boneless)	170 and 230°C (high and low fan)	Brick, 0.04 ×0.02 ×0.07m; finite element method	Energy for evaporation at the boundary is not considered; anisotropic properties
van der Sman (2007a, 2007b)	Beef *semimembranosus* muscles, 15 × 8 × 5.5 cm^3	175, 225°C	Rectangle; finite volume method	Anisotropic thermal conductivity and permeability
van der Sman (2013)	Chicken breast filet	45–150°C	3D irregular; finite volume method	Water in different phases, and water absorbed by solid

roasted, which requires longer cooking times, the evaporation front can move below the surface. In a few works, this situation has been considered. For instance, Holtz and Skjöldebrand (1986) developed a one-dimensional model with a moving evaporation front, which can deal with different heat transfer rates. They studied a $20 \times 5 \times 5$ cm^3 minced meat loaf (containing starch and spices) roasted at 175 and 200°C, with air velocities from 3 to 9 m/s. Townsend et al. (1989) developed a model considering the possibility of melting and boiling, but indicated that boiling was confined to the surface. High heat transfer rates are present commonly in contact cooking, where crust formation can be observed. Some authors have worked on the modeling of this heating method, for example, Sprague and Colvin (2011), Zorrilla and Singh (2000, 2003). In particular, during cooking of frozen hamburgers, two moving front problems are present simultaneously, that is, the melting of ice and the evaporation of water; additionally, melting of fat can be considered. Although these models were not developed for roasting, they could be used in situations where heat transfer determines that the evaporation front cannot remain at the surface. A possible approach is to utilize an apparent heat capacity to account for the evaporation of water or enthalpy jump. For instance, we have discussed this type of formulation in section 3.3.2 for modeling of baking (diffusive phenomenological models).

3.5.6 General porous media models

The main advantage of adopting a general (mechanistic) approach rely on a better understanding of the governing transport mechanisms. The resulting model is multiphasic in nature, since liquid and vapor water are considered separately, and a mechanism for water evaporation rate is incorporated. As a consequence, such model can handle situations where evaporation front remains at the surface, as well as cases where evaporation front moves toward inner regions of the product. On the other hand, the formulation and solving of these models are difficult tasks, involving solid knowledge of physics and expertise in numerical methods.

Datta (2007) presented a general description of the so-called hygroscopic porous media problem and discussed the different heat and mass transfer mechanisms and properties in food processing, including the distributed and sharp evaporation formulations to deal with phase change. Later on, Datta (2016) proposed different general frameworks for modeling food process, quality and safety aspects, based on the deformable, hygroscopic, porous media paradigm. Besides, Khan et al. (2016) reviewed the application of porous media modeling approach in food processes.

Regarding meat cooking, Dhall et al. (2012) implemented a multiphase and multicomponent porous media model for double-sided contact (intensive) heating of hamburger patties from frozen initial condition, including phase change for fat and ice melting, and water evaporation. Furthermore, for a similar problem, Dhall and Datta (2011) applied a comprehensive poromechanics-based modeling framework including deformation, assuming that patty is in a soft and rubbery state and evaporation stays on the surface during the entire cooking process. For deformation modeling, the authors utilized a modified Neo-Hookean constitutive model to account for the volume change due to moisture loss. More recently, Hernández-Alhambra et al. (2024) developed a model for pan frying of hamburger patty with large deformation mechanics, assuming also that the evaporation front remains at the surface in contact with the pan during cooking. In addition to shrinkage due to moisture loss, the authors incorporated the influence of protein denaturation (i.e., protein shrinkage) into the model. Besides the mentioned articles, the reader is referred to previous section 3.3.3, where this modeling framework was discussed for the case of baking.

3.5.7 Safety and quality modeling during meat roasting

Often the main interest of modeling is to obtain processing metrics/indexes or properties of roasted food, for which is mandatory to have accurate temperature predictions. For instance, Datta (2016) provided a general modeling framework for the prediction of quality features from temperature and moisture profiles. In the case of meat cooking, safety calculations are of paramount interest; for such studies, a kinetic model to predict the variation of local pathogenic or spoilage microorganisms' population (N, CFU/g) with processing time can be coupled to the cooking model. For this aim, the simplest formulation is based on the analogy with a first-order chemical reaction kinetics (Gil et al., 2006):

$$\frac{dN}{dt} = -k_{\mathrm{D}}N \tag{3.46}$$

Where k_D is the thermal death constant, which can be defined by an Arrhenius equation to establish temperature-dependence and thus coupling with the cooking model:

$$k_{\mathrm{D}} = k_0 e^{\frac{-E_{\mathrm{a}}}{RT}} \tag{3.47}$$

E_a is the activation energy (J/(mol K)), and k_0 is the velocity constant (1/s); both are specific values for a certain microorganism. The objective is to reduce several orders of magnitude of N, usually computed as integers of $-\log(N/N_0)$ (N_0 is the initial microorganism population). Alternatively, the thermal death time (F) can be defined as a function of the lethality (L):

$$F = \int_0^t L\,dt \tag{3.48}$$

$$L = 10^{\left(\frac{T-T_{\mathrm{R}}}{z}\right)} \tag{3.49}$$

Where T_R is a reference temperature and z is the thermal resistance constant (°C), which are specific for a given microorganism. The goal is to increase the F value (starting from 0) to a given integer of decimal reduction value D_R (e.g., $6D_R$ or $12D_R$).

More realistic kinetics for microorganism destruction can be used, as discussed by Gil et al. (2006) and Valdramidis et al. (2005). Additionally, cook values can be computed for relevant components (like vitamins) in the same way as for thermal death time (C values in analogy with F values).

Safety calculations can be applied in two main situations. First, the aim can be the decontamination of meat surface, since during slaughtering and handling, microorganisms present on the animal skin and viscera can contaminate the carcass surface. Since inner regions are not affected, short thermal treatments are commonly applied to sanitize the sample surface (Gil et al., 2006; Kondjoyan et al., 2006; Kondjoyan and Portanguen, 2008; Valdramidis et al., 2005). Second, when a product is made from minced or comminuted meat, the possible superficial contamination can be distributed over the whole product. In these situations, the cooking process must ensure the thermal death of microorganisms in the whole sample. Certainly, the associated paradigm is the cooking of hamburgers (Ou and Mittal, 2006, 2007; Zorrilla et al., 2003), and other comminuted products (Santos et al., 2008).

Besides safety, other aspects are relevant for meat cooking, for example, the variation of different quality features (hardness, color, etc.) are related to the operating conditions. These variations can be described also via kinetic models which are coupled to the roasting models. For instance, hardness is an important quality (texture) feature which determines consumer acceptance. Califano et al. (1997) described the hardness variation considering two simultaneous and opposite behaviors: collagen denaturation reducing hardness, and myosin and actin denaturation increasing hardness. Depending on the temperature range, one or another mechanism dominates and determines the entire behavior. Rabeler and Feyissa (2018) proposed a variable order kinetic model to describe the hardness, gumminess and chewiness of chicken breast during roasting.

On the other hand, the color of roasted meat represents the first contact of the product with consumers and determines its perception of doneness. Therefore, prediction of surface and inner color of different meats have been coupled with different roasting models. For instance, Yu et al. (2014) developed a model for protein denaturation and surface color (in $L^*a^*b^*$ space) during fish grilling. Rabeler et al. (2019) proposed a model for inner and surface lightness (L^*) prediction during chicken roasting, considering a whitening and a browning stage. Moya et al. (2021a) built kinetic models to predict color ($L^*a^*b^*$) during double-sided pan cooking of beef steak. Goñi (2010) and Goñi and Salvadori (2011) developed kinetic models to describe inner color during roasting of beef *semitendinosus* muscle.

Finally, the kinetics of harmful components generation has been coupled to cooking models as well. For example, Ngadi and Hwang (2007) developed a kinetic model for heterocyclic-amine formation during one-sided pan-frying of meat patties. Similarly, Sprague and Colvin (2011) coupled the heterocyclic-amine formation kinetics with a frying model for beef patty. Kondjoyan et al. (2014) analyzed the formation of heterocyclic aromatic amines and color variation during meat cooking.

It is worth noting that for all the above-mentioned safety and quality characteristics, the kinetic models are expressed as first order ordinary differential equations, which can be coupled and solved simultaneously with the cooking model, without increasing complexity in a significant way. Ultimately, all technological, safety and quality features can be considered in the multi-objective optimization of the roasting process.

3.5.8 Roasting models for coffee, cocoa, nuts, grains, and vegetables

Besides meat, roasting is applied to other products such as seeds, grains, nuts, in order to provide special characteristics (Sruthi et al., 2021). Models for roasting of these materials are usually the same than for drying but using higher temperatures. In the roasting process, the moisture content is reduced, but additional goals are pursued, for example, the development of flavor, color, etc. For instance, Castro et al. (2018) reviewed the mathematical modeling of convective drying of fruits, and Baidhe and Clementson (2024) discussed about the use of modeling and simulation on grain and seed drying. Since the heating of this type of foods is very fast, the drying front can quickly move to the inner zones of the products (Fadai et al., 2017). Table 3.5 shows some articles related to roasting of different foodstuffs.

3.6 Conclusion

In this chapter we discussed two basic and interrelated aspects of modeling considering the baking and roasting processes: (i) transport phenomena and other relevant changes (the "physics" of the processes) occurring in the products, based on experimental studies and fundamental concepts, and (ii) modeling approaches and frameworks to describe and predict the observed behavior. On the one hand, these processes involve complex, simultaneous and coupled phenomena. This is explained by the complex structure and composition of food materials. In bread and cake baking, the dispersed gas bubbles or pores in the dough and dough-crumb-crust transition play an essential role, in the same way as protein denaturation, swelling pressure and liquid flow in meat roasting. That is, macroscale or product scale observations and metrics are derived from phenomena and changes occurring at mesoscale. Consequently, the (traditional or "by-default") diffusive mechanisms of heat and mass transfer do not describe alone the observed behavior during baking and roasting.

On the other hand, different modeling approaches can be applied considering that different simplifications can be proposed, depending on either the specific objective and requirements of a given task, or the available resources and capability to develop a certain formulation. We should also consider the state of the art in modeling of food processing to understand the evolution of such approaches. In any case, it is worth recalling the inherent trade-off in modeling between complexity, dependence on experimental data, and level of universality physics involved in formulation and outputs of the models. Through previous sections we aimed at providing an overview of modeling approaches that have been applied in baking and roasting, from simple phenomenological to comprehensive physics-based frameworks. Such overview may be treated as a basic (incomplete) guide for modeling different aspects of these

Table 3.5 Articles about modeling and simulation of roasting of coffee, grains, seeds, vegetables, etc.

References	Product	Roasting conditions	Geometry and solution method	Additional details
Bottazzi et al. (2012)	Coffee	Fluidised-bed hot air laboratory roaster	Lumped And distributed parameters approach	High-temperature short-time roasting process at 533 K; low-temperature long-time process at 493 K
Fabbri et al. (2011)	Coffee	Roasting, 200°C for 10 min	Semielliptical and 3D irregular; finite element method	Fourier's law, Fick's law, evaporation at surface
Fadai et al. (2017)	Coffee	Roasting, 200°C	1D sphere; finite volume method and line method	Model 1: Fourier's law, Fick's law, evaporation at surface; model 2: Porous media approach, which predicts sharp evaporation front
Fadai et al. (2019)	Coffee	Roasting, 200°C	1D, sphere; finite difference method and line method	Deformation (poroviscoelastic model) and porous media approach
Jha (2005)	Gorgon nut	Roasting and popping, pan at 335°C	Macroscopic balances, empirical and theoretical approach	Kernel and shell structure, internal pressure, moisture content, volume change and temperature prediction
Llave et al. (2016)	Eggplant	Roasting, 250°C for 15 min	2D finite cylinder; finite element method	Fourier's law, Fick's law, evaporation inside the sample, deformation
Oliveros et al. (2017)	Coffee	Roasting, 470–533 K	3D irregular; fluent	Convection-diffusion Equation, airflow in the bed
Onwude et al. (2018)	Sweet potato	Combined infrared and hot-air drying, low temperature	2D axisymmetric geometry; finite element method	Fourier's law, Fick's law, evaporation at the surface
Robbins and Fryer (2003)	Barley	Spouted-bed roasting, 151, 188 and 216°C, 2–2.5 h	1D sphere; Crank-Nicolson method	Fourier's law, Fick's law, evaporation at the surface, equilibrium moisture at the surface

unit operations. Surely, new (or improved) frameworks will be applied in the near future, for example, physics-informed machine learning (Purlis, 2024b).

References

Agafonkina, I.V., Korolev, I.A., Sarantsev, T.A., 2019. The study of thermal denaturation of beef, pork, chicken and turkey muscle proteins using differential scanning calorimetry. Theory Pract. Meat Process. 4 (3), 19–23. https://doi.org/10.21323/2414-438X-2019-4-3-19-23.

Ahmad, S., Khan, M.A., Kamil, M., 2015. Mathematical modeling of meat cylinder cooking. LWT 60 (2), 678–683. https://doi.org/10.1016/j.lwt.2014.10.061.

Baidhe, E., Clementson, C.L., 2024. A review of the application of modeling and simulation to drying systems for improved grain and seed quality. Comput. Electron. Agric. 222, 109094. https://doi.org/10.1016/j.compag.2024.109094.

Bell, J.W., Farkas, B.E., Hale, S.A., Lanier, T.C., 2006. Effect of thermal treatment on moisture transport during steam cooking of skipjack tuna (*Katsuwonas pelamis*). Food Eng. Phys. Prop. 66 (2), 307–313. https://doi.org/10.1111/j.1365-2621.2001.tb11337.x.

Bengtsson, N.E., Jakobsson, B., Dagerskog, M., 1976. Cooking of beef by oven roasting: a study of heat and mass transfer. J. Food Sci. 41 (5), 1047–1053. https://doi.org/10.1111/j.1365-2621.1976.tb14386.x.

Bertram, H.C., Wu, Z., van den Berg, F., Andersen, H.J., 2006. NMR relaxometry and differential scanning calorimetry during meat cooking. Meat Sci. 74, 684–689. https://doi.org/10.1016/j.meatsci.2006.05.020.

Bikard, J., Coupez, T., Della Valle, G., Vergnes, B., 2012. Simulation of bread making process using a direct 3D numerical method at microscale: analysis of baking step. Int. J. Material Form. 5, 11–24. https://doi.org/10.1007/s12289-010-1018-3.

Bisceglia, B., Brasiello, A., Pappacena, R., Vietri, R., 2013. Food cooking process. Numerical simulation of the transport phenomena. In: Proceedings of the 2013 COMSOL Conference in Rotterdam, the Netherlands. https://www.comsol.com/paper/download/182043/bisceglia_paper.pdf. (Accessed 8 March 2025).

Blikra, M.J., Skipnes, D., Feyissa, A.H., 2019. Model for heat and mass transport during cooking of cod loin in a convection oven. Food Control 102, 29–37. https://doi.org/10.1016/j.foodcont.2019.03.001.

Bonacina, C., Comini, G., Fasano, A., Primicerio, M., 1973. Numerical solution of phase-change problems. Int. J. Heat Mass Tran. 16 (10), 1825–1832. https://doi.org/10.1016/0017-9310(73)90202-0.

Bottani, E., Volpi, A., 2009. An analytical model for cooking automation in industrial steam ovens. J. Food Eng. 90, 153–160. https://doi.org/10.1016/j.jfoodeng.2008.06.031.

Bottazzi, D., Farina, S., Milani, M., Montorsi, L., 2012. A numerical approach for the analysis of the coffee roasting process. J. Food Eng. 112, 243–252. https://doi.org/10.1016/j.jfoodeng.2012.04.009.

Boulet, M., Marcos, B., Dostie, M., Moresoli, C., 2010. CFD modeling of heat transfer and flow field in a bakery pilot oven. J. Food Eng. 97 (3), 393–402. https://doi.org/10.1016/j.jfoodeng.2009.10.034.

Bounton, P.E., Harris, P.V., Shorthose, W.R., 1975. Changes in shear parameters of meat associated with structural changes produced by aging, cooking and myofibrillar contraction. J. Food Sci. 40, 1122–1126. https://doi.org/10.1111/j.1365-2621.1975.tb01032.x.

Burfoot, D., Self, K.P., 1989. Predicting the heating times of beef joints. J. Food Eng. 9 (4), 251–274. https://doi.org/10.1016/0260-8774(89)90035-6.

Califano, A.N., Bertola, N.C., Bevilacqua, A.E., Zaritzky, N.E., 1997. Effect of processing conditions on the hardness of cooked beef. J. Food Eng. 34, 41–54. https://doi.org/10.1016/S0260-8774(97)00066-6.

Castro, A.M., Mayorga, E.Y., Moreno, F.L., 2018. Mathematical modelling of convective drying of fruits: A review. J. Food Eng. 223, 152e167. https://doi.org/10.1016/j.jfoodeng.2017.12.012.

Cepeda, J.F., Weller, C.L., Negahban, M., Subbiah, J., Thippareddi, H., 2013. Heat and mass transfer modeling for microbial food safety applications in the meat industry: A review. Food Eng. Rev. 5, 57–76. https://doi.org/10.1007/s12393-013-9063-6.

Cevoli, C., Panarese, V., Catalogne, C., Fabbri, A., 2020. Estimation of the effective moisture diffusivity in cake baking by the inversion of a finite element model. J. Food Eng. 270, 109769. https://doi.org/10.1016/j.jfoodeng.2019.109769.

Chang, H.C., Carpenter, J.A., Toledo, R.T., 1998. Modeling heat transfer during oven roasting of unstuffed turkeys. J. Food Sci. 63 (2), 257–261. https://doi.org/10.1111/j.1365-2621.1998.tb15721.x.

Chapwanya, M., Misra, N.N., 2015. A mathematical model of meat cooking based on polymer–solvent analogy. Appl. Math. Model. 39, 4033–4043. https://doi.org/10.1016/j.apm.2014.12.015.

Chen, H., Marks, B.P., Murphy, R.Y., 1999. Modeling coupled heat and mass transfer for convection cooking of chicken patties. J. Food Eng. 42 (3), 139–146. https://doi.org/10.1016/S0260-8774(99)00111-9.

Chhanwal, N., Anishaparvin, A., Indrani, D., Raghavarao, K.S.M.S., Anandharamakrishnan, C., 2010. Computational fluid dynamics (CFD) modeling of an electrical heating oven for bread-baking process. J. Food Eng. 100 (3), 452–460. https://doi.org/10.1016/j.jfoodeng.2010.04.030.

Cuq, B., Abecassis, J., Guilbert, S., 2003. State diagrams to help describe wheat bread processing. Int. J. Food Sci. Technol. 38 (7), 759–766. https://doi.org/10.1046/j.1365-2621.2003.00748.x.

Dalvi-Isfahan, M., 2023. Mathematical modeling for investigating the effect of single-sided flipping on moisture, fat content, and safety of hamburger patty. J. Food Process. Eng. 46, e14450. https://doi.org/10.1111/jfpe.14450.

Datta, A.K., 2007. Porous media approaches to studying simultaneous heat and mass transfer in food processes. I: Problem formulations. J. Food Eng. 80, 80–95. https://doi.org/10.1016/j.jfoodeng.2006.05.013.

Datta, A.K., 2016. Toward computer-aided food engineering: Mechanistic frameworks for evolution of product, quality and safety during processing. J. Food Eng. 176, 9–27. https://doi.org/10.1016/j.jfoodeng.2015.10.010.

Datta, A.K., van der Sman, R., Gulati, T., Warning, A., 2012. Soft matter approaches as enablers for food macroscale simulation. Faraday Discuss. 158, 435–459. https://doi.org/10.1039/C2FD20042B.

de Cindio, B., Correra, S., 1995. Mathematical modelling of leavened cereal goods. J. Food Eng. 24 (3), 379–403. https://doi.org/10.1016/0260-8774(95)90052-D.

Dhall, A., Datta, A.K., 2011. Transport in deformable food materials: A poromechanics approach. Chem. Eng. Sci. 66 (24), 6482–6497. https://doi.org/10.1016/j.ces.2011.09.001.

Dhall, A., Halder, A., Datta, A.K., 2012. Multiphase and multicomponent transport with phase change during meat cooking. J. Food Eng. 113, 299–309. https://doi.org/10.1016/j.jfoodeng.2012.05.030.

El Helou, P., Le Bideau, P., Fuentes, A., Glouannec, P., 2024. Experimental and numerical investigations of cake baking in mold. J. Food Eng. 382, 112215. https://doi.org/10.1016/j.jfoodeng.2024.112215.

Fabbri, A., Cevoli, C., Alessandrini, L., Romani, S., 2011. Numerical modeling of heat and mass transfer during coffee roasting process. J. Food Eng. 105 (2), 264–269. https://doi.org/10.1016/j.jfoodeng.2011.02.030.

Fadai, N.T., Melrose, J., Please, C.P., Schulman, A., Van Gorder, R.A., 2017. A heat and mass transfer study of coffee bean roasting. Int. J. Heat Mass Tran. 104, 787–799. https://doi.org/10.1016/j.ijheatmasstransfer.2016.08.083.

Fadai, N.T., Please, C.P., Van Gorder, R.A., 2019. Modelling structural deformations in a roasting coffee bean. Int. J. Non Lin. Mech. 110, 123–130. https://doi.org/10.1016/j.ijnonlinmec.2018.12.006.

Fan, J., Mitchell, J.R., Blanshard, J.M.V., 1999. A model for the oven rise of dough during baking. J. Food Eng. 41 (2), 69–77. https://doi.org/10.1016/S0260-8774(99)00070-9.

Feyissa, A.H., Adler-Nissen, J., Gernaey, K.V., 2009. Model of heat and mass transfer with moving boundary during roasting of meat in convection-oven. In: Proceedings of the 2009 COMSOL Conference Milan, Italy. https://www.comsol.com/paper/download/45309/Feyissa.pdf. (Accessed 8 March 2025).

Feyissa, A.H., Gernaey, K.V., Adler-Nissen, J., 2013. 3D modelling of coupled mass and heat transfer of a convection-oven roasting process. Meat Sci. 93 (4), 810–820. https://doi.org/10.1016/j.meatsci.2012.12.003.

Fukuoka, M., Kimura, M., Yamaguchi, S., Sakai, N., 2007. Heat and mass transfer analysis during cooking of meat. In: Proceedings of 2007 CIGR Section VI International Symposium on Food and Agricultural Products: Processing and Innovations (Italy).

Gil, M.M., Pereira, P.M., Brandão, T.R.S., Silva, C.L.M., Kondjoyan, A., Valdramidis, V.P., et al., 2006. Integrated approach on heat transfer and inactivation kinetics of microorganisms on the surface of foods during heat treatments—software development. J. Food Eng. 76 (1), 95–103. https://doi.org/10.1016/j.jfoodeng.2005.05.020.

Godsalve, E.W., Davis, E.A., Gordon, J., Davis, H.T., 1977. Water loss rates and temperature profiles of dry cooked bovine muscle. J. Food Sci. 42 (4), 1038–1045. https://doi.org/10.1111/j.1365-2621.1977.tb12662.x.

Goñi, S.M., 2010. Simulación y optimización de la cocción de productos cárneos en hornos convectivos. PhD Thesis. Universidad Nacional de Quilmes, Argentina. http://ridaa.unq.edu.ar/handle/20.500.11807/4711.

Goñi, S.M., Purlis, E., Salvadori, V.O., 2008. Evaluating a simple model for meat cooking simulation. In: Proceedings CIGR-International Conference of Agricultural Engineering 2008, Brazil.

Goñi, S.M., Salvadori, V.O., 2010. Prediction of cooking times and weight losses during meat roasting. J. Food Eng. 100 (1), 1–11. https://doi.org/10.1016/j.jfoodeng.2010.03.016.

Goñi, S.M., Salvadori, V.O., 2011. Kinetic modelling of colour changes during beef roasting. Procedia Food Sci. 1, 1039–1044. https://doi.org/10.1016/j.profoo.2011.09.155.

Goñi, S.M., Salvadori, V.O., 2012. Model-based multi-objective optimization of beef roasting. J. Food Eng. 111 (1), 92–101. https://doi.org/10.1016/j.jfoodeng.2012.01.022.

Goñi, S.M., Salvadori, V.O., 2014. Energy consumption estimation during oven cooking of food. In: Reiter, S. (Ed.), Energy Consumption: Impacts of Human Activity, Current and Future Challenges, Environmental and Socio-Economic Effects. Nova Science Publishers, pp. 99–116.

Hernández-Alhambra, E., Gufu, P., Cabeza-Gil, I., Ferrer-Mairal, A., Martínez, M.A., Calvo, B., et al., 2024. Towards domestic cooking efficiency: A case study on burger pan frying using experimental and computational results. J. Food Eng. 363, 111783. https://doi.org/10.1016/j.jfoodeng.2023.111783.

Ho, Q.T., Carmeliet, J., Datta, A.K., Defraeye, T., Delele, M.A., Herremans, E., et al., 2013. Multiscale modeling in food engineering. J. Food Eng. 114 (3), 279–291. https://doi.org/10.1016/j.jfoodeng.2012.08.019.

Holtz, E., Skjöldebrand, C., 1986. Simulation of the temperature of a meat loaf during the cooking process. J. Food Eng. 5 (2), 109–121. https://doi.org/10.1016/0260-8774(86)90011-7.

Huang, E., Mittal, G.S., 1995. Meatball cooking – modeling and simulation. J. Food Eng. 24 (1), 87–100. https://doi.org/10.1016/0260-8774(94)P1610-A.

Hung, C.C., Davis, E.A., Gordon, J., Davis, H.T., 1978. Mechanisms of water loss of bovine semitendinosus muscle dry cooked from the frozen state. J. Food Sci. 43, 1191–1195. https://doi.org/10.1111/j.1365-2621.1978.tb15266.x.

Ishiwatari, N., Fukuoka, M., Sakai, N., 2013. Effect of protein denaturation degree on texture and water state of cooked meat. J. Food Eng. 117 (3), 361–369. https://doi.org/10.1016/j.jfoodeng.2013.03.013.

Isleroglu, H., Keymak-Ertekin, F., 2016. Modelling of heat and mass transfer during cooking in steam-assisted hybrid oven. J. Food Eng. 181, 50–58. https://doi.org/10.1016/j.jfoodeng.2016.02.027.

Jha, S.N., 2005. Mathematical simulation of roasting of grain. J. Food Eng. 71 (3), 304–310. https://doi.org/10.1016/j.jfoodeng.2005.03.006.

Khan, MdI.H., Joardder, M.U.H., Kumar, C., Karim, M.A., 2016. Multiphase porous media modelling: A novel approach to predicting food processing performance. Crit. Rev. Food Sci. Nutr. 58 (4), 528–546. https://doi.org/10.1080/10408398.2016.1197881.

Khatir, Z., Thompson, H., Kapur, N., Toropov, V., Paton, J., 2013. Multi-objective Computational Fluid Dynamics (CFD) design optimisation in commercial bread-baking. Appl. Therm. Eng. 60 (1–2), 480–486. https://doi.org/10.1016/j.applthermaleng.2012.08.011.

Kondjoyan, A., Kohler, A., Realini, C.E., Portanguen, S., Kowalski, R., Clerjon, S., et al., 2014. Towards models for the prediction of beef meat quality during cooking. Meat Sci. 97, 323–331. https://doi.org/10.1016/j.meatsci.2013.07.032.

Kondjoyan, A., McCann, M.S., Rouaud, O., Havet, M., Foster, A.M., Swain, M., et al., 2006. Modelling coupled heat-water transfers during a decontamination treatment of the surface of solid food products by a jet of hot air-II. Validations of product surface temperature and water activity under fast transient air temperature conditions. J. Food Eng. 76, 63–69. https://doi.org/10.1016/j.jfoodeng.2005.05.015.

Kondjoyan, A., Oillic, S., Portanguen, S., Gros, J.-B., 2013. Combined heat transfer and kinetic models to predict cooking loss during heat treatment of beef meat. Meat Sci. 95, 336–344. https://doi.org/10.1016/j.meatsci.2013.04.061.

Kondjoyan, A., Portanguen, S., 2008. Prediction of surface and ''under surface'' temperatures on poultry muscles and poultry skins subjected to jets of superheated steam. Food Res. Int. 41, 16–30. https://doi.org/10.1016/j.foodres.2007.07.006.

Larson, E.M., Holm, E.T., Marchello, M.J., Slanger, W.D., 1992. Physical and sensory characteristics of fresh pork leg roasts cooked at low temperatures. J. Food Sci. 57 (6), 1300–1303. https://doi.org/10.1111/j.1365-2621.1992.tb06841.x.

Lepetit, J., 2008. Collagen contribution to meat toughness: Theoretical aspects. Meat Sci. 80, 960–967. https://doi.org/10.1016/j.meatsci.2008.06.016.

Llave, Y., Takemori, K., Fukuoka, M., Takemori, T., Tomita, H., Sakai, N., 2016. Mathematical modeling of shrinkage deformation in eggplant undergoing simultaneous heat and mass transfer during convection-oven roasting. J. Food Eng. 178, 124–136. https://doi.org/10.1016/j.jfoodeng.2016.01.013.

Lucas, T., Doursat, C., Grenier, D., Wagner, M., Trystram, G., Flick, D., 2015. Modeling of bread baking with a new, multi-scale formulation of evaporation–condensation–diffusion and evidence of compression in the outskirts of the crumb. J. Food Eng. 149, 24–37. https://doi.org/10.1016/j.jfoodeng.2014.07.020.

Moya, J., Lorente-Bailo, S., Ferrer-Mairal, A., Martínez, M.A., Calvo, B., Grasa, J., et al., 2021a. Color changes in beef meat during pan cooking: kinetics, modeling and application to predict turn over time. Eur. Food Res. Technol. 247, 2751–2764. https://doi.org/10.1007/s00217-021-03821-y.

Moya, J., Lorente-Bailo, S., Salvador, M.L., Ferrer-Mairal, A., Martínez, M.A., Calvo, B., et al., 2021b. Development and validation of a computational model for steak double-sided pan cooking. J. Food Eng. 298, e110498. https://doi.org/10.1016/j.jfoodeng.2021.110498.

Nelson, H., Deyo, S., Granzier-Nakajima, S., Puente, P., Tully, K., Webb, J., 2020. A mathematical model for meat cooking. Eur. Phys. J. Plus 135, 322. https://doi.org/10.1140/epjp/s13360-020-00311-0.

Ngadi, M.O., Hwang, D.-K., 2007. Modelling heat transfer and heterocyclic amines formation in meat patties during frying. Agric. Eng. Int. CIGR J. 9. Ms BC 04 004. https://cigrjournal.org/index.php/Ejounral/article/view/884/878.

Niamnuy, C., Devahastin, S., Soponronnarit, S., Raghavan, G.S.V., 2008. Modeling coupled transport phenomena and mechanical deformation of shrimp during drying in a jet spouted bed dryer. Chem. Eng. Sci. 63 (22), 5503–5512. https://doi.org/10.1016/j.ces.2008.07.031.

Nicolas, V., Glouannec, P., Ploteau, J.-P., Salagnac, P., Jury, V., 2017. Experiment and multiphysic simulation of dough baking by convection, infrared radiation and direct conduction. Int. J. Therm. Sci. 115, 65–78. https://doi.org/10.1016/j.ijthermalsci.2017.01.018.

Nicolas, V., Salagnac, P., Glouannec, P., Ploteau, J.-P., Jury, V., Boillereaux, L., 2014. Modelling heat and mass transfer in deformable porous media: Application to bread baking. J. Food Eng. 130, 23–35. https://doi.org/10.1016/j.jfoodeng.2014.01.014.

Nicolas, V., Vanin, F., Grenier, D., Lucas, T., Doursat, C., Flick, D., 2016. Modeling bread baking with focus on overall deformation and local porosity evolution. AIChE J. 62 (11), 3847–3863. https://doi.org/10.1002/aic.15301.

Obuz, E., Powell, T.H., Dikeman, M.E., 2002. Simulation of cooking cylindrical beef roasts. LWT 35 (8), 637–644. https://doi.org/10.1006/fstl.2002.0940.

Oillic, S., Lemoine, E., Gros, J.-B., Kondjoyan, A., 2011. Kinetic analysis of cooking losses from beef and other animal muscles heated in a water bath — Effect of sample dimensions and prior freezing and ageing. Meat Sci. 88, 338–346. https://doi.org/10.1016/j.meatsci.2011.01.001.

Oliveros, N.O., Hernández, J.A., Sierra-Espinosa, F.Z., Guardián-Tapia, R., Pliego-Solórzano, R., 2017. Experimental study of dynamic porosity and its effects on simulation of the coffee beans roasting. J. Food Eng. 199, 100–112. https://doi.org/10.1016/j.jfoodeng.2016.12.012.

Onwude, D.I., Hashim, N., Abdan, K., Janius, R., Chen, G., Kumar, C., 2018. Modelling of coupled heat and mass transfer for combined infrared and hot-air drying of sweet potato. J. Food Eng. 228, 12–24. https://doi.org/10.1016/j.jfoodeng.2018.02.006.

Ou, D., Mittal, G.S., 2006. Double-sided pan frying of unfrozen/frozen hamburgers for microbial safety using modelling and simulation. Food Res. Int. 39 (2), 133–144. https://doi.org/10.1016/j.foodres.2005.06.009.

Ou, D., Mittal, G.S., 2007. Single-sided pan frying of frozen hamburgers with flippings for microbial safety using modeling and simulation. J. Food Eng. 80 (1), 33–45. https://doi.org/10.1016/j.jfoodeng.2006.03.033.

Ousegui, A., Moresoli, C., Dostie, M., Marcos, B., 2010. Porous multiphase approach for baking process – explicit formulation of evaporation rate. J. Food Eng. 100 (3), 535–544. https://doi.org/10.1016/j.jfoodeng.2010.05.003.

Palka, K., Daun, H., 1999. Changes in texture, cooking losses, and myofibrillar structure of bovine *M. semitendinosus* during heating. Meat Sci. 51, 237–243. https://doi.org/10.1016/S0309-1740(98)00119-3.

Pan, Z., Singh, R.P., Rumsey, T.R., 2000. Predictive modeling of contact-heating for cooking a hamburger patty. J. Food Eng. 46 (1), 9–19. https://doi.org/10.1016/S0260-8774(00)00063-7.

Pathare, P.B., Roskilly, A.P., 2016. Quality and energy evaluation in meat cooking. Food Eng. Rev. 8, 435–447. https://doi.org/10.1007/s12393-016-9143-5.

Purlis, E., 2010. Browning development in bakery products – A review. J. Food Eng. 99 (3), 239–249. https://doi.org/10.1016/j.jfoodeng.2010.03.008.

Purlis, E., 2014. Optimal design of bread baking: Numerical investigation on combined convective and infrared heating. J. Food Eng. 137, 39–50. https://doi.org/10.1016/j.jfoodeng.2014.03.033.

Purlis, E., 2016. Heat and mass transfer during baking. In: Calderón-Domínguez, G., Gutiérrez-López, G.F., Niranjan, K. (Eds.), Advances in Heat Transfer Unit Operations: Baking and Freezing in Bread Making. CRC Press, pp. 173–188. https://doi.org/10.1201/9781315374543.

Purlis, E., 2019a. Modelling convective drying of foods: A multiphase porous media model considering heat of sorption. J. Food Eng. 263, 132–146. https://doi.org/10.1016/j.jfoodeng.2019.05.028.

Purlis, E., 2019b. Simple methods to predict the minimum baking time of bread. Food Control 104, 217–223. https://doi.org/10.1016/j.foodcont.2019.04.021.

Purlis, E., 2020. Simple models for predicting water loss of bread during baking. J. Food Process. Eng. 43 (11), e13526. https://doi.org/10.1111/jfpe.13526.

Purlis, E., 2023. Simple modelling of time-temperature profiles in food during baking. Appl Food Res 3 (1), 100271. https://doi.org/10.1016/j.afres.2023.100271.

Purlis, E., 2024a. Digital twin methodology in food processing: Basic concepts and applications. Curr Nutr Rep 13, 914–920. https://doi.org/10.1007/s13668-024-00584-2.

Purlis, E., 2024b. Physics-informed machine learning: the next big trend in food process modelling? Curr. Food Sci. Technol. Rep. 2, 1–6. https://doi.org/10.1007/s43555-023-00012-6.

Purlis, E., Cevoli, C., Fabbri, A., 2021. Modelling volume change and deformation in food products/processes: an overview. Foods 10 (4), 778. https://doi.org/10.3390/foods10040778.

Purlis, E., Salvadori, V.O., 2009a. Bread baking as a moving boundary problem. Part 1: Mathematical modelling. J. Food Eng. 91 (3), 428–433. https://doi.org/10.1016/j.jfoodeng.2008.09.037.

Purlis, E., Salvadori, V.O., 2009b. Bread baking as a moving boundary problem. Part 2: Model validation and numerical simulation. J. Food Eng. 91 (3), 434–442. https://doi.org/10.1016/j.jfoodeng.2008.09.038.

Purlis, E., Salvadori, V.O., 2010. A moving boundary problem in a food material undergoing volume change – Simulation of bread baking. Food Res. Int. 43 (4), 949–958. https://doi.org/10.1016/j.foodres.2010.01.004.

Rabeler, F., Feyissa, A.H., 2018. Modelling the transport phenomena and texture changes of chicken breast meat during the roasting in a convective oven. J. Food Eng. 237, 60–68. https://doi.org/10.1016/j.jfoodeng.2018.05.021.

Rabeler, F., Skytte, J.L., Feyissa, A.H., 2019. Prediction of thermal induced color changes of chicken breast meat during convective roasting: A combined mechanistic and kinetic modelling approach. Food Control 104, 42–49. https://doi.org/10.1016/j.foodcont.2019.04.018.

Rinaldi, M., Chiavaro, E., Gozzi, E., Massini, R., 2011. Simulation and experimental validation of simultaneous heat and mass transfer for cooking process of Mortadella Bologna PGI. Int. J. Food Sci. Technol. 46, 586–593. https://doi.org/10.1111/j.1365-2621.2010.02521.x.

Robbins, P.T., Fryer, P.J., 2003. The spouted-bed roasting of barley: development of a predictive model for moisture and temperature. J. Food Eng. 59 (2–3), 199–208. https://doi.org/10.1016/S0260-8774(02)00459-4.

Sakin, M., Kaymak-Ertekin, F., Ilicali, C., 2007. Simultaneous heat and mass transfer simulation applied to convective oven cup cake baking. J. Food Eng. 83 (3), 463–474. https://doi.org/10.1016/j.jfoodeng.2007.04.007.

Santos, M.V., Zaritzky, N.E., Califano, A., 2008. Modeling heat transfer and inactivation of Escherichia coli O157:H7 in precooked meat products in Argentina using the finite element method. Meat Sci. 79 (3), 595–602. https://doi.org/10.1016/j.meatsci.2007.12.014.

Sargolzaei, J., Abarzani, M., Aminzadeh, R., 2011. Modeling and simulation of hamburger cooking process using finite difference and CFD methods. Int. J. Ind. Chem. 2 (1), 52–62.

Schillaci, E., Gràcia, A., Capellas, M., Rigola, J., 2022. Numerical modeling and experimental validation of meat burgers and vegetarian patties cooking process with an innovative IR laser system. J. Food Process. Eng. 45 (9), e14097. https://doi.org/10.1111/jfpe.14097.

Schönfeldt, H.C., Strydom, P.E., 2011. Effect of age and cut on cooking loss, juiciness and flavour of South African beef. Meat Sci. 87, 180–190. https://doi.org/10.1016/j.meatsci.2010.10.007.

Seranthian, K., Datta, A.K., 2023. Dynamics of cupcake baking: Coupled multiphase heat and mass transport in a deformable porous material. Chem. Eng. Sci. 277, 118802. https://doi.org/10.1016/j.ces.2023.118802.

Singh, N., Akins, R.G., Erickson, L.E., 1984. Modeling heat and mass transfer during the oven roasting of meat. J. Food Process. Eng. 7 (3), 205–220. https://doi.org/10.1111/j.1745-4530.1984.tb00303.x.

Skjöldebrand, C., Hallström, B., 1980. Convection oven frying: heat and mass transport in the product. J. Food Sci. 45 (5), 1347–1353. https://doi.org/10.1111/j.1365-2621.1980.tb06552.x.

Sommier, A., Chiron, H., Colonna, P., Della Valle, G., Rouillé, J., 2005. An instrumented pilot scale oven for the study of French bread baking. J. Food Eng. 69 (1), 97–106. https://doi.org/10.1016/j.jfoodeng.2004.07.015.

Sprague, M.A., Colvin, M.E., 2011. A mixture-enthalpy fixed-grid model for temperature evolution and heterocyclic-amine formation in a frying beef patty. Food Res. Int. 44, 789–797. https://doi.org/10.1016/j.foodres.2011.01.011.

Sruthi, N.U., Premjit, Y., Pandiselvam, R., Kothakota, A., Ramesh, S.V., 2021. An overview of conventional and emerging techniques of roasting: Effect on food bioactive signatures. Food Chem. 348, 129088. https://doi.org/10.1016/j.foodchem.2021.129088.

Thorvaldsson, K., Skjöldebrand, C., 1996. Water transport in meat during reheating. J. Food Eng. 29, 13–21. https://doi.org/10.1016/0260-8774(95)00052-6.

Tornberg, E., 2005. Effect of heat on meat proteins – implications on structure and quality of meat products. Meat Sci. 70 (3), 493–508. https://doi.org/10.1016/j.meatsci.2004.11.021.
Tornberg, E., 2013. Engineering processes in meat products and how they influence their biophysical properties. Meat Sci. 95 (4), 871–878. https://doi.org/10.1016/j.meatsci.2013.04.053.
Townsend, M.A., Gupta, S., Pitts, W.H., 1989. The roast: nonlinear modeling and simulation. J. Food Process. Eng. 11 (1), 17–42. https://doi.org/10.1111/j.1745-4530.1989.tb00020.x.
Valdramidis, V.P., Belaubre, N., Zuniga, R., Foster, A.M., Havet, M., Geeraerd, A.H., et al., 2005. Development of predictive modelling approaches for surface temperature and associated microbiological inactivation during hot dry air decontamination. Int. J. Food Microbiol. 100 (1–3), 261–274. https://doi.org/10.1016/j.ijfoodmicro.2004.10.025.
van der Sman, R.G.M., 2007a. Moisture transport during cooking of meat: An analysis based on Flory–Rehner theory. Meat Sci. 76 (4), 730–738. https://doi.org/10.1016/j.meatsci.2007.02.014.
van der Sman, R.G.M., 2007b. Soft condensed matter perspective on moisture transport in cooking meat. AIChE J. 53 (11), 2986–2995. https://doi.org/10.1002/aic.11323.
van der Sman, R.G.M., 2013. Modeling cooking of chicken meat in industrial tunnel ovens with the Flory-Rehner theory. Meat Sci. 95, 940–957. https://doi.org/10.1016/j.meatsci.2013.03.027.
van der Sman, R.G.M., 2022. MULTICUBED: Multiscale-multiphysics simulation of food processing. Food Struct. 33, 100278. https://doi.org/10.1016/j.foostr.2022.100278.
Wagner, M.J., Lucas, T., Le Ray, D., Trystram, G., 2007. Water transport in bread during baking. J. Food Eng. 78 (4), 1167–1173. https://doi.org/10.1016/j.jfoodeng.2005.12.029.
Yang, W., Long, L., Zhang, L., Xu, K., Huang, Z., Ye, H., 2025. Heat and mass transfer and deformation during chiffon cake baking. J. Food Eng. 388, 112361. https://doi.org/10.1016/j.jfoodeng.2024.112361.
Yu, T.-Y., Morton, J.D., Clerens, S., Dyer, J.M., 2017. Cooking-induced protein modifications in meat. Compr. Rev. Food Sci. Food Saf. 16, 141–159. https://doi.org/10.1111/1541-4337.12243.
Yu, X., Llave, Y., Fukuoka, M., Sakai, N., 2014. Estimation of color changes in fish surface at the beginning of grilling based on the degree of protein denaturation. J. Food Eng. 129, 12–20. https://doi.org/10.1016/j.jfoodeng.2013.12.030.
Zhang, J., Datta, A.K., 2006. Mathematical modeling of bread baking process. J. Food Eng. 75 (1), 78–89. https://doi.org/10.1016/j.jfoodeng.2005.03.058.
Zhang, J., Datta, A.K., Muhkerjee, S., 2005. Transport processes and large deformation during baking of bread. AIChE J. 51 (9), 2569–2580. https://doi.org/10.1002/aic.10518.
Zorrilla, S.E., Banga, J.R., Singh, R.P., 2003. Dynamic optimization of double-sided cooking of meat patties. J. Food Eng. 58 (2), 173–182. https://doi.org/10.1016/S0260-8774(02)00342-4.
Zorrilla, S.E., Singh, R.P., 2000. Heat transfer in meat patties during double-sided cooking. Food Sci. Technol. Res. 6, 130–135. https://doi.org/10.3136/fstr.6.130.
Zorrilla, S.E., Singh, R.P., 2003. Heat transfer in double-sided cooking of meat patties considering two-dimensional geometry and radial shrinkage. J. Food Eng. 57 (1), 57–65. https://doi.org/10.1016/S0260-8774(02)00273-X.

Modeling of frying processes

4

E. Hernández-Alhambra[1], P. Guíu[2], A. Ferrer-Mairal[2], M.A. Martínez[3], B. Calvo[3], J. Grasa[3] and M.L. Salvador[2]
[1]Aragón Institute of Engineering Research (i3A), Universidad de Zaragoza, Zaragoza, Spain; [2]Instituto Agroalimentario de Aragon IA2, Universidad de Zaragoza-CITA, Zaragoza, Spain; [3]Centro de Investigación Biomédica en Red en Bioingeniería, Biomateriales y Nanomedicina (CIBER-BBN), Madrid, Spain

4.1 Introduction

Frying represents one of the most established and extensively applied cooking methods, both within industrial food processing and services, as well as in household practices. The technique involves immersing food in direct contact with oil or fat at temperatures exceeding the boiling point of water, typically ranging between 150 and 190°C (Farkas et al., 1996c). Frying modifies the food surface, resulting in a golden color, a crispy texture and characteristic flavor and taste, but also to the internal structure of the food, which are highly appreciated by consumers (Wang et al., 2023).

Food undergoes complex chemical and physical changes during frying (Ziaiifar et al., 2008). The processes involved encompass a wide range of physicochemical transformations, such as the reduction of moisture content, absorption of oil, protein denaturation, starch gelatinization, and the development of color and flavor through Maillard reactions and caramelization. Additional phenomena include the formation of a crust, structural shrinkage, and modifications in the mechanical properties that define texture (Gazmuri and Bouchon, 2009; Moya et al., 2021a; Hernández-Alhambra et al., 2024). Another notable effect of frying is the preservative capacity, which results from the microbial and enzymatic destruction associated with the heat treatment itself, together with a reduction of water associated with the dehydration process (Frakolaki et al., 2023). Food safety is particularly important when frying meat and meat products (Ou and Mittal, 2006, 2007; Dalvi-Isfahan, 2023) where inactivation of pathogenic microorganisms must be ensured. Nevertheless, the frying process must be carefully controlled to avoid compromising product quality. Excessive Maillard reactions, the generation of undesirable compounds such as acrylamide, polycyclic aromatic hydrocarbons (PAHs), and heterocyclic amines (HAAs), together with oil oxidation and nutrient degradation, represent critical risks that should be minimized through appropriate process management (Bassama et al., 2012; Xu et al., 2023).

Frying is characterized by the concurrent occurrence of heat transfer and mass transport processes. In terms of mass transfer, the process encompasses both moisture reduction and oil uptake. The internal water content of the product undergoes partial

Modeling of Unit Operations in the Food Industry. https://doi.org/10.1016/B978-0-12-822180-8.00004-3

evaporation and migrates outward into the surrounding medium. Once the temperature reaches the boiling point of water, a vapourization front develops at the surface and progressively advances toward the core of the food matrix. As a result, two different regions can be differentiated: the dried region called the crust, and the core. Frying is often considered a drying operation. Oil absorption predominantly occurs during the cooling stage of frying, primarily as a consequence of vacuum-driven effects (van Koerten et al., 2015; Liberty et al., 2019). Additional mechanisms have also been proposed, including the water-replacement phenomenon associated with moisture loss and pore development (Liberty et al., 2019; Xie et al., 2022), as well as the surface-active hypothesis, which attributes oil uptake to the generation of amphiphilic compounds formed during the frying process (Dana and Saguy, 2006).

Thermal energy is conveyed from the surrounding oil or fat medium to the outer surface of the product primarily through convective mechanisms. Subsequently, the heat penetrates inward from the surface toward the core by means of conductive transfer (Sahin et al., 1999). Key parameters influencing heat transfer during frying include thermal diffusivity, thermal conductivity, and heat transfer coefficients. These properties play a decisive role in governing the rate and efficiency of energy transmission within the food matrix throughout the cooking process (Dash et al., 2022).

The main factors influencing the frying process are dependent on both the process (type of frying, temperature, time, pressure, equipment, frying medium, …) and the food to be processed (composition, moisture content, geometric shape, surface to volume ratio, porosity, chemical reactions involved, …) (Dash et al., 2022; Hernández-Alhambra et al., 2026).

Traditional frying processes are generally classified into two different types (Fellows, 2022): deep-fat and pan-frying. In deep-fat frying, the food is completely (or mostly) immersed in the frying medium. The frying medium, composed of oil or fat, functions as an efficient carrier of thermal energy, facilitating its transfer to the product. Beyond its role in heat transmission, the medium also contributes to structural modifications and the development of characteristic sensory attributes such as texture and flavor. It is the most important way of frying on an industrial scale, and it is applied to a wide variety of food such as French fries, potato chips, donuts, chicken nuggets, etc. The deep-frying process is usually divided into three stages (Safari et al., 2018). During the initial phase of frying, the product is submerged in hot oil, where convective heat transfer occurs across the boundary layer surrounding the food. Consequently, the surface temperature rapidly approaches 100°C within a few seconds. This stage is characterized by vigorous boiling, driven by the evaporation of surface moisture and the release of steam bubbles. As frying progresses, water is gradually eliminated from the interior, a crust develops, and the overall temperature of the product increases. At this point, reduced fluid turbulence and a decline in the heat transfer coefficient are typically observed.

In pan or griddle frying, only a limited quantity of oil or fat is applied, forming a relatively uniform layer that separates the food from direct contact with the cooking surface. This technique is particularly suitable for products with a high surface-area-to-volume ratio, as well as for meat items such as steaks, bacon, or patties (Fellows, 2022). Unlike deep-fat frying, the heat distribution across the food surface is not homogeneous, which often results in variations in color and overall appearance. Moreover, the thickness of the oil layer is influenced by surface irregularities of the product, while steam bubbles generated during moisture evaporation contribute to localized temperature differences and non-uniform browning. Turning the food during frying reduces the frying time and contributes to homogeneity of the results (Thiffeault, 2022; Hernández-Alhambra et al., 2025). During pan-frying processing, two stages are considered regarding mass transfer. In the first stage, the temperature of the product rises to the boiling point of water (100°C) and the amount of oil absorbed is negligible (Moreira, 2014). Most of the oil is absorbed through the pores of the food surface, during the second stage or cooling stage, due to the increase in capillary pressure (Moreira and Barrufet, 1998; Mir-Bel et al., 2009). With respect to heat transfer, the initial stage is dominated by convective exchange between the product surface and the surrounding medium. This is subsequently accompanied by the evaporation of water. As the process advances, the temperature of the food rises, leading to the development of a rigid crust at the surface. Ultimately, heat penetrates toward the center, where the temperature progressively increases until the product reaches full doneness.

However, several authors distinguish up to four types of frying techniques, classified according to the relative proportion of oil employed. These include product frying, pan frying sauté/stir frying, shallow frying and deep-fat frying (Devi et al., 2021), each characterized by different levels of oil-to-food ratio and distinct thermal and mass transfer conditions. In recent years, several innovative frying technologies have been introduced, including hot-air frying, microwave-assisted frying, ultrasound-enhanced frying, high-pressure deep frying, and vacuum frying (Rani et al., 2023; Paul, 2023). These approaches are primarily aimed at shortening processing times, lowering oil absorption, and improving the overall quality of fried products. Additionally, surface modification through battering or breading has been extensively investigated as an effective strategy to limit oil uptake during frying (Liberty et al., 2019).

Whatever the frying system is, it is essential to control the conditions of time, temperature, water loss and fat uptake in order to attain the desired sensory effects, while contributing to the preservation of the food. Mathematical models are useful tools for these purposes.

Several models have been developed during the last years to improve the efficiency and sustainability of food processing, while contributing to the quality of the product (Datta et al., 2022; Erdogdu, 2023). Farkas et al. (1996a,b) were pioneers in the development and verification of a model used for deep-frying. Since then, different models have been developed to predict some properties of fried food, including internal and surface temperature (Cernela et al., 2015; van Koerten et al., 2017; Ghaderi et al., 2018; Gholamibozanjani et al., 2021; Moya et al., 2021b; Hernández-Alhambra et al., 2024), the rate of moisture loss (Troncoso and Pedreschi, 2009; Dehghan Nasiri et al., 2011; Cortés et al., 2014; Cernela et al., 2015; van Koerten et al., 2017; Naghavi et al., 2018; Moya et al., 2021b; Ran et al., 2023; Hernández-Alhambra et al., 2024), the formation of acrylamide (Bassama et al., 2012; Chan, 2020; Knight et al., 2021), microbial safety (Ou and Mittal, 2006, 2007), oil uptake (Bouchon and Pyle, 2005a, 2005b; Troncoso and Pedreschi, 2009; Cortés et al., 2014; Naghavi et al., 2018) or color (Krokida et al., 2001; Bansal et al., 2015; Ağçam, 2022).

However, models concerning the pan-frying processing of food products are still scarce in comparison to other types of thermal processing such as deep-frying (Dash et al., 2022) or baking (Purlis, 2023).

4.2 Heat, mass transfer and deformation during frying

Food engineers have been developing mathematical models for food processing for many years. Such models have substantially advanced the scientific understanding of the principal physical phenomena that govern these processes, including fluid dynamics, heat and mass transfer, and mechanical deformation (Farkas et al., 1996b; Erdogdu et al., 2005; Datta et al., 2022; Dash et al., 2022; Hernández-Alhambra et al., 2024; Rani et al., 2023). They have also been used to design new processes and optimize existing ones (Krokida et al., 2000; Peñalver Soto et al., 2022; Ran et al., 2023). During deep-fat frying (Fig. 4.1), two primary mass transfer phenomena occur: fat uptake and water loss. These processes significantly modify the microstructure and surface properties of the food. The resulting changes in microstructure, including pore development and cellular collapse, define textural attributes and overall quality of the fried product.

To improve the understanding and optimization of these complex phenomena, mathematical models have been proposed to simulate the heat and mass transfer processes during frying. These models help in predicting product behavior under operational conditions and allow for the design of processes that enhance the quality, efficiency, and safety of fried foods. By accurately describing the thermal and mass transfer dynamics, these models can be used to optimize critical parameters including frying time, temperature, and oil composition, ultimately leading to improved product quality and consistency.

4.2.1 Heat transfer

In the frying process, convective heat transfer governs the transfer of thermal energy from the oil to the food surface, whereas conductive mechanisms dominate the

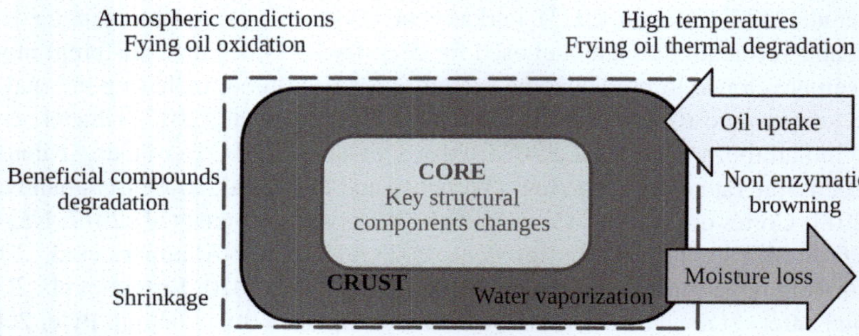

Figure 4.1 Heat transfer, mass transfer and shrinkage during deep fat frying.

movement of heat from the surface into the core. The rate of heat transmission between the frying medium and the surface of the food is controlled by the temperature difference between them and depends on the convective heat transfer coefficient. These coefficients are usually determined experimentally (Mir-Bel et al., 2012; Sandhu et al., 2016), and there are few general expressions that establish their behavior (Mekprayoon and Tangduangdee, 2012), with semiempirical expressions being common (Alvis et al., 2009; Bouchon and Pyle, 2005a). Specific heat, thermal conductivity, and thermal diffusivity are the parameters that greatly affect conductive heat transfer in foods. Their values can be estimated from the mass composition of the main constituents (water, protein, fat, carbohydrates, and ash) and the respective specific heats, densities, and conductivities of each of these components, which in turn depend on the temperature (Dash et al., 2022). Although there are many variations depending on the physicochemical nature of the product being fried, a series of common stages can be generalized for any food subjected to frying (Asokapandian et al., 2020; Farkas et al., 1996b).

4.2.1.1 Non-boiling stage

Before the food is introduced into the frying medium, the oil or fat used must reach a certain temperature, which decreases when the food, at a lower temperature, is submerged. These differences generate density profiles that cause heat transfer in the frying medium to be primarily through natural convection. The heat transferred through the medium affects the product as sensible heat. Since most foods tend to have low thermal diffusivities (Choi and Okos, 1986), temperature gradients begin to form in this stage, with the surface being at a higher temperature than the interior. In this stage, secondary or phase-change reactions, such as thawing, fat melting or starch gelatinization (Contardo et al., 2016), begin to occur. This stage is short in duration compared to the total frying time.

4.2.1.2 Boiling stage

During this stage, the heat supplied from the surrounding medium is largely used as latent energy for water evaporation. The steam bubbles generated increase the agitation in the frying medium, significantly improving heat transfer to the product and shifting the controlling mechanism to forced convection. The convective heat transfer coefficients change during the process showing a bell-shaped curve that depends on the amount of water evaporated, the bubble diameter, the steam retention time in the frying medium, and the turbulence generated; reaching its maximum value at high a rate of moisture loss (Safari et al., 2018; Sahasrabudhe et al., 2019). If excessive bubble formation occurs, it can cause a shielding effect that may reduce the convective efficiency of the process during this stage or even make it highly dependent on the geometry of the product and its position in the medium (Safari et al., 2024). The nature of the frying medium will have a substantial influence on the heat transfer of the system, especially the viscosity of the oil or fat used. Throughout the process, the viscosity of the oil is progressively modified by polymerization reactions and by changes occurring on the product's surface (Dana and Saguy, 2006; Ghaderi et al., 2024). The following sub-stages can be distinguished within this stage.

4.2.1.3 Surface boiling stage

The phase change of water occurs only at the food surface, which therefore remains at a constant temperature corresponding to the boiling point of water under the working conditions. Meanwhile, in the interior of the product, the temperature continues increasing, transferring liquid water to the surface, and resulting in a stage analogous to the constant-rate period in drying operations. When the surface is fully depleted of free water, its physicochemical properties change, forming a thin, porous and rigid crust with a structure significantly different to that in the food's core (Isik et al., 2016; Kalogianni and Papastergiadis, 2014; Lioumbas et al., 2012; Moreno and Bouchon, 2013; Onipe et al., 2020; van Koerten et al., 2015). This crust layer subsequently engages in toasting and non-enzymatic browning reactions, giving the product the characteristic sensory properties typical of fried foods (Amiryousefi et al., 2014).

4.2.1.4 Falling rate stage

The crust temperature exhibits a continued rise, approaching that of the frying medium, while inside the product, heat is still consumed to evaporate water. As the vapourization front advances, the crust thickens and the rate of vapor removal declines, causing the system to converge to a near-steady thermal condition, with the core stabilizing at a temperature slightly exceeding the boiling point. This small rise in temperature is attributed to the dissolved compounds in the water and by the accumulation of retained steam, which raises the internal pressure of the food (Budzaki and Seruga, 2005). The non-homogeneous temperature distribution may cause changes in the food structure (formation of cavities, tissue disintegration, etc.), which can sometimes make it difficult to distinguish between the crust and the core. As a function of the homogeneity and elastic properties of the material, structural tensions may develop, leading to the initiation of surface cracking or even the collapse of the outer layers (Liu et al., 2021a).

4.2.1.5 Bubble end point

Once most of the free water in the product has been removed, the decrease in moisture and the corresponding production of bubbles are reduced to a minimum. Some authors distinguish between two fractions of water: one that is semi-free and another more strongly bound to the food matrix, which is the fraction that would be eliminated in this stage (Vitrac et al., 2002; Grenier et al., 2007). Analysis of the relative evolution of the evaporation front and the crust thicknessing suggests an approximate correlation that holds up to the termination of the boiling regime. Beyond this stage, the phase-change front advances more rapidly than the dehydrated layer thickness (Lioumbas and Karapantsios, 2012).

4.2.2 Mass transfer

During deep-fat frying, mass transfer phenomena are central to the transformation of the food matrix (Sivaranjani et al., 2024). Water (as vapor) migrates outward, while oil simultaneously penetrates the product in the opposite direction. These concurrent fluxes lead to moisture loss, oil uptake, and the development of sensory attributes favored by consumers, such as the characteristic fried flavor, yellow-brown hue, and crispness of the final product.

4.2.2.1 Moisture loss

Upon reaching the saturation temperature, part of the internal water evaporates and is released from the food into the frying environment. Therefore, vapor and liquid water coexist within the food during frying. Intensive evaporation of surface water induces the development of a crust layer, which subsequently slows down further internal moisture evaporation (Ran et al., 2023; Vitrac et al., 2003; Zhang et al., 2020). Some studies consider that there is an evaporation front located at the crust-core interface that shifts inward as the crust thickness increases (Farkas et al., 1996a; Farid, 2002; Tangduangdee et al., 2004; Lioumbas and Karapantsios, 2012). Thus, water transport during frying is sometimes approached as a particular case of drying. Most foods can be considered hygroscopic and porous media, and therefore, the main mechanisms by which water can be transported are: molecular diffusion of water vapor, capillary-driven diffusion of liquid water, and pressure-driven convection of both liquid water and vapor (Datta, 2007). Diffusion is usually described by Fick's law. Due to differences in capillary attraction between areas of different moisture concentrations, a flow of liquid water occurs from areas of higher concentration to areas of lower concentration. The transport of liquid water and vapor due to pressure can be modeled using Darcy's law. However, the pressure in the liquid phase and in the vapor phase is not the same, as capillary attraction results in a negative pressure on the liquid, generating a water flux due to capillarity which is subsequently determined by the concentration gradients of moisture and temperature. Once water is removed from the product in the form of vapor, it contributes to oil degradation reactions, such as the de-esterification of triglycerides. The hydrolytic byproducts, due to their polar nature, promote oil absorption and the retention of vapor bubbles on the surface of the food, which causes hydrolysis to "autocatalyze" (Yildiz et al., 2024).

4.2.2.2 Oil uptake

The absorption of oil is complex, and its mechanisms are not fully understood. The extent of oil uptake during frying is influenced by multiple parameters, such as the initial water content of the product, characteristics of its external surface, product geometry, duration and temperature of the frying process, the type and compositional properties of the frying oil, its degradation state, and the application of treatments either prior to or following frying (Dehghannya and Ngadi, 2021). However, numerous studies reveal the importance of structural changes occurring in the food for oil penetration. The fundamental mechanisms that lead to oil uptake are water replacement, the effect associated with the cooling stage, and the formation of surfactants (Dana and Saguy, 2006). During frying, the oil replaces the evaporated water, with a direct relationship between moisture loss and lipid absorption. Heat transfer to the food matrix promotes the phase transition of internal moisture into vapor, generating an elevated internal vapor pressure. This pressure gradient drives vapor migration toward the surface, where it is discharged through structural discontinuities such as fissures, pores, and capillary pathways formed because of thermally induced microstructural transformations. As these voids expand and coalesce, especially in starchy and battered foods, they provide accessible pathways through which frying oil can infiltrate the newly formed macro- and microvoids. Most of the oil gained by the food during frying is limited to a superficial region (Khalilian et al., 2021; Al Faruq et al., 2022). This oil gradually penetrates the crust during the cooling phase after frying and thus determines the final oil content of the food, highlighting the importance of its modeling (Bouchon and Pyle, 2005a; Dehghannya and Ngadi, 2021; Debnath et al., 2009; He et al., 2012; Del Rosario-Santiago et al., 2022; Touffet et al., 2020). Most of these models propose that oil uptake during cooling occurs due to pressure-driven flow caused by capillary forces. When the food begins to cool, the water vapor inside condenses, leading to a reduction in internal pressure that causes the suction of oil. Moreover, during frying, the degradation of oil occurs, involving hydrolytic reactions favored by the high temperatures at which the bonds between glycerol and fatty acids break, generating compounds with surface-active characteristics, such as mono- and diacylglycerols. Consequently, the interfacial tension at the food-oil interface is reduced, which increases contact between these phases and leads to a higher oil uptake (Dana and Saguy, 2006; Liu et al., 2021b). Furthermore, polymerization reactions lead to an increase in oil viscosity, which complicates the drainage of oil after frying (Dehghannya and Ngadi, 2021; Ghaderi et al., 2024).

4.2.3 Shrinkage

The volumetric changes occurring during frying have been investigated less extensively than other quality-related attributes. However, changes in volume affect desirable attributes of fried products, such as crust development, textural properties, characterized by a brittle outer layer and a tender internal matrix, and visual quality (distinctive shapes). In addition, product volume is a critical parameter for subsequent operations such as filling, packaging, and distribution.

Shrinkage in frying processes is primarily driven by moisture evaporation, which causes food to lose mass and volume, and by heat-induced protein denaturation, which leads to alterations in the structural framework and textural properties of the food. In both pan frying and deep-frying processes, significant structural shrinkage of food materials occurs due to three primary physicochemical mechanisms.

4.2.3.1 Water loss

When exposed to typical frying temperatures, rapid water evaporation of surface moisture occurs, and as the outer layers dehydrate and heat up, the evaporation front gradually shifts. Liquid water migrates toward the surface mainly through capillary-driven flow, while vapor formed within the matrix diffuses outward. This phase change from liquid water to vapor results in substantial mass reduction and consequent volumetric shrinkage (Krokida et al., 2000). The dehydration rate is particularly influenced by pore formation dynamics and crust development, which create a moisture gradient between the core and surface regions.

4.2.3.2 Thermal and mechanical stresses

The creation of steep temperature gradients generates complex stress fields within the food matrix (Roshani et al., 2021). These stresses arise from differential thermal expansion coefficients between food components. Moreover, internal vapor pressure buildup from evaporating water and viscoelastic relaxation of biopolymer networks lead to either contraction or expansion of cellular structures, depending on the mechanical properties of the specific tissue. In most plant and animal tissues, this results in net shrinkage due to cell wall collapse and intercellular space reduction.

4.2.3.3 Protein denaturation and gelation

In meat and plant-based products, proteins denature and form a gel, contributing to changes in food structure and firmness, which in turn causes shrinkage (Shah and Takhar, 2022). Initial myofibrillar protein denaturation commences at relatively low temperatures, followed by progressive protein aggregation that establishes a continuous gel network. As this reorganization occurs, the contracting protein matrix forcibly expels moisture from its structure. These inter-connected phenomena of denaturation, gelation, and syneresis, collectively produce pronounced structural shrinkage, with muscle fibers exhibiting particularly dramatic dimensional changes. The extent of these transformations shows strong dependence on both thermal processing parameters (notably heating rate and final core temperature) and the intrinsic properties of the developed gel matrix, which ultimately govern critical quality attributes including final product texture and oil absorption behavior.

Additionally, this shrinkage alters the effective moisture diffusivity, density, and pore structure, among other thermophysical properties, which in turn influence thermal and mass transport phenomena (Baik and Mital, 2005; Estévez-Sánchez et al., 2021).

4.3 Modeling of frying processes

The progress achieved in food process modeling has been strongly influenced by the development of efficient numerical methods, initiated in the 1970s, which enabled the resolution of complex coupled spatio-temporal differential equations (Verboven et al., 2020). These advances provided researchers with the ability to simulate and analyze the intricate dynamics of heat transfer, mass transport, and other phenomena occurring in food systems across both time and space.

Among the most impactful approaches, finite element and finite volume methods have been central to the creation of highly efficient and user-friendly computational tools for solving model equations within computer-aided design (CAD) environments (Erdogdu et al., 2017). By integrating advanced numerical solvers, these techniques have facilitated accurate simulations of complex food processing phenomena, thereby streamlining the modeling of heat transfer, mass transfer, and structural transformations. This has significantly improved the capacity to design, optimize, and predict food processes.

For food engineers, such advancements marked a transition toward addressing the fundamental physics and associated properties of food materials, as well as the three-dimensional geometry of processing systems, rather than devoting extensive effort to the manual numerical implementation of algorithms (Datta, 2016). The availability of sophisticated modeling software has allowed professionals to focus on critical aspects of food processing, including recipe optimization, product quality improvement, and enhanced process efficiency.

4.3.1 Mathematical background

Lumped models provide a simpler, approximate solution when food size is small, and detailed internal processes are not critical. In contrast, multiphase models offer a detailed, spatially resolved solution that captures the complex interactions between heat, moisture, oil, and vapor, making them ideal for more comprehensive analysis of the deep-frying process.

4.3.1.1 Lumped models

Lumped models simplify the frying process by assuming that certain properties (like temperature or material parameters) are uniform throughout the food item. They treat the entire object as a single entity or "lump", avoiding spatial variations or simplifying as much as possible the problem physics.

Among others, the assumptions made in these models are.

- **Uniform Temperature:** The entire food is assumed to have a uniform temperature, meaning there are no temperature gradients within the food.
- **Simplified Heat Transfer:** In this approach, thermal transport within the food is represented by a single overall rate, frequently based on the lumped-capacity assumption. This simplification is applicable when the Biot number (defined as the ratio of internal to external thermal resistance) is sufficiently small, ensuring that temperature gradients inside the product can be neglected.
- **Single Phase Focus:** Typically, lumped models don't differentiate between different phases (solid, liquid, or vapor) within the food.

Basically, the primary application of lumped models is usually those related to simulating small food items where heat conduction is fast compared to external heat transfer. They can predict the average temperature rise but don't capture spatial temperature differences.

4.3.1.2 Multiphase porous media models

These approaches provide a more comprehensive representation of the intricate interactions among the solid, liquid, and gaseous phases present in food during frying. They explicitly account for the coupling between heat and mass transfer processes with moisture evaporation, vapor generation, and oil absorption, thereby offering a detailed framework for analyzing the dynamics of frying systems.

This coupling is captured through the following key aspects.

- **Phase Differentiation:** These models recognize and simulate different phases (solid food, liquid water, liquid oil, and vapourized moisture) and their interactions.
- **Mass and Heat Transfer:** Both mass transfer (moisture loss, oil absorption) and heat transfer (temperature changes) are modeled in a spatially resolved manner, meaning they account for temperature and concentration gradients within the food.
- **Evaporation and Condensation:** Multiphase models capture the phase changes occurring during frying, particularly moisture evaporation from the food's interior and vapor movement.
- **Porous Media Representation:** Food is conceptualized as a porous medium through which oil and vapor migrate via interconnected pores. This representation directly influences the description of heat and mass transfer phenomena, as the movement of fluids within the porous structure governs both thermal transport and moisture exchange during frying.
- **Detailed Spatial Modeling:** These models are computationally intensive, requiring solving coupled mass and heat transfer equations along with phase change phenomena (evaporation of water, absorption of oil).

Porosity is quantitatively expressed as the fraction of the total material volume that is occupied by pores (Halder et al., 2007):

$$\phi = \frac{\sum_{i=w,g,o} \Delta V_i}{\Delta V} \tag{4.1}$$

where V_i is the volume that the ith phase occupies in an elemental volume ΔV. These phases are: oil, water and gas. The saturation of each phase is defined as:

$$S_i = \frac{\Delta V_i}{\phi \, \Delta V} \tag{4.2}$$

The equation for heat transfer that incorporates mass diffusion can be expressed as:

$$\rho \, C_p \frac{\partial T}{\partial t} = \nabla \cdot (k \nabla T) + Q + \rho \, D \, h_m \nabla^2 C \tag{4.3}$$

where T is the temperature, t is time, ρ is the density, C_p is the specific heat capacity, k is the thermal conductivity, Q is the heat source term, D is the diffusion coefficient, h_m is the enthalpy contribution from mass diffusion, C is the concentration of the diffusing substance.

The dynamics of water removal during frying have been extensively characterized, providing a detailed insight into the phenomenon using Fick's diffusion law. Diffusion theory, grounded in concentration gradients, explains how liquids move across solid matrices. As thermal energy is transferred, moisture evaporates from the food's surface, establishing a gradient that propels water from the inner regions toward the exterior, generating a continuous release of vapor. In multiphase representations of water behavior in porous structures, two diffusion mechanisms are considered: binary diffusion and capillary diffusion.

The second Fick's law (diffusion equation) is given by:

$$\frac{\partial C}{\partial t} = D\nabla^2 C \tag{4.4}$$

where C is the concentration of the diffusing substance, D is the diffusion coefficient, and ∇^2 is the Laplace operator.

In the solid phase, stress is driven by temperature gradients and fluid pressures acting on the porous skeleton; enforcing equilibrium of stresses, pore pressure, and body forces, the linear momentum balance equation in differential form is:

$$\rho \frac{d\boldsymbol{v}}{dt} = \nabla \cdot \boldsymbol{\sigma} + \boldsymbol{b} \tag{4.5}$$

where ρ is the density of the material, v is the velocity vector field, $\frac{d}{dt}$ is the material derivative, $\nabla \cdot \boldsymbol{\sigma}$ is the divergence of the Cauchy stress tensor, $\boldsymbol{b}$ is the body force per unit volume. Assuming a quasi-steady deformation regime, where the acceleration term vanishes, the momentum balance in the solid results in an overall stress field that is divergence-free (Dhall and Datta, 2011):

$$\nabla \cdot \boldsymbol{\sigma} = \mathbf{0} \tag{4.6}$$

4.3.2 Modeling deep frying

4.3.2.1 Heat transfer

The Biot number (Bi) is a nondimensional parameter that relates a body's internal thermal resistance to its external convective resistance. It plays a key role in assessing the applicability of the lumped capacitance approach for transient heat transfer analysis. Mathematically, the Biot number is defined as the following ratio:

$$Bi = \frac{h\,L}{k} \tag{4.7}$$

where h is the convective heat transfer coefficient, L is the characteristic length, typically the ratio of volume to surface area $L = V/A$ and k is the thermal conductivity of the solid. When $Bi \ll 1$, the internal resistance to heat conduction is negligible compared to the external resistance due to convection. Therefore, the body can be assumed to have uniform temperature (lumped capacitance model applies). If $Bi \gg 1$, the internal resistance to heat conduction is significant, meaning temperature gradients exist within the solid. Finally, if $Bi \approx 1$, both internal and external resistances are comparable, requiring full transient heat conduction analysis.

In this framework, when adopting a lumped thermal capacity model, convection occurs from the surrounding fluid toward the solid's surface. Given that thermal conductivity largely exceeds convective effects, the solid's internal temperature can be regarded as spatially uniform. The heat transferred from the bath to the frying material is directly related to the variation in its internal energy. Based on the energy balance principle over an infinitesimal time interval "dt", the net heat entering the solid through its boundaries must equal the corresponding change in its internal energy during that period (Dash et al., 2022). This relationship can be written as:

$$h\,A\,(T_\infty - T)\,dt = C_p\,\rho\,V\,dT \tag{4.8}$$

where T_∞ is the temperature of the fluid, A is the outer surface area exposed to the fluid and V is its volume. By rearranging both sides of the equation and integrating them, one can analyze the temperature evolution over a finite time period, starting from $t = 0$ at temperature T_0 to time t at temperature T_t:

$$\int_{T_0}^{T_t} \frac{dT}{T_\infty - T} = \frac{h\,A}{C_p\,\rho\,V} t \tag{4.9}$$

which results in:

$$\frac{T_\infty - T_t}{T_\infty - T_0} = exp\left(\frac{-h\;A}{\rho\,C_p\;V} t\right) \tag{4.10}$$

Eq. (4.10) shows that the difference of temperature between the food and the surrounding frying oil decays exponentially over time. A larger $h \cdot A$ results in faster heat exchange, while a higher $\rho \cdot C_p \cdot V$ increases the thermal inertia of the food, leading to a slower temperature change.

When the lumped capacitance model can no longer be applied, several analytical expressions found in the literature can be used to solve the transient heat transfer equation under different basic geometrical assumptions.

Considering an infinite slab with uniform properties, Dash et al. (2022) propose that the heat conduction can be approximated to be one-dimensional:

$$\frac{\partial T}{\partial t} = \frac{k}{\rho C_p} \frac{\partial^2 T}{\partial x^2} \tag{4.11}$$

By imposing suitable boundary conditions (namely negligible surface resistance, a uniform initial temperature, and a constant surface temperature) the solution can be formulated as:

$$\frac{T_\infty - T(x,t)}{T_\infty - T_0} = \sum_{m=1}^{\infty} \frac{2(-1)^{m+1}}{\lambda_m} cos\left(\lambda_m \frac{x}{L}\right) exp\left(-\lambda_m^2 F_o\right) \tag{4.12}$$

where:

$$\lambda_m = (2m-1)\frac{\pi}{2}, \; and \; F_o = \frac{kt}{\rho C_p L^2} \tag{4.13}$$

being L half of the thickness of the slab. Different particular expressions could be found from this general equation to find the temperature evolution at different specific characteristic points as the center of the food (Dash et al., 2022).

For a cylindrical geometry of external radius R, the governing partial differential equation for heat transfer can be derived under the assumption of infinite length (eliminating axial conduction effects) (Dash et al., 2022; Köse, 2022):

$$\frac{\partial T}{\partial t} = \frac{k}{\rho C_p}\left(\frac{1}{4}\frac{\partial}{\partial r}\left(r\frac{\partial T}{\partial r}\right)\right) \tag{4.14}$$

In the same way, proper boundary and initial conditions lead to:

$$\frac{T_\infty - T(r,t)}{T_\infty - T_0} = \sum_{n=1}^{\infty}\frac{2}{\beta_n J_1(\beta_n)} J_0\left(\beta_n \frac{r}{R}\right) exp\left(-\beta_n^2 F_{o_c}\right) \tag{4.15}$$

Where $F_{o_c} = \frac{k\,t}{\rho\,C_p\,R^2}$ and β_n is the n th positive root of $J_0(\beta_n) = 0$, with the Bessel function defined as:

$$J_\alpha(x) = \sum_{m=0}^{\infty}\frac{(-1)^m\left(\frac{x}{2}\right)^{2m}}{m!\Gamma(m+\alpha+1)} \tag{4.16}$$

and the gamma function $\Gamma(m + \alpha + 1) = (m + \alpha)!$.

For a spherical geometry of a given radius R:

$$\frac{\partial T}{\partial t} = \frac{1}{r^2}\frac{\partial}{\partial r}\left(r^2\frac{\partial T}{\partial r}\right) \tag{4.17}$$

the solution:

$$\frac{T_\infty - T(r,t)}{T_\infty - T_0} = \frac{2R}{\pi r}\sum_{n=1}^{\infty}\frac{(-1)^{n+1}}{n} sin\left(\frac{n\pi r}{R}\right) exp\left(-n^2\pi^2 F_{o_s}\right) \tag{4.18}$$

which allows finding different particular solutions for the temperature in the food, such as at the center of the sphere, as follows:

$$\frac{T_\infty - T(0,t)}{T_\infty - T_0} = 2\,exp\left(-\pi^2 F_{o_s}\right) \tag{4.19}$$

In the context of frying a porous medium, the energy conservation principle is formulated as the sum of heat accumulation in each phase of the system together with the corresponding heat flux of each phase, which collectively equates to the conductive heat flux within those phases.

$$\rho_{\text{eff}}\, C_{p,\text{eff}} \frac{\partial T}{\partial t} = \nabla\left(k_{\text{eff}} \nabla T\right) + \nabla(\boldsymbol{n}_v h_v + \boldsymbol{n}_o h_o + \boldsymbol{n}_a h_a + \boldsymbol{n}_w h_w) + \rho_w L_v \Gamma_{evap} \tag{4.20}$$

The terms $\boldsymbol{n}_{\text{v}}$, $\boldsymbol{n}_{\text{o}}$, $\boldsymbol{n}_a$, and $\boldsymbol{n}_{\text{w}}$ represent the mass fluxes of vapor, oil, air, and water, respectively, while h_α represents the corresponding enthalpy of each phase, which accounts for the energy transported by the movement of each phase. L_v is the latent heat of vapourization and Γ_{evap} is the evaporation rate (mass of water evaporating per unit time and volume). The effective heat capacities and thermal conductivities are obtained as:

$$\rho_{\text{eff}}\, C_{p,\text{eff}} = \phi\, S_w \rho_w C_{p,w} + \phi\, S_o \rho_o C_{p,o} + (1-\phi)\rho_s C_{p,s} \tag{4.21}$$

$$k_{\text{eff}} = \left(k_s(1-\phi) + k_w \phi\, S_w + k_o \phi\, S_o + k_g \phi(1 - S_w - S_o)\right) \tag{4.22}$$

4.3.2.2 Mass transfer

During frying, oil and moisture contents within the product vary with position and time. Initially, moisture migrates primarily in liquid form, followed by a shift to vapor-phase transfer, with liquid-phase diffusion occurring more slowly. Moisture loss is controlled by diffusion, dictated by the concentration gradient, and in the case of a slab it can be represented through the solution of Fick's second law for transient, one-dimensional mass transfer (Pedreschi et al., 2005):

$$\frac{c_w(t) - c_{w,eq}}{c_{w0} - c_{w,eq}} = \frac{8}{\pi^2} \sum_{n=1}^{\infty} \left(\frac{1}{(2n+1)^2} exp\left(-\frac{(2n+1)^2 \pi^2}{4} \frac{D_{eff}}{L^2} t \right) \right) \tag{4.23}$$

where $c_w(t)$ represents the time-dependent moisture concentration, $c_{w,eq}$ is the moisture content when the equilibrium is reached, and c_{w0} denotes the initial moisture concentration and D_{eff} is the effective diffusivity that can be modeled as:

$$D_{\text{eff}} = D_0 \left(1 + \frac{D_0}{L^2} t \right)^b \tag{4.24}$$

with D_0 the diffusivity at the initial time and b a dimensionless parameter.

In modeling food as a porous medium, multiple transport mechanisms act concurrently: diffusion of vapor and air through the porous structure, capillary flow of liquid phases such as water and oil, and bulk movement of liquid, vapor, and air induced by overall pressure gradients. Under these conditions, the flux of air and vapor is expressed as (Yamsaengsung and Moreira, 2002a):

$$\boldsymbol{n}_a = -\rho_a \frac{k_i k_{gr}}{\mu_g} \nabla P - \frac{C^2}{\rho_g} M_a M_v D_{\text{eff,g}} \nabla y_a \tag{4.25}$$

$$\boldsymbol{n}_v = -\rho_v \frac{k_i k_{gr}}{\mu_g} \nabla P - \frac{C^2}{\rho_g} M_a M_v D_{\text{eff,g}} \nabla y_v \tag{4.26}$$

where k_{gr} is the gas relative permeability, c is the gas mixture molar density, y_a and y_v are the molar fraction of air and vapor, respectively.

And water and oil fluxes:

$$\boldsymbol{n_w} = -\rho_w \frac{k_i k_{wr}}{\mu_w} \nabla P - \rho_w \frac{k_i k_{wr}}{\mu_w} \frac{\partial p_c}{\partial S_w} \nabla S_w - \rho_w \frac{k_i k_{wr}}{\mu_w} \frac{\partial p_c}{\partial T} \nabla T \tag{4.27}$$

$$\boldsymbol{n_o} = -\rho_o \frac{k_i k_{or}}{\mu_o} \nabla P - \rho_o \frac{k_i k_{or}}{\mu_o} \frac{\partial p_c}{\partial S_o} \nabla S_o - \rho_o \frac{k_i k_{or}}{\mu_o} \frac{\partial p_c}{\partial T} \nabla T \tag{4.28}$$

The conservation of water mass, considering both liquid and vapor phases, establishes that the accumulation rate within the system, together with the variations arising from convective and diffusive fluxes, must balance with the rate of water evaporation:

$$\frac{\partial}{\partial t}(\phi(\rho_w S_w + \rho_v S_v)) + \nabla \boldsymbol{n_w} + \nabla \boldsymbol{n_v} = \Gamma_{\text{evap}} \tag{4.29}$$

The mass conservation of oil states that the accumulation rate of oil within the system is balanced by the combined effects of convective and diffusive fluxes:

$$\frac{\partial}{\partial t}(\phi \rho_o S_o) + \nabla \boldsymbol{n}_o = 0 \tag{4.30}$$

Finally, the mass conservation of air within the system is the due to diffusive fluxes:

$$\frac{\partial}{\partial t}(\phi \rho_a S_a) + \nabla \boldsymbol{n_a} = 0 \tag{4.31}$$

4.3.3 Modeling contact frying

Modeling contact frying shares fundamental similarities with deep frying, as both involve heat transfer, oil absorption, and moisture evaporation. However, a key distinction lies in the boundary conditions: in deep frying, the food is fully submerged in oil, leading to relatively uniform heating that often justifies, as seen in the previous section, a lumped capacitance approach where the Biot number is small. In contrast, contact frying features asymmetric heating, with intense conductive heat transfer at the pan-food interface and slower convective/radiative cooling on the opposite side.

This results in significant internal temperature gradients, making the Biot number too large to assume a lumped system.

Additionally, the thin, uneven oil layer in the contact frying region further complicates heat transfer, requiring a spatially resolved model that accounts for varying thermal resistances. Thus, contact frying demands a more rigorous treatment of boundary conditions and thermal gradients due to its inherently non-uniform heating. Modeling contact frying can be approached similarly to deep frying by treating the food as a porous medium, where mass and heat transfer are governed by the coupled equations previously presented (energy balance, oil conservation, and evaporation dynamics).

4.3.4 Finite element models

In this section, the key advancements in finite element method (FEM) models tailored for simulating deep frying and pan-frying processes are explored. FEM serves as a powerful numerical technique to solve the analytical heat and mass transfer equations presented in earlier sections, enabling detailed computational analysis of these complex cooking systems.

The pioneering works of Ikediala et al. (1996) and Ngadi et al. (1997) marked the initial application of the finite element method (FEM) to address the constitutive equations that describe heat and moisture transfer in frying operations. Ikediala et al. (1996) were the first to model heat transfer in meat patties subjected to single-sided pan-frying, providing important understanding of temperature profiles within the food matrix. Similarly, Ngadi et al. (1997) advanced the approach by examining moisture migration during deep-fat frying of chicken drumsticks. Their FEM models not only solved the heat transfer equations but also accurately captured the dynamics of moisture loss in complex frying environments. Building upon these foundational studies, Yamsaengsung and Moreira (2002b) refined the application of FEM by formulating a two-dimensional model capable of predicting heat and mass transfer throughout the frying and subsequent cooling of tortilla chips. Their work also introduced structural changes in the product (such as expansion and shrinkage) into the model, providing a more comprehensive simulation of the frying process. These pioneering efforts set the stage for subsequent advancements in frying simulations by integrating heat, mass transfer, and structural changes, greatly improving the understanding of food processing through computational methods.

Halder et al. (2007) focused on developing a porous media-based model to better simulate heat and mass transfer during deep-fat frying. This model emphasizes the non-equilibrium nature of the evaporation process, which is crucial for more accurately capturing oil absorption, crust formation, and acrylamide development. The key novelty in their approach is the ease of implementation in commercial CFD (computational fluid dynamics) software, making it practical for industrial applications. Additionally, their model includes input parameters that were validated against experimental data to ensure accuracy across different frying phases, from the initial non-boiling phase to surface boiling. Later, Arias-Mendez et al. (2013) proposed a dynamic optimization approach where the goal is to balance the textural quality of potato chips with minimizing the formation of harmful acrylamide. Their model integrates a multiphase transport system for heat and moisture, treating the potato as a porous medium, but it innovatively uses an optimization technique to suggest multi-zone temperature profiles that enhance both quality and safety. Their results showed that varying temperature profiles, starting with high heat and gradually reducing, achieve better results than constant temperature frying, which opens up new methods for controlling frying processes with regard to both time efficiency and chemical safety.

4.4 Case studies

This section presents the numerical simulation of meat pan-frying of two types of beef products: burgers and steaks. We will begin by describing the experimental setup, the recording of the main variables that allow us to define the state of the food and the products to be cooked. Subsequently, the numerical model to simulate the pan-frying is particularized, the finite element model is defined, as well as the initial, boundary conditions and parameters. Finally, two different simulations are presented: a 2D model to simulate burger cooking, applied to define the cooking time necessary to reach the temperature as a function of the number of turns; and a 3D model to simulate beef steak cooking, applied to study the effect of the pan temperature and the thickness of the steak on the cooking process.

4.4.1 Experimental setup

The experimental system used in all cooking tests is shown in Fig. 4.2. It consisted of an induction hob equipped with an automatic temperature control system (frying sensor BOSCH PXY675DW4E/01 model, BSH, Munich, Germany), placed on a balance (DS30K0.1 L, Kern and Sohn, Balinger-Frommern, Germany) to record the weight of the food every 5 s with an accuracy of 0.1 g using the KERNBC4 software. Cooking was performed with a multi-layer non-stick pan (teflon-steel-aluminum-steel) with a diameter of 210 mm. The pan temperature was measured at the base using a 1.5 mm diameter K-type thermocouple inserted into a hole specifically engineered for this measurement, and at the surface using a K-type RS PRO thermocouple (RS, London, UK), positioned 30 mm from the pan's edge. The food was placed in the pan when the surface temperature matched the target temperature with a deviation of $\pm 3°C$. All cooking tests were conducted with the temperature control system activated at 215°C or 140°C. A 1.5-mm-diameter T-type penetration thermocouple was used to record the temperature evolution at the core of the product, while the evolution of the surface temperature of the food was tracked using thermographic images captured regularly during cooking with an infrared thermal camera (875–2 model, Testo, Lenzkirch, Germany). All temperatures were recorded by a data logger (TC-08 Series, Farnell Components, Barcelona, Spain). The monitoring of the recorded data was performed using the LabTech v6.0.1.5 software (ConnectWise, Tampa, Florida, USA). Digital analysis using the free software ImageJ of images captured at regular intervals during cooking with a mobile phone (Samsung Galaxy A8) allowed for tracking shrinkage as the percent reduction in the top-surface area compared to the raw product.

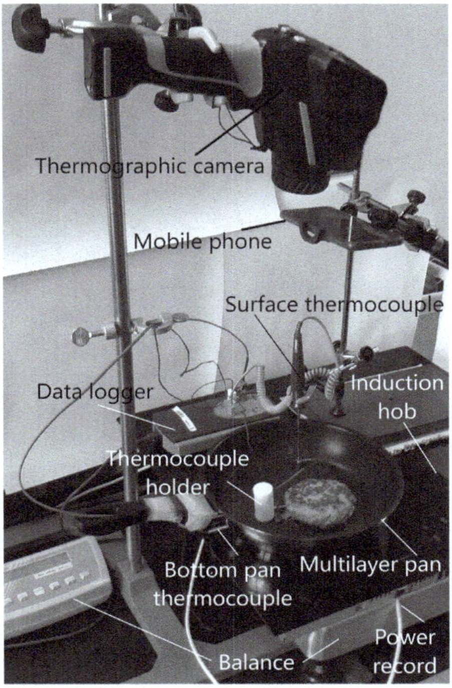

Figure 4.2 Experimental setup for temperature, weight loss and shrinkage measurement during the cooking process.

The cooking tests were conducted with two food matrices: burgers and steaks. In both cases, the middle part of beef loin (*Longissimus dorsi* muscle) was used. Steaks of two thicknesses (20 and 34 mm) were cut perpendicular to the longitudinal axis, and then trimmed to pieces of 12 × 8 cm. The burgers were made by grinding the meat with a 3 mm grinder plate and shaping it with a manual press, obtaining cylinders 100 mm in diameter and 20 mm in height. Both products were kept refrigerated until cooking in LDPE zip bags at 4°C for less than 48 h. Subsequently, they were tempered in a temperature-controlled chamber (MIR-153 Incubator, Sanyo, Osaka, Japan) at 20°C. The burgers were cooked with a different number of flips (Hernández-Alhambra et al., 2025) to assess the influence of this action, using a pan temperature of 140°C, and cooking until the product's core temperature reached 71°C, considered the minimum for ground meat to ensure food safety (USDA, 2025): 800 s (1 flip), 680 s (3 flips), 603 s (5 flips), and 603 s (7 flips). These times were the same as those used in cooking at a pan temperature of 215°C, since previous studies demonstrated that the pan temperature barely influenced the evolution of the core temperature of the product (Hernández-Alhambra et al., 2024). For the steaks, cooking was performed at 140°C and 215°C for 603 s with three flips for both thicknesses, as the objective was to assess the influence of thickness on temperature, weight loss and product shrinkage. The burgers and the steaks were cooked without adding oil.

The model requires as input the temperature dependency of the loss modulus, storage modulus, and water-holding capacity. The experimental methodology for their determination and the results obtained are those described in Moya et al. (2021b) for beef meat and in Hernández-Alhambra et al. (2024) for ground beef meat.

4.4.2 Modeling pan frying of meat

Pan-frying meat can be interpreted as a coupled flow and transport problem within a deforming solid matrix subjected to thermal processing. In this study, a mathematical framework was established under several assumptions. The beef sample is treated as a porous medium composed of a solid matrix saturated with liquid water. This biphasic material is assumed to be homogeneous, with its mechanical response described as hyper-elastic, meaning that the stress–strain relationship is nonlinear. During contact heating, both phases are considered to share the same temperature, while the pressure gradient generated by the contraction of connective tissue drives a flow governed by Darcy's law. Beyond these assumptions, additional hypotheses were introduced. One addresses the volumetric change of the meat: unlike earlier models that equated it solely to moisture loss, this work (supported by experimental evidence) also accounts for protein shrinkage. Furthermore, when liquid water reaches the lower surface of the beef in contact with the pan, it evaporates, removing heat from the system and thereby affecting the temperature of both the product and the pan.

To establish the mathematical model, the volume fraction ϕ_α corresponding to each phase is defined as:

$$\phi_\alpha = \lim_{V \to 0} \frac{V_\alpha}{V}, \quad \alpha = s, w \tag{4.32}$$

where V_α represents the volume occupied by the α phase (solid and water), the total volume is given by $V = V_s + V_w$, and the volume fractions satisfy $\phi_s + \phi_w = 1$.

The deformation kinematics of meat are characterized through a framework typically employed in solids experiencing finite strains. To capture distinct physical transformations, a sequence of fictitious states is introduced, leading to a multiplicative breakdown of the strain deformation gradient $\boldsymbol{F}$. If the deformation of the hamburger induced by thermal effects is considered negligible, the strain deformation gradient may be represented as the product of the deformation linked to water volume reduction ($\boldsymbol{F}_w$), protein shrinkage ($\boldsymbol{F}_p$), and the elastic deformation of the solid phase ($\boldsymbol{F}_e$):

$$\boldsymbol{F} = \boldsymbol{F}_w \cdot \boldsymbol{F}_p \cdot \boldsymbol{F}_e \tag{4.33}$$

The volume change resulting from moisture loss, J_M, can be determined by relating it to the variation in the volume fraction, ϕ_w, of the liquid water phase:

$$J_M = \frac{1 - \phi_{w,0}}{1 - \phi_w} \tag{4.34}$$

The moisture deformation gradient is defined by:

$$\boldsymbol{F}_w = J_M^{1/3}\boldsymbol{I} \tag{4.35}$$

where $\boldsymbol{I}$ denotes the 3x3 identity matrix, and the Jacobian is raised to the power of 1/ 3 so that $\det(\boldsymbol{F}_w) = J_M$.

During pre-portioning, it is assumed that the anisotropy linked to muscle fibers in minced meat disappears. Consequently, the deformation gradient related to protein shrinkage in the hamburger is considered isotropic and can be expressed as:

$$\boldsymbol{F}_{p,\text{local}} = \boldsymbol{F}_p = \begin{bmatrix} \lambda(T) & 0 & 0 \\ 0 & \lambda(T) & 0 \\ 0 & 0 & \lambda(T) \end{bmatrix} \tag{4.36}$$

Where $\lambda(T)$ denotes a shortening stretch that incorporates the temperature-dependent nature of protein shrinkage and is defined as:

$$\lambda(T) = \begin{cases} 1 \quad T \leq 30\,^{\circ}C \\ (\lambda_s - 1)(0.02T - 0.6) + 1 \quad 130\,^{\circ}C < T < 80\,^{\circ}C \\ \lambda_s \quad 80\,^{\circ}C \leq T \end{cases} \tag{4.37}$$

Where λ_s is a parameter representing the maximum protein shrinkage.

By disregarding body forces and inertial effects, the conservation of linear momentum leads to the quasi-static equilibrium equation $\nabla\boldsymbol{\sigma} = 0$ for the material, where $\boldsymbol{\sigma}$ denotes the total Cauchy stress. The conservation of angular momentum imposes the symmetry of this stress tensor, which can be represented as the sum of the partial stresses of the solid and fluid phases:

$$\boldsymbol{\sigma} = \widehat{\boldsymbol{\sigma}}_s + \widehat{\boldsymbol{\sigma}}_w \tag{4.38}$$

where

$$\widehat{\boldsymbol{\sigma}}_s = \phi_s\, \boldsymbol{\sigma}_e \text{ and } \widehat{\boldsymbol{\sigma}}_w = \phi_w\, \boldsymbol{\sigma}_w = -\phi_w\, p_w\, \boldsymbol{I} \tag{4.39}$$

Here, $\boldsymbol{\sigma}_e$ is the Cauchy elastic stress of the meat, while p_w represents the volume-averaged pressure exerted by the liquid water.

To simulate the elastic response of meat, the constitutive framework adopted corresponds to an isotropic, compressible Neo-Hookean material. In this formulation, the strain energy function is expressed in a decoupled form as:

$$\Psi(\mathbf{C}_e) = \Psi_{vol}(J_e) + \overline{\Psi}(\overline{\boldsymbol{C}}_e) = \frac{K}{2}(J_e - 1)^{\mathbf{2}} + \frac{G'}{2}\left(\overline{I_1} - 3\right) \tag{4.40}$$

where K and G' are the bulk and the shear elastic modulus, $J_e = \det(\boldsymbol{F}_e)$, and $\overline{I_1} = \boldsymbol{tr}(\overline{\boldsymbol{C}}_e)$ is the first invariant of the modified (deviatoric) right Cauchy-Green tensor $\overline{\boldsymbol{C}}_e = \overline{\boldsymbol{F}}_e^T \overline{\boldsymbol{F}}_e$, with $\overline{\boldsymbol{F}}_e = J_e^{-1/3}\boldsymbol{F}_e$.

In a non-dissipative process, the second Piola–Kirchhoff stress tensor is derived from the strain energy function as its corresponding derivative:

$$\boldsymbol{S}_e = 2\frac{\partial\Psi(\boldsymbol{C}_e)}{\partial\boldsymbol{C}_e} = \boldsymbol{S}_{e,vol} + \overline{\boldsymbol{S}}_e = J_e\, p\, \boldsymbol{C}_e^{-1} + J_e^{-\frac{2}{3}}\left(\mathbb{I} - \frac{1}{3}\boldsymbol{C}_e^{-1} \otimes \boldsymbol{C}_e\right) : \widetilde{\boldsymbol{S}}_e \tag{4.41}$$

where $\boldsymbol{S}_{e,vol}$ and $\overline{\boldsymbol{S}}_e$ are the volumetric and deviatoric parts of the second Piola-Kirchhoff stress tensor, p is the hydrostatic pressure, and $\widetilde{\boldsymbol{S}}_e$ is the fictitious second Piola-Kirchhoff stress tensor:

$$p = \frac{d\Psi_{vol}(J_e)}{dJ_e} \quad \widetilde{\boldsymbol{S}}_e = 2\frac{\partial\overline{\Psi}(\overline{\boldsymbol{C}}_e)}{\partial\overline{\boldsymbol{C}}_e} \tag{4.42}$$

The Cauchy stress tensor $\boldsymbol{\sigma}_e$ is $1/J_e$ times the push-forward of $\boldsymbol{S}_e$ ($\boldsymbol{\sigma}_e = J_e^{-1}\chi_*(\boldsymbol{S}_e)$). In the Neo-Hookean material formulation, the explicit expression is defined as a function of the invariant: $\overline{I_1}$:

$$\boldsymbol{\sigma}_e = p\,\boldsymbol{I} + \frac{2}{J_e} dev\left[\overline{\boldsymbol{F}}_e \frac{\partial\overline{\Psi}(\overline{\boldsymbol{C}}_e)}{\partial\overline{\boldsymbol{C}}_e}\overline{\boldsymbol{F}}_e^T\right] = p\,\boldsymbol{I} + \frac{2}{J_e}\left(\frac{\partial\overline{\Psi}}{\partial\overline{I}_1}\overline{\boldsymbol{b}}_e - \frac{1}{3}\frac{\partial\overline{\Psi}}{\partial\overline{I}_1}\overline{I}_1\boldsymbol{I}\right) \tag{4.43}$$

With $\boldsymbol{I}$ being the second-order identity tensor, $\mathrm{dev}(.) = (.) - \frac{1}{3}[(.) : \boldsymbol{I}]\boldsymbol{I}$, and $\overline{\boldsymbol{b}}_e = \overline{\boldsymbol{F}}_e\overline{\boldsymbol{F}}_e^T$, the modified left Cauchy-Green tensor.

For liquid water, the mass conservation equation is:

$$\frac{\partial c_w}{\partial t} + \nabla\cdot\boldsymbol{n}_w = 0 \tag{4.44}$$

Where $c_w = \phi_w\rho_w$ represents the concentration of liquid water, and $\boldsymbol{n_w}$ denotes the absolute flux of liquid water:

$$\boldsymbol{n}_w = \boldsymbol{n}_{w,G} - D_{wT}\nabla T \tag{4.45}$$

Here, D_{wT} is the diffusivity due to the temperature gradient, and the net flux of liquid water with respect to the ground frame, $\boldsymbol{n}_{w,G}$, is described by:

$$\boldsymbol{n}_{w,G} = D_w\nabla c_w + c_w\vec{\boldsymbol{v}}_{s,G} \tag{4.46}$$

In this expression, D_w is the diffusivity due to the water concentration gradient, and $\vec{\boldsymbol{v}}_{s,G}$ is the solid velocity.

Under the assumption of thermal equilibrium between the two phases, the corresponding energy balance equation is expressed as:

$$\rho_{\mathrm{eff}}\, C_{p,\mathrm{eff}}\, \frac{\partial T}{\partial t} + \sum_{w} \left(\boldsymbol{n}_{w,G} \cdot \nabla \left(C_{\mathrm{eff},w} T \right) \right) = \nabla \cdot \left(k_{\mathrm{eff}} \nabla T \right) \tag{4.47}$$

Material properties. The properties of the product, ρ_{eff}, $C_{p,\mathrm{eff}}$, k_{eff}, are obtained as functions of temperature and composition:

$$\rho_{\mathrm{eff}}(T) = (1 - \phi_w)\overline{\rho}_s + \phi_w \overline{\rho}_w \tag{4.48}$$

$$C_{p,\mathrm{eff}}(T) = x_s C_{p,s} + x_w C_{p,w} \tag{4.49}$$

$$k_{\mathrm{eff}}(T) = (1 - \phi_w) k_s + \phi_w k_w \tag{4.50}$$

where x_s is the mass fraction of the solid phase and x_w is the mass fraction of water.

For meat, the density of the solid phase depends on the mass fractions of protein (x_{prot}) and fat (x_{fat}). The overall meat density is then calculated as a function of temperature and composition as follows:

$$\overline{\rho}_s(T) = \left(\frac{x_{\mathrm{prot}}}{\overline{\rho}_{\mathrm{prot}}(T)} + \frac{x_{\mathrm{fat}}}{\overline{\rho}_{\mathrm{fat}}(T)} \right)^{-1} \tag{4.51}$$

The specific heat of the product $C_{p,s}(T)$ is defined for each component and then calculated using a mass fraction average mixing rule:

$$C_{p,s}(T) = x_{\mathrm{prot}} C_{p,\mathrm{prot}}(T) + x_{\mathrm{fat}} C_{p,\mathrm{fat}}(T) \tag{4.52}$$

The product is assumed to exhibit isotropic thermal conductivity, bounded between two limiting cases. The lower bound corresponds to a perpendicular model, in which all constituents are arranged in layers oriented perpendicular to the direction of heat flow:

$$\frac{1}{k_{\perp}(T)} = \sum_{\mathrm{i}} \frac{\phi_{\mathrm{i}}}{k_i(T)} \tag{4.53}$$

The upper bound corresponds to the parallel model, where the constituents are organized into layers aligned parallel to the direction of heat flow.

$$k_{\parallel}(T) = \sum_{i} \phi_i k_i(T) \tag{4.54}$$

The thermal conductivity of the product is estimated as:

$$k_p(T) = \mathrm{g}k_{\perp}(T) + (1-\mathrm{g})k_{\parallel}(T) \tag{4.55}$$

where g is a number between zero (fully parallel fibers) and one (fully perpendicular fibers).

4.4.3 Finite element model of meat pan frying

A finite element model was developed to simulate the meat cooking process, explicitly incorporating the geometry of the aluminum pan (210 mm in diameter, 5 mm thick) and the meat product. The burger has a diameter of 100 mm and a thickness of 20 mm; in this case, a 2D axisymmetric model was built. The steak has a fixed width and length of 120 and 80 mm; however, the thickness has been varied between 14 and 34 mm.

In both models, rectangular elements with quadratic interpolation were employed to approximate mass transfer, temperature, and deformation. A mesh sensitivity analysis was conducted to identify the optimal discretization, yielding a configuration of 504 nodes and 459 elements for the burger case, and 175 nodes with 96 elements for the steak case. To simulate the flipping of the product, two pans were modeled in accordance with the approach of Moya et al. (2021b), under the assumption that gravity has no influence on water diffusion. From the perspective of solid mechanics, deformation across the product thickness is negligible (Hernández-Alhambra et al., 2024), allowing the burger to be considered confined between the two pan surfaces. The influence of the lower and upper pan domains, illustrated in Fig. 4.3, on the cooking process is incorporated or omitted by multiplying the corresponding boundary conditions with a specific function defined as:

$$f_{surf\_flip}\begin{cases} 1 & \text{for } t_{act} \leq (t \cdot mod \cdot 2\tau) < \tau \\ 0 & \text{for } \tau \leq (t \cdot mod \cdot 2\tau) < 2\tau \end{cases} \tag{4.56}$$

Here, t_{act} is the time instant when the boundary condition activates, while $t \cdot mod \cdot 2\tau$ gives the position within the current period (2τ). Here, *mod* represents the mathematical operation that computes the remainder of a division, and 2τ corresponds to the period or total cooking time of both product sides. The software provides a smoothed version of this function, permitting up to two continuous derivatives. As illustrated in Fig. 4.3 the two step functions regulate the boundary conditions when the product undergoes a single flip.

4.4.4 Initial and boundary conditions

The pan was initially set at a temperature of 25°C, while the product exhibited uniform initial values of 23.4°C in the burger case and 21.4°C in the steak case. To reproduce the experimental conditions, a power control strategy was incorporated into the model to regulate the heat input necessary for raising the pan temperature. This process is described by the following equation:

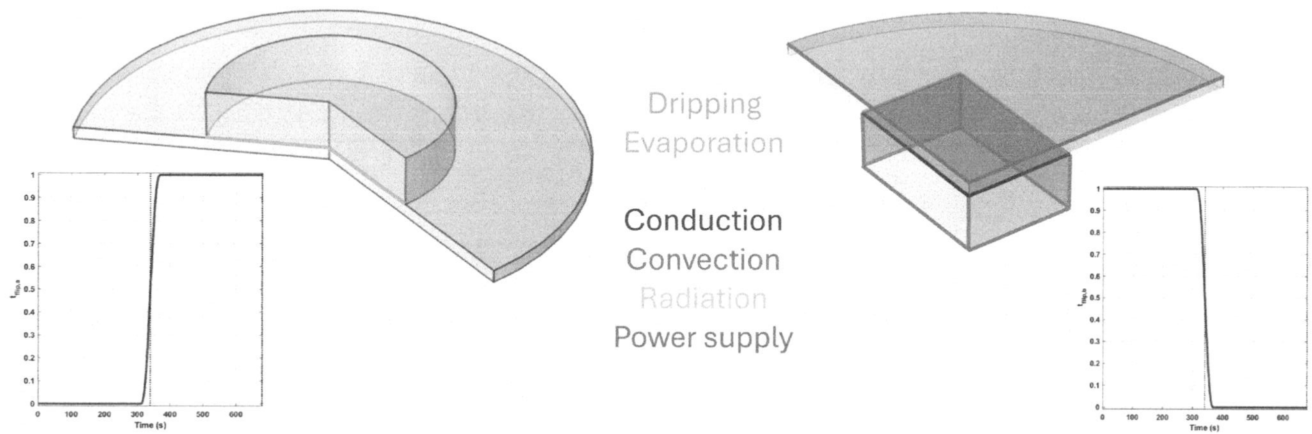

Figure 4.3 Mass and heat transfer boundary conditions applied in the contour of the product (mass transfer in the burger and heat transfer in the steak) and the pan governed by the two step functions to simulate the flipping process.

$$Power = \begin{cases} \kappa\left(T^{o}_{pan} - T^{*}_{pan}\right) & if\ \left(T^{o}_{pan} - T^{*}_{pan}\right) < \dfrac{W_{max}}{K} \\ W_{max} \quad if\ \left(T^{o}_{pan} - T^{*}_{pan}\right) \geq \dfrac{W_{max}}{K} \end{cases} \tag{4.57}$$

where $\kappa = 84$ W/K is a control parameter, T^{o}_{pan} denotes the target pan temperature, T^{*}_{pan} is the measured pan temperature at the central point, and W_{max} W represents the maximum power that the induction hob can deliver.

The heat transfer occurring at the interface of the two contacting surfaces can be formulated as:

$$q_{surf,cont} = H_c\ (T_{pan} - T_{surf})f_{surf,flip} \tag{4.58}$$

where T_{pan} and T_{surf} represent the temperatures of the pan and the meat at their contact interface, respectively, and H_c represents the thermal conductance at the interface of the two surfaces, governing the heat flux absorbed by the product.

The heat flux applied by the induction hob to the bottom surface of the pan is:

$$q_{surf,ind} = \begin{cases} \dfrac{\kappa}{A}\left(T^{o}_{pan} - T^{*}_{pan}\right)f_{surf,flip} & if\ \left(T^{o}_{pan} - T^{*}_{pan}\right) < W_{max}/\kappa \\ (W_{max}/A)f_{surf,flip} & if\ \left(T^{o}_{pan} - T^{*}_{pan}\right) \geq W_{max}/\kappa \end{cases} \tag{4.59}$$

in the expression, A is the surface of the pan $[m^2]T^{*}_{pan}$. The cooking process starts when the pan has reached the goal temperature. $T^{o}_{pan} = 215T$. Consistent with the experimental procedure, the initial 175 s of the simulation were dedicated to heating the pan to the target temperature. From this point onward, computational results are presented excluding this time interval.

The energy uptake by water during evaporation at the contact interface was modeled as:

$$q_{surf,evap} = \sigma_{evap}\ \overline{\rho}_w L_{evap}\ e\ f_v\ \left(\ T_{surf}\right)f_{surf,flip} \tag{4.60}$$

where σ_{evap} [1/s] is the evaporation rate constant, L_{evap} [J/kg] is the water latent heat of vapourization, $e = 1 \cdot 10^{-3}$ m is the assumed thickness of the liquid water layer and $f_v = \frac{1}{1 + e^{0.107\left(-T_{surf} + 121\right)}}$

A convective heat transfer boundary condition was imposed on the other exposed surfaces of both the product and the pan:

$$q_{surf,conv} = h\left(T_{amb} - T_{surf}\right)f_{surf,flip} \tag{4.61}$$

considers a surrounding air temperature T_{amb} [°C] and a convection coefficient h [W/(m^{2}K)].

A homogeneous distribution of water concentration is assumed as the initial condition of the simulation $c_{w0} = 44{,}085$ mol/m^3 for the burger case and $c_{w0} = 44{,}088$ mol/m^3 for the steak case. The boundary condition governing the liquid water equation accounts for both evaporation and drip fluxes. The evaporation flux is quantified as the product of the mass transfer coefficient h_m [m/s] and the difference in vapor density at the surface $\rho_{v,surf}$ [kg/m^3] and the boundary $\rho_{v,amb}$ [kg/m^3].

$$n_{w,surf,e} = h_m \left(\rho_{v,surf} - \rho_{v,amb}\right) f_{surf,flip} \tag{4.62}$$

Dripping takes place when the surface moisture concentration exceeds the water holding capacity. Consequently, the drip loss is defined as the difference between the total moisture flux arriving at the surface and the evaporation flux:

$$n_{w,surf,d} = \left(\boldsymbol{n_w} \cdot \boldsymbol{N} - n_{w,surf,e}\right) f_{surf,flip} \tag{4.63}$$

where $\boldsymbol{N}$ denotes the unit normal vector to the surface.

As deformation across the product thickness is disregarded, the normal displacements at the upper and lower surfaces are constrained for the entire duration of the simulation:

$$\boldsymbol{u} \cdot \boldsymbol{N} = 0 \tag{4.64}$$

Furthermore, in the case of steak, symmetry conditions are imposed because only a quarter of the real steak was simulated.

4.4.5 Model parameters

The parameters used in the model, obtained through experimental measurements or from literature, are listed in Table 4.1. Some parameters were optimized to fit the experimental results.

4.4.6 Modeling beef burger cooking

In this case study, our aim is to illustrate the capability of numerical modeling to predict the cooking time required for the geometric center of a burger to reach a microbiologically safe temperature, thereby ensuring the inactivation of pathogenic microorganisms present in raw meat, as a function of flipping frequency. Previously, this model was validated, for which it is necessary to cook burgers. Numerically, we will reproduce the experimental test conditions.

Computations were carried out until the central point of the burger reached 71.1°C, irrespective of the number of flips tested (3, 5, or 7). For the 3-flip condition, a total

Table 4.1 Model parameters used in the finite element simulation.

Name and description	Value for the burger	Value for the steak	Source
Model parameters			
κ Controller parameter [W/K]	84		Fitted
W_{max} Maximum Induction Hob Power [W]	2200		Measured
T_{amb} Ambient Temperature [°C]	25		Measured
T^{o}_{pan} Target pan temperature [°C]	140 or 210		Measured
P_{amb} Ambient Pressure [kPa]	101.3		Measured
H_c Thermal Conductance of Pan–Meat Contact [W/(m^2·K)]	200	75	Fitted
g Thermal conductivity parameter	0	0.45	Fitted
h convection coefficient [W/(m^2·K)]	5	10.22	Fitted
Water properties			
$\overline{\rho}_w$ Water density [kg/m^3]	997.2		Choi and Okos (1986)
D_w Water diffusivity [m^2/s]	$2.1044 \cdot 10^{-9}$	$1.25 \cdot 10^{-7}$	Fitted
$D_{w,T}$ Water diffusivity due to temperature gradient [kg/(m·s·K)]	$D_w \cdot \frac{\partial \rho_{f,eq}}{\partial T}$		Fitted
$C_{p,w}$ Water specific heat [kJ/(kg·K)]	$4.1289 - 9.0864 \cdot 10^{-5} \cdot T + 5.4731 \cdot 10^{-6} \cdot T^2$		Choi and Okos (1986)
k_w Water thermal conductivity [W/(m·K)]	0.57		Choi and Okos (1986)
L_{evap} Latent heat of vapourization [J/kg]	$2.26 \cdot 10^6$		Straub (1985)
h_m Mass Transfer coefficient [m/s]	$1.688 \cdot 10^{-3}$	$8.9 \cdot 10^{-3}$	Fitted
σ_{evap} evaporation rate constant [1/s]	$1.604 \cdot 10^{-3}$		Fitted

Meat properties			
$\overline{\rho}_p$ Water density [kg/m^3]	1330		Choi and Okos (1986)
$C_{p,s}$ Meat specific heat [J/(kg·K)]	$2.0082 + 1.2089 \cdot 10^{-3} \cdot T - 1.3129 \cdot 10^{-6} \cdot T^2$		Choi and Okos (1986)
k_p Product thermal conductivity [W/(m·K)]	$0.17881 + 1.1958 \cdot 10^{-3} \cdot T - 2.7178 \cdot 10^{-6} \cdot T^2$		Choi and Okos (1986)
WHC Water holding capacity [kg water/kg dry material]	Hernández-Alhambra et al. (2024)	Moya et al. (2021b)	Measured
G' Storage modulus [kPa]	Hernández-Alhambra et al. (2024)	Moya et al. (2021b)	Measured
σ_y yield stress [Pa]	Hernández-Alhambra et al. (2024)	Moya et al. (2021b)	Measured
$\dot{\gamma}_c$ Critical deformation velocity [1/s]	Hernández-Alhambra et al. (2024)	Moya et al. (2021b)	Measured
T_{ini} Initial product temperature [°C]	23.4	9.6 or 11.7	Measured
$T_{ini,pan}$ Initial pan temperature [°C]	25		Measured
e product thickness [m]	0.02	0.02; variable	Measured
d product size [m]	Diameter = 0.1	With = 0.08; length = 0.12	Measured
V heat Speed [m/s]	1.31		Measured

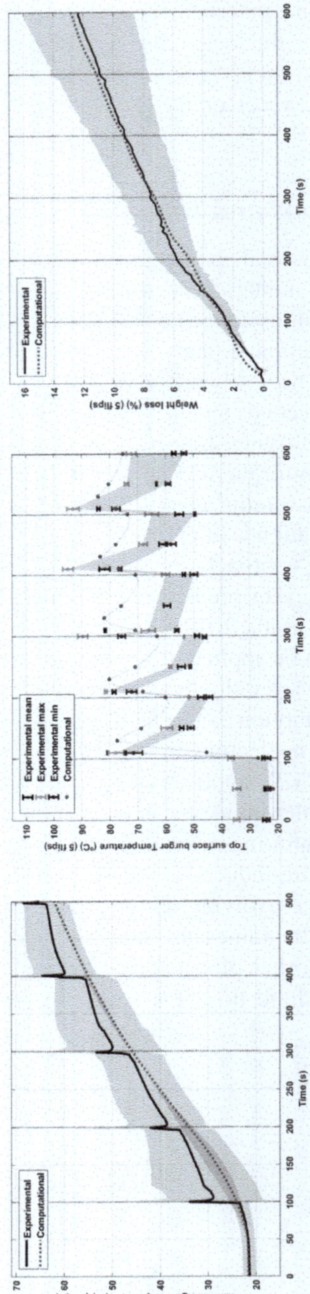

(a) Central temperature

(b) Upper temperature

(c) Weight loss

Figure 4.4 Experimental and numerical central temperature, upper temperature and weight loss for five flips.

cooking time of 680 s was required to achieve this target temperature. In contrast, both the 5-flip and 7-flip configurations reached the same safe internal temperature within 603 s, with turning operations performed at uniform time intervals. Increasing the flipping frequency thus led to a reduction in the total heating time needed to attain the desired core temperature, while simultaneously limiting cooking-induced weight loss and dimensional shrinkage.

Fig. 4.4 shows the evolution for five flips and compares it with the values recorded experimentally.

Fig. 4.5 presents the simulated temperature distribution and the corresponding deformation patterns immediately after the turnover for three, five and seven flips. Examination of the core region reveals that increasing the number of flips promotes higher central temperature and reduces internal thermal gradients, thereby yielding a more uniform degree of doneness throughout the burger.

At the end of the cooking, a clear color change and a noticeabe reduction in burger dimensions were observed (Fig. 4.6).

The textural properties of burgers are critical factors influencing consumer acceptance. Table 4.2 reports moisture and cooking losses, volumetric shrinkage, and instrumental textural parameters of burgers cooked with three, five and seven flips until core temperature of 71.1°C was reached. The data indicate that both water loss and overall cooking loss decrease as flipping frequency increases, demonstrating that multiple flipping enhances moisture retention ($P < .05$). Furthermore, burgers cooked with five flips exhibit significantly lower hardness and chewiness compared with those cooked with three flips ($P < .05$). However, increasing the number of flips beyond five does not provide additional reductions in these textural attributes.

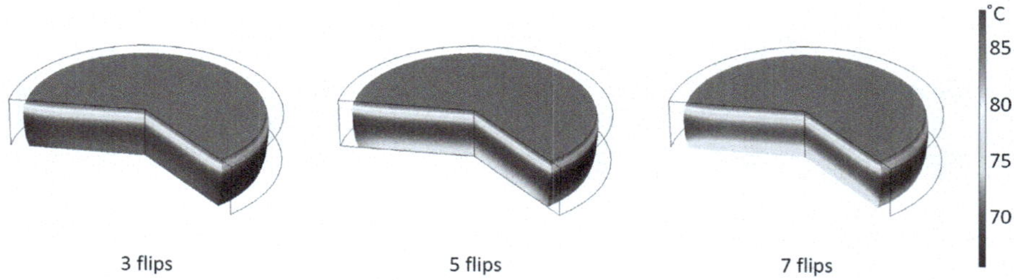

Figure 4.5 Map of temperatures and shrinkage of different number of flips.

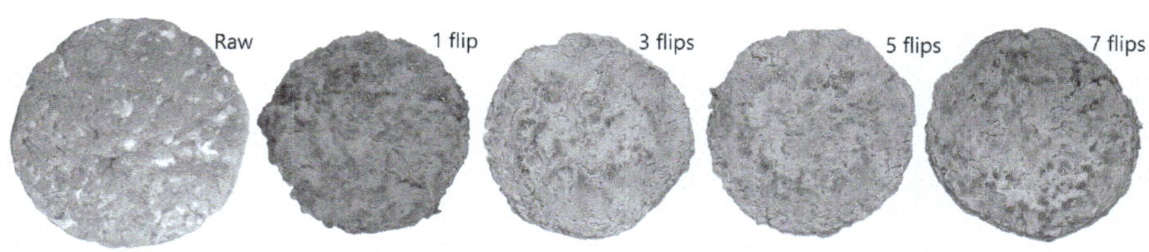

Figure 4.6 Final appearance of the burger as a function of the number of flips applied during cooking.

Table 4.2 Cooking at a pan temperature of 140°C. Data are expressed as means ± standard deviations. Values followed by different letters within the same row indicate significant difference ($P < .05$) between number of flips.

Parameters		Cooking conditions (number of flips – total cooking time)		
		3 flips – 680 s	**5 flips – 603 s**	**7 flips – 603 s**
Central temperature (°C)		67.95	69.35	71.72
Moisture loss (%)		14.23 ± 0.94^{b}	15.72 ± 0.68^{b}	11.54 ± 1.45^{a}
Cooking loss (%)		15.55 ± 2.46^{b}	13.84 ± 4.66^{b}	11.57 ± 1.86^{a}
Shrinkage (%)		20.17 ± 1.08^{a}	22.87 ± 1.45^{a}	21.54 ± 0.66^{a}
Texture parameters TPA	Hardness (N)	73.19 ± 3.09^{b}	60.78 ± 5.78^{a}	65.52 ± 9.37^{ab}
	Elasticity	0.70 ± 0.05^{a}	0.67 ± 0.03^{a}	0.68 ± 0.03^{a}
	Cohesiveness	0.61 ± 0.06^{b}	0.57 ± 0.03^{ab}	0.60 ± 0.04^{ab}
	Gumminess (N)	45.57 ± 4.15^{b}	34.77 ± 2.70^{a}	38.52 ± 6.07^{a}
	Chewiness (N)	33.20 ± 5.36^{b}	23.05 ± 1.62^{a}	26.13 ± 4.44^{a}

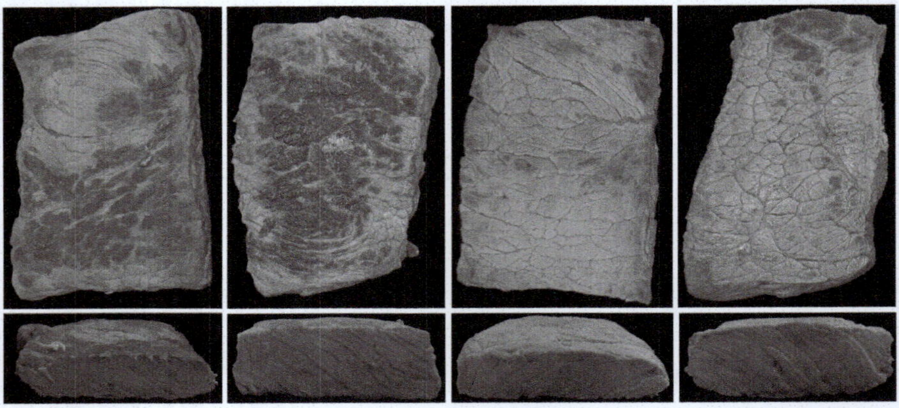

Figure 4.7 Final appearance of the beef steak as a function of the pan temperature and steak thickness.

4.4.7 Modeling beef steak cooking

In the case of beef steak cooking, the thickness varied between 14 and 34 mm, leaving the cooking time and the number of flips fixed at 680 s and three flips. Fig. 4.7 shows

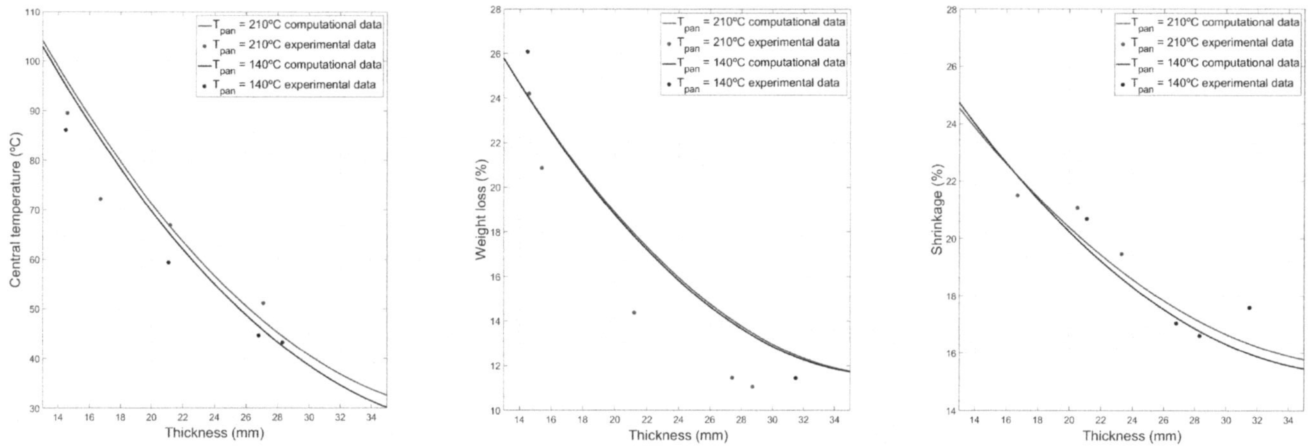

(a) Central temperature (b) Weight loss (c) Shrinkage

Figure 4.8 Evolution of central temperature, weight loss and shrinkage for several thicknesses for two different temperatures set in the pan.

the external and internal appearance of the fillets cooked at two different pan temperatures. The influence of thickness on the progression of core temperature, weight loss, and shrinkage is presented in Fig. 4.8. The data follow quadratic trends, and the high R^2 values in all cases indicate an excellent fit between the numerical data and the models, confirming the consistency of these relationships.

4.5 Conclusions

A 2D axisymmetric and a 3D computational model were developed to simulate pan cooking, explicitly accounting for coupled heat and moisture transport and meat deformation. The modeling approach reliably predicts the behavior of both burger and steak products.

The incorporation of flipping into the numerical simulations enabled accurate reproduction of the experimentally observed time–temperature histories, moisture and cooking losses, and shrinkage patterns in beef burgers and steaks. Experimental validation demonstrated that, when burgers are cooked to the same target core temperature, applying three or more flips substantially shortens the cooking time. This reduction in cooking duration enhances moisture retention, thereby decreasing cooking losses and dimensional shrinkage. Consequently, frequent flipping is strongly advised for burger cooking, although performing more than five flips does not yield further meaningful reductions in cooking time.

Caution is warranted when extrapolating these findings to different cooking scenarios. Nevertheless, based on previous observations (Hernández-Alhambra et al., 2024) it is expected that similar trends may arise across a wide range of pan surface temperatures (140–215°C), as the internal temperature profiles remain largely comparable except in a thin outer layer. Therefore, the cooking times required to achieve the same final core temperature are unlikely to diverge substantially from the values reported here. In contrast, for burgers of smaller thickness, where thermal penetration is more uniform, the advantage of multiple flips relative to a single flip may be less pronounced. The thickness analysis performed with the steak model led to consistent conclusions at both simulated pan temperatures. Increasing product thickness resulted in lower core temperatures, reduced mass loss, and diminished shrinkage. These outcomes align with the movement of the evaporation front: despite changes in pan temperature, the internal thermal gradient remains essentially constant, leading thicker samples to experience slower heating and reduced moisture loss.

References

Ağçam, E., 2022. Modeling of the changes in some physical and chemical quality attributes of potato chips during frying process. Appl. Food Res. 2, 100064. https://doi.org/10.1016/j.afres.2022.100064.

Al Faruq, A., Khatun, M.H.A., Azam, S.M.R., Sarker, M.S.H., Mahomud, M.S., Jin, X., 2022. Recent advances in frying processes for plant-based food. Food Chem. Adv. 1, 100086. https://doi.org/10.1016/j.focha.2022.100086.

Alvis, A., Vélez, C., Rada-Mendoza, M., Villamiel, M., Villada, H.S., 2009. Review: heat transfer coefficient during deep-fat frying. Food Control 20, 321–325. https://doi.org/10.1016/j.foodcont.2008.05.016.

Amiryousefi, M.R., Mohebbi, M., Khodaiyan, F., 2014. Applying an intelligent model and sensitivity analysis to inspect mass transfer kinetics, shrinkage and crust color changes of deep-fat fried ostrich meat cubes. Meat Sci. 96, 172–178. https://doi.org/10.1016/j.meatsci.2013.06.018.

Arias-Mendez, A., Warning, A., Datta, A.K., Balsa-Canto, E., 2013. Quality and safety driven optimal operation of deep-fat frying of potato chips. J. Food Eng. 119, 125–134. https://doi.org/10.1016/j.jfoodeng.2013.05.001.

Asokapandian, S., Swamy, G.J., Hajjul, H., 2020. Deep fat frying of foods: a critical review on process and product parameters. Crit. Rev. Food Sci. Nutr. 60, 3400–3413. https://doi.org/10.1080/10408398.2019.1688761.

Baik, O.O., Mital, G., 2005. Heat and moisture transfer and shrinkage simulation of deep-fat tofu frying. Food Res. Int. 28, 183–191. https://doi.org/10.1016/j.foodres.2004.10.003.

Bansal, H.S., Takhar, P.S., Alvarado, C.Z., Thompson, L.D., 2015. Transport mechanisms and quality changes during frying of chicken nuggets-hybrid mixture theory based modeling and experimental verification. J. Food Sci. 80, E2759–E2773. https://doi.org/10.1111/1750-3841.13082.

Bassama, J., Brat, P., Boulanger, R., Günata, Z., Bohuon, P., 2012. Modeling deep-fat frying for control of acrylamide reaction in plantain. J. Food Eng. 113, 156–166. https://doi.org/10.1016/j.jfoodeng.2012.04.004.

Bouchon, P., Pyle, D., 2005a. Modelling oil absorption during post-frying cooling: I: model development. Food Bioprod. Process. 83, 253–260. https://doi.org/10.1205/fbp.05115.

Bouchon, P., Pyle, D., 2005b. Modelling oil absorption during post-frying cooling: II: solution of the mathematical model, model testing and simulations. Food Bioprod. Process. 83, 261–272. https://doi.org/10.1205/fbp.05114.

Budzaki, S., Seruga, B., 2005. Determination of convective heat transfer coefficient during frying of potato dough. J. Food Eng. 66, 307–314. https://doi.org/10.1016/j.jfoodeng.2004.03.023.

Cernela, J., Heyd, B., Keller, S., Bailleul, J., Maillard, M., Bonazzi, C., Broyart, B., 2015. Experimental study of heat and mass transfer phenomena during the contact heating of solid food models. J. Food Eng. 146, 99–106. https://doi.org/10.1016/j.jfoodeng.2014.09.007.

Chan, D.S., 2020. Computer simulation with a temperature-step frying approach to mitigate acrylamide formation in French fries. Foods 9, 200. https://doi.org/10.3390/foods9020200.

Choi, Y., Okos, M.R., 1986. Effects of temperature and composition on thermal properties of foods. J. Food Process Appl. 1, 93–101.

Contardo, I., Parada, J., Leiva, A., Bouchon, P., 2016. The effect of vacuum frying on starch gelatinization and its in vitro digestibility in starch–gluten matrices. Food Chem. 197, 353–358. https://doi.org/10.1016/j. foodchem.2015.10.028.

Cortés, P., Badillo, G., Segura, L., Bouchon, P., 2014. Experimental evidence of water loss and oil uptake during simulated deep-fat frying using glass micromodels. J. Food Eng. 140, 19–27. https://doi.org/10.1016/j.jfoodeng.2014.04.005.

Dalvi-Isfahan, M., 2023. Mathematical modeling for investigating the effect of single-sided flipping on moisture, fat content, and safety of hamburger patty. J. Food Process. Eng. 46. https://doi.org/10.1111/jfpe.14450.

Dana, D., Saguy, I.S., 2006. Review: mechanism of oil uptake during deep-fat frying and the surfactant effect-theory and myth. Adv. Colloid Interface Sci. 128–130, 267–272. https://doi.org/10.1016/j.cis.2006.11.013.

Dash, K.K., Sharma, M., Tiwari, A., 2022. Heat and mass transfer modeling and quality changes during deep fat frying: a comprehensive review. J. Food Process. Eng. 45, e13999. https://doi.org/10.1111/jfpe.13999.

Datta, A., 2007. Porous media approaches to studying simultaneous heat and mass transfer in food processes. i: problem formulations. J. Food Eng. 80, 80–95. https://doi.org/10.1016/j.jfoodeng.2006.05.01305.013.

Datta, A., 2016. Toward computer-aided food engineering: mechanistic frameworks for evolution of product, quality and safety during processing. J. Food Eng. 176, 9–27. https://doi.org/10.1016/j.jfoodeng.2015.10.010.

Datta, A., Nicolaï, B., Vitrac, O., Verboven, P., Erdogdu, F., Marra, F., Sarghini, F., Koh, C., 2022. Computer-aided food engineering. Nat. Food 3, 894–904. https://doi.org/10.1038/s43016-022-00617-5.

Debnath, S., Rastogi, N.K., Krishna, G., Lokesh, B.R., 2009. Oil partitioning between surface and structure of deep-fat fried potato slices: a kinetic study. LWT Food Sci. Technol. (Lebensm.-Wiss.-Technol.) 42, 1054–1058. https://doi.org/10.1016/j.lwt.2009.01.006.

Dehghan Nasiri, F., Mohebbi, M., Tabatabaee Yazdi, F., Haddad Khodaparast, M.H., 2011. Kinetic modeling of mass transfer during deep fat frying of shrimp nugget prepared without a pre-frying step. Food Bioprod. Process. 89, 241–247. https://doi.org/10.1016/j.fbp.2010.11.009.

Dehghannya, J., Ngadi, M., 2021. Recent advances in microstructure characterization of fried foods: different frying techniques and process modeling. Trends Food Sci. Technol. 116, 786–801. https://doi.org/10.1016/j.tifs.2021.03.033.

Del Rosario-Santiago, J., López-Méndez, E.M., Ruiz-Espinosa, H., Ochoa-Velasco, C.E., Escobedo- Morales, A., Ruiz-López, I.I., 2022. Modelling post-frying oil absorption, water loss, and cooling of potato cylinders. Int. Food Res. J. 30, 1142–1158. https://doi.org/10.47836/ifrj.30.5.05.

Devi, S., Zhang, M., Ju, R., Bhandari, B., 2021. Recent development of innovative methods for efficient frying technology. Crit. Rev. Food Sci. Nutr. 61, 3709–3724. https://doi.org/10.1080/10408398.2020.1804319.

Dhall, A., Datta, A.K., 2011. Transport in deformable food materials: a poromechanics approach. Chem. Eng. Sci. 66, 6482–6497. https://doi.org/10.1016/j.ces.2011.09.001.

Erdogdu, F., 2023. Mathematical modeling of food thermal processing: current and future challenges. Curr. Opin. Food Sci. 51, 101042. https://doi.org/10.1016/j.cofs.2023.101042.

Erdogdu, F., Zorrilla, S., Singh, R., 2005. Effects of different objective functions on optimal decision variables: a study using modified complex method to optimize hamburger cooking. LWT Food Sci. Technol. 38, 111–118. https://doi.org/10.1016/j.lwt.2004.05.010.

Erdogdu, F., Sarghini, F., Marra, F., 2017. Mathematical modeling for virtualization in food processing. Food Eng. Rev. 9, 295–313. https://doi.org/10.1007/s12393-017-9161-y.

Estévez-Sánchez, K.H., Ruiz-Espinosa, H., Corona-Jiménez, E., López-Méndez, E.M., Cortés-Zavaleta, O., Ochoa-Velasco, C.E., Ruiz-López, I.I., 2021. Water diffusivity estimation in air-dried complex- shaped foods by the method of slopes: application to oblate spheroid geometry. Comput. Electron. Agric. 181, 105949. https://doi.org/10.1016/j.compag.2020.105949.

Farid, M., 2002. The moving boundary problems from melting and freezing to drying and frying of food. Chem. Eng. Process. Process Intensif. 41, 1–10. https://doi.org/10.1016/S0255-2701(00)00153-7.

Farkas, B., Singh, R., Rumsey, T., 1996a. Modeling heat and mass transfer in immersion frying. I, model development. J. Food Eng. 29, 211–226. https://doi.org/10.1016/0260-8774(95)00072-0.

Farkas, B., Singh, R., Rumsey, T., 1996b. Modeling heat and mass transfer in immersion frying. II, model solution and verification. J. Food Eng. 29, 227–248. https://doi.org/10.1016/0260-8774(95)00048-8 00048-8.

Farkas, B.E., Hubbard, L.J., Escher, F.E., 1996c. Analysis of moisture loss and oil absorption during frying of potato slices. J. Food Sci. 61, 679–685. https://doi.org/10.1111/j.1365-2621.1996.tb12179.x.

Fellows, P.J., 2022. Food Processing Technology: Principles and Practice, fifth ed. Woodhead Publishing.

Frakolaki, G., Kekes, T., Bizymis, A.P., Giannou, V., Tzia, C., 2023. Fundamentals of food frying processes. In: Jafari, S.M. (Ed.), High-Temperature Processing of Food Products. Unit Operations and Processing Equipment in the Food Industry. Woodhead Publishing, Cambridge, pp. 227–291. https://doi.org/10.1016/B978-0-12-818618-3.00001-X.

Gazmuri, A., Bouchon, P., 2009. Analysis of wheat gluten and starch matrices during deep-fat frying. Food Chem. 115, 999–1005. https://doi.org/10.1016/j.foodchem.2009.01.020.

Ghaderi, A., Dehghannya, J., Ghanbarzadeh, B., 2018. Momentum, heat and mass transfer enhancement during deep-fat frying process of potato strips: influence of convective oil temperature. Int. J. Therm. Sci. 134, 485–499. https://doi.org/10.1016/j.ijthermalsci.2018.08.035.

Ghaderi, A., Dehghannya, J., Ghanbarzadeh, B., 2024. Effect of oil type on momentum, heat, and mass transfer during deep-fat frying of potato strips: numerical and experimental study. Potato Res. https://doi.org/10.1007/s11540-024-09768-3.

Gholamibozanjani, G., Leong, S.Y., Oey, I., Bremer, P., Silcock, P., Farid, M., 2021. Heat and mass transfer modeling to predict temperature distribution during potato frying after pre-treatment with pulsed electric field. Foods 10, 1679. https://doi.org/10.3390/foods10081679.

Grenier, D., Bohuon, P., Méot, J., Lecomte, D., Baillères, H., 2007. Heat and mass transfer in fry drying of wood. Dry. Technol. 25, 511–518. https://doi.org/10.1080/07373930601184114.

Halder, A., Dhall, A., Datta, A., 2007. An improved, easily implementable, porous media based model for deep-fat frying: Part i: model development and input parameters. Food Bioprod. Process. 85, 209–219. https://doi.org/10.1205/fbp07033.

He, D.B., Xu, F., Hua, T.C., Song, X.Y., 2012. Oil absorption mechanism of fried food during cooling process. J. Food Process. Eng. 36, 412–417. https://doi.org/10.1111/j.1745-4530.2012.00681.x.

Hernández-Alhambra, E., Guíu, P., Cabeza-Gil, I., Ferrer-Mairal, A., Martínez, M., Calvo, B., Grasa, J., Salvador, M.L., 2024. Towards domestic cooking efficiency: a case study on burger pan frying using experimental and computational results. J. Food Eng. 363, 111783. https://doi.org/10.1016/j.jfoodeng.2023.111783.

Hernández-Alhambra, E., Guíu, P., Ferrer-Mairal, A., Martínez, M., Calvo, B., Grasa, J., Salvador, M.L., 2025. Predicting optimal flipping conditions during burger pan cooking with a numerical model. J. Food Eng. 387, 112315. https://doi.org/10.1016/j.jfoodeng.2024.112315.

Hernández-Alhambra, E., Guiu, P., Ferrer-Mairal, A., Martínez, M.A., Calvo, B., Grasa, J., Salvador, M.L., 2026. FEM modeling of hamburger pan cooking: Fat content influence and neural network-based prediction of fat loss. Curr. Res. Food Sci. 12. https://doi.org/10.1016/j.crfs.2026.101327.

Ikediala, J., Correia, L., Fenton, G., Ben-Abdallah, N., 1996. Finite element modeling of heat transfer in meat patties during single-sided pan-frying. J. Food Sci. 61, 796–802. https://doi.org/10.1111/j.1365-2621.1996.tb12205.x.

Isik, B., Sahin, S., Sumnu, G., 2016. Pore development, oil and moisture distribution in crust and core regions of potatoes during frying. Food Bioprocess Technol. 9, 1653–1660. https://doi.org/10.1007/s11947-016-1748-4.

Kalogianni, E.P., Papastergiadis, E., 2014. Crust pore characteristics and their development during frying of French-fries. J. Food Eng. 120, 175–182. https://doi.org/10.1016/j.jfoodeng.2013.07.029.

Khalilian, S., Mba, O.I., Ngadi, M.O., 2021. g-Frying of eggplant (*Solanum melongena* L.). J. Food Eng. 293, 110358. https://doi.org/10.1016/j.jfoodeng.2020.110358.

Knight, M., McWilliam, S., Peck, S., Koutsidis, G., Chope, G., Puddephat, I., Wedzicha, B., 2021. Kinetic modelling of acrylamide formation during the frying of potato chips. Food Chem. 352, 129305. https://doi.org/10.1016/j.foodchem.2021.129305.

Köse, Y.E., 2022. Kinetic modeling of heat and mass transfer during deep fat frying of churro. Bilge Int. J. Sci. Tech. Res. 6, 46–51. https://doi.org/10.30516/bilgesci.1064191.

Krokida, M.K., Oreopoulou, V., Maroulis, Z.B., 2000. Effect of frying conditions on shrinkage and porosity of fried potatoes. J. Food Eng. 43, 147–154. https://doi.org/10.1016/S0260-8774(99)00143-0.

Krokida, M., Oreopoulou, V., Maroulis, Z., Marinos-Kouris, D., 2001. Colour changes during deep fat frying. J. Food Eng. 48, 219–225. https://doi.org/10.1016/S0260-8774(00)00161-8.

Liberty, J.T., Dehghannya, J., Ngadi, M.O., 2019. Effective strategies for reduction of oil content in deep-fat fried foods: a review. Trends Food Sci. Technol. 92, 172–183. https://doi.org/10.1016/j.tifs.2019.07.050.

Lioumbas, J.S., Karapantsios, T.D., 2012. Evaporation front compared with crust thickness in potato deep- fat frying. J. Food Sci. 77, E17–E25. https://doi.org/10.1111/j.1750-3841.2011.02472.x.

Lioumbas, J.S., Kostoglou, M., Karapantsios, T.D., 2012. On the capacity of a crust–core model to describe potato deep-fat frying. Food Res. Int. 46, 185–193. https://doi.org/10.1016/j.foodres.2011.12.012.

Liu, Y., Tian, J., Duan, Z.H., Li, J.W., Fan, L.P., 2021a. Effect of oil surface activity on oil absorption behavior of potato strips during frying process. Food Chem. 365, 130427. https://doi.org/10.1016/j.foodchem.2021.130427.

Liu, Y., Tian, J., Hu, B., Yu, P., Fan, L., 2021b. Relationship between crust characteristics and oil uptake of potato strips with hot-air pre-drying during frying process. Food Chem. 360, 130045. https://doi.org/10.1016/j.foodchem.2021.130045.

Mekprayoon, R., Tangduangdee, C., 2012. Influence of nusselt number on weight loss during chilling process. Procedia Eng. 32, 90–96. https://doi.org/10.1016/j.proeng.2012.01.1241.

Mir-Bel, J., Oria, R., Salvador, M.L., 2009. Influence of the vacuum break conditions on oil uptake during potato post-frying cooling. J. Food Eng. 95, 416–422. https://doi.org/10.1016/j.jfoodeng.2009.06.001.

Mir-Bel, J., Oria, R., Salvador, M.L., 2012. Deep-fat frying under moderate vacuum of potato cylinders. Int. J. Food Sci. Technol. 47, 1369–1377. https://doi.org/10.1111/j.1365-2621.2012.02982.x.

Moreira, R.G., 2014. Vacuum frying versus conventional frying – an overview. Eur. J. Lipid Sci. Technol. 116, 723–734. https://doi.org/10.1002/ejlt.201300272.

Moreira, R.G., Barrufet, M.A., 1998. A new approach to describe oil absorption in fried foods: a simulation study. J. Food Eng. 35, 1–22. https://doi.org/10.1016/S0260-8774(98)00020-X.

Moreno, M.C., Bouchon, P., 2013. Microstructural characterization of deep-fat fried formulated products using confocal scanning laser microscopy and a non-invasive double staining procedure. J. Food Eng. 118, 238–246. https://doi.org/10.1016/j.jfoodeng.2013.03.007.

Moya, J., Lorente-Bailo, S., Ferrer-Mairal, A., Martínez, M.A., Calvo, B., Grasa, J., Salvador, M.L., 2021a. Color changes in beef meat during pan cooking: kinetics, modeling and application to predict turn over time. Eur. Food Res. Technol. 247, 2751–2764. https://doi.org/10.1007/s00217-021-03821-y.

Moya, J., Lorente-Bailo, S., Salvador, M., Ferrer-Mairal, A., Martínez, M., Calvo, B., Grasa, J., 2021b. Development and validation of a computational model for steak double-sided pan cooking. J. Food Eng. 298, 110498. https://doi.org/10.1016/j.jfoodeng.2021.110498.

Naghavi, E.A., Dehghannya, J., Ghanbarzadeh, B., 2018. 3D computational simulation for the prediction of coupled momentum, heat and mass transfer during deep-fat frying of potato strips coated with different concentrations of alginate. J. Food Eng. 235, 64–78. https://doi.org/10.1016/j.jfoodeng.2018.04.026.

Ngadi, M., Watts, K., Correia, L., 1997. Finite element method modelling of moisture transfer in chicken drum during deep-fat frying. J. Food Eng. 32, 11–20. https://doi.org/10.1016/S0260-8774(97)00095-2.

Onipe, O.O., Beswa, D., Jideani, A.I.O., 2020. Confocal laser scanning microscopy and image analysis for elucidating crumb and crust microstructure of bran-enriched south african fried dough and batter. Foods 9, 605. https://doi.org/10.3390/foods9050605.

Ou, D., Mittal, G., 2006. Double-sided pan-frying of unfrozen/frozen hamburgers for microbial safety using modelling and simulation. Food Res. Int. 39, 133–144. https://doi.org/10.1016/j.foodres.2005.06.009.

Ou, D., Mittal, G., 2007. Single-sided pan frying of frozen hamburgers with flippings for microbial safety using modeling and simulation. J. Food Eng. 80, 33–45. https://doi.org/10.1016/j.jfoodeng.2006.03.033.

Paul, I.D., 2023. Alternate Frying Technologies for Healthier and Better Fried Foods. CRC Press, pp. 1–24. https://doi.org/10.1201/9781003329244-10 (chapter 10).

Pedreschi, F., Hernández, P., Figueroa, C., and, P.M., 2005. Modeling water loss during frying of potato slices. Int. J. Food Prop. 8, 289–299. https://doi.org/10.1081/JFP-200059480.

Peñalver Soto, J.L., Muñoz Guillermo, M., Garre, A., Iguaz, A., Fernández, P.S., Egea, J.A., 2022. Multiobjective optimization of a frying process balancing acrylamide formation and quality: solution analysis and uncertainty propagation. Foods 11, 3689. https://doi.org/10.3390/foods11223689.

Purlis, E., 2023. Simple modelling of time-temperature profiles in food during baking. Appl. Food Res. 3, 100271. https://doi.org/10.1016/j.afres.2023.100271.

Ran, X., Lin, D., Zheng, L., Li, Y., Yang, H., 2023. Kinetic modelling of the mass and heat transfer of a plant-based fishball alternative during deep-fat frying and air frying and the changes in physicochemical properties. J. Food Eng. 350, 111457. https://doi.org/10.1016/j.jfoodeng.2023.111457.

Rani, L., Kumar, M., Kaushik, D., Kaur, J., Kumar, A., Oz, F., Proestos, C., Oz, E., 2023. A review on the frying process: methods, models and their mechanism and application in the food industry. Food Res. Int. 172, 113176. https://doi.org/10.1016/j.foodres.2023.113176.

Roshani, F., Movahhed, S., Chenarbon, H.A., 2021. Modelling shrinkage in deep-fried satina potato slices pretreated with ultrasound. Potato Res. 64, 257–265. https://doi.org/10.1007/s11540-020-09475-9.

Safari, A., Salamat, R., Baik, O.D., 2018. A review on heat and mass transfer coefficients during deep-fat frying: determination methods and influencing factors. J. Food Eng. 230, 114–123. https://doi.org/10.1016/j.jfoodeng.2018.01.022.

Safari, A., Dehghannya, J., Ghanbarzadeh, B., 2024. Two-phase (solid–fluid) coupled transfer phenomena modeling during frying of potato slices: effect of the product surface-to-volume ratio. J. Therm. Anal. Calorim. 149, 3181–3196. https://doi.org/10.1007/s10973-024-12929-8.

Sahasrabudhe, S.N., Chaudhari, S.S., Farkas, B.E., 2019. Experimental measurement of factors affecting dynamics of bubble growth from a submerged orifice: applications to the frying process. J. Food Eng. 251, 36–44. https://doi.org/10.1016/j.jfoodeng.2019.02.005.

Sahin, S., Sastry, S., Bayindirli, L., 1999. Heat transfer during frying of potato slices. LWT Food Sci. Technol. (Lebensm.-Wiss.-Technol.) 32, 19–24. https://doi.org/10.1006/fstl.1998.0494.

Sandhu, J., Parikh, A., Takhar, P.S., 2016. Experimental determination of convective heat transfer coefficient during controlled frying of potato discs. LWT Food Sci. Technol. 65, 180–184. https://doi.org/10.1016/j.lwt.2015.08.007.

Shah, Y., Takhar, P.S., 2022. Pressure development and volume changes during frying and post-frying of potatoes. LWT Food Sci. Technol. 172, 114243. https://doi.org/10.1016/j.lwt.2022.114243.

Sivaranjani, S., Joshi, J.T., Mukta, S.S., Rao, P.S., 2024. A comprehensive review of the mechanism, changes, and effect of deep fat frying on the characteristics of restructured foods. Food Chem. 450, 139393. https://doi.org/10.1016/j.foodchem.2024.139393.

Straub, J., 1985. In: Haar, Von L.Gallagher, J.S.Kell, G.S. (Eds.), NBS/NRC Steam Tables. Hemisphere Publishing Corp., Washington https://doi.org/10.1002/cite.330570931.

Tangduangdee, C., Bhumiratana, C., Tia, S., 2004. The role of moisture movement and crust thermal property on heat and mass transfer process during deep-fat frying. Int. Commun. Heat Mass Tran. 31, 73–84. https://doi.org/10.1016/S0735-1933(03)00203-3.

Thiffeault, J.L., 2022. The mathematics of burger flipping. Phys. Nonlinear Phenom. 439, 133410. https://doi.org/10.1016/j.physd.2022.133410.

Touffet, M., Trystam, G., Vitrac, O., 2020. Revisiting the mechanisms of oil uptake during deep-frying. Food Bioprod. Process. 123, 14–30. https://doi.org/10.1016/j.fbp.2020.06.007.

Troncoso, E., Pedreschi, F., 2009. Modeling water loss and oil uptake during vacuum frying of pre-treated potato slices. LWT Food Sci. Technol. (Lebensm.-Wiss.-Technol.) 42, 1164–1173. https://doi.org/10.1016/j.lwt.2009.01.008.

USDA, 2025. Food Safety and Inspection Service. https://www.fsis.usda.gov/food-safety/safe-food-handling-and-preparation/food-safety-basics/safe-temperature-chart (Last Accessed 21 May 2026).

van Koerten, K., Schutyser, M., Somsen, D., Boom, R., 2015. Crust morphology and crispness development during deep-fat frying of potato. Food Res. Int. 78, 336–342. https://doi.org/10.1016/j.foodres.2015.09.022.

van Koerten, K.N., Somsen, D., Boom, R.M., Schutyser, M.A.I., 2017. Modelling water evaporation during frying with an evaporation dependent heat transfer coefficient. J. Food Eng. 197, 60–67. https://doi.org/10.1016/j.jfoodeng.2016.11.007.

Verboven, P., Defraeye, T., Datta, A.K., Nicolai, B., 2020. Digital twins of food process operations: the next step for food process models? Curr. Opin. Food Sci. 35, 79–87. https://doi.org/10.1016/j.cofs.2020.03.002.

Vitrac, O., Dufour, D., Trystram, G., Raoult-Wack, A., 2002. Characterization of heat and mass transfer during deep-fat frying and its effect on cassava chip quality. J. Food Eng. 53, 161–176. https://doi.org/10.1016/S0260-8774(01)00153-4.

Vitrac, O., Trystram, G., Raoult-Wack, A., 2003. Continuous measurement of convective heat flux during deep-frying: validation and application to inverse modelling. J. Food Eng. 60, 111–124. https://doi.org/10.1016/S0260-8774(03)00024-4.

Wang, X., McClements, D.J., Xu, Z., Meng, M., Qiu, C., Long, J., Jin, Z., Chen, L., 2023. Recent advances in the optimization of the sensory attributes of fried foods: appearance, flavor, and texture. Trends Food Sci. Technol. 138, 297–309. https://doi.org/10.1016/j.tifs.2023.06.012.

Xie, D., Guo, D., Guo, Z., Hu, X., Luo, S., Liu, C., 2022. Reduction of oil uptake of fried food by coatings: a review. Int. J. Food Sci. Technol. 57, 3268–3277. https://doi.org/10.1111/ijfs.15266.

Xu, X., Liu, X., Zhang, J., Liang, L., Wen, C., Li, Y., Shen, M., Wu, Y., He, X., Liu, G., Xu, X., 2023. Formation, migration, derivation, and generation mechanism of polycyclic aromatic hydrocarbons during frying. Food Chem. 425, 136485. https://doi.org/10.1016/j.foodchem.2023.136485.

Yamsaengsung, R., Moreira, R., 2002a. Modeling the transport phenomena and structural changes during deep fat frying: Part I: model development. J. Food Eng. 53, 1–10. https://doi.org/10.1016/S0260-8774(01)00134-0.

Yamsaengsung, R., Moreira, R., 2002b. Modeling the transport phenomena and structural changes during deep fat frying: Part II: model solution and validation. J. Food Eng. 53, 11–25. https://doi.org/10.1016/S0260-8774(01)00135-2.

Yildiz, A., Echegaray, N., Öztekin, S., Lorenzo, J., 2024. Quality and stability of frying oils and fried foods in ultrasound and microwave–assisted frying processes and hybrid technologies. Compr. Rev. Food Sci. Food Saf. 23, e13405. https://doi.org/10.1111/1541-4337.13405.

Zhang, W., Chen, J., Yue, Y., Zhu, Z., Liao, E., Xia, W., 2020. Modelling the mass transfer kinetics of battered and breaded fish nuggets during deep-fat frying at different frying temperatures. J. Food Qual. 2020, 8874163. https://doi.org/10.1155/2020/8874163.

Ziaiifar, A.M., Achir, N., Courtois, F., Trezzani, I., Trystram, G., 2008. Review of mechanisms, conditions, and factors involved in the oil uptake phenomenon during the deep-fat frying process. Int. J. Food Sci. Technol. 43, 1410–1423. https://doi.org/10.1111/j.1365-2621.2007.01664.x.

Modeling of microwave and radio-frequency processes in the food industry

5

Samira Beikzadeh[1] *and Seid Mahdi Jafari*[2]
[1]Department of Nutrition and Food Sciences, Maragheh University of Medical Sciences, Maragheh, Iran; [2]Department of Food Materials and Process Design Engineering, Faculty of Food Science and Technology, Gorgan University of Agricultural Sciences and Natural Resources, Gorgan, Iran

5.1 Introduction

Advances in high-frequency engineering have been shaped by the parallel development of electronic device technologies and the sustained global need for greater communication capacity to support speech, digital data, and multimedia transmission since the late twentieth century. For much of its early history, microwave (MW) technology was predominantly associated with military and defense-related systems. This situation changed substantially with the rapid expansion of civilian communication applications, including wireless paging systems, cellular communication, video broadcasting, and both wired and wireless computer networking. Today, such systems operate across a wide range of environments, encompassing commercial office spaces, industrial and manufacturing sites, municipal infrastructure, and residential settings. The diversity of these applications, combined with large-scale production demands, has driven major improvements in cost-efficient manufacturing techniques for MW and radio-frequency (RF) components. These developments have, in turn, reduced the overall cost of deploying numerous RF and MW-based services, enabling widespread access to technologies such as portable global positioning system (GPS) navigation devices, automotive radar systems for collision prevention, and high-speed broadband digital communication. The inherently high operating frequencies of MW systems make them particularly suitable for modern communication and sensing tasks, as they allow both the allocation of multiple independent channels and the availability of substantial bandwidth within each channel for high-rate data transfer (Golio and Golio, 2018; Beikzadeh et al., 2020) Fig. 5.1 indicates coupled electromagnetic-thermal-mass transfer model.

MW signals, operating within the frequency range of 300 to 300,000 MHz, together with RF waves spanning 0.003–300 MHz, occupy distinct portions of the electromagnetic (EM) spectrum. When MW and RF energy interact with dielectric materials, such as food products, thermal energy is generated primarily through dipole

Modeling of Unit Operations in the Food Industry. https://doi.org/10.1016/B978-0-12-822180-8.00013-4

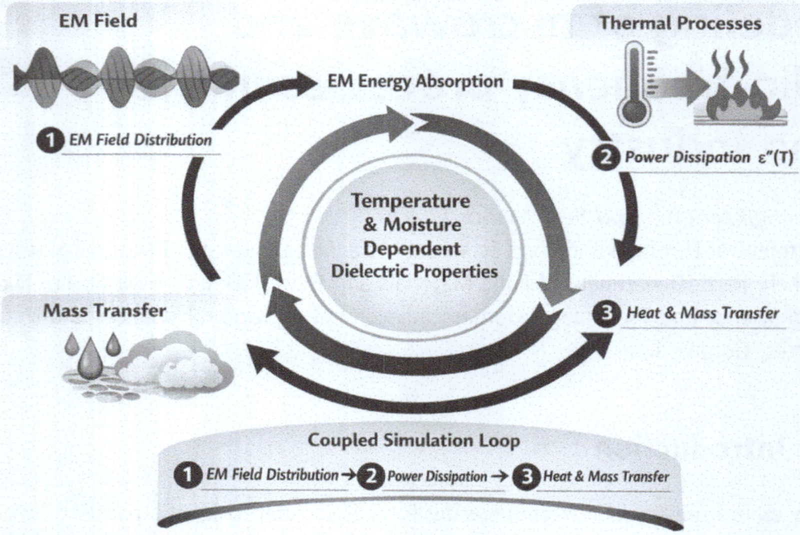

Figure 5.1 Coupled electromagnetic-thermal-mass transfer model.

rotation and ionic polarization mechanisms. Compared with conventional thermal processing techniques, the application of MW and RF heating offers notable advantages, including shorter processing durations, reduced equipment footprint, and enhanced product quality (Ramaswamy and Tang, 2008). RF and MW heating methods (commonly described as dielectric or dielectric-loss heating) produce heat internally as a result of molecular friction caused by rapidly alternating electric fields. Owing to characteristics such as rapid energy deposition and volumetric heating, dielectric heating has found widespread application across various industries. In food-processing operations, this approach has been successfully implemented for processes including drying, baking, and the thawing of frozen meat products (Beikzadeh et al., 2020).

The primary factor distinguishing RF engineering from MW engineering lies in the operating frequency of the signals involved and, consequently, their associated wavelengths (λ). This distinction originates from the finite speed at which EM waves propagate, which governs the temporal and spatial behavior of voltages and currents within electronic systems. In free space, the wavelength is defined by $\lambda = c/f$, where f denotes the signal frequency and c represents the speed of light. In low-frequency and conventional RF circuits (excluding specific elements such as antennas) the signal wavelength is typically much larger than the physical dimensions of the circuit components. By contrast, in MW systems, component dimensions are often on the same order of magnitude as the wavelength. When component lengths are significantly smaller than the wavelength ($\ell < \lambda/10$), phase variations across the component remain relatively minor. As component dimensions increase to become comparable to or exceed the wavelength, however, phase differences between opposite ends of the component grow increasingly significant.

This behavior forms the basis for differing analytical and design approaches in RF and MW engineering. In RF circuit design, components and interconnections are generally small enough relative to the wavelength to justify the use of lumped-element models, for which Kirchhoff's voltage and current laws remain applicable at every instant. Although parasitic inductance and capacitance must be included to capture frequency-dependent behavior, these effects can usually be represented accurately through equivalent lumped-element circuits. A commonly accepted guideline for the validity of this approximation is that component dimensions should not exceed one-10th of the wavelength at the operating frequency. At MW frequencies, where component sizes often surpass this limit, the finite propagation time of EM waves constitutes a significant fraction of the signal period. Under such conditions, lumped-element descriptions no longer provide sufficient accuracy, and distributed-element models become necessary to properly characterize electrical behavior. Accounting explicitly for propagation delays and distributed effects is therefore a defining aspect of MW engineering methodology (Golio and Golio, 2018).

5.2 Fundamentals of MW and RF heating

5.2.1 Principles of electromagnetic heating

Wide range of modern and widely used technologies rely on EM phenomena, including systems for radio communication, medical imaging based on X-rays, and optical signal processing. The present study focuses specifically on the RF and MW portion of the EM spectrum, which is defined here as extending from 3 MHz to 300 GHz. Similar to other wave-related physical phenomena, EM waves represent variations that evolve over both space and time. Unlike mechanical waves, however, EM waves possess the unique ability to propagate not only through material media but also through free space, enabling transmission in a vacuum (Balanis, 2011). The theoretical foundation of EM wave behavior was established in the mid-nineteenth century by Maxwell (1865), who unified the empirical laws formulated by Faraday, Ampère, and Gauss into a comprehensive system of coupled differential equations, now universally referred to as Maxwell's equations. These equations describe how EM radiation originates from the mutual interaction between time-dependent electric and magnetic fields. In an EM wave, the electric and magnetic field components oscillate orthogonally to each other and are both perpendicular to the direction in which energy propagates (Balanis, 2011). Maxwell's framework provides a complete description of EM wave propagation, including general cases of elliptically polarized waves. Technologies operating within the RF and MW frequency ranges encompass numerous applications, such as radar systems, radio and television broadcasting, mobile communication devices, GPS antennas, sensing technologies, and domestic MW heating equipment.

5.2.1.1 Maxwell's equations and electromagnetic wave propagation

EM fields are capable of being generated and transmitting energy through both empty space and material environments. The behavior of EM phenomena is fundamentally described by Maxwell's equations, together with the constitutive relations that characterize the response of the surrounding medium. When expressed in contemporary mathematical form and formulated within the Giorgi, or equivalently the MKS, system of units, these equations provide a consistent framework for analyzing EM field behavior (Ghosh, 2002), The differential formulation of Maxwell's equations may be expressed as follows:

$$\nabla \times E = -\partial \mathrm{B}/\partial \mathrm{t} \tag{5.1}$$

$$\nabla \times \mathrm{H} = \partial \mathrm{D}/\partial \mathrm{t} + j \tag{5.2}$$

$$\nabla .\ \mathrm{D} = p \tag{5.3}$$

$$\nabla \cdot \mathrm{B} = 0 \tag{5.4}$$

In these expressions, vector variables are represented using boldface notation, where.

E (V/m) denotes the electric field
D (C/m^2) represents the electric displacement
H (A/m) corresponds to the magnetic field
B (Wb/m^2) denotes the magnetic induction
J (A/m^2) represents the electric current density
p (C/m^2) represents the electric charge density.

In general, all of the aforementioned quantities vary with both position and time. By taking the divergence of Eq. (5.2) (noting that the divergence of a curl is always zero) and combining it with Eq. (5.3), one arrives at the continuity equation governing the electric current:

$$\nabla \times\ J = -\partial p/\partial \mathrm{t} \tag{5.5}$$

The continuity equation can be understood as a local, differential expression of the principle of electric charge conservation. Specifically, the rate at which charge density decreases within an infinitesimal volume is exactly balanced by the current flowing out of that volume. While the equations presented above represent Maxwell's laws in differential form, their integral counterparts are often preferred for computational analyses or for deriving numerical formulations, typically through the application of Stokes' theorem (Sorrentino and Bianchi, 2010):

$$\oint_{\mathrm{c}} \mathrm{E}\cdot \mathrm{dl} = -\ \partial/\partial \mathrm{t} \int_{\mathrm{S}} \mathrm{B}\cdot \mathrm{dS} \tag{5.6}$$

$$\oint_{\mathrm{c}} \mathrm{H.dl} = \partial/\partial \mathrm{t} \int_{\mathrm{S}} \mathrm{D.\ dS} + \int_{\mathrm{S}} \mathrm{J.\ dS} \tag{5.7}$$

$$\oint_{\mathrm{S}} \mathrm{D.\ dS} = \int_{\mathrm{V}} p.\mathrm{dV} \tag{5.8}$$

$$\oint_{\mathrm{S}} \mathrm{B.\ dS} = 0 \tag{5.9}$$

Likewise, by applying the divergence theorem, the continuity equation can be expressed in its integral form (Sorrentino and Bianchi, 2010):

$$\oint_S \text{J. dS} = -\partial/\partial t \int_V p \, dV$$

Maxwell's Eqs. (5.1–5.4) must be complemented by the constitutive relations that describe the EM characteristics of the medium in which the fields exist. For linear, isotropic materials (including vacuum) these constitutive relations reduce to simple scalar, linear expressions linking the electric displacement to the electric field and the magnetic induction to the magnetic field:

$$D = \varepsilon E \tag{5.10}$$

$$B = \mu H \tag{5.11}$$

A region without sources, which is typical in waveguides and transmission lines

$$\nabla \cdot \mathbf{E} = 0 \tag{5.12}$$

$$\nabla \cdot \mathbf{H} = 0 \tag{5.13}$$

$$\nabla \times \mathbf{E} = -j\omega\mu\mathbf{H} \tag{5.14}$$

$$\nabla \times \mathbf{H} = j\omega\varepsilon\mathbf{E} \tag{5.15}$$

The resulting wave equation for RF and MW applications

$$\nabla^2\mathbf{E} + k^2\mathbf{E} = 0 \tag{5.16}$$

$$\nabla^2 H + k^2 H = 0 \tag{5.17}$$

where

$$k = \omega\sqrt{\mu\varepsilon} \tag{5.18}$$

5.2.1.2 Frequency ranges and penetration depth comparison (MW vs. RF)

In general, a significant portion of an EM wave is reflected at the interface in an optical manner, while another substantial portion enters the material, changing direction due to refraction the same principle responsible for the apparent displacement of objects when viewed through water. Once inside the material, the transmitted energy is gradually absorbed or attenuated, causing the wave amplitude to diminish with increasing depth. This attenuation can be quantified by the penetration depth, D (also referred to as the skin depth), which is defined here as the depth at which the wave's power decreases by 86.5% (equivalent to $1/e^2$). In other words, 86.5% of the wave's power is absorbed between the surface and this depth. At distances significantly greater than D (e.g., beyond 2D), the wave becomes very weak, and minimal heating occurs, as observed in applications like diathermy.

The extent of both reflection and penetration depends strongly on the dielectric properties of the material. Two key parameters characterize the MW behavior of a material at a given frequency: the dielectric constant (more precisely, the relative dielectric constant, ε) and the loss tangent (tan δ). The dielectric constant provides critical information about the material's reflection characteristics and the wavelength of the wave within it. The wavelength inside the material is reduced from its free-space value, λ_0, by a factor of $\sqrt{\varepsilon}$, that is,

$$\lambda = \lambda_0 / \sqrt{\varepsilon} \tag{5.19}$$

or biological tissues, the square root of the dielectric constant, $\sqrt{\varepsilon}$, can be relatively high (sometimes reaching values up to 9) resulting in significantly reduced wavelengths within the tissue. For instance, at a frequency of 2450 MHz, the free-space wavelength of $\lambda_0 = 12.25$ cm is shortened to approximately $\lambda = 2$ cm inside the tissue. When the wave strikes the surface normally (perpendicularly), the fraction of power reflected relative to the incident power can be estimated as follows:

$$P_{reff}/P_{ine} = ((1-\sqrt{\varepsilon}) / (1+ \sqrt{\varepsilon}))^2 \tag{5.20}$$

For biological tissues, the reflected portion of the wave can reach values as high as 60%. The loss tangent, tan δ, provides a measure of the material's penetration.

$$D = 0.225\ \lambda_0 / \sqrt{\varepsilon} \sqrt{\ (\sqrt{\ (1 + \tan^2\delta) - 1)})} \tag{5.21}$$

or for small tanδ << 1

$$D = 0.318\ \lambda_0 / \sqrt{\varepsilon} \tan\delta \tag{5.22}$$

A higher value of tan δ corresponds to a reduced penetration depth, while an increased dielectric constant also decreases D. As the EM wave is attenuated within the material, the absorbed energy results in localized heating, quantified by the specific absorption rate (SAR), which is expressed as follows:

$$SAR = (1/2\rho)\ \omega\ \varepsilon_0\ \varepsilon \tan\delta\ E^2_{in} \text{ watts/kg.} \tag{5.23}$$

where:

SAR represents the power delivered to the absorbing material by the electric field within the body.

p denotes the mass density of the body, in kg/m^3
ε_0 denotes the permittivity of free space, in farads/m.
ε denotes the relative dielectric constant
tan δ denotes the loss tangent
ω denotes the radian frequency given by $\omega = 2\pi$ f

E_{in} the electric field (in volts per meter) at a specific location within the body, where the subscript "in" indicates that this internal field differs from the incident field outside the body.

Properties of various materials at the commonly used MW frequency of 2450 MHz are well documented. While materials such as paper and plastics are largely transparent to these waves, food items and biological tissues act as strong absorbers. With these fundamental concepts of EM waves in mind, it becomes useful to consider the typical magnitudes of key parameters across the EM spectrum. The spectrum classifies different types of energy according to their frequency (in Hz) and wavelength (in cm), where frequency and wavelength are inversely related, and either parameter alone can describe the wave. For instance, alternating current (household electricity) operates at 60 Hz, corresponding to 60 cycles per second. In this case, the extremely long wavelength, approximately 5×10^8 cm (5000 km), is less intuitive than the concept of 60 oscillations per second. At the opposite end of the nonionizing spectrum lies visible light, with frequencies around 10^{15} Hz. While such high frequencies are difficult to visualize, the corresponding wavelength, roughly 1/100,000 of a centimeter (about one-thousandth the diameter of a human hair) is easier to appreciate. Between these extremes lie MWs, which combine aspects of both ends of the spectrum: both electric and magnetic field components, along with radiation parameters such as power flux density (mW/cm^2), are relevant in describing their behavior.

Although optical phenomena, similar to visible or infrared (IR) light, still play a minor role, MW interactions also require consideration of lower-frequency concepts such as voltage, field, and current. Applications such as television broadcasting fall within this frequency range, while MW ovens operate at 915 MHz and 2450 MHz, two of the standard ISM bands (industrial, scientific, and medical). The boundaries of the MW portion of the EM spectrum are somewhat arbitrary, but can be defined based on the frequency range in which EM energy couples most effectively to macroscopic objects of practical relevance. Using this approach, MWs can be considered to span from roughly 10 MHz ($\lambda \approx 30$ m) to 100 GHz ($\lambda \approx 3$ mm). The biological and heating effects of nonionizing radiation are strongly influenced by the penetration depth of the waves, which typically reaches a maximum within the MW range. At low frequencies, EM fields are largely reflected or shunted by the conductive properties of biological tissue, while the dielectric constant also increases significantly. Conversely, at high frequencies, energy is absorbed within a shallow skin depth. Consequently, substantial internal field strengths occur only within a finite frequency window, which can be designated as the MW region. This range is dependent on the size of the organism, being approximately 100 MHz for humans and around 3 GHz for small animals such as mice (Osepchuk, 1979).

5.2.2 Dielectric properties of food materials

RF and MW heating have emerged as advanced thermal-processing techniques in the food engineering sector. These methods are primarily applied to dielectric materials, which inherently possess low electrical conductivity. Typically, materials with poor electrical conductivity also exhibit low thermal conductivity, leading to slow heat transfer when conventional methods such as conduction or convection are used (Awuah et al., 2014). In practical terms, heating dielectric materials using an external heat source requires considerable time, as heat must move from the surface to the interior through temperature gradients. In contrast, dielectric heating offers rapid, volumetric energy deposition, making it highly effective for such materials. Since most food products are dielectric in nature, this technique holds substantial promise for the food-processing industry. RF and MW energy occupy different portions of the EM spectrum, resulting in distinct wavelengths. RF waves generally span from 3 kHz to 300 MHz, whereas MWs cover the range from 300 MHz to 300 GHz. Compared with MWs, RF waves have lower frequencies and longer wavelengths. Both RF and MW bands overlap with radar frequencies and may interfere with communication systems such as mobile phone networks. Consequently, regulatory agencies restrict the use of certain frequencies for industrial, scientific, and medical (ISM) applications. In the United States, ISM usage is limited to 13.56 MHz, 27.12 MHz, and 40.68 MHz for RF, and 915 MHz and 2450 MHz for MWs. As frequency increases, the corresponding wavelength decreases, following the inverse relationship:

$$c = \lambda f \tag{5.24}$$

Here, c represents the speed of light ($c = 3.0 \times 10^8$ m/s), λ is the wavelength (in meters), and f is the frequency of the EM wave (in Hz). During dielectric heating, dielectric materials, such as food products, are exposed to an alternating electric field. Many food components contain polar molecules, such as water, which possess electric dipole moments meaning the centers of positive and negative charges do not coincide. When subjected to an electric field, these polar molecules attempt to align with the field, causing their positive and negative charge centers to separate and induce polarization. In an alternating electric field, the molecules rotate continuously to follow the changing field direction, a process known as dipole rotation (Marra et al., 2009; Sheikhi et al., 2020a, 2020b). The friction between molecules during this motion converts EM energy into heat, raising the temperature of the material. In addition to dipole rotation, dissociated ions in food also respond to the applied electric field. Their movement, which reverses direction as the field alternates, generates heat through collisions with surrounding molecules. This process is termed ionic conduction (Tang et al., 2002). Both dipole rotation and ionic conduction serve as the primary mechanisms underlying RF and MW heating (Piyasena and Dussault, 2003), and their efficiency depends on both temperature and the frequency of the applied field. Table 5.1 summarizes the dielectric properties (ε' and ε'') of various food materials as a

Table 5.1 Dielectric properties of food materials at different moisture contents and temperatures.

Material/Food matrix	Moisture content (%)	Temperature (°C)	Dielectric constant (ε')	Dielectric loss factor (ε'')	Key observations	Reference
Potato starch	20–50	20–80	4–12	0.2–1.5	ε' and ε'' increase with moisture	Zhu and Guo (2017)
Fruit by-products	10–60	25–90	5–15	0.3–2.0	ε'' shows nonlinear temperature dependence at high moisture	Solyom et al. (2020)
Vegetable processing by-products	15–55	30–85	6–14	0.4–2.2	High fiber reduces ε'' due to limited free water	Solyom et al. (2020)

function of moisture content and temperature. The data highlight the dependence of dielectric behavior on water content, temperature, and food composition, illustrating trends such as increased ε' and ε'' with higher moisture and the nonlinear effects of temperature at high moisture levels. These properties are crucial for accurate modeling and simulation of RF and MW heating processes.

5.2.2.1 Definition and significance (ε', ε'', loss factor)

The EM properties of a material play a crucial role in determining the effectiveness of dielectric heating treatments and serve as essential input parameters for computational modeling of the process. These properties primarily include permeability and permittivity. For the majority of dielectric materials, the contribution of permeability to heating is negligible or very small, and as a result, it is often disregarded in analyses (Zhang and Datta, 2001), Permittivity is a complex parameter that characterizes a material's dielectric behavior, including both the reflection of EM waves at interfaces and the attenuation of wave energy as it propagates through the material. Mathematically, permittivity can be expressed as:

$$\varepsilon = \varepsilon_0\,\varepsilon_r = \varepsilon_0\,(\varepsilon_r' - j\,\varepsilon_r'') = \varepsilon' - j\,\varepsilon'' \tag{5.25}$$

where ε denotes the permittivity of a material, and ε_0 represents the relative complex permittivity, and, $\varepsilon_0 = 8.854 \times 10^{-12}$ F/m. The components ε_r', and ε_r'', are the relative dielectric constant and relative dielectric loss factor, respectively. ε' and ε'' are commonly called the dielectric constant and dielectric loss factor, respectively, and $j = \sqrt{-1}$. ε_r', or ε' is related to the ability of the material to store electric energy that would affect the electric field distribution and the phase of waves traveling through the material. ε_r'' or ε'' describes the ability of materials to dissipate energy due to various polarization mechanisms. This ability would influence energy absorption and attenuation that finally gives rise to heat generation in the material. In practical applications, the relative dielectric constant and relative dielectric loss factor are the most frequently used parameters. Another key property is the loss tangent, which quantifies how efficiently a material dissipates energy (i.e., its power loss). The loss tangent, denoted as tan δ, is defined as the ratio of the dielectric loss factor to the dielectric constant, or equivalently, as the ratio of the relative dielectric loss factor to the relative dielectric constant. This relationship is expressed mathematically in the following equation (Piyasena and Dussault, 2003):

$$\tan\delta = \varepsilon''/\,\varepsilon' = \varepsilon_r''/\varepsilon_r' \tag{5.26}$$

Within the EM wave spectrum spanning 0.1 MHz to 1000 GHz, the contributions to the dielectric loss factor arise from both polarization and electrical conduction phenomena. Polarization can be further categorized into several types, including dipolar, atomic, electronic, and the Maxwell–Wagner mechanisms (Metaxas and Meredith,

1983; Kuang and Nelson, 1998). The relative significance of each polarization or conduction mechanism depends on factors such as the operating frequency, temperature, electrical conductivity, moisture content, and the size of the polar molecules. Among these mechanisms, ionic conduction and dipole rotation are the primary contributors to heating in RF and MW applications.

$$\varepsilon_r'' = \varepsilon_{rd}'' + \varepsilon_{r\sigma}'' = \varepsilon_{rd}'' + \sigma/(\varepsilon_0\ \omega) \tag{5.27}$$

Here, the subscripts d and σ denote the contributions from dipole rotation and ionic conduction, respectively. The parameter σ represents the ionic conductivity of the material (in S/m), $\omega = 2\pi f$ corresponds to the angular frequency of the waves (Hz). In aqueous solutions, the overall dielectric loss factor results from the combined effects of dipole rotation and ionic conduction across the relevant frequency ranges. At elevated temperatures, both the ionic conductivity contribution and the relaxation peak of the solution shift toward higher frequencies. This behavior arises from the decreased viscosity and increased mobility of ions and dipoles at higher temperatures (Mudgett, 1986).

5.2.2.2 Temperature, moisture, and composition dependence

The dielectric characteristics of food materials, including the dielectric constant (ε') and dielectric loss factor (ε''), are highly sensitive to temperature, moisture content, and overall composition. In RF and MW heating processes, these dependencies play a crucial role, as they directly govern the absorption of EM energy and its conversion into heat within the material. Experimental investigations across different food systems have consistently demonstrated that both ε' and ε'' change with variations in moisture and temperature, primarily due to alterations in water mobility and molecular interactions within (Zhu and Guo, 2017; Solyom et al., 2020).

Moisture content plays a crucial role in determining the dielectric properties of materials, as water is a highly polar molecule exhibiting a large dielectric constant and pronounced dipole relaxation. With increasing moisture levels, both ε' and ε'' generally rise because a greater amount of free water is available to interact with the EM field. This behavior has been consistently observed in a variety of materials, including fruit and vegetable processing by-products and potato starch, where higher water content leads to elevated ε' and ε'' values due to enhanced dipolar polarization and ionic conduction (Zhu and Guo, 2017; Solyom et al., 2020).

The influence of temperature on dielectric properties is intricate and often interacts with moisture content. At moderate moisture levels, raising the temperature generally leads to increases in both ε' and ε'', as thermal agitation enhances molecular mobility and lowers water viscosity, facilitating the alignment of dipoles and the movement of ions with the alternating electric field. However, at very high moisture contents, the temperature effect on ε' can be reversed. For example, in studies on fruit and vegetable by-products, ε' increased with temperature below a critical moisture threshold but decreased at higher moisture levels. This behavior arises from the competing contributions of bound and free water, as well as temperature-dependent changes in ionic mobility (Solyom et al., 2020; Bai et al., 2022).

The composition of food, including constituents such as salts, sugars, fats, and fibers, has a substantial impact on dielectric properties. Ionic species, such as dissolved salts, enhance the contribution of ionic conduction to ε'' and can intensify the temperature dependence of the dielectric loss factor. In contrast, foods with high fiber or lipid content limit the availability of free water, leading to lower dielectric losses. For instance, variations in dielectric behavior observed among different food waste materials have been linked to their chemical composition, with high-fiber or high-lipid materials exhibiting reduced ε'' due to decreased dipolar polarization and restricted ionic mobility (Solyom et al., 2020). Furthermore, the interplay among temperature, moisture, and food composition results in nonlinear effects on the dielectric response.

At lower moisture levels, bound water molecules contribute more significantly to dielectric loss, causing ε'' to rise with temperature due to enhanced dipole rotation. Conversely, at higher moisture contents or in materials with substantial ionic content, ionic conduction may become the dominant mechanism, leading to different temperature-dependent trends in ε'' that are influenced by both moisture and composition (Ling et al., 2018). A thorough understanding of these relationships is critical for accurate modeling and simulation of RF and MW heating processes. Incorporating precise dielectric measurements that reflect variations in temperature, moisture, and composition enhances predictions of energy absorption, heating uniformity, and overall process optimization (Tıraş et al., 2019).

5.2.3 Heat and mass transfer in electromagnetic fields

5.2.3.1 Coupled electromagnetic–thermal modeling

Coupled EM–thermal modeling integrates the simulation of EM wave propagation with the resulting heat generation and transport within materials. In RF and MW heating, this interaction is essential because the absorption of EM energy by a dielectric medium produces heat, which in turn alters the local temperature, modifies dielectric properties, and affects subsequent EM wave propagation and energy absorption. Accurate modeling requires simultaneously solving Maxwell's equations and the heat conduction equation, incorporating temperature-dependent dielectric properties such as $\varepsilon'(T)$ and $\varepsilon''(T)$, to predict the spatial and temporal evolution of temperature within the material (Thuéry and Grant, 1992).

Numerical approaches for these coupled EM–thermal problems commonly use either the Finite Element Method (FEM) or the Finite-Difference Time-Domain (FDTD) method. FEM discretizes the spatial domain into irregular meshes and iteratively solves the coupled partial differential equations (PDEs), allowing complex geometries, inhomogeneous materials, and temperature-dependent properties to be considered. FEM has been successfully applied in one-, two-, and three-dimensional configurations to simulate heating of materials such as concrete, zirconia, alumina, meat, water, SiC, fat, and muscle tissue.

In contrast, FDTD discretizes both space and time to directly simulate the dynamic evolution of EM fields and thermal responses (Torres and Jecko, 2002; Iskander et al., 1994). The coupling between EM and thermal fields is typically solved through an iterative three-step process (1) computation of the EM field distribution (2) evaluation of local power dissipation using the electric field and $\varepsilon''(T)$, and (3) solution of the heat conduction equation to update the temperature. This loop continues until convergence, ensuring consistent application of temperature-dependent dielectric properties (Loyd et al., 1994; Paulsen et al., 2002).

In materials undergoing moisture movement, evaporation, or phase change, mass transfer can also be incorporated. For example, during food drying or in lignocellulosic foams, moisture transport affects both dielectric and thermal behavior, influencing EM absorption and heat generation. Including mass transfer provides more realistic simulations when heat and moisture interactions are significant (Huang et al., 2018; Tauhiduzzaman et al., 2021). Temperature, moisture, and composition dependencies are critical in coupled EM–thermal modeling. Dielectric properties such as ε' and ε'' vary considerably with moisture content, temperature, and chemical composition. Water content, ionic species, sugars, fats, and fibers modulate dipole rotation and ionic conduction, producing nonlinear effects on dielectric losses. Accounting for these dependencies is essential to accurately predict heating patterns and process outcomes (Solyom et al., 2020). Computational complexity is a major challenge in coupled simulations.

Because EM fields evolve much faster than heat diffusion, approximations like the quasi-steady-state assumption are often employed, assuming the EM field reaches steady state within each thermal time step. Additional techniques to accelerate simulations include temperature-distribution corrections and reduced time-scale methods (Torres and Jecko, 2002; Sekkak et al., 2002). Understanding coupled EM-thermal–mass transfer phenomena is crucial for reliable modeling, sensitivity analyses, and optimization of RF and MW processes. Incorporating temperature- and moisture-dependent dielectric properties, along with composition effects, enables precise predictions of energy absorption, heating uniformity, and process efficiency (Huang et al., 2018).

5.2.3.2 Boundary conditions and heat distribution

For the EM component, common boundary conditions include impedance boundaries and port excitations. Impedance boundary conditions model metallic cavity walls as surfaces with negligible EM field penetration, effectively representing the cavity as either a perfect or lossy conductor. These conditions mathematically link the tangential components of the electric and magnetic fields at the boundary, enforcing either reflection or absorption characteristics. They are particularly crucial for simulating enclosed MW systems where field leakage is limited (Tauhiduzzaman et al., 2021; Li and Zhou, 2024). Port boundary conditions define the incoming EM energy at interfaces such as waveguides or antennas. In rectangular waveguides, which are frequently employed in MW heating setups, the dominant TE_{10} mode is typically applied at the port. This ensures controlled power input into the simulation domain and guarantees physically consistent entry and exit of EM energy, directly influencing field distribution and subsequent heat generation within the material (Tauhiduzzaman et al., 2021; Li and Zhou, 2024).

For the thermal field, boundary conditions may include thermally insulating walls, specified temperature surfaces, convective heat transfer, or radiative heat loss, depending on the physical system. In many MW cavity simulations, walls are assumed to be thermally insulating, implying zero heat flux across them. This assumption simplifies calculations by thermally isolating the material from its surroundings, though more realistic models can incorporate convective cooling or radiation. When modeling heat conduction with phase changes or interactions with an external environment, boundary conditions involving convection or prescribed temperature gradients are applied. These are especially relevant for processes such as food heating or thawing, where layered structures and phase transitions (e.g., ice melting) introduce moving boundaries that require dynamic definitions in the heat equation (Li and Zhou, 2024).

Mixed boundary conditions**,** where different types of conditions are applied across various parts of the domain, can enhance the accuracy of heat distribution predictions. Data-driven reduced models that combine fixed temperature and heat flux conditions have been shown to improve computational efficiency while maintaining reliable temperature estimations. Overall, correctly specifying boundary conditions for both EM and thermal fields is essential for accurate simulation of heat distribution in RF and MW heating. These conditions govern energy flow across domain boundaries and determine the internal temperature evolution, thereby affecting the coupled interactions between EM energy absorption and heat transfer (Zhang et al., 2020; Zhong and Liang, 2021).

5.3 Theoretical modeling frameworks of MW and RF processes

5.3.1 Coupled Maxwell–heat transfer equations

The theoretical foundation for simulating RF and MW heating relies on the coupled solution **of Maxwell's equations** for the EM fields and the heat transfer equation that governs temperature changes within the material. In these models, the EM field solution provides the spatial distribution of the electric field, which is then employed to determine the local power dissipation due to dielectric losses. This power dissipation serves as the source term in the heat transfer equation, driving temperature evolution in the medium. In a linear and isotropic medium, Maxwell's curl equations can be expressed in their classical form as follow:

$$-B\partial/t\partial = E \times \nabla \tag{5.28}$$

$$D\partial/t\partial + J = H \times \nabla \tag{5.29}$$

Here, E and H denote the electric and magnetic field intensities, while D and B represent the electric and magnetic flux densities, and J corresponds to the current density. The coupling between the EM and thermal fields is achieved through temperature-dependent dielectric properties, $\varepsilon'(T)$ and $\varepsilon''(T)$, which affect the EM field distribution and consequently determine the local heat generation within the material (Alpert and Jerby, 1999; Pitchai et al., 2012).

The thermal response of the material is described by the heat conduction equation, which includes a source term accounting for the power dissipated by EM fields:

$$Q_{EM} + (k\nabla T) \cdot \nabla = T\partial/t\partial\ {}_{\rho}C_{p} \tag{5.30}$$

In this equation, ρ represents the material density, C_p is the specific heat capacity, k denotes thermal conductivity, T is the local temperature, and Q_{EM} corresponds to the volumetric heat source resulting from EM energy absorption, which is typically calculated using either the Poynting vector or the dielectric loss term (Tauhiduzzaman et al., 2021; Zhang et al., 2020).

5.3.2 Governing equations and simplifications

The coupled system of EM and thermal phenomena is formulated as a set of PDEs that are typically addressed through numerical methods because of their nonlinear and complex nature. Maxwell's equations define how EM fields vary across the simulation domain, while the heat conduction equation governs how temperature evolves spatially and temporally. Due to the vastly different characteristic time scales (EM oscillations occurring within nanoseconds and thermal diffusion over seconds or longer) modeling often incorporates simplifying strategies. One commonly adopted method is the quasi-steady-state assumption for the EM field, in which Maxwell's equations are solved in the frequency domain assuming a harmonic steady field at each thermal time increment, while the thermal equation is integrated in the time domain to capture transient temperature changes. This approach significantly reduces computational load while preserving reliable accuracy in predicting the coupled EM–thermal behavior (Alpert and Jerby, 1999; Pitchai et al., 2012).

Additional simplifications can be applied depending on the system conditions. For scenarios in which temperature varies slowly compared to the period of EM oscillations, a steady-state thermal assumption can be employed, emphasizing the spatial distribution of temperature while neglecting transient effects. In contrast, for rapid heating processes or pulsed RF/MW applications, it may be necessary to solve the fully time-dependent forms of both Maxwell's and heat transfer equations. These transient analyses are essential for capturing dynamic phenomena, including the propagation of thermal waves, temperature-induced changes in dielectric properties, and the feedback interactions between the evolving temperature field and material response (Tauhiduzzaman et al., 2021). In numerous industrial and scientific applications, the coupled EM–thermal framework is often extended to include additional governing equations for mass transport **or** fluid flow, particularly in cases involving moisture migration, phase changes, or convective phenomena. Despite these extensions, the core formulation remains centered on the coupled Maxwell and heat transfer equations, incorporating temperature- and composition-dependent material properties, along with time-scale approximations tailored to the specific EM and thermal dynamics of the system (Tauhiduzzaman et al., 2021).

5.4 Numerical methods for modeling of MW and RF processes

5.4.1 Finite element method

The FEM is among the most widely employed numerical techniques in engineering disciplines. Originally developed for civil and aerospace engineering, its hallmark is the formulation of frequency-domain EM problems in terms of an integral equation, often expressed as the minimization of a functional, rather than directly using the differential form of Maxwell's equations. In contrast to finite-difference (FD) approaches, which discretize space using a Cartesian grid based on the discretized Maxwell's equations, FEM divides the domain into discrete elements volumetric tetrahedra in three dimensions or triangles in two dimensions. Within each element, the unknown quantity whether a potential or a field component is approximated by a polynomial function, with coefficients defined in terms of the values at the element's nodes or vertices. Minimizing the functional then results in a linear system of equations, where the unknowns correspond to the function values at the nodes. One of the main advantages of FEM over FD methods is its flexibility in handling complex geometries, since the tetrahedral mesh can conform to irregular 3D shapes and focus refinement only where necessary. In its basic form, FEM can generate spurious, nonphysical solutions, a challenge that has been mitigated through specialized mathematical techniques. Like other discretization methods such as FD frequency-domain (FDFD), FEM requires solving very large equation systems, which can become computationally demanding for extensive structures. Nevertheless, FEM remains highly effective and is widely implemented in commercial simulation software (Sorrentino and Bianchi, 2010).

5.4.2 Finite difference time domain

The FDTD technique is a widely utilized numerical approach for modeling how EM waves interact with materials in RF and MW applications. Being a time-domain method, FDTD solves Maxwell's curl equations by discretizing both spatial and temporal dimensions on a structured computational grid, allowing explicit updating of electric and magnetic field components over successive time steps. Originally developed by Kane S. Yee, the method has gained popularity due to its conceptual simplicity and its capability to capture transient phenomena and broadband responses without the need for matrix inversion. FDTD provides detailed field distributions across the entire simulation domain, making it especially suitable for complex dielectric materials and cavity structures commonly encountered in RF and MW heating scenarios. A primary strength of FDTD in MW modeling lies in its direct computation of time-dependent electric (E) and magnetic (H) fields, which allows visualization of EM wave propagation and identification of resulting heat sources for multiphysics coupling. In coupled EM–thermal simulations, the FDTD-derived power absorption density (e.g., $\varepsilon''|E|^2$) serves as the source term in the heat conduction equation, enabling accurate prediction of temperature distributions and potential hot spots in dielectric materials exposed to MW radiation.

5.4.3 Computational fluid dynamics (CFD) integration

The FDTD approach has been effectively employed in **hybrid simulation frameworks**, where an EM solver based on FDTD is combined with external thermal or Computational Fluid Dynamics (CFD) tools to model heat transfer and fluid motion within the heated domain (Chen et al., 2006). For instance, Kopyt and Celuch (2006) developed a hybrid simulation system integrating a commercial FDTD EM solver with a thermodynamic or CFD solver to capture heat diffusion and validate predictions against experimental measurements. This combined approach demonstrated improved accuracy in modeling MW heating. CFD integration becomes especially critical when convective heat transfer and fluid movement significantly affect the process. In industrial MW systems or reactor designs, where heating generates temperature gradients and fluid flow (e.g., in liquids or suspensions), the spatial EM power distribution obtained from FDTD analysis is mapped as a volumetric heat source in CFD simulations, enabling simultaneous computation of temperature and flow fields. This allows for realistic prediction of multiphysics interactions, such as natural convection induced by MW heating (Kopyt and Celuch, 2006).

In practical multi-physics applications, commercial platforms like COMSOL Multiphysics or ANSYS allow seamless coupling of FDTD or other EM solvers with CFD and thermal modules. These tools facilitate co-solution of EM fields, heat transfer, and fluid flow while maintaining feedback between modules. This capability is essential for analyzing coupled phenomena, such as temperature-dependent changes in dielectric properties and heat-driven fluid motion in MW processes. Despite its advantages, FDTD coupled with CFD and thermal solvers can be computationally intensive, particularly for three-dimensional domains with fine meshes or intricate geometries. Techniques such as domain decomposition, adaptive meshing, and parallel computing are commonly used to reduce simulation time while maintaining solution accuracy. In summary, integrating FDTD EM modeling with CFD provides a robust and comprehensive numerical framework for simulating MW and RF heating processes that involve coupled field, thermal, and fluid dynamics. This multi-physics methodology is increasingly applied in food engineering, chemical reactor design, and materials processing, where EM energy simultaneously drives heat and mass transfer alongside fluid flow.

5.5 Model validation and experimental techniques for MW and RF processes

5.5.1 Temperature mapping, thermal imaging, and fiber-optic sensors

Research on **numerical modeling of MW heating** began in the early 1970s using **FD methods** to simulate MW heating in large slabs of salted ham and beef, with predicted temperature distributions closely matching experimental measurements (Ohlsson and Bengtsson, 1971). Later studies applied FD modeling to high-moisture foods shaped as cylinders and rectangular blocks, achieving good agreement between calculated temperature profiles and experimental data for gel samples with varying moisture content and salt concentrations (Swami, 1982). Additional investigations examined heat and mass transfer during MW heating of materials such as bread, muffins, biscuits, and sandstone, with predicted temperatures generally within 5% of experimental values. FEM provides several advantages over finite difference methods (Zhou et al., 1995) and has been widely applied in MW food processing. FEM simulations have shown that in cylindrical gels or cylinder-shaped particulates, central regions tend to reach higher temperatures compared to the periphery due to volumetric heating effects. Incorporation of heat generation terms into axisymmetric FEM models has accurately captured pronounced heating along central axes.

FEM has also been applied to MW thawing of water and saline cylinders, with the EM field described by Maxwell's equations. Non-uniform temperature distributions were observed, demonstrating the inherent challenges in achieving homogeneous heating. Uniform heating in MW ovens is often enhanced using rotating carousels or mode stirrers; however, modeling these mechanisms mathematically is complex (Zhou et al., 1995). The EM field was described by Maxwell's equations. Non-uniformity of temperature distribution in test samples was also observed. The extent of uniform heating in an MW oven can be provided by either rotating the food on a carousel or using mode stirrers. However, this is considered to be too complex to model mathematically (Zhou et al., 1995). In general, the temperature distribution within a food sample depends on three primary factors: the sample size relative to MW penetration depth, the initial and boundary conditions, and the geometric shape of the sample.

By combining the continuity equation with the thermal energy equation, one obtains the coupled equations governing heat and mass transfer. These equations describe how temperature evolution and material transport interact within the medium, accounting for energy absorption, diffusion, and any phase-change effects:

$$pC_p \partial T/\partial t = \nabla .\ (K\nabla T) + Q \tag{5.31}$$

$$\partial M/\partial t = \nabla .\ (D\ \nabla M) \tag{5.32}$$

The boundary conditions for the coupled heat and mass transfer equations can be divided into two categories: those related to thermal transport and those associated with mass transfer. For the thermal boundary conditions, heat loss may occur due to convection and evaporative processes at the surface of the material. These conditions define the interaction between the material and its surrounding environment, influencing temperature distributions and energy removal from the system. Boundary Condition 1:

$$K\nabla Tn = -h\,(T_a - Ts) - \dot{m}_v/A\ \Delta H_v \text{ for } t > 0 \tag{5.33}$$

Boundary Condition 2: Loss of moisture resulting from evaporation:

$$D\nabla M.n = -h_m\,(M_a - M_S) \text{ for } t > 0 \tag{5.34}$$

The initial temperature and moisture content can vary with position and are expressed as follows:

$$T = T_0\,(x, y, z) \text{ at } t = 0 \tag{5.35}$$

$$M = M_0\,(x, y, z) \text{ at } t = 0 \tag{5.36}$$

5.5.2 Heat generation (Q)

The power absorbed by food during MW heating is described as the volumetric heat generation, denoted by Q. This heat generation depends on the local temperature and moisture content at a given position (x, y, z). To date, the mechanisms governing the power absorbed by food materials are not fully understood. Typically, Maxwell's equations and Lambert's law are employed to estimate the MW power intensity. Because of the inherent complexity in calculating the EM field inside an MW oven (Zhou et al., 1995), Lambert's law, as presented in the following equation, is often applied:

$$P = P_0 \exp\,(-2ad) \tag{5.37}$$

$$\alpha = 2\pi/\lambda\ [\varepsilon'\,(1+\tan^2\delta)^{1/2} - 1]/2]^{1/2} \tag{5.38}$$

$$\delta = \tan^{-1}\,(\varepsilon''/\,\varepsilon') \tag{5.39}$$

Where, a represents the attenuation factor, which depends on the dielectric constant ε' and the dielectric loss factor ε''. The volumetric heat generation Q can be calculated from the power intensity along the three orthogonal directions, as expressed in the following equation (Zhou et al., 1995):

$$Q = \partial P_x/\partial V + \partial P_y/\partial V + \partial Pz/\partial V \tag{5.40}$$

These equations are nonlinear and transient, representing the variations of temperature and moisture in food during MW heating. Due to their complexity, general analytical solutions cannot be obtained. In this study, FEM was employed because of its flexibility and capability to efficiently handle nonlinear problems (Zhou et al., 1995). The uniformity of MW heating is strongly influenced by factors such as the product's shape, size, composition, as well as the design and scale of the processing plant. Larger products tend to absorb MW energy from the surface inward, whereas smaller items of similar geometry are often heated from the center outward. Medium-sized products may develop hot spots located between the core and the outer surface. Regarding shape, spherical products exhibit different temperature distributions compared to brick-shaped items, due to focusing and edge effects (Dinani et al., 2020a, 2020b).

Processing parameters including MW frequency, phase, power, product movement, number of MW inlets, and their positions significantly affect the heat distribution during MW processing. Additionally, product characteristics such as load, placement and arrangement within the MW chamber, power-to-size ratio, shape, material homogeneity, composition, and moisture content contribute to variability in heating performance (Datta, 2001). Uneven temperature distribution represents a major obstacle to the industrial use of MW technology in food processing, primarily because of the occurrence of localized hot and cold regions. Areas that remain insufficiently heated may compromise product safety by allowing incomplete microbial inactivation or subsequent microbial growth in moisture-rich zones after drying. In contrast, excessively heated regions can lead to overprocessing, causing deterioration of sensitive components and a consequent decline in product quality (Chandrasekaran et al., 2013).

Ensuring a uniform temperature field throughout the sample volume (or at least reducing temperature non-uniformity as much as possible) is a critical requirement in MW-based processes. When uneven heating is identified, it can be mitigated to some extent through modification of operating conditions and product characteristics. Such adjustments may include changes in excitation frequency, power level, movement of the sample within the EM field, container design, sample geometry, placement, dimensions, and material composition. To evaluate the effectiveness of these corrective strategies and to compare different processing approaches, dependable techniques for measuring and quantifying heating uniformity are essential. These methods form the basis for assessing whether newly developed MW processing concepts lead to improved temperature homogeneity (Atuonwu and Tassou, 2018).

IR thermography is a widely applied technique for temperature measurement in MW-based processes. In recent years, this method has been increasingly employed to evaluate heating uniformity as well as drying behavior in food materials. As a non-contact and non-invasive approach, IR thermography utilizes IR-sensitive cameras to map temperature distributions without disturbing the EM field inside the MW cavity. The achievable spatial resolution is mainly governed by the camera's optical characteristics and its distance from the target surface. Despite these advantages, IR thermography has inherent limitations, as it only captures surface temperatures and therefore cannot directly provide information about internal or cross-sectional temperature distributions. Furthermore, because IR cameras are sensitive to both heat and EM radiation, their electronic components must be protected typically by an optical window when real-time (in-line) measurements are performed inside an MW chamber. Alternatively, samples may be removed from the cavity and measured externally (at-line). Such procedures can alter the observed temperature patterns due to internal heat conduction within the material or thermal losses to the surrounding environment through radiation (Knoerzer et al., 2009).

To obtain information about internal temperature distributions, products can be mechanically modified, for example by slicing, preferably with a minimal delay immediately after MW processing, followed by IR imaging of the exposed surfaces. However, this manipulation itself may induce additional temperature redistribution within the material. Consequently, temperature measurements should be carried out as quickly as possible after sample removal from the MW cavity, while maintaining identical processing duration, camera settings, and sample positioning. This consistency is essential to enable reliable quantitative comparison between different samples and experimental runs. A clear separation between the temperature measurement system and the EM field is particularly beneficial in multimode MW cavities. In such systems, even minor disturbances caused by measurement equipment can significantly alter the EM field configuration, potentially changing the excited modes and, as a result, the heating pattern within the product. Thermal imaging operates by sensing IR radiation emitted from heated surfaces.

The emitted radiation depends on both the surface temperature and the wavelength. For an ideal blackbody radiator, the spectral distribution of emitted energy follows Planck's radiation law. Because the mathematical expression of this relationship is complex, blackbody radiation behavior is commonly illustrated using a set of curves that describe the energy density as a function of wavelength, with temperature acting as a defining parameter, as typically presented in standard radiation diagrams. To evaluate internal temperature distributions using IR techniques, the product may be physically altered, for example by cutting it shortly after the heating process with minimal delay. Such manipulation can itself induce temperature redistribution inside the material. For this reason, IR measurements should be carried out immediately after the sample is taken out of the MW system, while maintaining identical processing duration, camera settings, and sample positioning, in order to ensure quantitative comparability between different samples and experimental runs.

Maintaining a physical distance between the temperature measurement system and the EM field is especially beneficial in multimode MW cavities. In these systems, even minor disturbances introduced by measuring instruments can modify the EM mode structure, which in turn alters the resulting heating pattern. Thermal imaging operates by sensing the IR radiation emitted from heated objects. The emitted radiation spectrum depends on both the surface temperature and the wavelength. For an ideal absorber, commonly referred to as a blackbody, the spectral distribution of radiant energy follows Planck's radiation theory. Because the mathematical formulation of this law is complex, blackbody radiation behavior is typically illustrated using graphical representations that show energy density as a function of wavelength, with temperature acting as a defining parameter (Monteiro et al., 2016). As temperature decreases, the magnitude of the emitted energy density is reduced, while the maximum of the spectrum moves toward longer wavelengths. As a result, objects at lower temperatures release weaker IR signals, which restricts the effectiveness of IR-based thermal imaging when samples are cool or when high measurement precision is required.

In practice, thermal imaging systems commonly operate within the wavelength interval of 8–14 μm. This spectral region corresponds to an atmospheric window in which absorption by air is minimal, while the radiation intensity remains sufficient for reliable detection by IR sensors. The radiation registered by the camera detector originates from multiple contributions. One component is the IR energy directly emitted by the object of interest. Another portion arises from radiation produced by surrounding surfaces, which is reflected by the target surface toward the detector. Additionally, the air layer between the object and the sensor both attenuates part of the transmitted radiation and emits its own IR energy, with the overall transmission characterized by a coefficient (τ) smaller than unity. For optimal temperature resolution, the temperature difference between the target and its environment should be at least 50 K, enabling measurement sensitivities in the range of approximately 0.05–1 K. Such conditions, however, are not maintained throughout the full duration of most MW heating operations in food processing applications.

Emissivity (ε) quantifies the thermal radiation emitted by a real object relative to that of an ideal blackbody. This parameter depends on material composition, surface morphology, observation angle, geometric factors, wavelength, and temperature-related phenomena such as phase transitions. Since real materials partially reflect and transmit IR radiation, emissivity values lie between zero and one. In food-related studies, an emissivity value close to that of water, typically around 0.96, is frequently assumed (Vollmer, 2021). This assumption is only reasonable for materials that contain a consistently high and stable amount of water. In products where water exists in different binding forms or where surface morphology varies, the emissivity can deviate markedly from that of pure water. This issue is particularly relevant in MW-assisted drying, where moisture removal continuously alters surface properties and, consequently, emissivity. Incorrect emissivity settings have been identified as one of the primary sources of error in IR temperature measurements. Nevertheless, many studies applying thermal imaging to heated food products do not explicitly state the emissivity values used during measurement or data analysis. It has been reported that, especially at low emissivity levels, IR-based temperature measurements may become unreliable. Only limited studies have addressed variable emissivity during MW drying, showing a clear decrease in emissivity as moisture content is reduced. In inline measurement configurations, additional attenuation caused by an optical window must also be considered and compensated through individual camera calibration. Such windows must transmit IR radiation while simultaneously preventing MW leakage, with quartz glass combined with metallic shielding being one possible solution.

To determine the effective transmission characteristics of optical windows, calibration approaches have been proposed in which IR temperature readings are correlated with reference temperatures obtained using contact sensors over a defined temperature range. Other calibration strategies have employed phase-change reference materials with known melting points. However, in some cases, the selected reference temperatures do not provide a sufficient temperature difference between the object and its surroundings to achieve optimal measurement accuracy. Moreover, important details such as emissivity values and optical properties of the container materials are often not documented. The use of closed glass containers further raises concerns, as the detected radiation originates primarily from the container surface rather than from the product itself. Open-container measurements may help to avoid such ambiguities.

Additional factors influencing IR temperature readings include steam formation and condensation during MW treatment, particularly when water vapor is not efficiently removed from the cavity. Despite the critical role of calibration and parameter selection, many publications provide limited or unclear information regarding the applied calibration procedures. Although IR thermography inherently captures only surface temperatures, this limitation can be mitigated by slicing samples immediately after processing to access internal temperature distributions. However, this offline approach requires manual handling, introduces time delays, and may result in temperature reductions compared to in-process conditions. To enhance the usefulness of surface temperature data, extrapolation methods from the surface into the product interior have been suggested, although experimental validation of such approaches remains limited. Heterogeneous material properties may further compromise the accuracy of extrapolated temperatures.

Combining surface imaging with internal access techniques appears promising, provided standardized procedures are established. When IR cameras are integrated directly into MW systems, real-time temperature monitoring becomes feasible. Due to their noncontact nature and high spatial resolution, thermographic cameras are well suited for automated, temperature-controlled MW processes and do not constrain product movement. Such systems have been successfully applied in controlled MW-assisted drying and liquid pasteurization processes, as well as for maintaining constant drying rates. Given the frequent occurrence of nonuniform temperature fields in MW applications, thermal imaging is particularly effective for capturing surface temperature distributions. However, accurate interpretation requires relatively uniform emissivity across the product surface. This condition may not be satisfied when major constituents such as water, fat, proteins, or polysaccharides are unevenly distributed. In cases where emissivity varies significantly, especially when high measurement accuracy is required, IR thermography alone cannot be considered sufficiently reliable. Consequently, combining IR imaging with complementary temperature measurement techniques, such as fiber-optic sensors or thermocouples, is strongly recommended (Ohlsson and Bengtsson, 1971; Minkina and Dudzik, 2009; Cuccurullo et al., 2018).

Temperature sensing using fiber optic technology can rely on several physical principles, including fluorescence decay measurement, Fabry–Pérot interference, and wavelength shifts in semiconductor-based transmission spectra. Fundamentally, optical fiber temperature measurement exploits the temperature dependence of Raman scattering occurring within a glass fiber, which is analyzed through a passive sensing element connected to a signal conditioning unit. Light pulses introduced into the fiber generated by sources such as lasers, xenon flash lamps, or light-emitting diodes travel along the fiber and interact with temperature-sensitive components integrated at the fiber tip, including phosphors or liquid crystal materials (Tang et al., 2002). When incident light excites these materials, photon–electron interactions occur through photoluminescence, producing scattered light whose characteristics depend on local temperature (Raman scattering). The returning optical signal comprises an elastically scattered component (Rayleigh scattering) and a temperature-related inelastic component (Raman scattering). The signal conditioner analyzes variations in backscattered light intensity and decay characteristics and converts them into temperature information. Specifically, the emission decay time and intensity change as a function of temperature-induced expansion and physical state of the sensor tip material.

In general, lower temperatures correspond to longer phosphor emission decay times (Tang et al., 2002). Fiber optic temperature probes are particularly suitable for MW environments because they do not perturb the EM field and remain unaffected by MW exposure. These sensors enable localized temperature measurements within samples across a broad range, from approximately −200 to 450°C. For food-related temperature ranges (−50–300°C), reported measurement accuracies reach approximately 0.1°C. Additionally, fiber optic sensors exhibit rapid response behavior, with typical response times between 0.05 and 2 s, enabling continuous inline acquisition of time–temperature data (Regier et al., 2017; Knoerzer et al., 2017). Despite these advantages, practical limitations exist. Fiber optic probes are delicate, as the sensing crystals are mechanically fragile, making handling challenging. Because the probe remains physically connected to an external signal processor outside the MW cavity, unrestricted sample movement is not feasible. For rotating platforms, continuous rotation may need to be replaced by oscillatory motion to maintain probe connection during heating. Accurate internal temperature readings require direct physical contact between the probe and the sample, and precise probe placement is therefore critical. As a result, measurements in highly porous or structurally fragile materials may lack reliability. Sample heterogeneity must also be considered. Furthermore, vapor bubble formation due to internal moisture evaporation can disrupt probe contact, displace the sensor, and compromise measurement accuracy (Dinani et al., 2020a, 2020b). Another limitation arises from the inherently point-based nature of conventional fiber optic measurements, resulting in poor spatial resolution.

Some commercially available systems address this issue by measuring temperature continuously along the fiber length with millimeter-scale resolution, using backscattered light along the fiber itself to infer local temperature and strain information. In such systems, local variations in refractive index alter the scattered signal, which is processed through interferometric comparison and fast Fourier transformation to associate frequency components with specific fiber positions. Although this approach increases the number of measurement points, it remains difficult to reconstruct fully resolved temperature fields because sensing is confined to regions immediately surrounding the fiber (Schwarz and Samiec, 2017). Given the prevalence of hot and cold spots and the likelihood of nonuniform temperature distributions during MW heating, fiber optic sensors alone may be insufficient for applications such as microbial inactivation, unless the coldest region is known in advance and the probe can be placed precisely at that location. Nevertheless, fiber optic sensors are highly valuable in processes requiring rapid thermal feedback, for example when integrated into magnetron control systems for real-time power adjustment.

Overall, challenges related to spatial resolution, probe positioning, and mechanical fixation limit their suitability for routine industrial process control, although they remain indispensable tools for process development and experimental optimization with respect to product geometry, size, and composition. In MW and RF heating systems, reliable temperature monitoring is essential because temperature data serve as feedback for process control and optimization. Conventional contact sensors such as thermocouples and thermistors are problematic in EM fields, as they can absorb MW energy, become damaged, or produce distorted readings. Noncontact techniques such as IR radiometry avoid EM interference by measuring temperature externally; however, these methods are restricted to surface temperature detection. In many applications, including polymer synthesis, surface measurements do not reliably represent internal temperatures. Therefore, contact sensors that are immune to EM fields are highly desirable.

Optical fiber sensors represent a promising alternative due to their small size, passive nature, environmental robustness, and resistance to EM interference. These properties allow placement at specific internal locations within a sample exposed to EM fields. Although numerous studies report successful use of optical fiber sensors in such environments, conventional systems remain limited to discrete sensing points. A small number of point sensors is often insufficient to capture full temperature profiles when strong gradients arise from uneven heating. Achieving adequate spatial coverage with a simple sensor configuration remains a key challenge. Recent advances in optical fiber sensing have enabled distributed temperature measurement along a single fiber. Techniques based on Raman scattering, Brillouin scattering, Rayleigh scattering, and fiber Bragg gratings (FBGs) allow continuous monitoring of temperature and strain at multiple locations. Such distributed sensing methods offer the potential to visualize spatial temperature distributions and identify heating nonuniformities using minimal sensor infrastructure.

Wada et al. (2015) demonstrated a distributed optical fiber sensing approach for monitoring temperature profiles during MW heating. A long fiber Bragg grating was employed as an electromagnetically immune sensor, and temperature distributions were interrogated using optical frequency domain reflectometry. In their study, water contained in a glass tube (820 mm length) was heated in an MW oven, and temperature profiles were measured at 5 mm intervals along the tube. IR measurements were used for comparison. The temporal and spatial evolution of temperature clearly revealed nonuniform heating patterns, demonstrating the effectiveness of distributed optical fiber sensing for MW heating applications (Wada et al., 2015). Table 5.2

Table 5.2 Temperature measurement methods in MW applications.

Measurement method	Temperature range (°C)	Spatial resolution	Advantages	Limitations
Infrared (IR) camera	–	Surface only	Non-contact, real-time	Measures only surface, emissivity-dependent
Fiber optic sensors	−200 to 450	Point/mm (distributed)	Immune to EM, can measure inside sample	Low spatial resolution, probe positioning critical
Thermocouple	–	Point	Low cost, widely available	EM interference, single point measurement
Pyrometer	–	Point	Non-invasive, fast	Measures spot temperature only

compares different temperature measurement techniques applied in MW and RF processes, including their operating range, spatial resolution, advantages, and limitations. This overview emphasizes the importance of method selection depending on the sample type, temperature range, and the need for contactless measurement to avoid perturbing the EM field.

5.5.3 Comparison between simulated and experimental data

MW technology is used for thermal treatment of both liquid and solid materials. When liquids are processed in continuous flow systems, turbulent flow regimes typically provide sufficient mixing, which homogenizes temperature variations in a manner similar to conventional heating approaches. Consequently, the need for techniques capable of identifying spatial temperature nonuniformities mainly concerns semisolid and solid food products, where pronounced temperature gradients and uneven heating across the product volume frequently occur. The techniques available for assessing temperature profiles or thermal load history in MW food processing vary widely with respect to their strengths, constraints, and inherent drawbacks. These approaches differ fundamentally in their measurement capabilities, and none of them alone is able to satisfy all necessary requirements. Therefore, employing a combination of complementary methods is often the most appropriate strategy. Certain techniques provide temperature information only at discrete probe locations or solely at the product surface.

Other approaches are capable of capturing spatial temperature distributions within the sample but fail to deliver comprehensive information on the overall thermal exposure history and associated time–temperature effects throughout the entire product volume. For this reason, combining multiple measurement strategies is the most effective way to meet mandatory safety and quality criteria by simultaneously monitoring temperature levels and quantifying heating effects. Techniques that enable three-dimensional temperature mapping, such as magnetic resonance imaging or MW radiometry, have limited practical relevance due to high equipment costs, operational complexity, and the inability to directly assess time–temperature histories. Not all conventional temperature sensing systems are suitable for use in MW environments. For instance, unshielded probes commonly applied in the analysis of transient heating processes cannot be reliably employed under MW conditions. Point-based temperature measurements, such as those obtained using fiber optic sensors, provide only localized information and offer limited insight into the global temperature field, as temperature patterns may vary unpredictably depending on product geometry, composition, and processing parameters.

When heating nonuniformity is expected and the locations of hot and cold regions are unknown, measuring the spatial temperature distribution becomes essential. This requires techniques capable of resolving temperature fields with sufficient spatial resolution. If the accuracy of the spatially resolved method alone is inadequate to guarantee food safety, it may be supplemented with localized temperature or effect measurements at identified critical regions. In industrial practice, pyrometers, thermocouples, and IR cameras are among the most widely adopted temperature measurement tools due to their ease of use and the availability of commercial data processing software. Thermocouples and pyrometers are particularly attractive because of their low cost and ready availability compared with more advanced alternatives. IR imaging systems benefit from flexible and customizable software solutions, which facilitates their integration into existing processing equipment and workflows. For real-time control of MW power input, pyrometers are often considered especially suitable, as they operate without physical contact, do not disturb the EM field, can be applied to a wide range of products, and can be readily incorporated into oven designs. Their primary limitation, however, lies in the fact that they provide only single-point temperature information and do not capture spatial temperature distributions or overall heating effects.

5.6 Sensitivity analysis and optimization of MW and RF processes

Analyzing sensitivity and optimizing MW and RF systems are essential for improving performance and guiding design, especially due to the intricate interactions among multiple design variables. At high frequencies, the behavior of MW components is strongly influenced by factors such as operating frequency, input power, and physical geometry. These effects occur because EM responses (including scattering characteristics, resonant behavior, and field distributions) are highly nonlinear and react sensitively to variations in both operational and structural parameters.

5.6.1 Sensitivity analysis framework

A thorough sensitivity analysis seeks to determine how changes in input parameters affect system behavior. For RF and MW components, such analyses are frequently integrated into global optimization strategies to facilitate effective design adjustments. A recent methodology, Fast Global Sensitivity Analysis (FGSA), identifies the directions in the parameter space that most strongly influence system responses. Unlike conventional dimension-specific metrics, such as Sobol indices, this technique emphasizes dominant directions that collectively explain most of the output variability, allowing for efficient dimensionality reduction and streamlined surrogate modeling for optimization purposes (Koziel and Pietrenko-Dabrowska, 2025). This fast sensitivity approach works by forming a relocation matrix from sampled points within the design space and then applying spectral decomposition to identify eigenvectors that represent the most influential directions. The magnitudes of these eigenvectors indicate the combined impact of specific parameter sets on electrical performance metrics, such as scattering parameters, allowing for focused design investigation within a lower-dimensional subspace (Koziel and Pietrenko-Dabrowska, 2024).

5.6.2 Influence of frequency on RF/MW behavior

The operating frequency is a critical factor in MW systems. Because RF and MW components often behave as resonant or dispersive structures, even small variations in frequency can significantly affect impedance matching, insertion loss, and the locations of resonant peaks. This frequency-dependent behavior is naturally reflected in EM simulation data used for sensitivity analysis, where frequency serves as a key input in the system's response or objective functions. Within surrogate-assisted optimization frameworks, frequency is inherently incorporated into the response surfaces that approximate device performance. Notable effects of frequency sensitivity include.

- **Resonance shifts:** In components such as couplers and filters, the resonant frequency strongly influences performance metrics like isolation and return loss.
- **Bandwidth effects:** The degradation of performance away from the center frequency is directly linked to how sensitive the geometry and material properties are to frequency variations.
- **Dispersion influences:** Changes in frequency alter phase velocity and dielectric behavior, which are especially important in transmission-line-based structures (Koziel and Pietrenko-Dabrowska, 2024).

5.6.3 Influence of power and nonlinear effects

Although the power level is not always treated as a direct design variable in linear RF analysis, it can strongly impact device behavior when operating near saturation or in nonlinear regimes. Power-related sensitivities may influence.

- **Thermal responses:** Elevated power levels increase conductor and dielectric losses, which can alter material properties and the overall sensitivity landscape.
- **Nonlinear distortions:** In components containing nonlinear elements, such as varactors or PIN diodes, the response may deviate from small-signal predictions as power increases.

While most sensitivity analyses primarily emphasize geometric and frequency parameters, the effects of power can be incorporated into extended cost functions within optimization routines, particularly when multiple performance objectives, such as efficiency and linearity, must be balanced.

5.6.4 Geometric parameter influence and optimization

Geometric dimensions, such as lengths, widths, and spacings, are among the most direct factors in MW design. They determine transmission line lengths, resonator dimensions, and the topology of matching networks, thereby shaping the EM field distribution and scattering responses across the operating frequency range. In surrogate-assisted optimization approaches, these geometric parameters form a high-dimensional vector that defines the design space of the system. Using the methodology proposed by Koziel and Pietrenko-Dabrowska, sensitivity analysis identifies combinations of geometric variables that have the largest influence on output variations. By focusing on these dominant directions, the dimensionality-reduced subspace streamlines the optimization process, allowing efficient global search using machine-learning models, followed by local refinement to achieve optimal performance across the entire design space (Koziel and Pietrenko-Dabrowska, 2025).

5.6.5 Optimization via sensitivity-guided machine learning

Combining rapid sensitivity analysis with surrogate-based machine-learning optimization frameworks offers substantial benefits in RF and MW system design. By focusing on the most influential directions in the parameter space, the dimensionality of the problem is significantly reduced, which greatly lowers the computational cost of global optimization. For instance, performing surrogate-assisted optimization within this reduced space can achieve near-optimal designs while requiring far fewer full-wave EM simulations, maintaining high accuracy in the results (Koziel and Pietrenko-Dabrowska, 2025). This two-step optimization approach, consisting of a global surrogate-based search followed by local fine-tuning, achieves a balance between computational efficiency and design precision. The global stage explores the most impactful directions in the design space, while the local stage allows refinement across all parameters, including those with minor contributions. Frequency critically influences resonance locations, bandwidth, and dispersive behavior of RF and MW components, making it a key factor in sensitivity analysis and performance optimization. Power affects nonlinear responses and thermal behavior, particularly in devices operating near saturation or under high RF levels. Geometric dimensions remain fundamental in shaping scattering, impedance matching, and resonant characteristics, and sensitivity-driven strategies can prioritize the most influential geometric combinations for design enhancement. Integrating rapid sensitivity assessment with surrogate-assisted machine-learning optimization substantially reduces computational cost while preserving high-quality design outcomes (Koziel & Pietrenko-Dabrowska, 2024, 2025).

5.6.6 Multi-objective optimization (MOD)

MOD is essential in MW and RF engineering, where system performance is often influenced by several competing criteria. Unlike single-objective optimization, which targets improvement of a single metric, MOO seeks to identify a set of optimal trade-off solutions that balance multiple objectives simultaneously, such as energy efficiency, heating uniformity, EM behavior, and robustness. These trade-offs are typically visualized through the Pareto front, where enhancement of one objective cannot occur without compromising at least one other (Koziel and Pietrenko-Dabrowska, 2023; Wei et al., 2022).

5.6.7 Energy efficiency as a multi-objective criterion

Energy efficiency is a key objective in the optimization of MW and RF systems. For RF components, higher efficiency corresponds to lower insertion losses, reduced power dissipation, and enhanced overall performance. In MW-driven physical processes, such as heating and material processing, energy efficiency directly affects operating costs and environmental impact. Consequently, it is frequently considered a primary goal in multi-objective optimization problems, often alongside constraints on EM performance (Wei et al., 2022). Surrogate-assisted multi-objective evolutionary algorithms (MOEAs) have shown strong effectiveness in optimizing energy efficiency in MW component design. By integrating full-wave EM simulations with surrogate models, these approaches allow efficient navigation of complex design spaces while considerably lowering computational demands. Decomposition-based MOEAs supported by surrogate modeling can generate Pareto-optimal solutions that achieve high energy efficiency while maintaining critical frequency-domain specifications in MW filters (Wei et al., 2022).

5.6.8 Uniformity optimization in MW processes

Uniformity represents a key objective in MW-based processes such as heating, drying, and material treatment. Uneven distributions of EM fields or temperature can lead to localized overheating, diminished product quality, and material degradation. Therefore, enhancing uniformity while maintaining satisfactory energy efficiency is a classical multi-objective optimization challenge in MW process engineering. Recent investigations have indicated that coordinated tuning of system parameters including power settings, source configuration, and operational strategies—can substantially improve uniformity without excessive energy consumption. Collaborative switching strategies in multi-source MW systems have been shown to effectively enhance temperature-field uniformity while simultaneously reducing unnecessary energy losses. These findings highlight that uniformity and energy efficiency are intrinsically linked objectives that require joint optimization rather than separate treatment (Yang et al., 2024).

5.6.9 Pareto-based multi-objective optimization framework

Contemporary multi-objective optimization in MW and RF design often employs Pareto-based evolutionary algorithms, including NSGA-II and MOEA/D, frequently combined with surrogate modeling approaches. These frameworks produce a set of non-dominated solutions, giving designers the flexibility to select trade-offs that align with application-specific requirements. Instead of providing a single optimal design, Pareto-based methods offer insight into the interactions and compromises between conflicting objectives, such as energy efficiency, uniformity, and system robustness (Koziel and Pietrenko-Dabrowska, 2023; Wei et al., 2022). A rapid multi-objective optimization strategy has been introduced that leverages knowledge-based surrogate models, allowing efficient investigation of trade-offs among several design objectives in MW components. This methodology highlights the integration of sensitivity analysis with multi-objective optimization to steer the design search toward physically relevant and computationally efficient solutions. The findings indicate that using sensitivity-informed surrogates in combination with Pareto-based optimization notably accelerates convergence while maintaining high-quality designs (Koziel and Pietrenko-Dabrowska, 2023). In conclusion, multi-objective optimization is a core strategy for modeling and improving MW and RF processes. Since objectives like energy efficiency and uniformity often conflict, Pareto-based optimization approaches are essential to identify practical trade-off solutions. Combining surrogate modeling, sensitivity analysis, and evolutionary multi-objective algorithms has proven to be an effective and widely adopted method for managing the computational challenges associated with complex MW and RF system optimization (Koziel and Pietrenko-Dabrowska, 2023; Wei et al., 2022; Yang et al., 2024). Summary of sensitivity analysis and optimization techniques in MW and RF processes are presented in Table 5.3.

Table 5.3 Summary of sensitivity analysis and optimization techniques in MW and RF processes.

Category	Optimization/Analysis focus	Key parameters	Methodology	Main objectives	Application Area	Reference
Sensitivity analysis	Fast global sensitivity analysis (FGSA)	Geometry, frequency	Eigenvector-based dominant direction analysis	Identification of most influential parameter combinations	MW passive components	Koziel and Pietrenko-Dabrowska (2024)
Sensitivity-Guided optimization	Dimensionality reduction for optimization	Geometry, EM responses	Surrogate modeling + reduced-order design space	Computational efficiency, design accuracy	MW passives	Koziel and Pietrenko-Dabrowska (2025)
Single-objective optimization	Energy-efficient EM performance	Geometry, frequency	Surrogate-assisted evolutionary optimization	Minimized loss, improved efficiency	MW filters	Wei et al. (2022)
Multi-objective optimization	Trade-off exploration	Geometry, frequency	MOEA/D with surrogate models	Energy efficiency versus EM performance	RF and MW components	Wei et al. (2022)
Multi-Criterial optimization	Performance–robustness trade-off	Geometry tolerances	Knowledge-based surrogate modeling	Performance stability, robustness	MW circuit design	Koziel and Pietrenko-Dabrowska (2023)
Multi-objective process optimization	Energy efficiency versus uniformity	Power level, source configuration	Coordinated power control and switching	Uniform temperature distribution, reduced energy waste	MW heating processes	Yang et al. (2024)
Multi-source optimization	Field/temperature uniformity	Power, timing, source coordination	Collaborative switching strategies	Uniformity improvement with energy constraints	Industrial MW systems	Yang et al. (2024)

5.7 Conclusion and future trends

The modeling and optimization of MW and RF heating processes require a comprehensive understanding of the interplay between EM fields, thermal responses, and material properties. Dielectric parameters, including ε' and ε'', are highly dependent on moisture content, temperature, and chemical composition, directly influencing energy absorption, heat generation, and spatial temperature distribution. Coupled EM–thermal simulations using FEM and FDTD methods, supplemented by appropriate boundary conditions and, when applicable, mass transfer models, have proven effective for predicting heating behavior in complex geometries and heterogeneous materials. Experimental temperature measurement techniques, such as IR thermography and fiber optic sensors, provide valuable validation and insight into heating uniformity, though challenges related to emissivity, probe positioning, and spatial resolution remain.

Recent advances in sensitivity analysis, surrogate-assisted modeling, and multi-objective optimization enable the efficient identification of influential design parameters and trade-offs between competing objectives, such as energy efficiency and uniformity. Pareto-based optimization frameworks, combined with machine-learning-driven surrogate models, allow rapid exploration of high-dimensional design spaces while reducing computational costs, offering practical pathways for both system design and process optimization. Looking forward, the integration of distributed temperature sensing, real-time feedback control, and adaptive MW power management holds promise for further improving heating uniformity, process safety, and energy efficiency. Future research should focus on standardizing measurement protocols, developing more accurate material property databases under dynamic conditions, and extending multi-physics models to incorporate chemical reactions, phase transitions, and moisture transport. Additionally, hybrid optimization strategies that combine global and local search methods, coupled with high-fidelity surrogate models, are expected to accelerate the design and control of industrial-scale MW and RF systems. These advancements will support scalable, energy-efficient, and high-quality processing in food, pharmaceutical, and materials engineering applications.

Abbreviations list

CFD	Computational fluid dynamics
EM	Electromagnetic
FDTD	Finite-difference time-domain
FEM	Finite element method
MOO	Multi-objective optimization
MW	Microwave
RF	Radio-frequency

References

Alpert, Y., Jerby, E., 1999. Coupled thermal-electromagnetic model for microwave heating of temperature-dependent dielectric media. IEEE Trans. Plasma Sci. 27 (2), 555–562.

Atuonwu, J.C., Tassou, S.A., 2018. Quality assurance in microwave food processing and the enabling potentials of solid-state power generators: a review. J. Food Eng. 234, 1–15.

Awuah, G.B., Ramaswamy, H.S., Tang, J., 2014. Radio-Frequency Heating in Food Processing: Principles and Applications. CRC Press.

Bai, S., Liu, L., Yu, H., Guan, X., Li, R., Hou, L., et al., 2022. Thermal and dielectric properties of wolfberries as affected by moisture content and temperature associated with radio frequency and microwave dehydrations. Foods 11 (23), 3796.

Balanis, C.A., 2011. Modern Antenna Handbook. John Wiley & Sons.

Beikzadeh, S., Ghorbani, M., Shahbazi, N., Izadi, F., Pilevar, Z., Mortazavian, A.M., 2020. The effects of novel thermal and nonthermal technologies on the properties of edible food packaging. Food Eng. Rev. 12 (3), 333–345.

Chandrasekaran, S., Ramanathan, S., Basak, T., 2013. Microwave food processing—a review. Food Res. Int. 52 (1), 243–261.

Chen, H., Tang, J., Liu, F., 2006. Coupled simulation of an electromagnetic heating process using the finite difference time domain method. J. Microw. Power Electromagn. Energy 41 (3), 50–68.

Cuccurullo, G., Giordano, L., Metallo, A., Cinquanta, L., 2018. Drying rate control in microwave assisted processing of sliced apples. Biosyst. Eng. 170, 24–30.

Datta, A.K., 2001. Mathematical modeling of microwave processing of foods: an overview. Food Processing Operations Modeling, pp. 150–190.

Dinani, S.T., Hasić, M., Auer, M., Kulozik, U., 2020a. Assessment of uniformity of microwave-based heating profiles generated by solid-state and magnetron systems using various shapes of test samples. Food Bioprod. Process. 124, 121–130.

Dinani, S.T., Kubbutat, P., Kulozik, U., 2020b. Assessment of heating profiles in model food systems heated by different microwave generators: solid-state (semiconductor) versus traditional magnetron technology. Innov. Food Sci. Emerg. Technol. 63, 102376.

Ghosh, S.N., 2002. Electromagnetic Theory and Wave Propagation. CRC Press.

Golio, M., Golio, J., 2018. RF and Microwave Applications and Systems. CRC Press.

Huang, J., Xu, G., Hu, G., Kizil, M., Chen, Z., 2018. A coupled electromagnetic irradiation, heat and mass transfer model for microwave heating and its numerical simulation on coal. Fuel Process. Technol. 177, 237–245.

Iskander, M.F., Smith, R., Andrade, A., Kimrey, H., Wal, L., 1994. FDTD simulation of microwave sintering of ceramics in multimode cavities. IEEE Trans. Microw. Theor. Tech. 42 (5), 793–800.

Knoerzer, K., Regier, M., Hardy, E., Schuchmann, H., Schubert, H., 2009. Simultaneous microwave heating and three-dimensional MRI temperature mapping. Innov. Food Sci. Emerg. Technol. 10 (4), 537–544.

Knoerzer, K., Regier, M., Schubert, H., 2017. Measuring temperature distributions during microwave processing. In: The Microwave Processing of Foods. Elsevier, pp. 327–349.

Kopyt, P., Celuch, M., 2006. Coupled electromagnetic-thermodynamic simulations of microwave heating problems using the FDTD algorithm. J. Microw. Power Electromagn. Energy 41 (4), 18–29.

Koziel, S., Pietrenko-Dabrowska, A., 2023. Rapid multi-criterial design of microwave components with robustness analysis by means of knowledge-based surrogates. Sci. Rep. 13 (1), 5953.

Koziel, S., Pietrenko-Dabrowska, A., 2024. Optimization of microwave components using machine learning and rapid sensitivity analysis. Sci. Rep. 14 (1), 31265.

Koziel, S., Pietrenko-Dabrowska, A., 2025. Globalized parameter tuning of microwave passives by dimensionality-reduced surrogates and multi-fidelity simulations. Sci. Rep. 15 (1), 21776.

Kuang, W., Nelson, S., 1998. Low-frequency Dielectric Properties of Biological Tissues: A Review with Some New Insights.

Li, G., Zhou, J., 2024. COMSOL-based simulation of microwave heating of Al2O3/SiC composites with parameter variations. Symmetry 16 (10), 1254.

Ling, B., Liu, X., Zhang, L., Wang, S., 2018. Effects of temperature, moisture, and metal salt content on dielectric properties of rice bran associated with radio frequency heating. Sci. Rep. 8 (1), 4427.

Hyperthermia of the prostate from a heat transfer point of view. In: Loyd, D., Karlsson, M., Ask, P., Erlandson, B.-E. (Eds.), 1994. Proceedings of 16th Annual International Conference of the. IEEE Engineering in Medicine and Biology Society.

Marra, F., Zhang, L., Lyng, J.G., 2009. Radio frequency treatment of foods: review of recent advances. J. Food Eng. 91 (4), 497–508.

Maxwell, J.C., 1865. VIII. A dynamical theory of the electromagnetic field. Phil. Trans. Roy. Soc. Lond. (155), 459–512.

Metaxas, A., Meredith, R.J., 1983. Industrial Microwave Heating. IET.

Minkina, W., Dudzik, S., 2009. Infrared Thermography: Errors and Uncertainties. John Wiley & Sons.

Monteiro, R.L., Carciofi, B.A., Laurindo, J.B., 2016. A microwave multi-flash drying process for producing crispy bananas. J. Food Eng. 178, 1–11.

Mudgett, R.E., 1986. Electrical properties of foods. In: Engineering Properties of Foods, 2, pp. 389–455.

Ohlsson, T., Bengtsson, N., 1971. Microwave heating profiles in foods.: a comparison between heating experiments and computer simulation. A research note. Microw. Energy Appl. Newsl. 4 (6), 3–8.

Osepchuk, J.M., 1979. Sources and basic characteristics of microwave/RF radiation. Bull. N. Y. Acad. Med. 55 (11), 976.

Paulsen, K., Lynch, D., Strahbehn, J., 2002. Three-dimensional finite, boundary, and hybrid element solutions of the Maxwell equations for lossy dielectric media. IEEE Trans. Microw. Theor. Tech. 36 (4), 682–693.

Pitchai, K., Birla, S., Subbiah, J., Jones, D., Thippareddi, H., 2012. Coupled electromagnetic and heat transfer model for microwave heating in domestic ovens. J. Food Eng. 112 (1–2), 100–111.

Piyasena, P., Dussault, C., 2003. Continuous radio-frequency heating of a model viscous solution: influence of active current, flow rate, and salt content on temperature rise. Can. Biosyst. Eng. 45 (3), 3–27.

Ramaswamy, H., Tang, J., 2008. Microwave and radio frequency heating. Food Sci. Technol. Int. 14 (5), 423–427.

Regier, C., Hoult, N.A., Moore, I.D., 2017. Laboratory study on the behavior of a horizontal-ellipse culvert during service and ultimate load testing. J. Bridge Eng. 22 (3), 04016131.

Schwarz, J., Samiec, D., 2017. B3. 4-Fiber-Optic measurement of temperature profiles. Proc. Sensor 2017, 212–216.

Sekkak, A., Pichon, L., Razek, A., 2002. 3-D FEM magneto-thermal analysis in microwave ovens. IEEE Trans. Magn. 30 (5), 3347–3350.

Sheikhi, Z., Farhoodi, M., Beikzadeh, S., Kazemian-Bazkiaee, F., Shojaee-Aliabadi, S., Mirmoghtadaie, L., 2020a. Effect of Using Cold Plasma Treatment on the Surface and Physicochemical Properties of Starch-Chitosan Composite Film.

Sheikhi, Z., Mirmoghtadaie, L., Khani, M., Farhoodi, M., Beikzadeh, S., Abdolmaleki, K., et al., 2020b. Physicochemical characterization of argon plasma-treated starch film. J. Agric. Sci. Technol. 22 (4), 999–1008.

Solyom, K., Lopez, P.R., Esquivel, P., Lucia, A., 2020. Effect of temperature and moisture contents on dielectric properties at 2.45 GHz of fruit and vegetable processing by-products. RSC Adv. 10 (28), 16783–16790.

Sorrentino, R., Bianchi, G., 2010. Microwave and RF Engineering. John Wiley & Sons.

Swami, S., 1982. Microwave Heating Characteristics of Simulated High Moisture Foods. University of Massachusetts.

Tang, J., Hao, F., Lau, M., 2002. Microwave heating in food processing. In: Advances in Bioprocessing Engineering. World Scientific, pp. 1–44.

Tauhiduzzaman, M., Hafez, I., Bousfield, D., Tajvidi, M., 2021. Modeling microwave heating and drying of lignocellulosic foams through coupled electromagnetic and heat transfer analysis. Processes 9 (11), 2001.

Thuéry, J., Grant, E.H., 1992. Microwaves: Industrial, Scientific, and Medical Applications. (No Title).

Tıraş, B., Dede, S., Altay, F., 2019. Dielectric properties of foods. Turk. J. Agric. Food Sci. Technol. 7 (11), 1805–1816.

Torres, F., Jecko, B., 2002. Complete FDTD analysis of microwave heating processes in frequency-dependent and temperature-dependent media. IEEE Trans. Microw. Theor. Tech. 45 (1), 108–117.

Vollmer, M., 2021. Physics of the electromagnetic spectrum. In: Electromagnetic Technologies in Food Science, pp. 1–32.

Wada, D., Sugiyama, J-i, Zushi, H., Murayama, H., 2015. An optical fiber sensing technique for temperature distribution measurements in microwave heating. Meas. Sci. Technol. 26 (8), 085105.

Wei, Y., Qi, G., Wang, Y., Yan, N., Zhang, Y., Feng, L., 2022. Efficient microwave filter design by a surrogate-model-assisted decomposition-based multi-objective evolutionary algorithm. Electronics 11 (20), 3309.

Yang, B., Zhao, Z., Zhang, H., Chen, Y., Chen, X., 2024. Temperature field optimization for multi-microwave sources based on collaborative switching under uncertain communication. Appl. Sci. 14 (17).

Zhang, H., Datta, A.K., 2001. Electromagnetics of microwave heating: magnitude and uniformity of energy absorption in an oven. In: Handbook of Microwave Technology for Food Application. CRC Press, pp. 33–68.

Zhang, M., Jia, X., Tang, Z., Zeng, Y., Wang, X., Liu, Y., et al., 2020. A fast and accurate method for computing the microwave heating of moving objects. Appl. Sci. 10 (8), 2985.

Zhong, J., Liang, S., 2021. A data-driven based spatiotemporal model reduction for microwave heating process with the mixed boundary conditions. Processes 9 (5), 827.

Zhou, L., Puri, V., Anantheswaran, R., Yeh, G., 1995. Finite element modeling of heat and mass transfer in food materials during microwave heating—model development and validation. J. Food Eng. 25 (4), 509–529.

Zhu, Z., Guo, W., 2017. Frequency, moisture content, and temperature dependent dielectric properties of potato starch related to drying with radio-frequency/microwave energy. Sci. Rep. 7 (1), 9311.

Modeling of high-pressure processes

6

Laura Otero
Institute of Food Science, Technology and Nutrition (ICTAN-CSIC). MALTA Consolider Team. Madrid, Spain

6.1 High-pressure food processing: General description

High-pressure processing (HPP) is, nowadays, a food preservation technology successfully implemented at an industrial level all over the world. Thus, in 2020, almost 600 industrial HPP units, dedicated to the commercial processing of food products, were operating in production worldwide. Of these, 50% were located in America, 25% in Europe, 19% in Asia, 5% in Oceania, and 1% in Africa (Houška et al., 2022). Current commercial applications include a wide variety of products such as juices, sauces and dressings, vegetable dips, fruit purees, meat, seafood, and dairy products, ready-to-eat meals, or even baby food. Moreover, new applications are constantly emerging and, as a result, the global HPP food market is growing rapidly. Thus, in 2024, it was valued at US$ 3228.6 million and is forecast to reach US$ 5267.4 million by 2030 (Valuates Reports, 2024).

During industrial HPP, foods are immersed in a fluid, usually water at room temperature, and pressurized up to 400–600 MPa for several minutes (generally <10 min). Treatments are typically performed in high-pressure machines that process prepacked products in batches; however, since 2019, there has also been HPP equipment available that can process beverages in bulk before bottling (Houška et al., 2022; Tonello-Samson et al., 2020). As illustrated in Fig. 6.1A, the main components of typical commercial-scale HPP equipment are: (1) conveyors used to load and unload carriers or baskets containing food products; (2) a cylindrical stainless-steel pressure vessel where foods are processed; (3) vessel closures, that is, plugs and wedges, restrained by a yoke; and (4) intensifiers that pump pressurizing fluid into the vessel to reach the target pressure (Balasubramaniam et al., 2016; Queirós et al., 2023). Although early industrial machines were equipped with vertically positioned pressure vessels, today more than 90% of them have a horizontal configuration. The largest and most automated HPP systems have pressure vessels of 525 L capacity and are capable of processing more than 3000 kg of product per hour (Houška et al., 2022; Tonello-Samson et al., 2020). A cycle begins by loading the prepacked products into the carriers, pushing them into the pressure vessel, and filling the vessel with water. Then, during the compression step, pressure intensifiers pump more water into the HPP vessel until the target pressure is reached (Fig. 6.1B). Pressure is subsequently maintained for the time required to achieve the intended HPP-induced effect on the

Modeling of Unit Operations in the Food Industry. https://doi.org/10.1016/B978-0-12-822180-8.00010-9

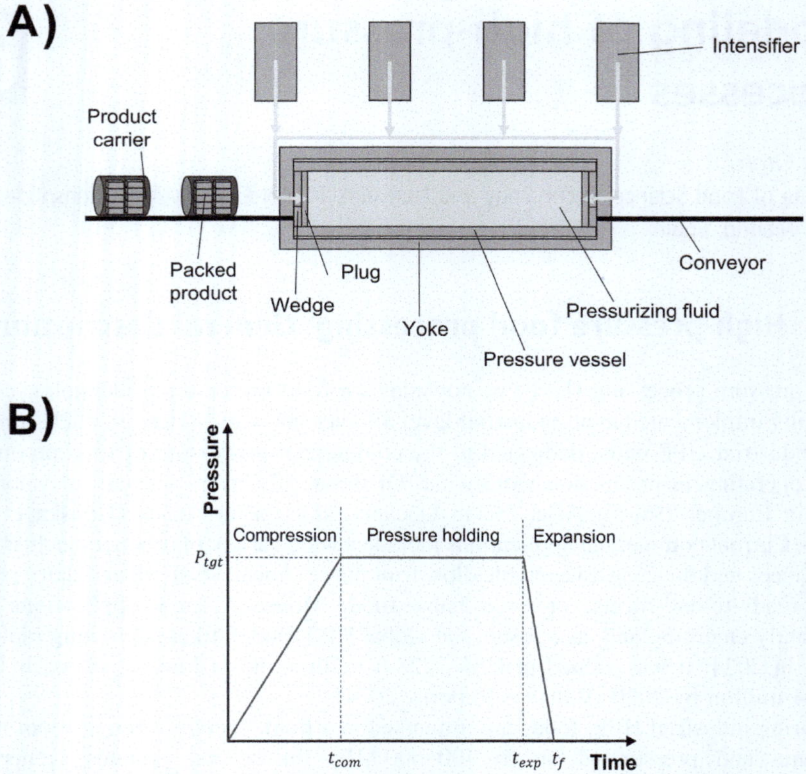

Figure 6.1 (A) Main components of typical commercial-scale HPP equipment. (B) Steps of a typical pressure cycle: compression, pressure-holding time, and expansion. P_{tgt}: target pressure, t_{com}: time needed to reach the target pressure during the compression step, t_{exp}: time at which expansion is performed, and t_f: time at which atmospheric pressure is reached after expansion.

product, typically less than 10 min. Then, decompression is performed by opening the release valves that allow water to flow out of the vessel, and the cycle ends when the HPP vessel is opened, and the processed product is uploaded (Houška et al., 2022; Queirós et al., 2023).

At room temperature and below, pressure processing can inactivate vegetative cells of foodborne pathogens and spoilage microorganisms present in foods, as well as reduce the activity of certain endogenous enzymes responsible for quality degradation, thus extending food shelf-life significantly (Aganovic et al., 2021; Houška et al., 2022). For this reason, HPP is mainly used in the food industry as a food preservation technology, although other interesting applications have also been commercially implemented, such as the use of HPP processing for shucking lobsters, crabs, or oysters, among others (Roobab et al., 2022). Unlike traditional thermal processing,

HPP at room temperature or below rarely affects vitamins, pigments, and aroma compounds and, therefore, the organoleptic and nutritional properties of foods are minimally affected after pressure processing. In addition, by increasing food safety and extending the shelf life of foods, HPP allows for a reduction in synthetic additives used in product formulation (Huang et al., 2017; Janahar and Balasubramaniam, 2022). All these advantages contribute to meeting the current consumer demand for safe, fresh-like, minimally processed, and additive-free foods and drive the global pressure-treated foods market, which is expected to grow at a compound annual growth rate of 8.5% between 2024 and 2030 (Valuates Reports, 2024).

Although often considered a nonthermal technology, pressure processing can also be performed at elevated temperatures from 40°C to more than 120°C; that is, high-pressure thermal processing (HPTP). Depending on the pressure level and temperature applied, HPTP can inactivate not only vegetative microorganisms but also bacterial spores to obtain microbiologically safe and stable foods (Gänzle, 2023). Overall, HPTP minimizes the negative impact of thermal processing on food quality by reducing the time required at elevated temperatures. Thus, nutritional and organoleptic properties are generally better preserved in HPTP foods than in those conventionally thermally treated, while the production of food processing contaminants during HPTP is also reduced (Sevenich et al., 2023). Despite all these advantages, the industrial application of HPTP remains a challenge, as it involves the development of economically viable equipment and the scale-up and validation of the results obtained under lab conditions (Queirós et al., 2023).

Other lesser-known high-pressure processes of potential interest to the food industry include high-pressure freezing (HPF) and high-pressure thawing (HPT), which involve subzero temperatures. Depending on the process conditions, pressure can accelerate phase transitions, thus improving the quality of high-pressure frozen and thawed foods (Otero, 2023). Although widely studied in the literature, these processes have not yet been exploited industrially, probably due to the difficulties associated with their implementation, which include the need for temperature control, inefficient heat exchange, and long processing times. In addition, the use of subzero temperatures requires pressurizing fluids other than water, with a freezing point lower than the subzero temperatures employed, and strict specifications for the materials used in the construction of the pressure vessel.

This chapter provides an overview of the current state of the art in modeling all the above-mentioned high-pressure processes, whether industrially applied or not. The complexity of the phenomena involved, current difficulties in building models, new trends, and recent advances are discussed in depth. This chapter should encourage further research to characterize, better understand, and evaluate high-pressure processes of potential interest to the food industry, improve process uniformity, move from lab to industrial scale, and optimize not only process parameters but also pressure equipment. Thus, it should contribute to the design of economically viable applications that expand the use of high-pressure technology.

6.2 Uniformity of high-pressure processes

Uniformity of high-pressure processes is a prerequisite to ensure the required safety and quality throughout the treated product. Since most pressure-induced effects are also temperature-dependent, maintaining not only uniform pressure but also uniform temperature conditions throughout the product during processing is essential to ensure food safety and quality.

Although the heterogeneity and anisotropy of food products make a homogeneous treatment strictly impossible, a homogeneous effect of pressure during HPP is commonly assumed due to the almost instantaneous velocity of pressure propagation (Grauwet et al., 2016; Nair et al., 2016; Sevenich and Rauh, 2021). Thus, according to the isostatic principle, pressure is considered to be transmitted uniformly and instantaneously throughout the product, regardless of its size, shape, or composition. In contrast, there is much evidence that significant temperature nonuniformities occur during high-pressure food processing that can lead to pronounced inhomogeneities in microbial and/or enzyme inactivation and nutritional and/or sensorial degradation induced in the processed food (Grauwet et al., 2016; Queirós et al., 2023). This is especially relevant in HPTP when pursuing commercial food sterility, as the reactions leading to spore inactivation are highly temperature-dependent.

6.2.1 *Adiabatic heat of compression, heat transfer, and fluid motion phenomena*

Temperature nonuniformities during HPP occur due to several causes, one of the most relevant being differences in the compression heating of the materials involved. During pressurization, the conversion of compressive work against intermolecular forces into internal energy produces a temperature increase in all materials inside the pressure vessel (food and packaging, product containers or carriers, and pressurizing fluid). Assuming adiabatic conditions, this temperature increase can be theoretically calculated by Eq. (6.1):

$$\frac{dTk}{dP} = \frac{\alpha(P,T)}{\rho(P,T) \cdot Cp(P,T)} \cdot Tk \tag{6.1}$$

where *Tk* denotes the temperature in K, *P* the pressure in Pa, ρ the density in $kg \cdot m^{-3}$, α the coefficient of thermal expansion in K^{-1}, and *Cp* the specific heat capacity in $J \cdot kg^{-1} \cdot K^{-1}$. Evidently, upon decompression, the process is reversed, and adiabatic cooling of the products inside the vessel occurs.

Eq. (6.1) reveals that temperature changes after compression or expansion depend not only on the initial temperature and the magnitude of the pressure change, but also on certain thermophysical properties (α, ρ, and *Cp*) which, in turn, depend on the composition of the compressed materials. This means that, depending on their composition, the temperature changes associated with adiabatic compression or expansion can be very different in the processed food, the packaging, the product carrier, and the pressurizing fluid, as clearly illustrated in Table 6.1. For example, rough estimations, made using $\Delta T/100$ MPa values given in Table 6.1, reveal that, after

Table 6.1 Quasiadiabatic temperature increases, ΔT/100 MPa (°C), observed in the literature in different food products, packaging and carrier materials, and pressurizing fluids after compression at different pressure and temperature conditions.

Product	Initial T (°C)	Pressure applied (MPa)	ΔT/100 MPa (°C)	References
Foods				
2% fat milk	25	n.r.	≈3.0	Ting et al. (2002)
Beef ground	25 ± 1	150.4	3.0 ± 0.1	Patazca et al. (2007)
	25 ± 1	324.8	3.3 ± 0.0	Patazca et al. (2007)
	25 ± 1	457.1	3.4 ± 0.0	Patazca et al. (2007)
	25 ± 1	540.7	3.2 ± 0.0	Patazca et al. (2007)
Chicken breast	25 ± 1	146.2	3.0 ± 0.1	Patazca et al. (2007)
	25 ± 1	322.6	3.0 ± 0.1	Patazca et al. (2007)
	25 ± 1	462.6	3.1 ± 0.0	Patazca et al. (2007)
	25 ± 1	542.3	3.0 ± 0.0	Patazca et al. (2007)
Chicken fat	25	n.r.	≈4.5	Ting et al. (2002)
Cream cheese	1 ± 1	157.5	3.4	Patazca et al. (2007)
	1 ± 1	289.7	3.0	Patazca et al. (2007)
	1 ± 1	453.2	3.0	Patazca et al. (2007)
	1 ± 1	585.0	3.1	Patazca et al. (2007)
	25 ± 1	146.4	4.9 ± 0.1	Patazca et al. (2007)
	25 ± 1	146.6	4.9	Patazca et al. (2007)
	25 ± 1	316.1	4.8 ± 0.0	Patazca et al. (2007)
	25 ± 1	317.1	4.8	Patazca et al. (2007)
	25 ± 1	447.6	4.6 ± 0.1	Patazca et al. (2007)
	25 ± 1	448.0	4.6	Patazca et al. (2007)
	25 ± 1	533.3	4.7 ± 0.0	Patazca et al. (2007)
	25 ± 1	586.2	4.4	Patazca et al. (2007)
	50 ± 1	159.2	5.1	Patazca et al. (2007)
	50 ± 1	292.4	4.8	Patazca et al. (2007)
	50 ± 1	459.7	4.9	Patazca et al. (2007)
	50 ± 1	590.2	4.8	Patazca et al. (2007)
	70 ± 1	159.7	5.7	Patazca et al. (2007)
	70 ± 1	295.6	5.3	Patazca et al. (2007)
	70 ± 1	446.2	5.0	Patazca et al. (2007)
	70 ± 1	588.2	4.7	Patazca et al. (2007)
Crude beef fat	25 ± 2	162	4.1 ± 0.1	Rasanayagam et al. (2003)
	25 ± 2	315	4.4 ± 0.3	Rasanayagam et al. (2003)
	25 ± 2	463	4.4 ± 0.5	Rasanayagam et al. (2003)
	25 ± 2	605	4.4 ± 0.8	Rasanayagam et al. (2003)
Egg white	25 ± 1	163.0	2.8 ± 0.1	Patazca et al. (2007)
	25 ± 1	332.5	2.8 ± 0.1	Patazca et al. (2007)
	25 ± 1	465.0	2.8 ± 0.1	Patazca et al. (2007)
	25 ± 1	541.2	2.9 ± 0.0	Patazca et al. (2007)

Continued

Table 6.1 Quasiadiabatic temperature increases, ΔT/100 MPa (°C), observed in the literature in different food products, packaging and carrier materials, and pressurizing fluids after compression at different pressure and temperature conditions.—cont'd

Product	Initial T (°C)	Pressure applied (MPa)	ΔT/100 MPa (°C)	References
Egg yolk	25 ± 1	161.4	4.5 ± 0.0	Patazca et al. (2007)
	25 ± 1	338.9	4.4 ± 0.0	Patazca et al. (2007)
	25 ± 1	472.0	4.3 ± 0.0	Patazca et al. (2007)
	25 ± 1	544.8	4.3 ± 0.0	Patazca et al. (2007)
Extracted beef fat	25	n.r.	≈6.3	Ting et al. (2002)
	25 ± 2	170	8.3 ± 0.7	Rasanayagam et al. (2003)
	25 ± 2	342	6.5 ± 0.4	Rasanayagam et al. (2003)
	25 ± 2	465	6.2 ± 0.4	Rasanayagam et al. (2003)
	25 ± 2	613	6.3 ± 0.4	Rasanayagam et al. (2003)
Gravy beef	25 ± 1	153.6	3.1 ± 0.1	Patazca et al. (2007)
	25 ± 1	307.7	3.0 ± 0.1	Patazca et al. (2007)
	25 ± 1	448.9	2.9 ± 0.0	Patazca et al. (2007)
	25 ± 1	534.5	3.0 ± 0.0	Patazca et al. (2007)
Hass avocado	25 ± 1	148.9	4.1 ± 0.2	Patazca et al. (2007)
	25 ± 1	316.5	4.0 ± 0.0	Patazca et al. (2007)
	25 ± 1	480.4	3.6 ± 0.0	Patazca et al. (2007)
	25 ± 1	580.0	3.7 ± 0.1	Patazca et al. (2007)
Honey	1 ± 1	157.7	3.0	Patazca et al. (2007)
	1 ± 1	296.9	2.9	Patazca et al. (2007)
	1 ± 1	444.5	2.7	Patazca et al. (2007)
	1 ± 1	580.5	2.7	Patazca et al. (2007)
	25 ± 1	157.5	3.5 ± 0.1	Patazca et al. (2007)
	25 ± 1	299.1	3.2 ± 0.0	Patazca et al. (2007)
	25 ± 1	449.3	3.1 ± 0.1	Patazca et al. (2007)
	25 ± 1	581.6	2.9 ± 0.0	Patazca et al. (2007)
	50 ± 1	161.8	3.8	Patazca et al. (2007)
	50 ± 1	300.6	3.6	Patazca et al. (2007)
	50 ± 1	437.7	3.5	Patazca et al., 2007
	50 ± 1	585.1	3.3	Patazca et al. (2007)
	70 ± 1	165.4	4.7	Patazca et al. (2007)
	70 ± 1	299.5	4.1	Patazca et al. (2007)
	70 ± 1	452.8	3.9	Patazca et al. (2007)
	70 ± 1	587.1	3.7	Patazca et al. (2007)
Mashed potato	25	n.r.	≈3.0	Ting et al. (2002)
Mayonnaise dressing	25 ± 1	149.3	7.2 ± 0.3	Patazca et al. (2007)
	25 ± 1	318.6	5.7 ± 0.0	Patazca et al. (2007)
	25 ± 1	468.0	5.0 ± 0.0	Patazca et al. (2007)
	25 ± 1	574.6	5.3 ± 0.1	Patazca et al. (2007)
Olive oil	25	n.r.	From 8.7 to <6.3	Ting et al. (2002)
	25 ± 2	173	8.7 ± 0.4	Rasanayagam et al. (2003)
	25 ± 2	325	7.3 ± 0.4	Rasanayagam et al. (2003)

Table 6.1 Quasiadiabatic temperature increases, ΔT/100 MPa (°C), observed in the literature in different food products, packaging and carrier materials, and pressurizing fluids after compression at different pressure and temperature conditions.—cont'd

Product	Initial T (°C)	Pressure applied (MPa)	ΔT/100 MPa (°C)	References
	25 ± 2	455	6.9 ± 0.4	Rasanayagam et al. (2003)
	25 ± 2	501	7.2 ± 0.2	Rasanayagam et al. (2003)
Orange juice	25	n.r.	≈3.0	Ting et al. (2002)
Salmon	25	n.r.	≈3.2	Ting et al. (2002)
	25 ± 2	159	2.8 ± 0.3	Rasanayagam et al. (2003)
	25 ± 2	321	2.9 ± 0.1	Rasanayagam et al. (2003)
	25 ± 2	463	3.0 ± 0.2	Rasanayagam et al. (2003)
	25 ± 2	611	3.0 ± 0.1	Rasanayagam et al. (2003)
Skim milk	25 ± 1	148.3	2.9 ± 0.2	Patazca et al. (2007)
	25 ± 1	314.6	3.0 ± 0.1	Patazca et al. (2007)
	25 ± 1	458.8	2.9 ± 0.0	Patazca et al. (2007)
	25 ± 1	544.8	3.0 ± 0.1	Patazca et al. (2007)
Soy oil	25 ± 2	n.r.	From 9.1 to <6.2	Ting et al. (2002)
	25 ± 2	172	8.3 ± 0.7	Rasanayagam et al. (2003)
	25 ± 2	312	7.5 ± 0.1	Rasanayagam et al. (2003)
	25 ± 2	452	6.9 ± 0.3	Rasanayagam et al. (2003)
	25 ± 2	617	6.3 ± 0.4	Rasanayagam et al. (2003)
Tomato salsa	25	n.r.	≈3.0	Ting et al. (2002)
Vegetable oil	1 ± 1	170.3	9.1	Patazca et al. (2007)
	1 ± 1	286.4	8.6	Patazca et al. (2007)
	1 ± 1	446.6	8.0	Patazca et al. (2007)
	1 ± 1	584.6	6.9	Patazca et al. (2007)
	25 ± 1	156.1	9.2 ± 0.3	Patazca et al. (2007)
	25 ± 1	293.8	8.0 ± 0.3	Patazca et al. (2007)
	25 ± 1	446.7	7.0 ± 0.0	Patazca et al. (2007)
	25 ± 1	586.0	6.6 ± 0.0	Patazca et al. (2007)
	50 ± 1	167.8	9.6	Patazca et al. (2007)
	50 ± 1	292.9	8.3	Patazca et al. (2007)
	50 ± 1	462.3	7.2	Patazca et al. (2007)
	50 ± 1	593.8	6.5	Patazca et al. (2007)
	70 ± 1	164.5	9.7	Patazca et al. (2007)
	70 ± 1	312.5	8.2	Patazca et al. (2007)
	70 ± 1	462.6	7.4	Patazca et al. (2007)
	70 ± 1	591.0	6.8	Patazca et al. (2007)
Whole egg	25 ± 1	142.1	3.2 ± 0.1	Patazca et al. (2007)
	25 ± 1	328.4	3.4 ± 0.1	Patazca et al. (2007)
	25 ± 1	470.7	3.5 ± 0.2	Patazca et al. (2007)
	25 ± 1	552.3	3.3 ± 0.0	Patazca et al. (2007)
Whole milk	25 ± 1	145.4	3.1 ± 0.1	Patazca et al. (2007)
	25 ± 1	326.6	3.2 ± 0.1	Patazca et al. (2007)
	25 ± 1	472.5	3.2 ± 0.0	Patazca et al. (2007)
	25 ± 1	556.8	3.2 ± 0.0	Patazca et al. (2007)

Continued

Table 6.1 Quasiadiabatic temperature increases, ΔT/100 MPa (°C), observed in the literature in different food products, packaging and carrier materials, and pressurizing fluids after compression at different pressure and temperature conditions.—cont'd

Product	Initial T (°C)	Pressure applied (MPa)	ΔT/100 MPa (°C)	References
Materials for food packaging, insulators, and product carriers				
High-density polyethylene (HDPE)	25	600	4.92	Casulli et al. (2017)
	50	600	5.42	Casulli et al. (2017)
	75	600	5.80	Casulli et al. (2017)
Low-density polyethylene (LDPE)	25	600	5.65	Casulli et al. (2017)
	50	600	6.44	Casulli et al. (2017)
	75	600	7.41	Casulli et al. (2017)
Polyamide (PA)	25	600	4.25	Casulli et al. (2017)
	50	600	4.74	Casulli et al. (2017)
	75	600	5.18	Casulli et al. (2017)
Polypropylene (PP)	25	600	4.29	Casulli et al. (2017)
	50	600	4.65	Casulli et al. (2017)
	75	600	5.01	Casulli et al. (2017)
Polystyrene (PS)	25	600	4.31	Casulli et al. (2017)
	50	600	4.39	Casulli et al. (2017)
	75	600	4.53	Casulli et al. (2017)
Pressurizing fluids				
Ethanol	25 ± 2	144	10.6 ± 0.6	Rasanayagam et al. (2003)
	25 ± 2	325	5.4 ± 0.6	Rasanayagam et al. (2003)
	25 ± 2	457	5.5 ± 0.2	Rasanayagam et al. (2003)
	25 ± 2	540	6.8 ± 0.4	Rasanayagam et al. (2003)
Propylene glycol	25 ± 2	164	5.8 ± 0.6	Rasanayagam et al. (2003)
	25 ± 2	325	5.3 ± 0.4	Rasanayagam et al. (2003)
	25 ± 2	470	5.1 ± 0.2	Rasanayagam et al. (2003)
	25 ± 2	548	5.1 ± 0.5	Rasanayagam et al. (2003)
Water	1 ± 1	156.9	1.3	Patazca et al. (2007)
	1 ± 1	292.9	1.7	Patazca et al. (2007)
	1 ± 1	439.0	1.9	Patazca et al. (2007)
	1 ± 1	594.5	1.9	Patazca et al. (2007)
	25	n.r.	≈3.0	Ting et al. (2002)
	25 ± 2	149	2.6 ± 0.1	Rasanayagam et al. (2003)
	25 ± 2	298	2.7 ± 0.2	Rasanayagam et al. (2003)
	25 ± 2	443	2.8 ± 0.1	Rasanayagam et al. (2003)
	25 ± 2	592	2.9 ± 0.3	Rasanayagam et al. (2003)
	25 ± 1	164.6	2.6	Patazca et al. (2007)
	25 ± 1	290.4	2.8	Patazca et al. (2007)
	25 ± 1	449.8	2.9	Patazca et al. (2007)
	25 ± 1	589.0	2.9	Patazca et al. (2007)
	50 ± 1	165.4	4.0	Patazca et al. (2007)

Table 6.1 Quasiadiabatic temperature increases, ΔT/100 MPa (°C), observed in the literature in different food products, packaging and carrier materials, and pressurizing fluids after compression at different pressure and temperature conditions.—cont'd

Product	Initial T (°C)	Pressure applied (MPa)	ΔT/100 MPa (°C)	References
	50 ± 1	294.1	3.7	Patazca et al. (2007)
	50 ± 1	443.6	3.7	Patazca et al. (2007)
	50 ± 1	588.9	3.7	Patazca et al. (2007)
	70 ± 1	160.5	4.8	Patazca et al. (2007)
	70 ± 1	302.0	4.5	Patazca et al. (2007)
	70 ± 1	448.3	4.3	Patazca et al. (2007)
	70 ± 1	595.5	4.3	Patazca et al. (2007)
Water/glycol (50/50)	25	n.r.	From 4.8 to <3.7	Ting et al. (2002)

n.r., Not reported.

pressurization up to 600 MPa, the temperature of water (typical pressurizing fluid), polypropylene (typical product carrier material), low-density polyethylene (typical product packaging material), and mayonnaise dressing (food product) would increase from 25°C to 42.4°C, 50.7°C, 58.9°C, and 56.8°C, respectively. Thus, these simple calculations demonstrate that, although frequently regarded as a nonthermal technology, HPP at room temperature can induce significant thermal effects in the processed products. These pressure-induced thermal effects are even more relevant in HPTP, as they are specifically used to heat and cool food products rapidly and uniformly during compression and expansion, respectively. Adiabatic heating during compression allows products at moderate initial temperatures to rapidly reach the high temperatures required to achieve the intended effects (e.g., commercial sterility), while adiabatic cooling after expansion results in immediate decrease in product temperature. In this way, HPTP makes it possible to reduce the exposure time of foods to high temperatures, thus minimizing the negative impacts of thermal processing on quality attributes (Gänzle, 2023; Queirós et al., 2023).

However, differences in the compression heating of the materials inside the pressure vessel are not the only cause of thermal nonuniformities during HPP. Due to the low compressibility of steel, the vessel walls and plugs experience negligible compression heating, and this also contributes to create thermal gradients. In addition, the pressurizing fluid pumped into the vessel to reach the target pressure, usually more than 15% of the equipment volume, is another relevant cause of temperature inhomogeneities (Queirós et al., 2023). All these factors jointly contribute to induce heat exchanges between the food products and packaging, the product carriers, the pressurizing fluid, and the external environment through the vessel walls, thus causing temperature gradients and fluid motion phenomena inside the pressure vessel during HPP. Moreover, other factors such as the equipment design (insulation, vessel orientation, among others), the compression rate, or the vessel filling ratio, for example, can significantly affect heat exchanges and fluid motion phenomena, also impacting the uniformity of high-pressure processes (Knoerzer et al., 2010a; Otero et al., 2007; Smith et al., 2014).

Abundant evidence of thermal gradients during HPP has been obtained in lab equipment by using thermocouples placed at different points of the high-pressure vessel, both in the sample and the pressure medium (Denys et al., 2000a, 2000b; Hartmann et al., 2004; Knoerzer et al., 2007; Otero et al., 2007). For example, Denys et al. (2000b) measured temperature differences higher than 5°C between the surface and center of tomato paste samples during HPP at 340 MPa and 20°C, while Knoerzer et al. (2007) recorded temperature differences higher than 10°C at different locations inside the product carrier when pressurizing water at 600 MPa and 90°C. More sophisticated techniques have also been used to reveal the existence of thermal gradients during HPP. Thus, Pehl et al. (2000) and Delgado and Hartmann (2003) employed temperature-sensitive liquid crystals to visualize temperature distribution and fluid motion phenomena in liquid food ingredients pressurized, up to 700 MPa, in a tempered high-pressure optical cell equipped with sapphire windows.

Many authors have demonstrated that the observed temperature differences throughout the high-pressure vessel may result in pronounced inhomogeneities in microbial and/or enzyme inactivation and nutritional and/or sensorial degradation within the processed product (Juliano et al., 2009c; Rauh et al., 2009). Thus, for example, Kilimann et al. (2006) processed *Lactobacillus lactis* cells in different buffer systems at 500 MPa and observed temperature differences of up to 6°C between the upper and the lower part of the pressure vessel that induced differences of more than one decimal reduction in the inactivation observed. Similarly, Denys et al. (2000b) measured residual activities between 48% and 73% in solutions of *Bacillus subtilis* α-amylase located at different positions in the pressure vessel and processed at an average pressure of 543 MPa for 25 min.

All the data available in the literature reveal that heat exchanges occurring during HPP can cause nonuniform temperature distributions throughout the treated foods, which can lead to significant inhomogeneities in the intended pressure-induced effect. Therefore, understanding the heat transfer phenomena that occur throughout the vessel during HPP is of paramount importance not only for ensuring process uniformity, but also for scaling up processes and designing industrial-scale equipment.

6.2.2 Mapping uniformity of high-pressure processes

Direct measurement of temperature at different points inside the pressure vessel would be, evidently, the most straightforward method to map thermal uniformity and "visualize" heat transfer phenomena during pressure processing; however, implementing temperature sensors in industrial-scale HPP equipment is today a technological challenge.

In vertically oriented, lab- and small pilot-scale equipment, temperature is generally monitored by one or more wired thermocouples that are usually fitted through the nonmobile bottom plug of the vessel. Implementing thermocouples in industrial-scale HPP machines is, however, much more complicated because they typically have horizontal configuration and both the vessel and the closure plugs are moving parts during normal operation. To avoid this inconvenience, several pressure-resistant, wireless systems have been developed for direct measurement and acquisition of temperature data under pressure. These systems basically consist of a temperature logger inside a water-tight, pressure-resistant shell that acquires temperature data directly measured by a temperature sensor located either inside the shell or through it (Knoerzer et al., 2010d; Queirós et al., 2023). The main problem with these devices is that several are needed to monitor temperature at different positions and, as they are relatively large, they can affect free fluid motion and act as heat sinks, thus affecting heat exchanges inside the pressure vessel.

Pressure-temperature-time indicators (pTTIs) have also been proposed for an indirect estimation of temperature heterogeneities inside the pressure vessel. These sensors can be defined as small, wireless devices that exhibit an easily, quantifiable, and irreversible response that depends on the pressure, temperature, and time conditions experienced. Thus, by locating pTTIs at different positions in the pressure vessel, temperature uniformity can be assessed. For example, Grauwet and coworkers used small, flexible microtubes filled with different active proteins in buffer solutions as pTTIs and, by measuring the residual activity, they estimated temperature uniformity during both HPP and HPTP (Grauwet et al., 2009, 2010a, 2010b, 2011). Although pTTIs are cheap and broadly employable systems, they also present limitations mainly related to their specific application window and temperature sensitivity. Moreover, it is important to note that pTTIs integrate the combined effect of pressure, temperature, and time and, therefore, the same readout can be obtained for different processing conditions (Grauwet et al., 2016).

A major drawback of both temperature sensors and pTTIs is that they provide measurements only at specific points. Thus, even though multiple sensors can be positioned at different locations in the pressure vessel, they cannot completely map temperature uniformity throughout the vessel. In this context, mathematical modeling, and more specifically computational fluid dynamics (CFD) modeling, can offer a valuable solution since it allows predicting temperature and flow distribution at any point inside the pressure vessel. These predictions are, in turn, essential to gain insight into how the equipment design and process variables can affect heat exchange and fluid motion phenomena occurring during high-pressure food processing and, in consequence, process uniformity. Moreover, coupling CFD models with kinetic models of microbial and/or enzyme inactivation or quality degradation allows evaluation of process uniformity in terms of food safety and quality and, therefore, the design of safe high-pressure processes with minimal impact on the product quality.

6.3 Thermophysical properties under pressure

One of the main difficulties in modeling heat transfer and simulating the temperature distribution within the pressure vessel during HPP is the lack of data on the pressure-dependent thermophysical properties of food products, packaging, product carriers, and pressurizing fluids other than water. To overcome this inconvenience, different theoretical, semiempirical, and empirical models have been proposed in the literature (Otero and Datta, 2026), many of them based on previous models used to predict thermophysical properties at atmospheric conditions (Guignon et al., 2008; Otero et al., 2010). As theoretical models are more general and can be applied to a wide variety of products and situations, in this chapter, only they are described in depth. The reader is directed to Table 6.2 to find a collection of references in which semiempirical and empirical models to predict thermophysical properties of several types of products at different processing conditions are described.

6.3.1 Thermophysical properties required to model heat transfer and fluid motion phenomena under pressure

Modeling and simulating **heat transfer** phenomena during HPP requires the knowledge of the density, specific heat capacity, and thermal conductivity (K) of the materials involved at the pressure and temperature conditions of the treatment. Although for a number of food products and fluids of interest, measurements of some of these properties have been performed at certain pressure and temperature conditions, and even semiempirical and empirical models have been, in some cases, derived (Table 6.2), thermophysical data are very limited or inexistent for most materials under pressure. The only exception is pure water that is the main component of most foods and, as previously mentioned, the pressurizing fluid most frequently used in high-pressure food processing. Thus, the International Association for the Properties of Water and Steam (IAPWS) provides accurate formulations for the calculation of $\rho(P,T), Cp(P,T)$, and $K(P,T)$ values for pure water in the stable fluid region from the melting-pressure curve to 1000°C and pressures up to 1000 MPa (IAPWS, 2011, 2018). However, for the rest of the materials, thermophysical properties have generally to be estimated when modeling heat transfer during HPP.

Since foods are heterogeneous materials made up of different components (proteins, fats, simple sugars, complex carbohydrates, and ash) and pressurizing fluids other than water frequently are mixtures of two or more liquids in variable proportions, it is evident that their thermophysical properties will depend on the relative proportion and intrinsic properties of each component. Therefore, when composition is known, additive models are often used to predict thermophysical properties of food products and pressurizing fluids, both at atmospheric and at high pressure (Choi and Okos, 1986; Guignon et al., 2010; Otero et al., 2010; Phinney et al., 2017):

$$\frac{1}{\rho(P,T)} = \sum_{j=1}^{n} \frac{m_j}{\rho_j(P,T)} \tag{6.2}$$

$$Cp(P,T) = \sum_{j=1}^{n} m_j(P,T) \cdot Cp_j(P,T) \tag{6.3}$$

Table 6.2 Semiempirical and empirical models available in the literature to predict thermophysical properties of foods and pressurizing fluids under pressure.

	Semiempirical models				Empirical models			
		Validity range				Validity range		
Property	**Product**	**P (MPa)**	**T (°C)**	**References**	**Product**	**P (MPa)**	**T (°C)**	**References**
Initial freezing point	General use	0.1–210	–	Guignon et al. (2008) Otero et al. (2006) Otero et al. (2010)	Aqueous solutions	0.1–210	–	Guignon et al. (2005)
						0.1–210		Guignon et al. (2008)
						0.1–350		Guignon et al. (2009)
					Vegetable products	0.1–111.4	–	Denys (2000)
						0.1–210		Schlüter and Knorr (2002)
						0.1–210		Zhu et al. (2006)
						0.1–210		Guignon et al. (2008)
					Fish products	0.1–210	–	Zhu et al. (2006)
								Guignon et al. (2008)
								Zhu et al. (2014)
					Meat products	0.1–210	–	Zhu et al. (2006)
					Others: Gels	0.1–210	–	Guignon et al. (2008)

Continued

Table 6.2 Continued

Property	Semiempirical models				Empirical models			
		Validity range				Validity range		
	Product	P (MPa)	T (°C)	References	Product	P (MPa)	T (°C)	References
Density/ Specific volume	Aqueous solutions	All P/T combinations at which $\rho_w(P,T)$ values are available		Werner et al. (2007)	Aqueous solutions	0.1–700	25	Min et al. (2009, 2010)
	Fats and oils	0.1–145	30–80	Acosta et al. (1996)	Fats and oils	0.1–45	10–90	Freitas et al. (2013)
						0.1–145	30–80	Acosta et al. (1996)
						0.1–350	15–55	Guignon et al. (2009)
						0.1–520	20–50	Kiełczyński et al. (2017)
						0.1–700	25	Min et al. (2009, 2010)
						0.1–1000	10–50	Rostocki et al. (2011)
						0.1–1000	20	Malanowski et al. (2013)
						0.1–1000	20	Pawlicki et al. (2022)
						0.1–1000	20–40	Rostocki et al. (2009)

	Milk products	–	–	–	Milk products		0.1–350	0–60	Guignon et al. (2014)
							0.1–700	25	Min et al. (2009, 2010)
	Vegetable products	0.1–400	−20.5–60	Denys (2000)	Vegetable products		0.1–350	−20–40	Guignon et al. (2012)
							0.1–350	0–50	Aparicio et al. (2011)
							0.1–700	25	Min et al. (2009, 2010)
	Fish products	–	–	–	Fish products		0.1–700	25	Min et al. (2009, 2010)
	Meat products	–	–	–	Meat products		0.1–700	25	Min et al. (2009, 2010)
	Other foods	–	–	–	Other foods	Honey	0.1–700	25	Min et al. (2010)
	Pressurizing fluids:	–	–	–	Pressurizing fluids:				
	Water	–	–	–	Water				
	Other than water				Other than water		0.1–350	0–40	Guignon et al. (2010)

Continued

Table 6.2 Continued

Property	Semiempirical models				Empirical models			
		Validity range				Validity range		
	Product	P (MPa)	T (°C)	References	Product	P (MPa)	T (°C)	References
Thermal expansivity	Fats and oils	0.1–350	15–55	Guignon et al. (2009)	–	–	–	–
		0.1–520	20–50	Kiełczyński et al. (2017)				
	Vegetable products	0.1–350	−20–40	Guignon et al. (2012)	–	–	–	–
Specific heat capacity	Aqueous solutions	–	–	–	Aqueous solutions	0.1–600	25	Nguyen et al. (2012)
	Fats and oils	–	–	–	Fats and oils	0.1–356	25	Leu et al. (2010)
						0.1–600	25	Nguyen et al. (2012)
	Vegetable products	–	–	–	Vegetable products	0.1–600	25	Nguyen et al. (2012)
	Others	–	–	–	Other foods: Honey	0.1–600	25	Nguyen et al. (2012)

Thermal conductivity	General use	All P/T combinations at which the required inputs are available		Otero et al. (2010)	Aqueous solutions	0.1–600	25	Nguyen et al. (2012)
	Liquids			Werner et al. (2007)				
	Liquids			Bridgman (1923)				
	Aqueous solutions			Werner et al. (2007)				
				Rastorguev and Ganiev (1968)				
				Wilke (1950)				
				Pandey and Mishra (2005)				
				El'darov (2003)				
		0.1–600	25	Ross et al. (1984)				
	Fats and oils	0.1–400	10–60	McLaughlin (1969)	Fats and oils	0.1–600	25	Nguyen et al. (2012)
		0.1–400	10–60	Werner et al. (2008)				
		0.1–600	25	Ross et al. (1984)		0.1–700	25	Ramaswamy et al. (2007)
	Milk products	–	–	–	Milk products	0.1–350	5	Zhu et al. (2007)
						0.1–350	5–25	Zhu et al. (2008)
						0.1–400	25	Sun et al. (2016)
						0.1–600	25	Nguyen et al. (2012)

Continued

Table 6.2 Continued

Property	Semiempirical models					Empirical models				
			Validity range					Validity range		
	Product		P (MPa)	T (°C)	References	Product		P (MPa)	T (°C)	References
	Vegetable products		0.1–600	25	Nguyen et al. (2012)	Vegetable products		0.1–350	5	Zhu et al. (2007)
								0.1–350	5–25	Zhu et al. (2008)
								0.1–400	25–55	Sun et al. (2018)
								0.1–400	30–65	Denys and Hendrickx (1999)
								0.1–600	25	Nguyen et al. (2012)
								0.1–700	25	Ramaswamy et al. (2007)
	Fish products		–	–	–	Fish products		0.1–350	5–25	Zhu et al. (2008)
	Meat products		–	–	–	Meat products		0.1–350	5–25	Zhu et al. (2008)
								0.1–400	25	Sun et al. (2016)
								0.1–400	25–55	Sun et al. (2018)
	Other foods	Honey	0.1–600	25	Ross et al. (1984)	Other foods	Flour dough	0.1–400	25–55	Sun et al. (2018)
							Egg white	0.1–400	25	Sun et al. (2016)
								0.1–400	25	Sun et al. (2016)
							Egg yolk	0.1–600	25	Nguyen et al. (2012)
							Honey	0.1–700	25	Ramaswamy et al. (2007)
							Honey			
							Corn syrup	0.1–700	25	Ramaswamy et al. (2007)

Viscosity	Aqueous solutions	0.1–700	5–60	Först et al. (2002)	Aqueous solutions	–	–	–
	Fats and oils	0.1–480	1–100	Hersey and Hopkins (1937)	Fats and oils	0.1–150	20	Schaschke et al. (2006)
		0.1–150	20	Schaschke et al. (2006)				
		0.1–300	25	Zolotykh et al. (1975)				
		0.1–330	40	Ohno et al. (1997)				
		0.1–350	20–80	Mawatari et al. (2013)				
		0.1–400	40	Mia et al. (2007)				
		0.1–500	20–80	Hö glund (1999)				
		0.1–1230	1–91	Biresaw and Bantchev (2013)				
		0.1–1230	2–91	Barlow et al. (1973)				

$$\frac{1}{K(P,T)}=\sum_{j=1}^{n}\frac{\varepsilon_j(P,T)}{K_j(P,T)} \tag{6.4}$$

$$K(P,T)=\sum_{j=1}^{n}\varepsilon_j(P,T)\cdot K_j(P,T) \tag{6.5}$$

$$K(P,T)=\prod_{j=1}^{n}K_j(P,T)^{\varepsilon_j(P,T)} \tag{6.6}$$

where $\rho(P,T)$ and $\rho_j(P,T), Cp(P,T)$ and $Cp_j(P,T)$, $K(P,T)$ and $K_j(P,T)$, are the pressure- and temperature-dependent densities, specific heat capacities, and thermal conductivities of the product and of each of its "n" components, respectively, while $m_j(P,T)$ and $\varepsilon_j(P,T)$ are the corresponding mass and volume fractions. As the thermal conductivity of foods depends not only on their composition, but also on their structure, series, Eq. (6.4), and parallel models, Eq. (6.5), defining the lower and upper extremes between which the effective thermal conductivity of any food should be, are here presented. Moreover, Eq. (6.6) represents the geometric-mean model that predicts K values intermediate between those calculated from the series and parallel models.

The main limitation of these additive models is the lack of data on the intrinsic properties under pressure of food components and pressurizing fluids other than water. Thus, only in a number of fluids, including ethanol, ethylene glycol, propylene glycol, castor oil, or silicon oil (Guignon et al., 2010; Mawatari et al., 2013; Werner et al., 2008; Yusa et al., 1977), some thermophysical properties have been measured at certain pressure and temperature conditions, and there hardly exist data for proteins, fats, simple sugars, complex carbohydrates, and ash. For this reason, when using additive models to predict thermophysical properties of foods under pressure, the intrinsic properties of food constituents are usually considered not pressure-dependent, but only temperature-dependent (Guignon et al., 2012; Otero et al., 2006, 2010).

If detailed food composition is not known, but only water and dry matter contents, Eqs. (6.2)–(6.6) can still be used if the food properties at atmospheric conditions, $\rho(P_{atm},T), Cp(P_{atm},T)$, and $K(P_{atm},T)$ are available (Otero et al., 2006, 2010). In this case, dry matter can be considered as a whole and $\rho_{dm}(P_{atm},T)$,$Cp_{dm}(P_{atm},T)$, and $K_{dm}(P_{atm},T)$ values can be derived from food properties at atmospheric conditions by using Eqs. (6.2)–(6.6).

When modeling HPF and HPT processes, it is important to note that not only liquid water but also ice must be considered among the food components in Eqs. (6.2)–(6.6). To estimate the ice fraction in frozen and partially frozen foods, the knowledge of the pressure-dependent initial freezing point, $T_{if}(P)$, is mandatory but, unfortunately, there hardly exist data for real foods under pressure. Therefore, in most cases, $T_{if}(P)$ values need to be estimated and this can be done by using theoretical models such as the Robinson–Stokes equation and the modified Raoult's law (Guignon et al., 2008).

The Robinson–Stokes equation, Eq. (6.7), links the initial freezing point of a food with its water activity, a_w, at this temperature:

$$-ln\ a_w(P, Tk_{if}) = \frac{L_m^0(P) - \Delta Cp_m^0(P) \cdot Tk_{if}^0(P)}{R} \cdot \left(\frac{1}{Tk_{if}(P)} - \frac{1}{Tk_{if}^0(P)} \right) + \frac{\Delta Cp_m^0(P)}{R} \cdot ln \frac{Tk_{if}^0(P)}{Tk_{if}(P)} \tag{6.7}$$

where $Tk_{if}^0(P)$ and $Tk_{if}(P)$ are the freezing points (K), at the target pressure, of pure water and food, respectively, $L_m^0(P)$ is the molar latent heat ($J{\cdot}mol^{-1}$) of pure water, $\Delta Cp_m^0(P)$ is the difference in molar heat capacities ($J{\cdot}mol^{-1}{\cdot}K^{-1}$) of liquid water and ice at the corresponding freezing point $Tk_{if}^0(P)$, and R is the universal gas constant. If $T_{if}(P_{atm})$ of the food product is known, the Robinson–Stokes equation can be used to predict $T_{if}(P)$ values at any pressure by considering that water activity is constant with pressure and temperature along the melting curve. In this case, $a_w(P_{atm}, Tk_{if})$ is first derived from Eq. (6.7) at atmospheric conditions and then, this value is used to predict $T_{if}(P)$ values at any pressure.

When $T_{if}(P_{atm})$ of the food product is unknown, compositional models based on the Clausius–Clapeyron relation and the Raoult's law, Eq. (6.8), can be employed provided the food composition is available:

$$Tk_{if}(P) = Tk_{if}^0(P) - \frac{R \cdot Tk_{if}^0(P)^2 \cdot \sum_{i=1}^{n} \frac{m_i}{M_i}}{L_m^0(P) \cdot \left(\frac{m_{tw} - m_{bw}}{M_w} + \sum_{i=1}^{n} \frac{m_i}{M_i} \right)} \tag{6.8}$$

where m_i is the mass fraction of the different solutes (ions, sugars, acids, and alcohols), M_i corresponds to their molar masses ($g{\cdot}mol^{-1}$), m_{tw} and m_{bw} represent the mass fraction of total and bound water, respectively, and M_w is the molar mass of water (Guignon et al., 2008; Otero et al., 2010).

Once $T_{if}(P)$ of the food product is estimated at the target pressure, the corresponding mass fraction of ice can be calculated at any temperature below $T_{if}(P)$ by this simple mass balance:

$$m_{ice}(P, T) = (m_{tw} - m_{ufzw}) - m_{ufw}(P, T) \tag{6.9}$$

where m_{ufzw} and m_{ufw} represent the mass fractions of unfreezable and unfrozen free water, respectively. For simplicity, unfreezable water is generally considered equal to the bound water and unaffected by pressure, while unfrozen free water must be estimated at each pressure and temperature condition. Assuming that foods can be treated as ideal solutions at low solute concentrations, the mole fraction of unfrozen free water, X_{ufw}, can be predicted by the Robinson–Stokes equation as:

$$ln\ X_{ufw}(P, Tk) = -\frac{L_m^0(P) - \Delta Cp_m^0(P) \cdot Tk_{if}^0(P)}{R} \cdot \left(\frac{1}{Tk} - \frac{1}{Tk_{if}^0(P)}\right)$$
$$-\frac{\Delta Cp_m^0(P)}{R} \cdot ln\frac{Tk_{if}^0(P)}{Tk} \tag{6.10}$$

and, by solving Eq. (6.11), $m_{ufw}(P, T)$ can be easily obtained (Otero et al., 2010):

$$X_{ufw}(P, T) = \frac{\frac{m_{ufw}(P, T)}{M_w}}{\frac{m_{ufw}(P, T)}{M_w} + \frac{m_{dm}}{M_{dm}}} \tag{6.11}$$

Alternatively, and for small excursions from the initial freezing point and for $T_{if}(P) \approx T_{if}^0(P)$, ice content in frozen and partially frozen foods can also be calculated as:

$$m_{ice}(P, T) = (m_{tw} - m_{ufzw}) \cdot \left(1 - \frac{Tk_{if}(P) - Tk_{if}^0(P)}{Tk - Tk_{if}^0(P)}\right) \tag{6.12}$$

Depending on the model formulation, simulation of heat transfer phenomena during HPF and HPT also requires the knowledge of either the apparent specific heat capacity, $Cp_{app}(P, T)$, or the enthalpy, $H(P, T)$. Otero et al. (2010) showed that these pressure-dependent additive properties can be estimated as:

$$Cp_{app}(P, T) = \sum_{n=1}^{j} m_j(P, T) \cdot Cp_j(P, T) - L^0(P, T_{if}(P)) \cdot \frac{\partial m_{ice}(P, T)}{\partial T} \tag{6.13}$$

$$H(P, T) = \int_{T_{ref}(P)}^{T} Cp_{app}(P, T)dT \tag{6.14}$$

where $L^0(P, T_{if}(P))$ is the latent heat of fusion of pure ice at the initial freezing point of the food considered:

$$L^0(P, T_{if}(P)) = L^0\left(P, T_{if}^0(P)\right) - \left(Cp_w\left(P, T_{if}^0\right) - Cp_{ice}\left(P, T_{if}^0\right)\right) \cdot \left(T_{if}^0(P) - T_{if}(P)\right) \tag{6.15}$$

and the variation of the ice content with temperature can be derived from Eq. (6.16) as:

$$\frac{\partial m_{ice}(P, T)}{\partial T} = \frac{(m_{tw} - m_{ufzw}) \cdot \left(T_{if}(P) - T_{if}^0(P)\right)}{\left(T - T_{if}^0(P)\right)^2} \tag{6.16}$$

Moreover, the reference temperature in Eq. (6.14), at which $H = 0$, can be calculated as:

$$T_{ref}(P) = T_{if}^0\ (P) - 40 \tag{6.17}$$

Apart from density, specific heat capacity, and thermal conductivity, modeling **fluid motion** phenomena under pressure also requires the knowledge of the dynamic viscosity, $\mu(P, T)$, of all the fluids involved. Although viscosity values at high pressure have been measured for a number of fluids in the literature (Först et al., 2002; Fukui et al., 2010; Mia et al., 2007; Yusa et al., 1977), the existing data are insufficient as they do not cover all the fluids and pressure and temperature conditions of interest to high-pressure food processing. Estimations for the dynamic viscosity of fluid foods and aqueous pressurizing fluids can be made by using the "ratio model," which takes into account the increase in dynamic viscosity with pressure, akin to the behavior observed in pure water:

$$\mu(P, T) = \frac{\mu(P_{atm}, T)}{\mu_w(P_{atm}, T)} \cdot \mu_w(P, T) \tag{6.18}$$

Since pressure- and temperature-dependent viscosity of water can be easily calculated from models available in the literature (Huber et al., 2009; IAPWS, 2008), the only input required by the ratio model is the viscosity of the product at atmospheric conditions.

Evidently, the ratio model can also be used to obtain rough estimations of density, specific heat capacity, and thermal conductivity of any unfrozen food or aqueous pressurizing fluid under pressure by simply substituting μ by the respective property (Knoerzer and Chapman, 2011). In fact, it has been successfully used in the literature to simulate heat transfer and fluid motion phenomena during HPP of a number of foods such as milk (Hartmann et al., 2003) and beef fat (Abdul Ghani and Farid, 2007). Moreover, Guignon et al. (2010) employed this model to predict the specific volume (i.e., the inverse of density) of binary mixtures used as pressuring fluids. They found that the ratio model predicts the specific volume of binary mixtures adequately, provided water is the main component.

6.3.2 Thermophysical properties required to model compression heating

When simulating high-pressure processes, the accurate prediction of the adiabatic heating experienced by all materials inside the pressure vessel during compression is essential because this adiabatic heating is the main cause of nonuniformities during treatments. As mentioned in Section 6.2.1, rough estimations of the temperature changes during compression and expansion can be made by considering ΔT/100 MPa values reported in Table 6.1. However, due to the nonlinear nature of these pressure- and temperature-dependent changes, such estimations could be far from reality (Knoerzer, 2023).

Accurate predictions of the compression heating can be performed by using Eq. (6.1). Solving this equation requires the knowledge of not only the density and specific heat capacity of the material, but also the corresponding thermal expansivity, α. This property describes how the volume of a material changes with temperature at a constant pressure:

$$\alpha = \frac{1}{v} \cdot \left(\frac{\partial v}{\partial T}\right)_P \tag{6.19}$$

where v represents the specific volume.

Unfortunately, data on the thermal expansivity of food products, packaging and carrier materials, and pressurizing fluids are very scarce in the literature, probably because their determination requires arduous measurements under well-controlled temperature and pressure conditions (Aparicio et al., 2011; Guignon et al., 2009). Ideally, simple additive models, similar to those described in Eqs. (6.2)–(6.6), could be employed to predict the pressure- and temperature-dependent thermal expansivity of both foods (Guignon et al., 2014) and pressurizing fluids made of a mixture of liquids (Guignon et al., 2010). However, the lack of data on the thermal expansivity, even at atmospheric conditions, of the major food components and pure pressurizing fluids makes it inviable. On the other hand, if derivable expressions for $v(P,T)$ are available, $\alpha(P,T)$ could be predicted from Eq. (6.19), but these expressions only exist for a very limited number of products, including diacylglycerol, sunflower and olive oils, and orange juice and concentrates (Guignon et al., 2009, 2012; Kiełczyński et al., 2017).

To overcome this inconvenience, a compression heating coefficient, $k_c(P,T)$, encompassing thermal expansivity, density, and specific heat capacity, can be defined:

$$k_c(P,T) = \frac{\alpha(P,T)}{\rho(P,T) \cdot Cp(P,T)} \tag{6.20}$$

Knoerzer and coworkers showed that $k_c(P, T)$ values for both solid products (Knoerzer et al., 2010c) and liquids (Knoerzer et al., 2010b) can be derived from the temperature profiles acquired in these products during HPP. Thus, by using a sophisticated methodology, these authors obtained $k_c(P, T)$ values for water, propylene glycol, and water/propylene glycol mixtures, and also for some insulating plastics commonly used for the manufacture of product carriers, such as high-density polyethylene (HDPE), polypropylene (PP), and polytetrafluoroethylene (PTFE). They observed that the effects of pressure and temperature on the compression heating coefficient depended on the molecular characteristics of the product and, therefore, experimental $k_c(P, T)$ data were fitted by more or less complex expressions depending on the material (Table 6.3).

These expressions can then be used to accurately predict temperature changes during compression or expansion in any high-pressure process:

$$\frac{dT}{dP} = \frac{\alpha(P,T)}{\rho(P,T) \cdot Cp(P,T)} \cdot Tk = k_c(P,T) \cdot Tk \tag{6.21}$$

Thus, Knoerzer et al. (2010b, 2010c) found very good agreement between measured and predicted temperatures, calculated according to the equations shown in Table 6.3, in all the pressurizing fluids and plastics investigated.

6.4 Numerical modeling of high-pressure processes and applications

Different approaches have been used in the literature to model high-pressure processes, including analytical and numerical methods, macroscopic models, and artificial neural networks (ANNs). All of these modeling approaches present different capabilities, advantages, and limitations, but numerical models are, undoubtedly, the most versatile, as they allow the prediction of process variables with high local resolution and in a wide variety of situations. Specifically, CFD modeling is especially interesting for HPP because it can predict the pressurizing fluid flow based on the governing equations of conservation of mass, momentum, and energy. Moreover, partial differential equations for flow and heat transfer can be coupled with those for microbial and/or enzyme inactivation and solved simultaneously to predict time-dependent distributions throughout the pressure vessel during processing (Juliano et al., 2009b). Nowadays, building CFD models for simulation of high-pressure processes is relatively easy, as there are many user-friendly CFD software packages available on the market. This section details the steps required to build CFD models describing high-pressure processes and also some examples of the use of these models in the literature.

Table 6.3 Equations and coefficients for prediction of the compression heating coefficient, $k_c(P,T)$, of some fluids and plastics of interest to high-pressure processing (Knoerzer et al., 2010b, 2010c).

PRESSURIZING FLUIDS	$\frac{k_c(P,T)}{Pa^{-1}} = a_1 + a_2\cdot\frac{P}{MPa} + a_3\cdot\frac{Tk}{K} + a_4\cdot\frac{P^2}{MPa^2} + a_5\cdot\frac{Tk^2}{K^2} + a_6\cdot\frac{P\cdot Tk}{MPa\cdot K} + a_7\cdot\frac{P^3}{MPa^3} + a_8\cdot\frac{Tk^3}{K^3} + a_9\cdot\frac{P^2\cdot Tk}{MPa^2\cdot K} + a_{10}\cdot\frac{P\cdot Tk^2}{MPa\cdot K^2} + a_{11}\cdot\frac{P^4}{MPa^4} + a_{12}\cdot\frac{Tk^4}{K^4} + a_{13}\cdot\frac{P^3\cdot Tk}{MPa^3\cdot K} + a_{14}\cdot\frac{P\cdot Tk^3}{MPa\cdot K^3} + a_{15}\cdot\frac{P^2\cdot Tk^2}{MPa^2\cdot K^2}$				
Coefficients	**Water**	**25% PG**	**50% PG**	**75% PG**	**100% PG**
a_1	$-1.5804\cdot10^{-10}$	$-1.3148\cdot10^{-10}$	$-6.6041\cdot10^{-10}$	$1.6177\cdot10^{-9}$	$1.2933\cdot10^{-9}$
a_2	$8.8676\cdot10^{-12}$	$1.4631\cdot10^{-11}$	$-5.6923\cdot10^{-13}$	$-1.3241\cdot10^{-12}$	$-1.0785\cdot10^{-12}$
a_3	$-8.0720\cdot10^{-12}$	$-1.8247\cdot10^{-11}$	$7.6012\cdot10^{-12}$	$-1.1040\cdot10^{-11}$	$-8.2096\cdot10^{-12}$
a_4	$-8.2422\cdot10^{-15}$	$-1.3209\cdot10^{-15}$	$1.1105\cdot10^{-15}$	$1.9398\cdot10^{-16}$	$7.1228\cdot10^{-16}$
a_5	$6.8470\cdot10^{-14}$	$1.6885\cdot10^{-13}$	$-2.2833\cdot10^{-14}$	$2.7505\cdot10^{-14}$	$2.1188\cdot10^{-14}$
a_6	$-5.3341\cdot10^{-14}$	$-1.2705\cdot10^{-13}$	$3.3747\cdot10^{-16}$	$6.6598\cdot10^{-15}$	$3.5162\cdot10^{-15}$
a_7	$3.1596\cdot10^{-18}$	$5.0109\cdot10^{-19}$	$-3.9688\cdot10^{-19}$	$-3.2535\cdot10^{-19}$	$-8.5059\cdot10^{-19}$
a_8	$-1.7692\cdot10^{-16}$	$-4.9836\cdot10^{-16}$	$2.3384\cdot10^{-17}$	$-2.1010\cdot10^{-17}$	$-1.7034\cdot10^{-17}$
a_9	$3.4467\cdot10^{-17}$	$4.8818\cdot10^{-18}$	$-1.3442\cdot10^{-18}$	$1.1304\cdot10^{-18}$	$1.6959\cdot10^{-18}$
a_{10}	$1.0872\cdot10^{-16}$	$3.7093\cdot10^{-16}$	$3.0288\cdot10^{-19}$	$-1.1585\cdot10^{-17}$	$-7.4571\cdot10^{-18}$
a_{11}	$1.8932\cdot10^{-22}$	$2.8259\cdot10^{-22}$	N/A	N/A	N/A
a_{12}	$1.5544\cdot10^{-19}$	$4.9163\cdot10^{-19}$	N/A	N/A	N/A
a_{13}	$-1.0089\cdot10^{-20}$	$-3.2461\cdot10^{-21}$	N/A	N/A	N/A
a_{14}	$-8.1633\cdot10^{-20}$	$-3.7172\cdot10^{-19}$	N/A	N/A	N/A
a_{15}	$-2.8612\cdot10^{-20}$	$1.9668\cdot10^{-21}$	N/A	N/A	N/A
PLASTICS	**HDPE** **PP**	$\frac{k_c(P,T)}{Pa^{-1}} = a(T)\cdot\left(\frac{P}{MPa}\right)^{b(T)}$ $a(T) = a_1 + a_2\cdot\frac{Tk}{K} + a_3\cdot\frac{Tk^2}{K^2} + a_4\cdot\frac{Tk^3}{K^3}$ $b(T) = b_1 + b_2\cdot\frac{Tk}{K} + b_3\cdot\frac{Tk^2}{K^2}$			

	PTFE	$\frac{k_c(P,T)}{Pa^{-1}} = a(T) \cdot e^{b(T) \cdot \frac{P}{MPa}} + c(T) \cdot e^{d(T) \cdot \frac{P}{MPa}}$ $a(T) = a_1 + a_2 \cdot \frac{Tk}{K} + a_3 \cdot \frac{Tk^2}{K^2} + a_4 \cdot \frac{Tk^3}{K^3}$ $b(T) = b_1 + b_2 \cdot \frac{Tk}{K} + b_3 \cdot \frac{Tk^2}{K^2} + b_4 \cdot \frac{Tk^3}{K^3}$ $c(T) = c_1 + c_2 \cdot \frac{Tk}{K} + c_3 \cdot \frac{Tk^2}{K^2} + c_4 \cdot \frac{Tk^3}{K^3}$ $d(T) = d_1 + d_2 \cdot \frac{Tk}{K} + d_3 \cdot \frac{Tk^2}{K^2} + d_4 \cdot \frac{Tk^3}{K^3}$

Coefficients	**HDPE**	**PP**	**PTFE**
a_1	$7.8600 \cdot 10^{-11}$	$-3.4187 \cdot 10^{-9}$	$-1.6461 \cdot 10^{-16}$
a_2	$6.5011 \cdot 10^{-12}$	$6.5859 \cdot 10^{-11}$	$1.5266 \cdot 10^{-13}$
a_3	$-2.3327 \cdot 10^{-14}$	$-2.2454 \cdot 10^{-13}$	$-4.6594 \cdot 10^{-11}$
a_4	$7.2730 \cdot 10^{-17}$	$2.3769 \cdot 10^{-16}$	$4.7694 \cdot 10^{-9}$
b_1	-0.0686	-1.1992	$2.1616 \cdot 10^{-10}$
b_2	-0.0015	0.0032	$-8.9724 \cdot 10^{-8}$
b_3	$4.4771 \cdot 10^{-7}$	$-3.9233 \cdot 10^{-6}$	$3.4022 \cdot 10^{-6}$
b_4	N/A	N/A	$2.3878 \cdot 10^{-7}$
c_1	N/A	N/A	$4.0761 \cdot 10^{-15}$
c_2	N/A	N/A	$-3.9907 \cdot 10^{-12}$
c_3	N/A	N/A	$1.2938 \cdot 10^{-9}$
c_4	N/A	N/A	$-1.3855 \cdot 10^{-7}$
d_1	N/A	N/A	$5.6995 \cdot 10^{-10}$
d_2	N/A	N/A	$-3.5614 \cdot 10^{-7}$
d_3	N/A	N/A	$3.0223 \cdot 10^{-6}$
d_4	N/A	N/A	$2.3878 \cdot 10^{-7}$

HDPE, High-density polyethylene; *P*, Pressure (MPa); *PG*, Propylene glycol; *PP*, Polypropylene; *PTFE*, Polytetrafluoroethylene; *Tk*, Temperature (K).

6.4.1 Building a CFD model

The workflow for CFD modeling usually implies the following steps: (1) defining the problem and the model assumptions, (2) building the geometry and setting the computational domain, (3) specifying material properties, (4) defining physics as governing equations, (5) stating process, initial, and boundary conditions, (6) generating the computational grid, and (7) running simulations. The simulation results are then analyzed and interpreted in (8) the postprocessing step by generating graphs and plots and extracting relevant data to validate the model.

6.4.1.1 Problem definition and model assumptions

As mentioned in Section 6.1, HPP at room temperature is a well-established technology in the food industry, but there are also other high-pressure processes of potential interest, including HPTP, HPF, and HPT. Simulation of all these processes involves different strategies to be defined at the different steps of the model-building process, as will be described below.

The complexity of both HPP and food materials means that certain assumptions are often made to simplify model building (Juliano et al., 2009a, 2011). Thus, 2D-axial symmetry is frequently assumed in vertical HPP systems, even though some nonaxisymmetric features, such as a noncentered pressurizing fluid inlet, can exist (Juliano et al., 2009c; Knoerzer et al., 2007). Furthermore, the pressure increase/decrease during compression/expansion is generally assumed to be linear and, consequently, the pressurization and depressurization rates are constant. In addition, the expansion of the pressure vessel, associated with fluid pumping during compression, is generally considered insignificant. Also, it is typically assumed that the composition, structure, size, and shape of foods and product carriers are not altered during pressure processing, thus presuming that structural and physicochemical changes due to compression are negligible. Food packaging is frequently assumed to be infinitely thin and, therefore, ignored. Fluids are typically considered Newtonian, compressible, and chemically inert, and it is usually assumed they are at rest at the beginning of the process. At this moment, the pressure vessel is usually supposed to be completely filled and at thermal equilibrium. In thermoregulated high-pressure systems, temperature fluctuations associated with the operation of the thermostat are usually ignored.

On the contrary, to achieve good accuracy, some apparently irrelevant process characteristics and variables should not be neglected (Juliano et al., 2011). Thus, even though convective currents created by the entrance of pressurizing fluid during compression are frequently neglected (Otero et al., 2007; Smith et al., 2014), some authors have demonstrated that they can significantly affect temperature uniformity during processing (Hartmann, 2002; Hartmann and Delgado, 2002; Knoerzer et al., 2007). In this sense, the temperature and velocity of the incoming fluid during compression should be considered, and the possibility of turbulent conditions evaluated. Since the incoming pressurizing fluid is usually at room temperature, this is especially important when modeling HPTP, as Khurana and Karwe (2009) demonstrated.

6.4.1.2 Geometry and computational domain

CFD modeling of high-pressure processes involves the definition of a computational domain to which the related physics and associated interactions are applied. The complexity of this domain has a substantial impact on both the accuracy and the computational costs of the model and, therefore, a compromise solution is usually adopted by accurately reproducing all geometric details that significantly affect the variables of interest and neglecting those that have no relevant effect.

Fig. 6.2 depicts two examples of geometries representing typical vertical and horizontal HPP systems. Although HPP equipment in the food industry is horizontally oriented, most simulations in the literature consider vertical configurations, probably because lab- and pilot plant equipment, usually employed to validate models, generally have vertical vessels. Vertically oriented devices often comprise axis-symmetric features that allow building two-dimensional (2D) axis-symmetric models (Hartmann and Delgado, 2002; Kilimann et al., 2006; Knoerzer et al., 2007; Otero et al., 2007). In contrast, for horizontal HPP systems, only axis-symmetric features no longer exist and, therefore, three-dimensional (3D) models are required. However, in this case, the computational domain can be halved by cutting the cylinder longitudinally, as shown in Fig. 6.2, thus reducing the number of degrees of freedom (Smith et al., 2014).

The number of subdomains defined in the model geometry depends on the complexity of the system considered and the accuracy expected, but typically, at least three are included, representing the food sample, the pressure medium, and the pressure vessel. More complex models may also include a number of packed foods, product carriers, and insulation and thermoregulating systems (e.g., Juliano et al., 2009c; Knoerzer and Chapman, 2011). However, increasing the level of detail in the computational domain usually means losing the symmetry of the system and, in consequence, the need for a 3D model (Abdul Ghani and Farid, 2007).

Another important point to consider when defining the computational domain is the compression system used by the HPP equipment to generate pressure. As mentioned, in commercial-scale equipment, pressurization is performed by using indirect compression systems, that is, pressure intensifiers that pump pressurizing fluid inside

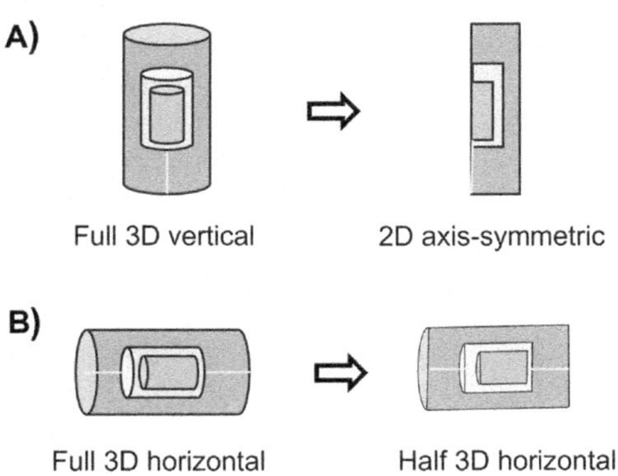

Figure 6.2 Geometries describing (A) vertically and (B) horizontally oriented HPP systems and computational simplifications commonly made to model high-pressure processes.

the vessel to reach the target pressure. However, in lab- and pilot plant equipment, this is not the unique option, and direct compression systems, using a movable piston at the bottom of the vessel to build up the pressure, can also be employed. In this case, the complexity of the computational domain significantly increases because the movement of the piston must be considered and all the calculations must be performed on a moving mesh (Rauh et al., 2009).

6.4.1.3 Material properties

As commented in Section 6.3, the pressure- and temperature-dependent thermophysical properties required to implement HPP models are unknown for most foods, pressurizing fluids, and packaging and carrier materials. Only water properties are well defined by the IAPWS formulations in a wide temperature and pressure range (IAPWS, 2008, 2011, 2018). However, these formulations are computationally expensive algorithms and, therefore, their implementation in the modeling software is difficult. For this reason, a number of authors have derived, from IAPWS data, more simple polynomial equations, fitted at the pressure/temperature conditions of interest to HPP processing (Table 6.4). These expressions can be easily incorporated in the software and provide predictions of the water properties accurate enough for modeling purposes (Chourot et al., 1997; Juliano et al., 2009c; Knoerzer et al., 2007, 2010a; Park & Jun, 2015; Salvi et al., 2017; Zhu et al., 2007, 2008).

For the rest of the materials, thermophysical properties are usually approximated by making simplifying assumptions. Thus, the properties of the product carriers and insulators are frequently considered not pressure dependent, and the corresponding values at atmospheric pressure are assumed (Juliano et al., 2009c; Knoerzer et al., 2007; Salvi et al., 2017). Similarly, in food products, the intrinsic properties of the dry matter, treated either as a whole or as the sum of its specific components, are usually considered not pressure-dependent, but only temperature-dependent (Guignon et al., 2012; Otero et al., 2006, 2010). For calculations, expressions such as those shown in Table 6.4 can be implemented in additive models (Choi and Okos, 1986; Phinney et al., 2017).

The ratio model, which assumes that changes in the product properties with pressure are similar to those observed in pure water, is also frequently used to estimate thermophysical properties of foods and aqueous pressurizing fluids under pressure (Abdul Ghani and Farid, 2007; Hartmann et al., 2003; Knoerzer and Chapman, 2011). Moreover, due to the lack of data on both the thermal expansivity and k_c values of most materials under pressure, simplifying assumptions are also made to simulate temperature changes during compression and expansion. Thus, if neither α nor k_c data are available, compression heating of both the vessel and the product carrier is generally assumed to be zero (Juliano et al., 2009c; Knoerzer et al., 2010a), while that of pressuring fluids other than water and foods is frequently assumed to be similar to that of water (Knoerzer and Chapman, 2011).

However, it is important to evaluate the impact of adopting these simplifying assumptions, as they can lead to nonnegligible inaccuracy in the model predictions. For example, Knoerzer and Chapman (2011) simulated heat transfer during HPP in

Table 6.4 Simple polynomial expressions proposed in the literature to be implemented in HPP models for the prediction of thermophysical properties of pure water and food components.

Property	Product/ Food component	Expression	Validity range		References
			P (MPa)	T (°C)	
Density ($kg\cdot m^{-3}$)	Water	$-0.5597 \cdot P + 2.679 \cdot 10^{-7} \cdot Tk + 1.1839 \cdot 10^3$	n.r.	n.r.	Knoerzer et al. (2007)
		$997 + 4.025 \cdot 10^{-1} \cdot P - 2 \cdot 10^{-4} \cdot P^2$	0.1–700	25	Park and Jun (2015)
		$-0.28 \cdot P - 0.5 \cdot Tk + 1161.37$	0.1–700	0–20	Salvi et al. (2017)
	Protein	$1314.4 - 0.51945 \cdot T$	–	10–150	Phinney et al. (2017)
	Fat	$924.5 - 0.40698 \cdot T$	–	10–150	Phinney et al. (2017)
	Simple sugar	$1704.2 - 0.69452 \cdot T$	–	10–150	Phinney et al. (2017)
	Complex carbohydrates	$1322.1 - 0.29996 \cdot T$	–	10–150	Phinney et al. (2017)
	Ash	$2423.8 - 0.28063 \cdot T$	–	10–150	Phinney et al. (2017)
Specific heat capacity ($J\cdot kg^{-1}\ K^{-1}$)	Water	$-0.1177 \cdot P - 6.393 \cdot 10^{-7} \cdot Tk + 4.134 \cdot 10^3$	n.r.	n.r.	Knoerzer et al. (2007)
		$4.181 \cdot 10^3 - 2.4 \cdot P + 5 \cdot 10^{-3} \cdot P^2 - 4 \cdot 10^{-6} \cdot P3$	0.1–700	25	Park and Jun (2015)
		$-9.08 \cdot 10^{-5} \cdot Tk + 5.47 \cdot 10^{-6} \cdot Tk^2 + 4176.2$	–	0–20	Salvi et al. (2017)
	Protein	$1.9861 + 1.375 \cdot 10^{-3} \cdot T$	–	10–150	Phinney et al. (2017)
	Fat	$2.0065 + 6.33 \cdot 10^{-4} \cdot T$	–	10–150	Phinney et al. (2017)
	Simple sugar	$1.4662 + 1.384 \cdot 10^{-3} \cdot T$	–	10–150	Phinney et al. (2017)
	Complex carbohydrates	$1.8631 + 9.16 \cdot 10^{-4} \cdot T$	–	10–150	Phinney et al. (2017)
	Ash	$1.1072 + 1.316 \cdot 10^{-3} \cdot T$	–	10–150	Phinney et al. (2017)

Continued

Table 6.4 Continued

Property	Product/ Food component	Expression	Validity range P (MPa)	Validity range T (°C)	References
Thermal conductivity ($W \cdot m^{-1} K^{-1}$)	Water	$1.294 \cdot 10^{-3} \cdot P + 3.671 \cdot 10^{-10} \cdot Tk + 2.209 \cdot 10^{-1}$	n.r.	n.r.	Knoerzer et al. (2007)
		$0.571 + 5.26 \cdot 10^{-4} \cdot P - 5.96 \cdot 10^{-7} \cdot P^2$	0.1–400	5	Zhu et al. (2007)
		$0.607 + 4.82 \cdot 10^{-4} \cdot P - 3.96 \cdot 10^{-7} \cdot P^2$	0.1–350	25	Zhu et al. (2008)
		$0.607 + 5 \cdot 10^{-4} \cdot P - 7 \cdot 10^{-7} \cdot P^2 + 6 \cdot 10^{-10} \cdot P^3$	0.1–700	25	Park and Jun (2015)
		$3 \cdot 10^{-4} \cdot P + 1.4 \cdot 10^{-3} \cdot Tk + 0.19$	0.1–700	0–20	Salvi et al. (2017)
	Protein	$0.1973 + 7.46 \cdot 10^{-4} \cdot T$	–	10–150	Phinney et al. (2017)
	Fat	$0.1811 - 3.00 \cdot 10^{-4} \cdot T$	–	10–150	Phinney et al. (2017)
	Simple sugar	$0.2267 + 6.62 \cdot 10^{-4} \cdot T$	–	10–150	Phinney et al. (2017)
	Complex carbohydrates	$0.2139 + 6.59 \cdot 10^{-4} \cdot T$	–	10–150	Phinney et al. (2017)
	Ash	$0.3404 + 9.09 \cdot 10^{-4} \cdot T$	–	10–150	Phinney et al. (2017)
Thermal expansivity (K^{-1})	Water	$1.754 \cdot 10^{-6} \cdot P - 3.513 \cdot 10^{-13} \cdot Tk - 6.665 \cdot 10^{-6}$	n.r.	n.r.	Knoerzer et al. (2007)
Viscosity (Pa·s)	Water	$-3.634 \cdot 10^{-6} \cdot P + 2.175 \cdot 10^{-7} \cdot Tk + 1.645 \cdot 10^{-3}$	n.r.	n.r.	Knoerzer et al. (2007)
		$4.23 \cdot 10^{-7} \cdot P - 9.7 \cdot 10^{-6} \cdot Tk + 0.0037$	0.1–700	0–20	Salvi et al. (2017)
Initial freezing point (°C)	Water	$-7.2192 \cdot 10^{-2} \cdot P - 1.55 \cdot 10^{-4} \cdot P^2$	0–210	–	Chourot et al. (1997)
Latent heat ($J \cdot kg^{-1}$)	Water	$333549.295 - 399.369 \cdot P - 0.388 \cdot P^2$	0–210	–	Chourot et al. (1997)
		$3.335 \cdot 10^5 - 3.379 \cdot 10^2 \cdot P - 1.292 \cdot P^2 + 3.114 \cdot 10^{-3} \cdot P^3$	0–210		Otero and Sanz (2006)

P, Pressure (MPa); *Tk*, Temperature (K), and *T*, Temperature (°C). n.r., Not reported.

a vessel filled with a 40% propylene glycol/water mixture and found that using compression heating properties of water instead of those of the real pressure medium induced significant errors in the temperature predictions. Therefore, evaluating the effect of the assumptions built into the model is crucial to quantify their impact on the reliability and accuracy of the resulting predictions.

6.4.1.4 Governing equations

During high-pressure food processing, heat transfer and fluid motion phenomena occur that must be considered together when modeling. Thus, in solid subdomains such as pressure vessels, product carriers, food packaging, and solid foods, heat transfer occurs primarily through conduction. In contrast, in fluid subdomains, which include pressurizing fluids and fluid foods, it is essential to take convection currents into account.

In the heat transfer balance, unsteady-state heat conduction during processing is described by the Fourier's law of heat diffusion, while adiabatic heating/cooling during compression/expansion is represented by a heat generation term, Q:

$$\frac{\partial}{\partial t}(\rho \cdot Cp \cdot T) = \nabla \cdot (K \nabla T) + Q \tag{6.22}$$

where t represents time. Assuming adiabatic conditions, Q can be expressed, according to Eq. (6.1), as:

$$Q = Tk \cdot \alpha \cdot \frac{dP}{dt} \tag{6.23}$$

Evidently, this Q term yields positive and negative values during compression and expansion, respectively, while it is zero during the pressure-holding time.

The thermo- and fluid-dynamic behavior of fluids during pressure processing can be described by the conservation equations of mass, momentum, and energy that account for convection currents inside the pressure vessel. Thus, the mass conservation equation describes how density changes with pressure and/or temperature during processing result in fluid motion:

$$\frac{\partial \rho}{\partial t} + \nabla \cdot \left(\rho \vec{u}\right) = 0 \tag{6.24}$$

the momentum conservation equation considers pressure, viscous, and gravity forces affecting fluid mechanics:

$$\rho\left(\frac{\partial \vec{u}}{\partial t} + \left(\vec{u} \cdot \nabla\right)\vec{u}\right) = -\nabla P + \nabla \cdot \left(\mu \cdot \nabla \vec{u}\right) + \rho g \tag{6.25}$$

while the energy conservation equation accounts for heat transfer through convection and conduction:

$$\frac{\partial(\rho \cdot Cp \cdot T)}{\partial t} + \nabla \cdot \left(\rho \vec{u} CpT\right) = \nabla \cdot (K \nabla T) + Q \tag{6.26}$$

where $\vec{u}$ represents the velocity vector and g is the gravity constant.

As mentioned in Section 6.4.1.1, turbulent flow conditions can arise in the pressurizing fluid during compression mainly depending on the geometry of the pressurizing fluid inlet and the fluid velocity. To take the contributions of turbulent eddies into account, several authors in the literature have applied the k-ε turbulence model that includes additional terms in Eqs. (6.25) and (6.26) by considering "turbulent viscosity," μ_T, and "turbulent thermal conductivity," K_T, respectively (Juliano et al., 2009c; Knoerzer et al., 2007, 2010a; Salvi et al., 2017):

$$\rho\left(\frac{\partial \vec{u}}{\partial t} + \left(\vec{u} \cdot \nabla\right)\vec{u}\right) = -\nabla P + \nabla \cdot \left((\mu + \mu_T) \cdot \nabla \vec{u}\right) + \rho g \tag{6.27}$$

$$\frac{\partial(\rho \cdot Cp \cdot T)}{\partial t} + \nabla \cdot \left(\rho \vec{u} CpT\right) = \nabla \cdot ((K + K_T)\nabla T) + Q \tag{6.28}$$

Furthermore, when modeling high-pressure processes with phase transitions, the governing energy equation and, if fluids are considered, also the momentum equation must be modified to deal with the latent heat release/absorption occurring during freezing/melting. To do so, different model formulations can be adopted, with the enthalpy formulation being the most commonly utilized (Kowalczyk et al., 2004; Norton et al., 2009; Smith et al., 2012). Moreover, pressure effects on the freezing point and enthalpy must be considered. In this sense, it is important to note that, as described in detail by Otero (2023), there are different types of HPF and HPT processes with very important differences between them. Thus, depending on the process, the phase transition can take place either at the target pressure, as in high-pressure-assisted freezing and in HPT, or after expansion, as in high-pressure shift freezing (HPSF). These particularities of each process must be taken into account to adequately formulate the corresponding models.

HPSF is of special interest because the freezing dynamics are very different from those of conventional freezing at atmospheric pressure and, therefore, particular considerations must be taken into account when modeling this process.

Fig. 6.3 shows a schematic of the process. In HPSF, the product to be frozen is compressed to the target pressure (path AB in Fig. 6.3) and then cooled under pressure to a subzero temperature in unfrozen state (path BC in Fig. 6.3). Once the desired temperature is reached throughout the product, pressure is quickly released. After expansion (path CD in Fig. 6.3), the product is at atmospheric pressure and at a temperature well below its freezing point, but still in an unfrozen state (point D in Fig. 6.3); that is, it is highly and uniformly supercooled. This large supercooling induces

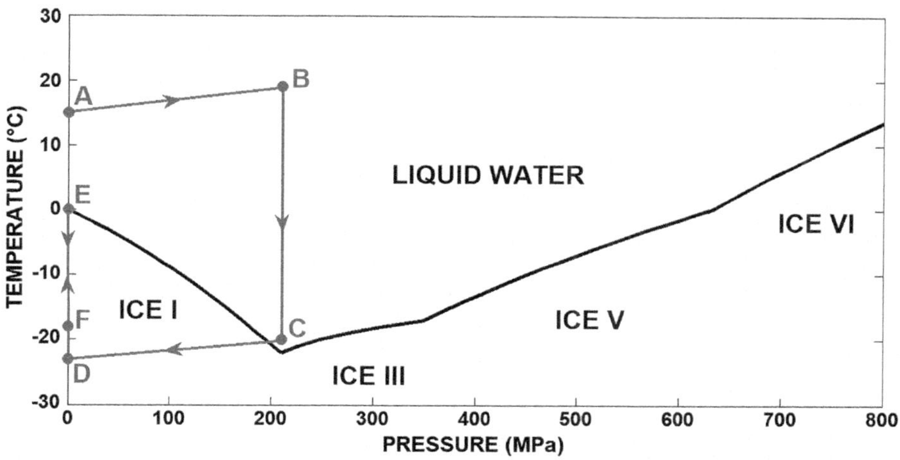

Figure 6.3 Schematic drawing of a high-pressure shift freezing process. AB: compression, BC: precooling under pressure, CD: expansion, DE: nucleation, EF: phase transition and tempering at atmospheric pressure. Bold lines represent the water phase diagram.
Credit: Reproduced with permission from Otero, L., 2023. Application of high pressure processing in freezing and thawing processes. In S. M. Jafari & N. Therdthai (Eds.), *Non-thermal Food Processing Operations* (pp. 359–405). Elsevier Inc. https://doi.org/10.1016/b978-0-12-818717-3.00005-6

quasiinstantaneous and uniform nucleation throughout the product and, thus, a significant amount of ice is instantaneously formed just after expansion. The release of the corresponding latent heat increases the product temperature to its freezing point (path DE in Fig. 6.3), and freezing is then completed at atmospheric pressure (path EF in Fig. 6.3). Therefore, modeling HPSF processes involves mathematically describing not only the supercooling achieved after expansion, but also the amount of ice instantaneously formed.

Supercooling achieved after expansion, ΔT_{sup}, can be readily estimated from Eq. (6.29):

$$\Delta T_{sup} = T_{if}(P_{atm}) - T_{min} \tag{6.29}$$

where $T_{if}(P_{atm})$ is the initial freezing point of the product at atmospheric conditions and T_{min} is the minimum temperature reached after expansion. If not known, $T_{if}(P_{atm})$ can be estimated from Eq. (6.7) or (6.8), while T_{min} values can be predicted from Eq. (6.1).

Furthermore, the amount of ice instantaneously formed can be predicted by considering the following heat balance (Otero and Sanz, 2006):

$$m^*_{ice} \cdot L^0(P_{atm}) = (m^*_{ice} \cdot \overline{Cp_{ice}(P_{atm})} + \left(1 - m^*_{ice}\right) \cdot \overline{Cp_w(P_{atm})} \cdot \Delta T_{sup} \tag{6.30}$$

where m_{ice}^* represents the mass of ice divided by the initial mass of water and $\overline{Cp_{ice}(P_{atm})}$ and $\overline{Cp_w(P_{atm})}$ are the specific heat capacities, at atmospheric pressure, of ice and water, respectively (mean values calculated at T_{min} and at $T_{if}(P_{atm})$).

6.4.1.5 Process, initial, and boundary conditions

The governing equations can be solved numerically based on process variables appropriate to the model, initial conditions specified in each subdomain, and boundary conditions describing how the subdomains in the modeled system interact with each other and with the environment. Depending on the complexity of the HPP system, the level of detail of the computational domain, and the assumptions made, different conditions can be established.

Fig. 6.4 illustrates an example of computational domain representing a vertical 2D axis-symmetric geometry in which typical subdomains, including a solid food product, Ω_F, the pressurizing fluid, Ω_{PF}, and the vessel, Ω_V, are depicted. Fig. 6.4 also details the internal (Γ_{F-PF}: food-pressurizing fluid boundary and Γ_{PF-V}: pressurizing fluid-vessel boundary) and the external boundaries (Γ_{inlet}: at the inlet tube and Γ_{up_V}, Γ_{r_V}, Γ_{low_V}: upper, right, and lower vessel boundaries).

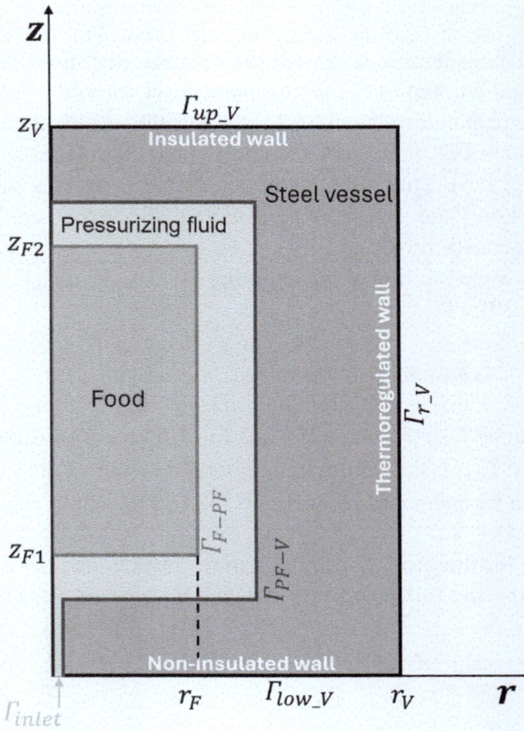

Figure 6.4 Typical computational domain representing a vertical 2D axis-symmetric geometry. Ω_F, Ω_{PF}, and Ω_V: subdomains occupied by the food product, the pressurizing fluid, and the steel vessel, respectively. Internal boundaries: Γ_{F-PF} (food-pressurizing fluid boundary, ▬) and Γ_{PF-V} (pressurizing fluid-vessel boundary, ▬). External boundaries: Γ_{inlet} (inlet tube boundary, ▬), Γ_{up_V}, Γ_{r_V}, and Γ_{low_V} (upper, right, and lower vessel boundaries, ▬).
Note that the schematic diagram is not to scale.

Process conditions are usually defined as input variables to the model and include the pressure-time profile. As mentioned in Section 6.4.1.1, pressure profiles are generally estimated by considering a linear increase/decrease in pressure during compression/expansion and a constant pressure during the holding time. However, it is important to note that actual pressure profiles are typically more complex, neither linear during compression and expansion nor constant during holding time. As the pressure profile directly affects the compression heating rate, the pressure- and temperature-dependent thermophysical properties, and the convective flow distribution throughout the liquid domains, the use of estimated pressure profiles may induce inaccurate model predictions. Thus, Knoerzer and Chapman (2011) compared temperatures measured at various points in the pressure vessel during HPTP with model predictions made by using either estimated or real pressure profiles and found that predictions from real pressure profiles were much more accurate. For this reason, these authors recommended collecting a number of real pressure profiles and obtaining representative average close-to-reality process conditions before modeling high-pressure processes in a particular device.

As for the initial conditions, at the beginning of the process, the pressure medium and the fluid food, if any, are typically considered to be at rest and, therefore, their initial velocity is generally assumed to be zero. If isothermal conditions are also assumed, then a $T_0 = cte$ should be defined in all the subdomains. Alternatively, different temperatures can be considered in different subdomains; for example, in the food sample, the pressure medium, or the pressure vessel (Infante et al., 2009; Salvi et al., 2017). This is especially interesting when modeling HPTP since the effects of vessel preheating can be analyzed.

Adequate boundary conditions must also be defined before solving the model, as they are crucial to ensuring that HPP simulations are both realistic and reliable. Symmetry boundary, inflow velocity boundary, pressure boundary, fluid–solid, and thermal boundary conditions are the most commonly applied in HPP modeling (Juliano et al., 2011; Knoerzer and Juliano, 2016).

Thus, in simplified 2D axis-symmetric models (Fig. 6.4), boundary conditions for symmetry on the axis ($r = 0$) are assumed:

$$- K\frac{\partial T}{\partial r}\bigg|_{r=0} = 0 \quad \forall z, \forall t > 0 \tag{6.31}$$

and

$$- \mu\frac{\partial \overrightarrow{u}}{\partial r}\bigg|_{r=0} \quad \forall z, \forall t > 0 \tag{6.32}$$

When modeling HPP equipment using pressure intensifiers as compression system, an inflow velocity boundary condition is generally stated at the pressurizing fluid inlet (Γ_{inlet} in Fig. 6.4) during compression to take into account the entrance of pressure medium from the pump (Knoerzer et al., 2007):

$$u_z = u_{in}, \forall t \in [0, t_{com}] \tag{6.33}$$

$$u_{in} = 0, \forall t > t_{com} \tag{6.34}$$

where u_{in} is the velocity of the incoming pressurizing fluid. Similarly, an outflow velocity boundary condition can be defined for the outlet tube to take into account the release of pressurizing fluid at expansion.

Pressure conditions are also imposed during compression, pressure holding, and expansion phases in this boundary:

$$\frac{\partial P}{\partial t} = p_{rate} \quad \forall t \in [0, t_f] \tag{6.35}$$

During compression:

$$p_{rate} = \frac{P_{tgt}}{t_{com}}, \forall t \in [0, t_{com}] \tag{6.36}$$

During the pressure-holding time:

$$p_{rate} = 0 \quad and \quad P = P_{tgt}, \forall t \in [t_{com}, t_{exp}] \tag{6.37}$$

During expansion:

$$p_{rate} = -\frac{P_{tgt}}{(t_f - t_{exp})} \quad \forall t \in [t_{exp}, t_f] \tag{6.38}$$

Fluid–solid boundary conditions define the flow behavior at any solid boundary. In Fig. 6.4, fluid–solid internal boundaries include Γ_{F-PF} and Γ_{PF-V} while, in more complex models, boundaries between the pressurizing fluid and the product carriers or between fluid foods and packaging, for example, should also be considered. Fluid–solid boundary conditions depend on whether laminar or turbulent flow is considered. If laminar flow is assumed, a no-slip condition is usually set, that is, fluid velocity is assumed to be zero due to the fluid adhesion to the walls (Otero et al., 2007; Smith et al., 2014). To account for solid walls subjected to turbulent flow, more sophisticated fluid–solid boundary conditions are required. For example, Knoerzer et al. (2007) successfully used the logarithmic wall function, an empirical relationship between the value of the velocity parallel to the wall and the wall friction.

Thermal boundary conditions describe temperature and/or heat flux at the boundaries of the computational domain and, therefore, they depend on the subdomains included in the model and the specific characteristics of the HPP device. For example, when the steel vessel is included in the model as a subdomain, as shown in Fig. 6.4, different boundary conditions can be assumed at the external vessel walls depending on whether they are thermoregulated, insulated, or neither. Thus, in thermoregulated walls and at the pressurizing fluid inlet (Γ_{r_V} and Γ_{inlet} in Fig. 6.4), constant temperatures can be assumed:

$$T|_{r=r_V} = T_{tm\_w}, \forall z \in [0, z_V], \forall t > 0 \quad (6.39)$$

$$T|_{z=0} = T_{in}, \forall r \in [0, r_{in}], \forall t > 0 \quad (6.40)$$

where T_{tm_w} and T_{in} are the constant temperatures of the thermoregulated wall and the incoming pressurizing fluid, respectively.

In insulated vessel walls, on the other hand, an adiabatic boundary should be defined. For example, at the upper vessel wall in Fig. 6.4, that is, on Γ_{up_V}:

$$-K\frac{\partial T}{\partial z}\bigg|_{z=z_V} = 0 \quad \forall r \in [0, r_V], \forall t > 0 \quad (6.41)$$

In noninsulated external walls and at all fluid–solid boundaries inside the pressure vessel (e.g., on Γ_{r_V}, Γ_{F-PF}, and Γ_{PF-V} in Fig. 6.4), a convective flux is usually considered:

$$-K_s \cdot \frac{\partial T}{\partial n} = h_{s-fl} \cdot \left(T - T_{fl}\right) \quad (6.42)$$

where subindexes *s* and *fl* represent the solid and fluid materials of the specific subdomains implied in each boundary, *n* is the normal boundary direction, and *h* is the heat transfer coefficient ($W \cdot m^{-2}\ K^{-1}$) between the solid and fluid materials implied. For example, on Γ_{F-PF} ,

$$-K_F \cdot \frac{\partial T}{\partial n} = h_{F-PF} \cdot (T - T_{PF}), \begin{cases} \forall r \in [0, r_F], \forall z = z_{F1}, \forall t \\ \forall r = r_F, \forall z \in [z_{F1}, z_{F2}], \forall t \\ \forall r \in [0, r_F], \forall z = z_{F2}, \forall t \end{cases} \quad (6.43)$$

In foods packaged in thin films, an equivalent heat-transfer coefficient is usually assumed to include the real heat-transfer coefficient and the thermal conductivity and thickness of the packaging material (Juliano et al., 2009c).

If very specific knowledge of the modeled system is available, time-dependent temperature conditions can also be stated. For example, Knoerzer et al. (2007) assumed that the temperature of the vessel walls increased linearly during compression, while decreased during the pressure-holding time:

$$T_{wall} = pt + q \quad (6.44)$$

where

$$\begin{cases} p > 0 & \forall t, 0 < t < t_{com} \\ p < 0 & \forall t, t_{com} < t < t_{exp} \\ p = 0 & \forall t, t_{exp} < t < t_f \end{cases} \tag{6.45}$$

Finally, continuity of heat flux is usually assumed at all fluid–solid and solid–solid boundaries. For example, in Fig. 6.4, this boundary condition can be stated on Γ_{F-PF} as:

$$n \cdot (K_F \nabla T) = n \cdot (K_{PF} \nabla T), \begin{cases} \forall r \in [0, r_F], \forall z = z_{F1}, \forall t \\ \forall r = r_F, \forall z \in [z_{F1}, z_{F2}], \forall t \\ \forall r \in [0, r_F], \forall z = z_{F2}, \forall t \end{cases} \tag{6.46}$$

6.4.1.6 Meshing, model solving, and simulation

During meshing, the complex computational domain is discretized into multiple elements where the governing equations are solved numerically to approximate the exact solution. These elements can assume a variety of shapes (e.g., quadrilateral or triangle elements in 2D) and sizes, thus enabling them to adapt to the specific characteristics of the domain to be modeled. Since the mesh density influences both the solution accuracy and the computational time, it is essential to adjust it not only to provide a nonmesh-dependent solution but also to achieve a balance between these two factors.

Various common discretization methods, such as the finite difference method, the finite element method, or the finite volume method, have been successfully employed in the literature for HPP modeling. Although each method has its own advantages and disadvantages, the choice of one or the other usually depends on the computational means available to solve the model. Thus, when self-developed mathematical codes are used, the finite difference method is the prevailing option, as it is the easiest method to implement in numerical models (Denys et al., 2000a, 2000c). Conversely, the finite element method and the finite volume method are employed when commercial software is utilized, as these are typically the methods that are incorporated into such software (Hartmann, 2002; Knoerzer et al., 2007; Otero et al., 2007).

By using discretization methods, partial differential equations for fluid flow and heat transfer are solved simultaneously. Moreover, if partial differential equations describing microbial or enzyme inactivation kinetics are available, they can be coupled with those of thermo-fluid dynamics to predict the inactivation level achieved throughout the product from the calculated temperature distribution. Coupling of equations can be done either internally, by simultaneously solving the thermo-fluid dynamics and inactivation kinetic equations within the same software package, or externally, by using another software to calculate the distribution of the microbial load or the enzyme activity throughout the product from the predicted pressure-temperature-time profiles (Juliano et al., 2009b, 2009c; Knoerzer et al., 2007). In both cases, inactivation data predicted at different locations can be represented versus time, and microbial and/or enzyme activity distributions can be plotted at specific times and even displayed as animations.

6.4.1.7 Model validation and sensitivity evaluation

Validation involves the comparison of the numerical solution calculated by the model with real experimental data. When modeling HPP in lab- and pilot-plant equipment, temperature validation is typically performed by comparing real temperature data, measured with thermocouples at specific locations inside the pressure vessel during processing, with those calculated during simulations at the same locations. During validation, attention should be paid to the design of the thermocouples' assembly, as it can significantly affect their dynamic response. For example, Khurana and Karwe (2009) observed that the high-pressure tubing surrounding the thermocouples significantly increased their response time and, therefore, they had to obtain a time constant for each thermocouple to correct the experimental data for the lag. When modeling HPP in industrial equipment, temperature validation is much more complicated because, as mentioned in Section 6.2.2, thermocouples cannot be implemented, and relatively large pressure-resistant, wireless systems must be used for temperature data acquisition, which can interfere with free fluid movement and act as heat sinks.

Fluid flow validation is also complicated, as real velocity measurements inside the pressure vessel during processing are not feasible. As previously commented, to date, only Pehl et al. (2000) and Delgado and Hartmann (2003) have visualized flow fields inside a 2-mL high-pressure optical cell by using encapsulated thermochromic liquid crystals and sapphire windows within the cell structure.

Model validation for other pressure- and/or temperature-dependent responses, such as microbial and enzyme inactivation or quality degradation, for example, requires consideration of a volume rather than selected points. In these cases, experimental measurements must be made, after pressure treatment, on packages located at specific points in the pressure vessel and containing a known initial amount of the substance of interest (e.g., microorganisms or enzymes). These measurements can then be compared with mean values calculated from model predictions in the vessel volume where the packages were located (Juliano et al., 2009b).

In addition to validating the model, it is also advisable to analyze its sensitivity. As mentioned in the previous sections, many simplifications and assumptions are made when building HPP models, which can affect the robustness of the model in different ways. For example, thermophysical properties of foods are usually estimated based on different assumptions, and pressure-time profiles are often simplified. Therefore, it is important to analyze how model predictions change when these parameters are varied in order to know which factors most influence model performance. To do this, perturbed models should be generated by randomly varying the considered variables (e.g., $\rho, \mu, T_0,$ or P, among others), and the mean errors calculated and compared. Although this is extremely useful for assessing model confidence, few studies in the literature have performed this type of analysis (Infante et al., 2009; Knoerzer and Chapman, 2011).

6.4.2 Applications of numerical models in high-pressure processing

Numerical models have been widely used in the literature to describe the heat transfer and fluid motion phenomena that occur inside the pressure vessel during processing. These models are fundamental to characterize high-pressure processes, both with and without phase transitions, and to truly understand how different variables, such as the type of pressurizing fluid, filling ratio, or product packaging, can affect the thermo-fluid dynamics of the system. In addition, numerical models are very useful for assessing the uniformity of pressure-induced microbial and enzyme inactivation. Thus, they allow the identification of possible cold spots or cold regions and the investigation of their causes. Only with this knowledge can appropriate strategies be designed to minimize thermal nonuniformities and, thus, inhomogeneities in the expected effects of the pressure treatment. Moreover, numerical models are extremely useful in the scale-up of high-pressure processes from the laboratory to industry, as they can predict the effects of changes in equipment dimensions and conditions on the process. In this way, numerical models and simulations can be used to improve, optimize, and scale up high-pressure processes from laboratory to industrial scale.

6.4.2.1 Process characterization, improvement, and optimization

Many authors have used CFD simulations to gain insight into the thermal and fluid dynamic phenomena that occur during the various stages of the pressure treatment, that is, compression, pressure-holding time, and expansion (Hartmann, 2002; Rauh et al., 2009). For example, Hartmann (2002) simulated the compression of water at 500 MPa in a 4-mL chamber, considering different inlet velocities of the pressurizing fluid, in this case water. They found that the fluid motion inside the vessel is dominated, at the beginning of the compression step, by forced convection in an area close to the pressurizing fluid inlet. The CFD analysis revealed that compression heating increases the temperature inside the vessel during pressurization, and the faster the velocity of the incoming pressurizing fluid, the faster the compression and the higher the temperature reached. The simulation showed how heat removal occurred through the colder vessel walls, thus generating a nonuniform temperature distribution inside the vessel which, in turn, created an inhomogeneous density distribution. This generated buoyancy-induced fluid motion and vortices near the walls. Thus, cold water flowed downward at the vessel walls and upward at the center of the pressure chamber. This free convection drove the fluid motion during the pressure-holding time, which decreased in intensity as the thermal gradients in the vessel disappeared.

By predicting time-dependent fluid velocity and temperature fields, CFD simulations allow a better understanding of how the specific characteristics of the HPP equipment and process variables affect thermal and fluid dynamic phenomena during processing. In this sense, Rauh et al. (2009) built a 3D CFD model to analyze thermal and fluid dynamic inhomogeneities in a 2.3-L vessel filled with water during processing at 700 MPa, either at 20°C or at 50°C, for 2 min. They observed that the initial temperature of the water and the temperature of the vessel wall significantly affected the temperature distribution in the water during pressure processing. The model revealed that when both temperatures are identical, the lower they are, the smaller the thermal gradient established between them after compression since, as shown in Eq. (6.1), the smaller the compression heating is. Thus, the thermal stratification predicted by the model at the end of the process was less significant in simulations at 20°C than in those at 50°C.

Otero et al. (2007) analyzed how the filling ratio in the vessel affects temperature uniformity during HPP. They found that convection currents in the pressurizing fluid significantly accelerate the removal of heat generated during compression and, therefore, the temperature uniformity of the process significantly improves when low filling ratios are used in the pressure vessel. Since low filling ratios can hinder the cost-effectiveness of the process, the authors used the CFD model to explore ways to improve process uniformity other than reducing the filling ratio. Numerical simulations revealed that high-pressure processes with high thermal uniformity can be achieved, regardless of the filling ratio, by anticipating the temperature rise caused by compression and adjusting the initial temperature of the food products and the pressurizing fluid accordingly. The effectiveness of this strategy was also confirmed by simulations performed by Rauh et al. (2009) and Infante et al. (2009).

Hartmann and Delgado (2003) investigated the effect of the packaging material on the inactivation induced in *Bacillus subtilis* α-amylase solutions during pressure processing. They found that the properties of the packaging material can significantly affect the inactivation rate and process uniformity. Thus, the CFD model showed that the lower the heat transfer coefficient for conduction through the packaging material, the longer the compression heating is maintained inside the package and the lower the temperature differences between packages. Therefore, the model revealed that an appropriate choice of the packaging material can maximize both the level of inactivation achieved in the product and the uniformity of the process.

As mentioned previously, thermal uniformity is particularly important in HPTP, as relatively small temperature differences can result in significant differences in spore inactivation. In this process, compression heating is used to rapidly heat the sample throughout its volume to the target temperature, which should be maintained throughout the entire process. However, depending on the process conditions, significant thermal inhomogeneities can occur during HPTP that threaten the effectiveness of the process. In this sense, Knoerzer et al. (2007) simulated temperature and flow distribution in a 35-L vertical vessel filled with water at 90°C during processing at 600 MPa for 5 min. After compression, they observed maximum temperatures of up to 110°C in the upper 2/3 of the pressure chamber, but at its bottom, the temperature was more than 10°C lower due to the cooling effect of the pressure medium entering the vessel. Moreover, during the pressure-holding time, thermal conduction through the vessel walls caused excessive cooling and, thus, bulk temperatures of only 96°C were predicted just before expansion.

Using a CFD model, Salvi et al. (2017) found that the temperature of the incoming fluid during compression significantly affects the thermal uniformity of HPTP. For example, they observed that the temperature distribution inside the vessel was more uniform when the temperature of the incoming fluid was the same as the initial vessel temperature. The model also allowed the identification of cold regions inside the vessel during HPTP. For example, the authors simulated the temperature distribution in a 350-L horizontal vessel filled with water and five cylindrical oil pouches, with an initial temperature of 50°C, during processing at 700 MPa for 13 min. They assumed that the water entering the vessel during compression was at 25°C and found significant temperature differences, of up to 12°C, both within a given oil pouch and from pouch to pouch. Due to the lower temperature of the incoming water, the pouches near the water inlets were the coldest.

All these results reveal the need to minimize the cooling effects of the incoming compression fluid and provide some thermal insulation to the system. To do so, Knoerzer et al. (2007) proposed the use of product carriers. They tested the effect of using two different cylindrical carriers, made either of metal or of insulating PTFE. Simulations illustrated how both carriers acted as a barrier to the incoming pressurizing fluid, thus minimizing its cooling effect during compression, but the PTFE carrier was much more effective in retaining heat and achieving a homogeneous sterilization temperature throughout its volume. In this sense, the authors estimated that, under the conditions tested, a reduction of more than 12 log units in *C. botulinum* spores would be achieved in 94.6% of the volume of the PTFE carrier, whereas no significant reduction would occur if no carrier was used. The effectiveness of the PTFE carrier in retaining heat and improving temperature uniformity was confirmed by Juliano et al. (2009c), who evaluated the temperature distribution inside the carrier in seven cylindrical packages filled with a solid food model. Thus, after pressurizing at 600 MPa/60°C, they found uniform and similar temperatures in all the packages during the pressure-holding time. On the other hand, Salvi et al. (2017) evaluated the impact of adding a PTFE insulation jacket on the inner wall of the pressure vessel on thermal uniformity during processing. The model revealed that the addition of insulation leads to a higher temperature inside the vessel, but to a less uniform temperature distribution compared to the absence of insulation.

Numerical models have also been used to characterize HPF and HPT processes. Thus, Kowalczyk et al. (2004) found that, since the density anomaly of water disappears under pressure, convective fluid motion in freezing and thawing processes under pressure was significantly different from that observed under atmospheric conditions. By predicting the temperature evolution during processing, numerical models allow identifying how different variables can affect the freezing and thawing times and, in consequence, the quality of the frozen and thawed products. For example, several authors have demonstrated that the pressure and temperature conditions before expansion determine the amount of ice instantaneously formed in HPSF and, therefore, significantly affect the phase transition time (Norton et al., 2009; Otero and Sanz, 2006; Smith et al., 2012).

Mathematical models and simulation can also be used to optimize equipment design and/or process parameters to maximize temperature uniformity, process yield, or food quality, for example. Models allow rapid evaluation of prospective changes and identification of optimal solutions, thus reducing expensive and labor-intensive trial-and-error approximations. For example, process yield can be maximized by reducing the wall thickness of the product carrier as long as thermal uniformity is not compromised. In this regard, Knoerzer et al. (2010) developed an iterative software routine that progressively and automatically changed the carrier wall thickness (0–70 mm), solved the model, and evaluated thermal uniformity and load capacity. The study showed that the wall thickness could be reduced from 28 mm to approximately 4 mm without compromising temperature retention, leading to an increase in usable volume from 12.5 to 24 L.

However, it is important to note that, during optimization processes, models must be solved numerous times with varying data sets. The computational time required for this can be considerable, especially in the case of highly complex models. In such cases, simplified models can offer a useful alternative, provided that they yield results within an acceptable margin of error. In this context, Infante et al. (2009) developed simplified models for HPP processing that take into account, for instance, constant thermophysical properties, which are maintained at their average value throughout the range of pressures and temperatures involved in the process. They compared temperature predictions made with simplified models with those calculated with full models in different scenarios and found that differences were acceptable, thus concluding that simplified models requiring significantly shorter computational times were a good alternative to the full models for optimization procedures.

6.4.2.2 Assessing uniformity of microbial and enzyme inactivation

Numerical modeling and simulation can also help to assess the uniformity of microbial and enzyme inactivation achieved throughout the product during processing and to identify possible causes of inhomogeneities. Thus, many authors in the literature have coupled partial differential equations either for conductive heat transfer only or for conductive and convective heat transfer and fluid motion with those for microbial and/or enzyme inactivation to jointly predict the transient distribution of the microbial load and/or the enzyme activity throughout the vessel during pressure processing (Ardia et al., 2004; Denys et al., 2000b; Hartmann et al., 2003; Infante et al., 2009; Juliano et al., 2009c; Kilimann et al., 2006; Knoerzer and Juliano, 2016; Salvi et al., 2017). Although only a few of these papers include validation experiments, the good agreement observed between experimental and predicted data confirms the validity of this approach (Denys et al., 2000b; Hartmann et al., 2003; Kilimann et al., 2006).

Evidently, the availability of reliable kinetic models, capable of accurately predicting the level of inactivation achieved based on combined temperature and pressure data, is a crucial prerequisite. In this regard, Juliano et al. (2009c) built a CFD model of a 35-L vessel, including seven cylindrical food packages inside a PTFE carrier, to compare the performance of four kinetic models (first-order kinetic model, Weibull model, an *n*th-order model, and a combined discrete log-linear *n*th-order model) in predicting the inactivation of *C. botulinum* throughout the food packages during processing. To do so, the authors externally coupled the kinetic models, expressed as ordinary differential equations, with the temperature predictions made by the CFD model. They found that, as with conventional thermal processes, the different inactivation models gave very different predictions. These results clearly show the importance of having kinetic models sufficiently validated over a wide range of pressure and temperature conditions in order to make reliable predictions.

If suitable kinetic models are available, fairly accurate spatiotemporal predictions can be achieved. For example, Kilimann et al. (2006) evaluated the effect of pressure, temperature, and holding time on the inactivation kinetics of *Lactococcus lactis* suspended in various buffer systems. From the data collected, they derived a logistic inactivation model capable of predicting viable cell counts, both injured and uninjured. Coupling this model to the equations of thermo-fluid dynamics under pressure, they predicted the inactivation of *L. lactis*, at the top and at the bottom of a 3.3-L high-pressure vessel, during pressure processing at 500 MPa and various temperatures between 28°C and 48°C. The model revealed significant temperature differences between the top and bottom of the vessel during the high-pressure treatment that resulted in differences of more than 1 log unit in *L. lactis* inactivation predictions. These predictions agreed very well with the experimental data, thus demonstrating the effectiveness of numerical modeling in predicting nonuniformities during pressure processing.

When simulating the inactivation of microbes and/or enzymes, it is crucial to remember that in solid foods, food particles remain stationary, while in liquid foods, they circulate throughout the food matrix. Therefore, in this case, the trajectory of each particle must be considered to obtain its temperature evolution during HPP and, in turn, the associated microbial or enzymatic inactivation. Otherwise, the predictions obtained may be quite inaccurate. For example, Infante et al. (2009) conducted a numerical study to analyze how the type of food, solid or liquid, affects the inactivation of *Bacillus subtilis* α-amylase, lipoxygenase, and pectin methylesterase during pressure processing and observed that, due to the mass transfer phenomenon, enzyme activity distributions were much more homogeneous in liquid foods.

It is also important to note that the more pressure- or temperature-sensitive the microbial population or enzyme is, the more accurate the model must be. Otherwise, small inaccuracies in simplified pressure profiles or in temperature predictions could induce large errors in those of residual microbial load or enzyme activity. Thus, to evaluate the robustness of the models, it is important to assess how the accuracy of the model variables (thermophysical properties, pressure profiles, and so on) affects the accuracy of the temperature predictions and, in turn, the accuracy of the inactivation results (Infante et al., 2009).

6.4.2.3 *Scaling-up high-pressure processes*

Moving from laboratory or pilot scale to commercial scale requires not only increasing the capacity of HPP equipment to meet market demand, but also considering the potential effects that changes in the vessel size can have on process uniformity and pressure-induced effects observed in foods. In this sense, Hartmann and Delgado (2002) used CFD modeling and numerical simulation together with a time-scale analysis to investigate the effect of the vessel size (0.8 mL, 0.8 L, and 6.3 L, representing micro-scale, lab-scale, and pilot-scale vessels, respectively) on the spatial and temporal evolution of temperature, velocity, and activity fields in a *Bacillus subtilis* α-amylase solution at 40°C during processing at 530 MPa for 20 min. They found that, depending on the size of the vessel, the efficiency of the heat removal and the relative importance of conductive and convective heat transport were very different during the inactivation process. Thus, in the 0.8-mL vessel, thermal and velocity gradients were damped out rapidly due to the large ratio between the heat-exchanging vessel surface and the volume and, thus, after only 50 s, isothermal conditions were reached. Therefore, practically the entire inactivation process was carried out at a uniform temperature. In contrast, in the 0.8-L and the 6.3-L vessels, removal of the compression heating was less effective due to the lower ratio between the heat-exchanging surfaces and the volume. In consequence, fluid motion and convective heat transfer were very intense during much of the inactivation process, thus dampening spatial temperature gradients. This means that, at the conditions tested by Hartmann and Delgado (2002), a geometrical scale-up resulted in more efficient enzyme inactivation, as temperature increases due to compressive heating subsisted longer in larger volumes and intensive fluid motion limited nonuniformities in the activity distribution.

Apart from the size of the vessel, there are many other factors related to the equipment design (vessel material, insulation, and orientation, among others) that can significantly affect thermal and fluid-dynamic phenomena during processing and must, therefore, be considered when scaling-up. For example, lab machines are frequently not fabricated from stainless steel and, therefore, they must use pressurizing fluids other than water, such as silicone oil, castor oil, or glycol solutions, among others. As shown in Table 6.1, quasiadiabatic temperature increase in oils and glycol solutions can be two or three times that of water and, since most food products are water-based, these pressure media may significantly heat the product upon compression. Therefore, effects observed in foods pressurized in lab machines could include unintended thermal effects that would not be replicated when performing experiments in industrial units using water as a pressurizing fluid (Ting et al., 2002).

The configuration of pressure equipment can also affect temperature uniformity during processing and, consequently, the transferability of lab results to an industrial scale. As previously mentioned, although most studies in the literature are carried out in lab- and pilot-scale vertical systems, industrial HPP equipment is usually designed with a horizontal orientation to avoid sublevel construction requirements, eliminate height and floor load restrictions, and facilitate loading and unloading of the product. Since the use of thermocouples to map the temperature distribution in horizontal industrial HPP equipment is not an option, CFD modeling is especially useful for evaluating the spatial and temporal evolution of the temperature and fluid velocity fields in this equipment. Thus, to compare temperature performance and uniformity in vertically and horizontally oriented HPP equipment, Smith et al. (2014) built 3D vertical and horizontal models and ran simulations using exactly the same process conditions. They found that, due to differences in the location of the colder vessel walls in relation to the sample (at the left and right or at the top and bottom in the vertical and horizontal equipment, respectively), flow distribution and temperature profiles during processing were very different for the two orientations. Thus, the simulations revealed that horizontal equipment allows a more uniform temperature distribution throughout the vessel than vertical HPP units. Therefore, the authors concluded that the research conducted on vertical systems is not directly transferable to commercial horizontal ones, and further research is needed to develop horizontal models that can assist in the design of strategies to ensure food quality and safety.

All these results highlight the value of CFD modeling and simulation in scaling up HPP from laboratory conditions to industry. Thus, CFD simulations can help in understanding the effect of scaling-up on key process parameters, such as temperature, that can significantly affect the efficiency of the high-pressure process.

6.5 Other modeling approaches

Apart from numerical models, other modeling approaches have also been used in the literature to describe high-pressure processes, including analytical models, macroscopic models, and ANNs (Juliano et al., 2009a, 2009b). Their advantages and limitations are described in the following sections.

6.5.1 Analytical models

Some authors have employed the analytical solutions of the governing heat transfer equations to simulate the temperature evolution in foods during high-pressure processes, both with and without phase transitions (Carroll et al., 2003; Sanz and Otero, 2000). Unlike numerical models, analytical models provide exact solutions, and they are faster and easier to apply. In general, analytical models give good results in predicting the thermal behavior of foods during pressure processing as long as the assumptions of the model (sample geometry, initial and boundary conditions, etc.) are met. Evidently, this limits the possible applications considerably.

For example, Sanz and Otero (2000) developed a model to describe HPSF in a cylindrical agar sample. To do so, they assumed that the total freezing time, t_{tot_fr}, was the sum of the precooling, phase change, and tempering times (t_{prec}, t_{ph_tr}, and t_{temp}, respectively):

$$t_{tot\_fr} = t_{prec} + t_{ph\_tr} + t_{temp} \tag{6.47}$$

By assuming a uniform initial temperature, T_0, in the sample and a constant temperature in the freezing medium, T_{fm} (i.e., the pressurizing fluid inside the pressure vessel) and considering that the unsteady-state heat transfer equation is, for a finite cylinder, the cross-product of those for an infinite slab and an infinite cylinder, they came up with the following set of equations to estimate the precooling and tempering times:

$$\left(\frac{T_{fm} - T}{T_{fm} - T_0}\right)_{fin_{cyl}} = \left(\frac{T_{fm} - T}{T_{fm} - T_0}\right)_{inf_{slab}} \cdot \left(\frac{T_{fm} - T}{T_{fm} - T_0}\right)_{inf_{cyl}} \tag{6.48}$$

$$\left(\frac{T_{fm}-T}{T_{fm}-T_0}\right)_{inf_{slab}}=2\cdot\sum_{n=1}^{\infty}\frac{\sin B_n\cdot exp\left[-B_n^2\cdot\left(\frac{K\cdot t}{\rho\cdot Cp\cdot l^2}\right)\right]}{B_n+\sin B_n\cdot\cos B_n} \tag{6.49}$$

$$\left(\frac{T_{fm}-T}{T_{fm}-T_0}\right)_{inf_{cyl}}=2\cdot\sum_{n=1}^{\infty}\frac{J_1(C_n)\cdot exp\left[\left(\frac{-C_n^2\cdot K\cdot t}{Cp\cdot rd^2}\right)\right]}{J_0^2(C_n)+J_1^2(C_n)\cdot C_n} \tag{6.50}$$

where l is half of the height of the infinite slab, rd is the radius of the infinite cylinder, and B_n and C_n are the n-roots of Eqs. (6.51) and (6.52), respectively:

$$m=\frac{h\cdot l}{K}=B_n\cdot\tan B_n \tag{6.51}$$

$$n=\frac{h\cdot rd}{K}=C_n\cdot\frac{J_1(C_n)}{J_0(C_n)} \tag{6.52}$$

while $J_0(C_n)$ and $J_1(C_n)$ are Bessel zero- and first-order functions of C_n, respectively.

Moreover, to estimate the phase transition time, the authors employed the following modified Planck's equation:

$$t_{ph\_tr}=\frac{\rho\cdot L\cdot m_{tw}\cdot\omega}{T_{if}-T_{fm}}\cdot\left(\frac{P_{Plank}\cdot D}{h}+\frac{R_{Plank}\cdot D^2}{K}\right) \tag{6.53}$$

where ω is the mean ice content (%) between those corresponding to the final freezing temperature and to T_{fm}, D is the diameter of the sample, and P_{Plank} and R_{Plank} are shape factors which depend on the sample geometry.

Unlike in conventional freezing at atmospheric conditions, model assumptions in HPSF can be well fulfilled during the precooling step, provided the sample is at a uniform initial temperature. Thus, as the model assumes, in HPSF, only sensible heat is eliminated throughout the product during this step and, therefore, accurate model predictions should be expected. In this sense, Sanz and Otero (2000) found good agreement between the temperatures recorded at the center of cylindrical agar samples during the precooling step of HPSF experiments performed at different conditions (92 MPa/−8°C, 130 MPa/−12°C, 180 MPa/−18°C, and 210 MPa/−21°C) and those estimated by the analytical model. In fact, mean deviations were lower than 1% in all the cases. The modified Plank's equation also predicted satisfactorily the phase transition times as long as the amount of instantaneous ice, produced just after adiabatic expansion, was considered. Thus, experimental and predicted t_{ph_tr} values fitted well with a maximum relative error of 11%. However, the model overestimated the tempering times as, in this phase of the process, model assumptions were not fulfilled. As occurs in conventional freezing, when the tempering step starts at the sample center in HPSF, the temperature at the sample surface is considerably below the freezing point. In consequence, model predictions are significantly less accurate during this step. For example, in HPSF experiments at 210 MPa/−21°C, Sanz and Otero (2000) found mean deviations of 15.5% between experimental and predicted temperatures at the sample center.

Analytical models have also been proposed to extract unknown thermal properties of materials under pressure from the temperature profiles recorded during pressure processing. Thus, Carroll et al. (2003) used an analytical solution for conductive heat transfer in an infinite cylinder to model the cooling of the product during the pressure-holding time. By using a curve-fitting algorithm, they proposed to employ this analytical solution to extract some thermal properties under pressure from the cooling curve recorded during real experiments. To test the reliability of the proposed method, they pressurized pure water to 600 MPa in a vessel thermoregulated at 51°C. By fitting the analytical solution and the temperatures recorded at the center of the sample, they extracted mean values for the thermal diffusivity and expansivity of pure water at 600 MPa and 50–60°C. They found that these values were 20% higher and 20%–25% lower than those provided from the NIST database, respectively, and attributed this misalignment to convective heat transfer occurring in water that is not considered in the model.

All these results reveal that analytical models, even though useful, should be treated with caution if their assumptions are not completely fulfilled. Otherwise, model predictions may lead to rather inaccurate conclusions. In this sense, it is important to note that strict requirements on sample geometry or uniform initial temperatures, for example, can rarely be met in real foods and complex processes and, therefore, this is a major limitation for the use of analytical models in HPP.

6.5.2 Macroscopic models

Macroscopic models have been used in the literature to offer a generalized representation of high-pressure systems. Thus, by using block diagrams and partial heat balances to represent heat exchanges in all the components of the system, Otero et al. (2002) built a macroscopic model describing a high-pressure system composed of a 2.35-L high-pressure vessel filled with water and thermoregulated by a surrounding coil connected to a heating–cooling bath (Fig. 6.5).

To simulate heat exchanges in the high-pressure system, the model took into account convective and conductive heat transfer between all the components of the system: between the heating–cooling bath and the environment, in the pipes between the bath and the coil surrounding the high-pressure vessel, between the vessel and the thermoregulating fluid in the coil, and between the water inside the vessel and the vessel walls (Fig. 6.5). Moreover, the model also included the calculation of the adiabatic compression heating/cooling occurring during compression/expansion. Otero et al. (2002) found that this model was able to accurately predict the evolution of temperature in different points of the system (in water inside the pressure vessel, at the entrance and exit of the surrounding coil, and inside the heating–cooling bath) during pressure processing, both after compression and expansion.

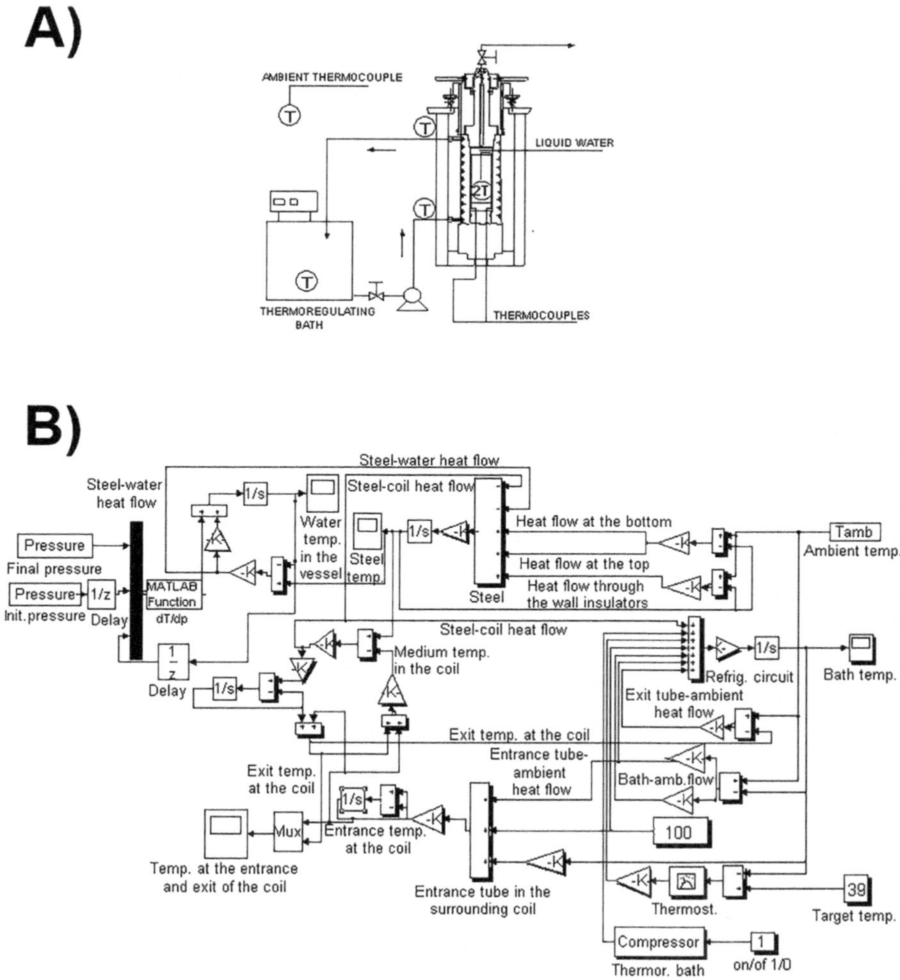

Figure 6.5 (A) High-pressure system composed of a 2.35-L pressure chamber thermoregulated by a surrounding coil connected to a heating–cooling bath. (B) Macroscopic model for the simulation of thermal exchanges throughout the high-pressure system. (T): Position of the thermocouples in the system.
Credit: Reproduced with permission from Otero, L., Molina-García, A. D., Ramos, A. M., & Sanz, P. D., 2002. A model for real thermal control in high-pressure treatment of foods. *Biotechnol. Prog.*, 18 (4), 904–908. https://doi.org/10.1021/bp010157a.

By assessing the influence of the different components on the thermal performance of the complete high-pressure system, this macroscopic model facilitates the identification of the key elements in the thermal regulation of high-pressure treatments. For example, the model revealed the huge importance that the steel mass of the pressure

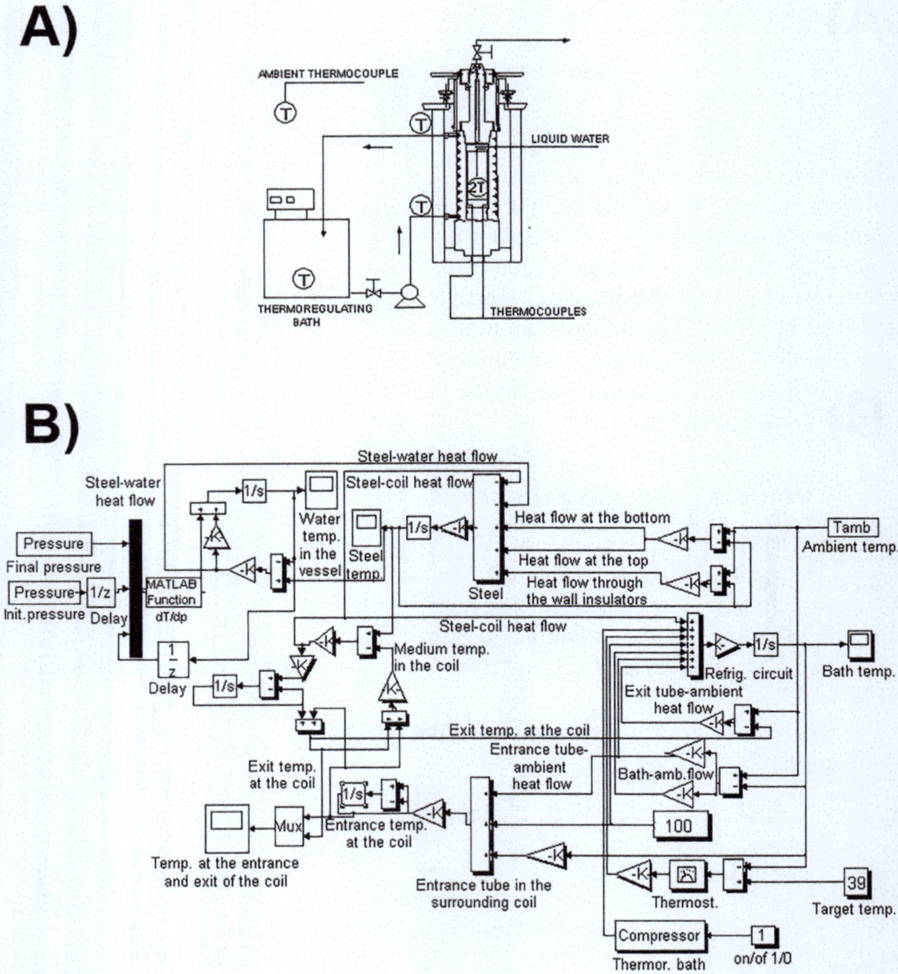

Figure 6.5 cont'd.

vessel has in the thermal behavior of the system, thus demonstrating that the initial temperature of the vessel is a key factor to consider in the process design. Furthermore, the model makes it possible to quickly evaluate how different variables (such as the temperature of the pressurizing fluid, the thermoregulating fluid, the flow rate, the heating/cooling capacity of the bath, or the number of baths, among others) affect the temperature inside the pressure vessel and, in this way, this type of model can be very useful for better thermal control of the treatments performed. However, it is important to note that this type of model does not reproduce the thermal gradients that occur within the pressure vessel and is, therefore, not useful for evaluating the uniformity of high-pressure processes.

6.5.3 Artificial neural networks

An ANN is a mathematical algorithm that, by using interconnected groups of nodes or artificial neurons, handles and processes information in a way similar to animal brains. Thus, ANNs learn from examples through iteration, by relating input and output parameters without requiring prior knowledge of the relationships between them.

The structure of ANNs is relatively simple and basically consists of sets of interconnected neurons typically arranged in an input layer, one or more hidden layers, and an output layer. As Fig. 6.6A illustrates, the input layer receives the information from

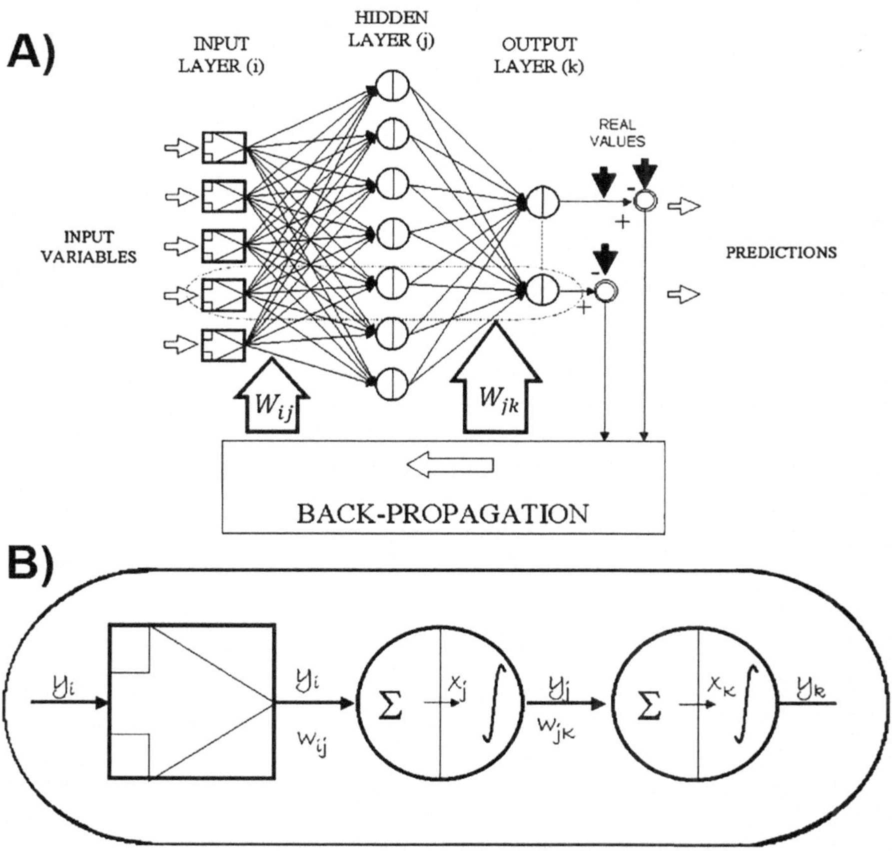

Figure 6.6 (A) Typical structure of a neural network model. A standard back-propagation algorithm is applied in the training step to optimize the connection weights. (B) Details of the function of a neuron: the inputs into a neuron (y) are multiplied by their corresponding connection weights (w) and summed together to yield x values. These x values are then transformed through the sigmoid function to produce a single output that may be passed on to other neurons.

Credit: Modified with permission from Torrecilla, J. S., Otero, L., & Sanz, P. D., 2005. Artificial neural networks: a promising tool to design and optimize high-pressure food processes. *J. Food Eng.*, 69 (3), 299–306. https://doi.org/10.1016/j.jfoodeng.2004.08.020.

an external source and passes it to the first hidden layer, where this information is processed by a combination of linear and nonlinear mathematical algorithms. The results are subsequently sent to the next hidden layer or, in neural network architectures with only one hidden layer, to the output layer that finally presents the predictions to the external receptor.

Connections between neurons are controlled by weighted links, w, that are optimized during the learning process, usually with real input and output data from the system to be modeled (supervised method). Fig. 6.6B depicts the calculation process in hidden and output neurons that consists of activation and transfer functions. The activation function multiplies the neuron inputs, y, by the corresponding weighted links, w, and adds together to yield x values, which are subsequently fed into a transfer function to produce a single y value, the output of the considered neuron.

ANNs have been successfully applied in food technology in a number of tasks related to process control, microbiology, sensory analysis, quality control, and classification (Bhagya Raj and Dash, 2022; Guiné, 2019; Nayak et al., 2020). In the HPP sector, ANNs have been employed to predict not only the thermal behavior of foods during processing (Torrecilla and Sanz, 2011) but also some critical parameters and properties for the design and optimization of high-pressure processes, such as the adiabatic compression heating temperature (Knoerzer, 2024) and the initial freezing point under pressure (Guignon et al., 2008).

The capability of ANNs to predict the thermal behavior of foods during HPP was demonstrated by Torrecilla et al. (2004, 2005, 2007). Thus, these authors built ANNs capable of predicting the maximum temperature reached inside the pressure vessel after compression, T_{max} (Torrecilla et al., 2005, 2007) and the time needed to dissipate compression heating and recover the thermal equilibrium, t_{eq} (Torrecilla et al., 2004, 2005, 2007). To train and test the ANNs, they used the data generated by the macroscopic model described in Section 6.5.2 (Otero et al., 2002). These data comprised a wide range of processing conditions, with proved influence on the thermal evolution of the system, including the applied pressure, P (250–450 MPa), the compression rate, p_{rate} (1–2 MPa s^{-1}), the room temperature, T_{room} (10–30°C), the target temperature, T_{tgt} (40–80°C), and the vessel temperature, T_V (between the room and the target temperatures) and also the corresponding T_{max} and t_{eq} values predicted by the macroscopic model. The authors grouped the data generated by the model in two totally independent sets and used them to train the ANN (training set) and to evaluate the competence of the trained model (test set). After testing different training algorithms (13 functions), topologies (from 1 to 13 neurons in the hidden layer), and values for the learning coefficient (0.001–1) and the learning coefficient decrease (0.1–1), they obtained an optimized feed-forward ANN consisting of three layers: an input layer with five nodes corresponding to the independent variables P, p_{rate}, T_{room}, T_{tgt}, and T_V, a hidden layer with 12 neurons, and an output layer with two neurons to present T_{max} and t_{eq} predictions. The ANN validation revealed the excellent predictive ability of the optimized network and, hence, the mean prediction errors were 0.62% and 0.4% for temperature and time predictions, respectively (Torrecilla et al., 2007). These results demonstrate that ANNs can be an interesting alternative to physically based models for control of high-pressure processes, especially when the thermophysical properties of the materials involved are not available under pressure. Unlike physically based models, ANNs do not require the knowledge of thermophysical properties under pressure to make predictions, although enough experimental observations and a proper training must be done for each new product and/or high-pressure system tested. However, once the network is optimized, many combinations of processing conditions can be evaluated after a short computing time.

Moreover, as previously mentioned, ANNs have also been used to predict critical parameters for the design and optimization of high-pressure processes and unknown properties under pressure. In this sense, Knoerzer (2024) has recently demonstrated the efficacy of ANNs in accurately predicting adiabatic compression heating of both liquid (water and propylene glycol) and solid materials (PP, HDPE, and PTFE) during HPP and HPTP. The author built a network composed of an input layer with two nodes corresponding to the pressure and the initial temperature, three hidden layers, each with 10 neurons, and an output layer with one node for the final temperature. To train the ANN, Knoerzer (2024) used temperature profiles recorded in experiments performed at a wide range of pressures, from 0.1 to 750 MPa. Moreover, to avoid potential temperature losses or gains, the temperature profiles used for building the ANN corresponded only to the pressure release, as this part of the process is very fast and ranged from about 90°C to below 10°C for each experiment. The optimized network accurately predicted the resulting compression heating temperature for all the materials tested and, hence, root mean squared errors were lower than 1 in all the cases.

The initial freezing point of foods and pressurizing fluids under pressure is an essential property for the design and optimization of high-pressure processes involving phase transitions, such as HPF and HPT. For this reason, Guignon et al. (2008) developed a feed-forward ANN to predict this property in aqueous systems under pressure. To do so, they used about 400 independent freezing data points of aqueous solutions, gels, and foods obtained either experimentally at the authors' laboratory or from the literature. The optimized ANN was composed of an input layer with two neurons corresponding to the pressure at which the initial freezing point is required and the system's water activity, a hidden layer with 6 neurons, and an output layer with one neuron for the predicted freezing temperature. This network was able to accurately predict the initial freezing temperature of all the products tested and, hence, experimental and predicted data differ by only 0.3°C.

All these findings demonstrate the significant potential of neural networks in modeling complex, high-pressure processes. However, it is also crucial to consider their limitations. For instance, neural networks function as black boxes and, therefore, they are ineffective in elucidating the mechanisms through which specific process parameters influence the variables of interest. Furthermore, to ensure robustness, they require a substantial amount of experimental data for training, and they typically exhibit limited capacity to adapt to disparate scenarios. Thus, a network trained with a specific high-pressure system is not applicable to a different unit unless they are identical and work under the same environmental conditions. All these limitations mean that, although they can be very useful, their use in high-pressure food processing is not yet widespread.

6.6 Concluding remarks

Different modeling approaches, including analytical and numerical models, macroscopic models, and ANNs, have been successfully used in the literature to describe high-pressure processes. Of these, numerical models and, more specifically, CFD models are, undoubtedly, the most versatile option. CFD models can accurately describe the fluid motion and heat transfer phenomena occurring inside the pressure vessel and make spatiotemporal predictions with high local resolution and in a wide variety of situations. Thus, they are powerful tools for understanding what happens inside the pressure vessel during processing and the effect of different process variables and equipment characteristics on thermo-fluid dynamics. Furthermore, by coupling the partial differential equations for thermo-fluid dynamics with those of microbial and/or enzyme inactivation, CFD models allow to assess the uniformity of the process in terms not only of temperature but also of food safety and quality. Thus, they can play a significant role in the characterization, improvement, optimization, and scale-up of high-pressure processes.

However, the widespread use of CFD models is currently hampered by a lack of knowledge on the thermophysical properties of foods, packaging and carrier materials, and pressurizing fluids under pressure. This often forces the use of simplifying assumptions that jeopardize the accuracy of model predictions. Furthermore, the scarcity of kinetic models capable of describing the inactivation of different microorganisms and enzymes of interest over a wide range of processing conditions (pressure, temperature, pH, water activity, etc.) limits the usefulness of CFD models for designing safe high-pressure processes with minimal impact on product quality. Therefore, to improve the accuracy and usefulness of CFD models, future research efforts should focus on obtaining thermophysical properties of materials under pressure and deriving suitable kinetic models to describe microbial and enzymatic inactivation under pressure.

Nomenclature

a_w	Water activity
CFD	Computational fluid dynamics
Cp	Specific heat capacity
Cp_{app}	Apparent specific heat capacity
D	Sample diameter
g	Gravity constant
H	Enthalpy
HDPE	High-density polyethylene
HPF	High-pressure freezing

HPP	High-pressure processing
HPSF	High-pressure shift freezing
HPT	High-pressure thawing
HPTP	High-pressure thermal processing
h	Heat transfer coefficient
IAPWS	International Association for the Properties of Water and Steam
K	Thermal conductivity
k_c	Compression heating coefficient
L	Latent heat
l	Half of the height of the infinite slab
M	Molar mass
m	Mass fraction
m^*_{ice}	Mass of ice divided by the initial mass of water
n	Normal boundary direction
P	Pressure
P_{atm}	Atmospheric pressure
P_{Plank}	Geometric shape factor in Plank's equation
PP	Polypropylene
PTFE	Polytetrafluoroethylene
p_{rate}	Compression/decompression rate
pTTIs	Pressure-temperature-time indicators
Q	Heat generation term
R	Universal gas constant
R_{Plank}	Geometric shape factor in Plank's equation
r	Horizontal position (radial direction)
rd	Radius of the infinite cylinder
T	Temperature
T_0	Initial temperature
T_{fm}	Freezing medium temperature
Tk	Temperature in K
T_{max}	Maximum temperature reached inside the pressure vessel after compression
T_{min}	Minimum temperature reached after expansion
T_{ref}	Reference temperature at which $H = 0$
T_{room}	Room temperature
T_{tm_w}	Temperature of thermoregulated wall
t	Time
t_{com}	Time needed to reach the target pressure during the compression step
t_{eq}	Time needed to dissipate compression heating and recover the thermal equilibrium
t_{exp}	Time at which expansion is performed
t_f	Time at which atmospheric pressure is reached after expansion
t_{ph_tr}	Phase transition time
t_{prec}	Precooling time
t_{temp}	Tempering time
t_{tot_fr}	Total freezing time
$\overline{u}$	Fluid velocity
v	Specific volume
w	Weighted link
X	Mole fraction

x	Result of the activation function
y	Neuron input/output
z	Vertical position (axial direction)

Greeks

α	Coefficient of thermal expansion
Γ_{F-PF}	Food-pressurizing fluid boundary
Γ_{inlet}	Inlet tube boundary
Γ_{low_V}	Lower vessel boundary
Γ_{PF-V}	Pressurizing fluid-vessel boundary
Γ_{r_V}	Right vessel boundary
Γ_{up_V}	Upper vessel boundary
ΔT_{sup}	Supercooling
ε	Volume fraction
μ	Dynamic viscosity
ρ	Density
ω	Mean ice content (%) between those corresponding to the final freezing temperature and to the temperature of the freezing medium
Ω_F	Food subdomain
Ω_{PF}	Pressurizing fluid subdomain
Ω_V	Vessel subdomain

Superindexes

0	Pure water

Subindexes

bw	Bound water
dm	Dry matter
F	Food
fin_{cyl}	Finite cylinder
fl	Fluid
ice	Ice
if	Initial freezing point
in	Pressurizing fluid inlet
inf_{cyl}	Infinite cylinder
inf_{slab}	Infinite slab
m	Molar
PF	Pressurizing fluid
s	Solid
T	Turbulent conditions

tgt	Target
tw	Total water
ufw	Unfrozen free water
ufzw	Unfreezable water
V	Pressure vessel
w	Water
wall	Vessel wall
z	Axial direction

References

Abdul Ghani, A.G., Farid, M.M., 2007. Numerical simulation of solid-liquid food mixture in a high pressure processing unit using computational fluid dynamics. J. Food Eng. 80 (4), 1031–1042. https://doi.org/10.1016/j.jfoodeng.2006.08.018.

Acosta, G.M., Smith, R.L., Arai, K., 1996. High-pressure PVT behavior of natural fats and oils, trilaurin, triolein, and n-tridecane from 303 K to 353 K from atmospheric pressure to 150 MPa. J. Chem. Eng. Data 41, 961–969. https://doi.org/10.1021/je960006x.

Aganovic, K., Hertel, C., Vogel, R.F., Johne, R., Schlüter, O., Schwarzenbolz, U., Jäger, H., Holzhauser, T., Bergmair, J., Roth, A., Sevenich, R., Bandick, N., Kulling, S.E., Knorr, D., Engel, K.H., Heinz, V., 2021. Aspects of high hydrostatic pressure food processing: perspectives on technology and food safety. Compr. Rev. Food Sci. Food Saf. 20 (4), 3225–3266. https://doi.org/10.1111/1541-4337.12763.

Aparicio, C., Guignon, B., Otero, L., Sanz, P.D., 2011. Thermal expansion coefficient and specific heat capacity from sound velocity measurements in tomato paste from 0.1 up to 350 MPa and as a function of temperature. J. Food Eng. 104 (3). https://doi.org/10.1016/j.jfoodeng.2010.12.026.

Ardia, A., Knorr, D., Heinz, V., 2004. Adiabatic heat modelling for pressure build-up during high-pressure treatment in liquid-food processing. Food Bioprod. Process. 82 (1), 89–95. https://doi.org/10.1205/096030804322985362.

Balasubramaniam, V.M., Barbosa-Cánovas, G.V., Lelieveld, H.L.M., 2016. High-pressure processing equipment for the food industry. Food Eng. Ser. 39–65. https://doi.org/10.1007/978-1-4939-3234-4_3.

Barlow, A.J., Harrison, G., Kim, M.G., Lamb, J., 1973. Pressure dependence of the viscoelastic properties of castor oil. J. Chem. Soc. Faraday Trans. 2 Mol. Chem. Phys. 69, 1446–1453.

Bhagya Raj, G.V.S., Dash, K.K., 2022. Comprehensive study on applications of artificial neural network in food process modeling. Crit. Rev. Food Sci. Nutr. 62 (10), 2756–2783. https://doi.org/10.1080/10408398.2020.1858398.

Biresaw, G., Bantchev, G.B., 2013. Pressure viscosity coefficient of vegetable oils. Tribol. Lett. 49, 501–512. https://doi.org/10.1007/s11249-012-0091-9.

Bridgman, P.W., 1923. The thermal conductivity of liquids under pressure. Proceeding Am. Acad. Arts Sci. 59, 141–169.

Carroll, T., Chen, P., Fletcher, A., 2003. A method to characterise heat transfer during high-pressure processing. J. Food Eng. 60 (2), 131–135. https://doi.org/10.1016/S0260-8774(03)00026-8.

Casulli, K.E., Dhakal, S., Sandeep, K.P., Balasubramaniam, V.M., 2017. Compression heating of selected polymers during high-pressure processing. J. Food Process. Eng. 40, 1–7. https://doi.org/10.1111/jfpe.12417.

Choi, Y., Okos, M.R., 1986. Effects of temperature and composition on the thermal properties of food. In: Maguer, M.L., Jelen, P. (Eds.), Food Engineering and Process Applications, Transport Phenomena, Vol 1. Elsevier Applied Science, pp. 93–101.

Chourot, J.-M., Boillereaux, L., Havet, M., Le Bail, A., 1997. Numerical modeling of high pressure thawing: application to water thawing. J. Food Eng. 34 (1), 63–75. https://doi.org/10.1016/S0260-8774(97)00067-8.

Delgado, A., Hartmann, C., 2003. Pressure treatment of food: instantaneous but not homogeneous effect. In: Winter, R. (Ed.), Advances in High Pressure Bioscience and Biotechnology, II. Springer Berlin Heidelberg, pp. 459–464.

Denys, S., Hendrickx, M.E., 1999. Measurement of the thermal conductivity of foods at high pressure. J. Food Sci. 64, 709–713. https://doi.org/10.1111/j.1365-2621.1999.tb15116.x.

Denys, S., 2000. Process calculations for design of batch high hydrostatic pressure processes without and with phase transitions. Ph. D. Dissertation no. 456 at the Faculty of Agricultural and Applied Biological Sciences, Katholieke Universiteit Leuven, Belgium.

Denys, S., Ludikhuyze, L.R., Van Loey, A.M., Hendrickx, M.E., 2000a. Modeling conductive heat transfer and process uniformity during batch high-pressure processing of foods. Biotechnol. Prog. 16 (1), 92–101. https://doi.org/10.1021/bp990123r.

Denys, S., Van Loey, A.M., Hendrickx, M.E., 2000b. A modeling approach for evaluating process uniformity during batch high hydrostatic pressure processing: combination of a numerical heat transfer model and enzyme inactivation kinetics. Innov. Food Sci. Emerg. Technol. 1 (1), 5–19. https://doi.org/10.1016/S1466-8564(99)00003-X.

Denys, S., Van Loey, A.M., Hendrickx, M.E., 2000c. Modeling conductive heat transfer during high-pressure thawing processes: determination of latent heat as a function of pressure. Biotechnol. Prog. 16 (3), 447–455. https://doi.org/10.1021/bp000018d.

El'darov, V.S., 2003. Thermal conductivity of aqueous solutions of KCl-NaCl-$CaCl_2$ system at high temperatures and pressures. High Temp 41, 327–331. https://doi.org/10.1023/A:1024282308625.

Freitas, S.V.D., E Silva, F.A., Pastoriza-Gallego, M.J., Piñeiro, M.M., Lima, Á.S., Coutinho, J.A.P., 2013. Measurement and prediction of densities of vegetable oils at pressures up to 45 MPa. J. Chem. Eng. Data 58, 3046–3053. https://doi.org/10.1021/je400474w.

Först, P., Werner, F., Delgado, A., 2002. On the pressure dependence of the viscosity of aqueous sugar solutions. Rheol. Acta 41 (4), 369–374. https://doi.org/10.1007/s00397-002-0238-y.

Fukui, K., Asakuma, Y., Maeda, K., 2010. Determination of liquid viscosity at high pressure by DLS. J. Phys. Conf. 215. https://doi.org/10.1088/1742-6596/215/1/012073.

Gänzle, M.G., 2023. Introduction to high pressure thermal processing and pressure assisted thermal sterilization. In: Knoerzer, K., Sevenich, R. (Eds.), High Pressure Thermal Processing. Elsevier Academic Press, pp. 1–15. https://doi.org/10.1016/B978-0-12-823872-1.00005-3.

Grauwet, T., Plancken, I. Van der, Vervoort, L., Hendrickx, M.E., Loey, A. Van, 2010a. Protein-based indicator system for detection of temperature differences in high pressure high temperature processing. Food Res. Int. 43 (3), 862–871. https://doi.org/10.1016/j.foodres.2009.12.002.

Grauwet, T., Van der Plancken, I., Vervoort, L., Hendrickx, M.E., Van Loey, A., 2010b. Solvent engineering as a tool in enzymatic indicator development for mild high pressure pasteurization processing. J. Food Eng. 97 (3), 301–310. https://doi.org/10.1016/j.jfoodeng.2009.10.018.

Grauwet, T., Van Der Plancken, I., Vervoort, L., Hendrickx, M.E., Van Loey, A., 2009. Investigating the potential of *Bacillus subtilis* α-amylase as a pressure-temperature-time indicator for high hydrostatic pressure pasteurization processes. Biotechnol. Prog. 25 (4), 1184–1193. https://doi.org/10.1002/btpr.170.

Grauwet, T., Van der Plancken, I., Vervoort, L., Hendrickx, M., Van Loey, A., 2016. High-pressure processing uniformity. In: Balasubramaniam, V.M., Barbosa-Cánovas, G.V., Lelieveld, H.L.M. (Eds.), High Pressure Processing of Food. Principles, Technology and Applications. Springer, New York, NY, pp. 253–268. https://doi.org/10.1007/978-1-4939-3234-4_13.

Grauwet, T., Van Der Plancken, I., Vervoort, L., Matser, A., Hendrickx, M., Van Loey, A., 2011. Temperature uniformity mapping in a high pressure high temperature reactor using a temperature sensitive indicator. J. Food Eng. 105 (1), 36–47. https://doi.org/10.1016/j.jfoodeng.2011.01.001.

Guignon, B., Aparicio, C., Sanz, P.D., 2009. Volumetric properties of sunflower and olive oils at temperatures between 15 and 55°C under pressures up to 350 MPa. High Press. Res. 29 (1), 38–45. https://doi.org/10.1080/08957950802659823.

Guignon, B., Aparicio, C., Sanz, P.D., 2010. Volumetric properties of pressure-transmitting fluids up to 350 MPa: water, ethanol, ethylene glycol, propylene glycol, castor oil, silicon oil, and some of their binary mixture. J. Chem. Eng. Data 55 (9), 3017–3023. https://doi.org/10.1021/je9010568.

Guignon, B., Aparicio, C., Sanz, P.D., Otero, L., 2012. Orange juice pvT-properties for high pressure processing and modeling purposes: importance of soluble solids concentration. Food Res. Int. 46 (1), 83–91. https://doi.org/10.1016/j.foodres.2011.12.002.

Guignon, B., Otero, L., Molina-García, A.D., Sanz, P.D., 2005. Liquid water-ice I phase diagrams under high pressure: sodium chloride and sucrose models for food systems. Biotechnol. Prog. 21, 439–445. https://doi.org/10.1021/bp049666d.

Guignon, B., Rey-Santos, I., Sanz, P.D., 2014. Determination, analysis and prediction of the volumetric behavior of milk at high pressure. Food Res. Int. 64, 336–347. https://doi.org/10.1016/j.foodres.2014.06.046.

Guignon, B., Torrecilla, J.S., Otero, L., Ramos, A.M., Molina-García, A.D., Sanz, P.D., 2008. The initial freezing temperature of foods at high pressure. Crit. Rev. Food Sci. Nutr. 48 (4), 328–340. https://doi.org/10.1080/10408390701347736.

Guiné, R.P.F., 2019. The use of artificial neural networks (ANN) in food process engineering. Int. J. Food Eng. 5 (1), 15–21. https://doi.org/10.18178/ijfe.5.1.15-21.

Hartmann, C., 2002. Numerical simulation of thermodynamic and fluid-dynamic processes during the high-pressure treatment of fluid food systems. Innov. Food Sci. Emerg. Technol. 3 (1), 11–18. https://doi.org/10.1016/S1466-8564(01)00060-1.

Hartmann, C., Delgado, A., 2002. Numerical simulation of convective and diffusive transport effects on a high-pressure-induced inactivation process. Biotechnol. Bioeng. 79 (1), 94–104. https://doi.org/10.1002/bit.10302.

Hartmann, C., Delgado, A., 2003. The influence of transport phenomena during high-pressure processing of packed food on the uniformity of enzyme inactivation. Biotechnol. Bioeng. 82 (6), 725–735. https://doi.org/10.1002/bit.10619.

Hartmann, C., Delgado, A., Szymczyk, J., 2003. Convective and diffusive transport effects in a high pressure induced inactivation process of packed food. J. Food Eng. 59 (1), 33–44. https://doi.org/10.1016/S0260-8774(02)00427-2.

Hartmann, C., Schuhholz, J.P., Kitsubun, P., Chapleau, N., Le Bail, A., Delgado, A., 2004. Experimental and numerical analysis of the thermofluid dynamics in a high-pressure autoclave. Innov. Food Sci. Emerg. Technol. 5 (4), 399–411. https://doi.org/10.1016/j.ifset.2004.07.003.

Hersey, M.D., Hopkins, R.F., 1937. Collected results on viscosity of lubricants under pressure. I. Fatty oils. J. Appl. Phys. 8, 560–566. https://doi.org/10.1063/1.1710341.

Höglund, E., 1999. Influence of lubricant properties on elastohydrodynamic lubrication. Wear 232, 176–184. https://doi.org/10.1016/S0043-1648(99)00143-X.

Houška, M., Silva, F.V.M., Evelyn, Buckow, R., Terefe, N.S., Tonello, C., 2022. High pressure processing applications in plant foods. Foods 11 (2). https://doi.org/10.3390/foods11020223.

Huang, H.W., Wu, S.J., Lu, J.K., Shyu, Y.T., Wang, C.Y., 2017. Current status and future trends of high-pressure processing in food industry. Food Control 72 (12), 1–8. https://doi.org/10.1016/j.foodcont.2016.07.019.

Huber, M.L., Perkins, R.A., Laesecke, A., Friend, D.G., Sengers, J.V., Assael, M.J., Metaxa, I. N., Vogel, E., Mareš, R., Miyagawa, K., 2009. New international formulation for the viscosity of H_2O. J. Phys. Chem. Ref. Data 38 (2), 101–125. https://doi.org/10.1063/1.3088050.

IAPWS, 2008. In: Release on the IAPWS Formulation 2008 for the Viscosity of Ordinary Water Substance. International Association for the Properties of Water and Steam, pp. 1–9.

IAPWS, 2011. In: Release on the IAPWS Formulation 2011 for the Thermal Conductivity of Ordinary Water Substance. http://www.iapws.org.

IAPWS, 2018. In: Revised Release on the IAPWS Formulation 1995 for the Thermodynamic Properties of Ordinary Water Substance for General and Scientific Use. https://doi.org/10.1063/1.1461829.

Infante, J.A., Ivorra, B., Ramos, Á.M., Rey, J.M., 2009. On the modelling and simulation of high pressure processes and inactivation of enzymes in food engineering. Math. Model Methods Appl. Sci. 19 (12), 2203–2229. https://doi.org/10.1142/S0218202509004091.

Janahar, J.J., Balasubramaniam, V.M., 2022. Application of High-Pressure-Based Technologies in the Food Industry. *FST-FABE-1001. Agriculture and Natural Resources*. The Ohio State University, pp. 1–9. https://ohioline.osu.edu/factsheet/fst-fabe-1001.

Juliano, P., Knoerzer, K., Barbosa-Cánovas, G.V., 2009a. High-pressure processes: thermal and fluid dynamic modeling applications. In: Simpson, R. (Ed.), Engineering Aspects of Thermal Processing. CRC Press, pp. 161–208.

Juliano, P., Knoerzer, K., Barbosa-Cánovas, G.V., 2009b. High-pressure thermal processes: thermal and fluid dynamic modeling principles. In: Simpson, R. (Ed.), Engineering Aspects of Thermal Processing. CRC Press, pp. 91–158.

Juliano, P., Knoerzer, K., Fryer, P.J., Versteeg, C., 2009c. *C. botulinum* inactivation kinetics implemented in a computational model of a high-pressure sterilization process. Biotechnol. Prog. 25 (1), 163–175. https://doi.org/10.1002/btpr.116.

Juliano, P., Knoerzer, K., Versteeg, C., 2011. Computational fluid dynamics applied in high-pressure high-temperature processes: spore inactivation distribution and process optimization. In: Knoerzer, K., Juliano, P., Roupas, P., Versteeg, C. (Eds.), Innovative Food Processing Technologies: Advances in Multiphysics Simulation, first ed. John Wiley & Sons, Ltd, pp. 75–100. https://doi.org/10.1002/9780470959435.ch5.

Khurana, M., Karwe, M.V., 2009. Numerical prediction of temperature distribution and measurement of temperature in a high hydrostatic pressure food processor. Food Bioprocess Technol. 2 (3), 279–290. https://doi.org/10.1007/s11947-008-0096-4.

Kiełczyński, P., Szalewski, M., Balcerzak, A., Wieja, K., Rostocki, A.J., Ptasznik, S., 2017. Evaluation of high-pressure thermophysical parameters of the diacylglycerol (DAG) oil using ultrasonic waves. Food Bioprocess Technol. 10 (2), 358–369. https://doi.org/10.1007/s11947-016-1827-6.

Kilimann, K.V., Kitsubun, P., Delgado, A., Gänzle, M.G., Chapleau, N., Le Bail, A., Hartmann, C., 2006. Experimental and numerical study of heterogeneous pressure-temperature-induced lethal and sublethal injury of *Lactococcus lactis* in a medium scale high-pressure autoclave. Biotechnol. Bioeng. 94 (4), 655–666. https://doi.org/10.1002/bit.20852.

Knoerzer, K., 2023. Adiabatic compression heating properties of liquids and solids. In: Knoerzer, K., Sevenich, R. (Eds.), High Pressure Thermal Processing. Elsevier Inc, pp. 17–40. https://doi.org/10.1016/B978-0-12-823872-1.00010-7.

Knoerzer, K., 2024. Leveraging artificial intelligence for simplified adiabatic compression heating prediction: comparing the use of artificial neural networks with conventional numerical approach. Innov. Food Sci. Emerg. Technol. 91, 103546. https://doi.org/10.1016/j.ifset.2023.103546.

Knoerzer, K., Buckow, R., Chapman, B., Juliano, P., Versteeg, C., 2010a. Carrier optimisation in a pilot-scale high pressure sterilisation plant - an iterative CFD approach employing an integrated temperature distributor (ITD). J. Food Eng. 97 (2), 199–207. https://doi.org/10.1016/j.jfoodeng.2009.10.010.

Knoerzer, K., Buckow, R., Sanguansri, P., Versteeg, C., 2010b. Adiabatic compression heating coefficients for high-pressure processing of water, propylene-glycol and mixtures - a combined experimental and numerical approach. J. Food Eng. 96 (2), 229–238. https://doi.org/10.1016/j.jfoodeng.2009.07.017.

Knoerzer, K., Buckow, R., Versteeg, C., 2010c. Adiabatic compression heating coefficients for high-pressure processing - a study of some insulating polymer materials. J. Food Eng. 98 (1), 110–119. https://doi.org/10.1016/j.jfoodeng.2009.12.016.

Knoerzer, K., Chapman, B., 2011. Effect of material properties and processing conditions on the prediction accuracy of a CFD model for simulating high pressure thermal (HPT) processing. J. Food Eng. 104 (3), 404–413. https://doi.org/10.1016/j.jfoodeng.2011.01.003.

Knoerzer, K., Juliano, P., 2016. Modeling high-pressure processes: equipment design, process performance evaluation, and validation. In: Balasubramaniam, V.M.B., Barbosa-Cánovas, G.V., Lelieveld, H.L.M. (Eds.), High Pressure Processing of Food. Springer Science + Bussiness Media, New York, pp. 217–252. https://doi.org/10.1007/978-1-4939-3234-4_12.

Knoerzer, K., Juliano, P., Gladman, S., Versteeg, C., Fryer, P.J., 2007. A computational model for temperature and sterility distributions in a pilot-scale high-pressure high-temperature process. AIChE J. 53 (11), 2996–3010. https://doi.org/10.1002/aic.11301.

Knoerzer, K., Smith, R., Juliano, P., Kelly, M., Steele, R., Sanguansri, P., Versteeg, C., 2010d. The thermo-egg: a combined novel engineering and reverse logic approach for determining temperatures at high pressure. Food Eng. Rev. 2 (3), 216–225. https://doi.org/10.1007/s12393-010-9025-1.

Kowalczyk, W., Hartmann, C., Delgado, A., 2004. Modelling and numerical simulation of convection driven high pressure induced phase changes. Int. J. Heat Mass Tran. 47 (5), 1079–1089. https://doi.org/10.1016/j.ijheatmasstransfer.2003.07.030.

Leu, B.M., Yavaş, H., Kantor, I., Prakapenka, V.B., 2010. Specific heat of olive oil to 356 MPa. JAOCS, J. Am. Oil Chem. Soc. 87, 1517–1520. https://doi.org/10.1007/s11746-010-1635-5.

Malanowski, A., Rostocki, A.J., Kiełczyński, P., Szalewski, M., Balcerzak, A., Kościesza, R., Tarakowski, R., Ptasznik, S., Siegoczyński, R.M., 2013. Viscosity and compressibility of diacylglycerol under high pressure. High Press. Res. 33, 178–183. https://doi.org/10.1080/08957959.2013.775649.

Mawatari, T., Fukuda, R., Mori, H., Mia, S., Ohno, N., 2013. High pressure rheology of environmentally friendly vegetable oils. Tribol. Lett. 51 (2), 273–280. https://doi.org/10.1007/s11249-013-0180-4.

McLaughlin, E., 1969. Theory of thermal conductivity of fluids. In: Tye, R.P. (Ed.), Thermal Conductivity. Academic Press, Massachusetts, USA, pp. 1–65.

Mia, S., Hayashi, S., Ohno, N., 2007. High pressure tribological behavior of vegetable oils. In: Proceedings of the International Conference on Mechanical Engineering.

Min, S., Sastry, S., Balasubramaniam, V.M., 2009. Variable volume piezometer for measurement of volumetric properties of materials under high pressure. High Press. Res. 29, 278–289. https://doi.org/10.1080/08957950902734021.

Min, S., Sastry, S.K., Balasubramaniam, V.M., 2010. Compressibility and density of select liquid and solid foods under pressures up to 700 MPa. J. Food Eng. 96, 568–574. https://doi.org/10.1016/j.jfoodeng.2009.09.003.

Nair, A., Maldonaldo, J.A., Miyazawa, Y., Cuitiño, A.M., Schaffner, D.W., Karwe, M., 2016. Numerical simulation of stress distribution in heterogeneous solids during high pressure processing. Food Res. Int. 84, 76–85. https://doi.org/10.1016/j.foodres.2016.03.016.

Nayak, J., Vakula, K., Dinesh, P., Naik, B., Pelusi, D., 2020. Intelligent food processing: journey from artificial neural network to deep learning. Comput. Sci. Rev. 38, 100297. https://doi.org/10.1016/j.cosrev.2020.100297.

Nguyen, L.T., Balasubramaniam, V.M., Sastry, S.K., 2012. Determination of in-situ thermal conductivity, thermal diffusivity, volumetric specific heat and isobaric specific heat of selected foods under pressure. Int. J. Food Prop. 15, 169–187. https://doi.org/10.1080/10942911003754726.

Norton, T., Delgado, A., Hogan, E., Grace, P., Sun, D.W., 2009. Simulation of high pressure freezing processes by enthalpy method. J. Food Eng. 91 (2), 260–268. https://doi.org/10.1016/j.jfoodeng.2008.08.031.

Ohno, N., Shiratake, A., Kuwano, N., Hirano, F., 1997. Behavior of some vegetable oils in EHL contacts. Tribol. Ser. 32, 243–251. https://doi.org/10.1016/s0167-8922(08)70453-x.

Otero, L., 2023. Application of high pressure processing in freezing and thawing processes. In: Jafari, S.M., Therdthai, N. (Eds.), Non-thermal Food Processing Operations. Elsevier Inc, pp. 359–405. https://doi.org/10.1016/b978-0-12-818717-3.00005-6.

Otero, L., Datta, A.K., 2026. Modeling key thermophysical properties for high-pressure food processing: Foundations for a digital knowledge base. J. Food Eng. https://doi.org/10.1016/j.jfoodeng.2025.112857.

Otero, L., Guignon, B., Aparicio, C., Sanz, P.D., 2010. Modeling thermophysical properties of food under high pressure. Crit. Rev. Food Sci. Nutr. 50 (4), 344–368. https://doi.org/10.1080/10408390802248817.

Otero, L., Molina-García, A.D., Ramos, A.M., Sanz, P.D., 2002. A model for real thermal control in high-pressure treatment of foods. Biotechnol. Prog. 18 (4), 904–908. https://doi.org/10.1021/bp010157a.

Otero, L., Ousegui, A., Guignon, B., Le Bail, A., Sanz, P.D., 2006. Evaluation of the thermophysical properties of tylose gel under pressure in the phase change domain. Food Hydrocoll. 20 (4), 449–460. https://doi.org/10.1016/j.foodhyd.2005.04.001.

Otero, L., Ramos, A.M., de Elvira, C., Sanz, P.D., 2007. A model to design high-pressure processes towards an uniform temperature distribution. J. Food Eng. 78 (4), 1463–1470. https://doi.org/10.1016/j.jfoodeng.2006.01.020.

Otero, L., Sanz, P.D., 2006. High-pressure-shift freezing: main factors implied in the phase transition time. J. Food Eng. 72 (4), 354–363. https://doi.org/10.1016/j.jfoodeng.2004.12.015.

Pandey, J.D., Mishra, R.K., 2005. Theoretical evaluation of thermal conductivity and diffusion coefficient of binary liquid mixtures. Phys. Chem. Liq. 43, 49–57. https://doi.org/10.1080/00319100420003O3554.

Park, S.H., Jun, S., 2015. Practical estimation of the in situ physical properties of foods under high pressure. Food Sci. Biotechnol. 24 (3), 777–782. https://doi.org/10.1007/s10068-015-0101-4.

Patazca, E., Koutchma, T., Balasubramaniam, V.M., 2007. Quasi-adiabatic temperature increase during high pressure processing of selected foods. J. Food Eng. 80, 199–205. https://doi.org/10.1016/j.jfoodeng.2006.05.014.

Pawlicki, L.T., Rostocki, A.J., Tefelski, D.B., Siegoczyński, R.M., Ptasznik, S., 2022. Mechanical properties of sunflower oil under pressure. Eur. Food Res. Technol. 248, 283–287. https://doi.org/10.1007/s00217-021-03880-1.

Pehl, M., Werner, F., Delgado, A., 2000. First visualization of temperature fields in liquids at high pressure using thermochromic liquid crystals. Exp. Fluid 29 (3), 302–304. https://doi.org/10.1007/s003480000111.

Phinney, D.M., Frelka, J.C., Heldman, D.R., 2017. Composition-based prediction of temperature-dependent thermophysical food properties: reevaluating component groups and prediction models. J. Food Sci. 82 (1), 6–15. https://doi.org/10.1111/1750-3841.13564.

Queirós, R.P., Serment-Moreno, V., Tonello-Samson, C., 2023. High pressure thermal processing systems. In: Knoerzer, K., Sevenich, R. (Eds.), High Pressure Thermal Processing. Elsevier Academic Press, pp. 205–241. https://doi.org/10.1016/B978-0-12-823872-1.00009-0.

Ramaswamy, R., Balasubramaniam, V.M., Sastry, S.K., 2007. Thermal conductivity of selected liquid foods at elevated pressures up to 700 MPa. J. Food Eng. 83, 444–451. https://doi.org/10.1016/j.jfoodeng.2007.04.006.

Rasanayagam, V., Balasubramaniam, V.M., Ting, E., Sizer, C.E., Bush, C., Anderson, C., 2003. Compression heating of selected fatty food materials during high-pressure processing. J. Food Sci. 68, 254–259. https://doi.org/10.1111/j.1365-2621.2003.tb14148.x.

Rastorguev, Y.L., Ganiev, Y.A., 1968. Thermal conductivity of solutions. Inzhenerno-Fizicheskii Zhurnal 14, 689–697.

Rauh, C., Baars, A., Delgado, A., 2009. Uniformity of enzyme inactivation in a short-time high-pressure process. J. Food Eng. 91 (1), 154–163. https://doi.org/10.1016/j.jfoodeng.2008.08.019.

Roobab, U., Fidalgo, L.G., Arshad, R.N., Khan, A.W., Zeng, X.-A., Bhat, Z.F., Bekhit, A.E.-D. A., Batool, Z., Aadil, R.M., 2022. High-pressure processing of fish and shellfish products: safety, quality, and research prospects. Compr. Rev. Food Sci. Food Saf. 21 (4), 3297–3325. https://doi.org/10.1111/1541-4337.12977.

Ross, R.G., Andersson, P., Sundqvist, B., Backstrom, G., 1984. Thermal conductivity of solids and liquids under pressure. Reports Prog. Phys. 47, 1347–1402. https://doi.org/10.1088/0034-4885/47/10/002.

Rostocki, A.J., Tefelski, D.B., Ptasznik, S., 2009. Compressibility studies of some vegetable oils up to 1 GPa. High Press. Res. 29, 721–725. https://doi.org/10.1080/08957950903457804.

Rostocki, A.J., Tefelski, D.B., Wieja, K., 2011. The equation of state of triolein up to 1GPa. High Press. Res. 31, 168–171. https://doi.org/10.1080/08957959.2010.541871.

Salvi, D., Khurana, M., Karwe, M.V., 2017. Prediction of temperature distribution in a horizontal high pressure food processing vessel and its impact on process uniformity. J. Food Process. Eng. 40 (5). https://doi.org/10.1111/jfpe.12547.

Sanz, P.D., Otero, L., 2000. High-presssure shift freezing. Part 2. Modeling of freezing times for a finite cylindrical model. Biotechnol. Prog. 16 (6), 1037–1043. https://doi.org/10.1021/bp0001213.

Schaschke, C.J., Allio, S., Holmberg, E., 2006. Viscosity measurement of vegetable oil at high pressure. Food Bioprod. Process. 84, 173–178. https://doi.org/10.1205/fpb.05122.

Schlüter, O., Knorr, D., 2002. Impact of the metastable state of water on the design of high pressure supported freezing and thawing processes. In: ASAE Meeting Paper No. 026024, ASABE Paper No. 026197. ASAE, St. Joseph, MI. https://doi.org/10.13031/2013.10852.

Sevenich, R., Nieder, S., Rauh, C., Olivier, S., Knoerzer, K., 2023. High-pressure thermal sterilization (HPTS) and its effect on production of food processing contaminants and quality-related properties in food in comparison to thermal-only processing. In: Knoerzer, K., Sevenich, R. (Eds.), High Pressure Thermal Processing. Elsevier Academic Press, pp. 103–182. https://doi.org/10.1016/B978-0-12-823872-1.00002-8.

Sevenich, R., Rauh, C., 2021. Modeling methods and numerical tools for process validation/control during high pressure processing. In: Knoerzer, K., Muthukumarappan, K. (Eds.), Innovative Food Processing Technologies: A Comprehensive Review. Elsevier Inc, pp. 141–159. https://www.scopus.com/inward/record.uri?eid=2-s2.0-85118417399&partnerID=40&md5=91f06be00907255f14b517f327ffe019.

Smith, N.A.S., Knoerzer, K., Ramos, A.M., 2014. Evaluation of the differences of process variables in vertical and horizontal configurations of High Pressure Thermal (HPT) processing systems through numerical modelling. Innov. Food Sci. Emerg. Technol. 22, 51–62. https://doi.org/10.1016/j.ifset.2013.12.021.

Smith, N.A.S., Peppin, S.S.L., Ramos, Á.M., 2012. Generalized enthalpy model of a high-pressure shift freezing process. Proc. R. Soc. A 468 (2145), 2744–2766. https://doi.org/10.1098/rspa.2011.0622.

Sun, W., Jianping, L., Xiaowei, Z., Songming, Z., Yong, Y., 2016. Ultra high pressure increasing thermal conductivity of food materials. Trans. Chinese Soc. Agric. Eng. 32, 291–296.

Sun, W., Zhu, S., Ramaswamy, H.S., Yu, Y., Li, J., 2018. Thermal conductivity of selected foods at high-pressure processing conditions. Trans. ASABE 61, 317–325. https://doi.org/10.13031/trans.12435.

Ting, E., Balasubramaniam, V.M., Raghubeer, E., 2002. Determining thermal effects in high-pressure processing. Food Technol. 56 (2), 31–35.

Tonello-Samson, C., Queirós, R.P., González-Angulo, M., 2020. Advances in high-pressure processing in-pack and in-bulk commercial equipment. In: Barba, F.J., Puértolas, E., Tonello-Samson, C., Lavilla (Eds.), Present and Future of High Pressure Processing: A Tool for Developing Innovative, Sustainable, Safe and Healthy Foods. Elsevier Inc, pp. 297–316. https://doi.org/10.1016/B978-0-12-816405-1.00013-3.

Torrecilla, J.S., Otero, L., Sanz, P.D., 2004. A neural network approach for thermal/pressure food processing. J. Food Eng. 62 (1), 89–95. https://doi.org/10.1016/S0260-8774(03)00174-2.

Torrecilla, J.S., Otero, L., Sanz, P.D., 2005. Artificial neural networks: a promising tool to design and optimize high-pressure food processes. J. Food Eng. 69 (3), 299–306. https://doi.org/10.1016/j.jfoodeng.2004.08.020.

Torrecilla, J.S., Otero, L., Sanz, P.D., 2007. Optimization of an artificial neural network for thermal/pressure food processing: evaluation of training algorithms. Comput. Electron. Agric. 56 (2), 101–110. https://doi.org/10.1016/j.compag.2007.01.005.

Torrecilla, J.S., Sanz, P.D., 2011. Neural networks: their role in high-pressure processing. In: Knoerzer, K., Juliano, P., Roupas, P., Versteeg, C. (Eds.), Innovative Food Processing

Technologies: Advances in Multiphysics Simulation. John Wiley & Sons, Ltd. and Institute of Food Technologists, pp. 39–55. https://doi.org/10.1002/9780470959435.ch3. https://onlinelibrary.wiley.com/doi/10.1002/9780470959435.ch3#:~:text=https%3A.

Valuates Reports, 2024. In: Global High Pressure Processing (HPP) Food Market Insights, Forecast to 2030. In: https://reports.valuates.com/market-reports/QYRE-Auto-0364/global-high-pressure-processing-hpp-food.

Werner, M., Baars, A., Eder, C., Delgado, A., 2008. Thermal conductivity and density of plant oils under high pressure. J. Chem. Eng. Data 53 (7), 1444–1452. https://doi.org/10.1021/je700685q.

Werner, M., Baars, A., Werner, F., Eder, C., Delgado, A., 2007. Thermal conductivity of aqueous sugar solutions under high pressure. Int. J. Thermophys. 28, 1161–1180. https://doi.org/10.1007/s10765-007-0221-z.

Wilke, C.R., 1950. A viscosity equation for gas mixtures. J. Chem. Phys. 18, 517–519. https://doi.org/10.1063/1.1747673.

Yusa, M., Mathur, G.P., Stager, R.A., 1977. Viscosity and compression of ethanol-water mixtures for pressures up to 40,000 psig. J. Chem. Eng. Data 22 (1), 32–35. https://doi.org/10.1021/je60072a020.

Zhu, S., Le Bail, A., Ramaswamy, H.S., 2006. High-pressure differential scanning calorimetry: Comparison of pressure-dependent phase transition in food materials. J. Food Eng. 75, 215–222. https://doi.org/10.1016/j.jfoodeng.2005.04.009.

Zhu, S., Marcotte, M., Ramaswamy, H., Shao, Y., Le-Bail, A., 2008. Evaluation and comparison of thermal conductivity of food materials at high pressure. Food Bioprod. Process. 86 (3), 147–153. https://doi.org/10.1016/j.fbp.2006.08.001.

Zhu, S., Ramaswamy, H.S., Marcotte, M., Chen, C., Shao, Y., Le Bail, A., 2007. Evaluation of thermal properties of food materials at high pressures using a dual-needle line-heat-source method. J. Food Sci. 72 (2), E49–E56. https://doi.org/10.1111/j.1750-3841.2006.00243.x.

Zhu, S.M., Su, G.M., He, J.S., Ramaswamy, H.S., Le Bail, A., Yu, Y., 2014. Water phase transition under pressure and its application in high pressure thawing of agar gel and fish. J. Food Eng. 125, 1–6. https://doi.org/10.1016/j.jfoodeng.2013.10.016.

Zolotykh, E.V., Kuznetsov, D.I., Krupina, A.N., 1975. Viscosities of some liquids at high pressures. J. Eng. Phys. 29, 1163–1166.

Modeling of food drying processes

7

Kyuya Nakagawa
Department of Chemical Engineering, Faculty of Engineering, Kyushu University, Fukuoka, Japan

7.1 Introduction

7.1.1 Food drying system

Drying is one of the techniques that has been used since ancient times to control the moisture content of food. Food stability is strongly related to moisture content. Food drying techniques include solar drying, smoking, salting, candying, pickling, frying, and roasting. Consider salted meat as an example. The osmotic pressure difference that occurs between the surface of the meat that comes in contact with the highly concentrated salt solution and the inside of the meat causes water inside the cell membrane to move to the outside. Smoking meat whose water content has been reduced by salting greatly reduces the water content on the surface of the meat. This also causes skin formation due to the denaturation of the protein, where the aromatic components of the smoke are absorbed, inhibit the growth of bacteria, and improve the flavor of the meat. During the storage of this meat in a cool, dark place for a long period, the water content decreases further. As the water content decreases, we can expect the crystallization of *umami* components and the death of viruses in meat. By appropriately managing the water content of food, we can ensure quality in terms of storage, taste, and safety. To reproduce this complex management system, which has been empirically passed down from generation to generation, it is important to correctly understand the mechanisms involved and predict how these mechanisms will interact with industrial operations. In the case of drying operations, this includes factors such as the applicable temperature, airflow velocity, pressure, and geometry of the product and equipment. This is the motivation for modeling the process.

There is high demand for dried food in the food market. In addition to traditional dried foods (e.g., products made by directly drying natural products, powdered products), the production of highly processed foods (such as instant foods and nutritional supplements) is increasing. For example, the production of freeze-dried foods (i.e., packaged products for individual servings) in Japan in 2022 exceeded 672 million servings and is expected to continue to grow in the long term (Japan Freeze-Drying Food Industry Association, production survey).

In recent years, the food industry has needed to address many issues to achieve the Sustainable Development Goals (SDGs). Drying accounts for approximately 5%–10% of the energy consumed in industrial processes (Mujumdar, 2006).

Modeling of Unit Operations in the Food Industry. https://doi.org/10.1016/B978-0-12-822180-8.00001-8

In addition, the current food system accounts for approximately 30% of global energy consumption (FAO, 2021). Most foods require drying during processing, and because food accounts for approximately 34% of the weight of the world's produce (Dev and Raghavan, 2012), reducing energy consumption during the drying process has a significant impact. However, because food has complex quality requirements, process optimization is complex and unique. The quality requirements of dried foods are summarized in Table 7.1. In many cases, reducing the water content of food increases its shelf life. By removing water and turning it into a crystalline state, it is possible to maximize the stability of the solid; however, this only applies to a limited number of foods, such as salt and monosodium glutamate. Most dried foods (solid foods) are thought to be in a glass or rubber state of the plastic matrix formed by the aggregation of natural macromolecules, rather than in a solid state.

Many food components are macromolecules or oligomers such as proteins and polysaccharides. Water behaves as a plasticizer for the aggregates of these moleculars. The viscosity of a plasticized polymer significantly changes due to the influence of water at the glass transition temperature. It can be assumed that the molecular motion in this glass phase is restricted depending on the viscosity. At temperatures below the glass transition point, viscosity increases rapidly and molecular motion is greatly suppressed, causing a slowdown in various reaction rates (e.g., quality changes, bacterial growth). Using the distance from the glass transition point (which is dependent on temperature and water content) as an indicator, the dynamics of various phenomena that occur within the system can be described.

In systems where molecular motion is extremely suppressed near the glass transition point, true equilibrium is not reached within a considerable period. Although glass systems are nonequilibrium systems, it is possible to deal with their apparent equilibrium properties, such as water sorption and desorption. The food stability map published by Labuza suggests that the shelf life of food products can be discussed in a unified manner using water activity (Labuza, 1970; Slade et al., 1991). The food

Table 7.1 Food quality, state, and moisture contents.

Moisture content	State of product		Example	Quality factor
Low ↑	Solid ↑	Crystalline solid	*salt, amino acid*	*Purity, Particle size, Melting temperature*
		Amorphous solid	*chocolate*	
		Glassy state	*sugar, powdered foods, instant noodle, pasta, biscuit*	*Handling property, Dissolution rate, Texture, Palatability, Storage stability, Nutritional value*
	↓	Rubbery state	*dried fruits, dried fish, meat jerky, cheese, dry-cured ham*	
	Liquid/Solid mixture	Emulsion, Gel	*souse, mayonnaise, jam, yogurt*	*Kinetic stability*
	↑	Colloidal suspension	*milk, soup, juice*	
↓ High	↓ Liquid	Solution	*juice, supplement drink*	*Homogeneity*

stability map shows that the chemical and biochemical reactions in food are strongly dependent on water activity. For example, even if the water content of food is the same, foods with lower water activity can be predicted to have higher storage stability. The relationship between the water content and water activity can be expressed by sorption isotherms. This is a way of thinking that considers water sorption at equilibrium. However, with the widespread use of concepts such as glass transition and water as a plasticizer, it has been recognized that water activity alone does not provide a correct understanding of the physical properties of water and food products. At the Faraday Discussion Conference held in Cambridge in 1985, the limitations of treating water activity as an equilibrium theory, the fallacy of old ideas such as bound water and antifreeze water, and the idea of water as a plasticizer were abandoned (Slade et al., 1991).

Many foods are dried to attain a glassy state. This drying process can be traced on the solid–liquid phase equilibrium diagram shown in Fig. 7.1. When dehydration occurs isothermally through contact with air, the wet product dries according to *line (1)*. As the solid concentration increases as drying progresses, crystallization may begin at point (a). In this case, the pure solid phase separates from the aqueous solution phase. However, it typically exists in a supersaturated state with increasing viscosity. The product gradually enters a rubbery state, then undergoes a glass transition at point (b) when the viscosity reaches 10^{12} Pas. As the product approaches the glass transition line (T_g curve), the diffusivity of water in the solution is significantly reduced. This affects dehydration dynamics. When drying is stopped in the region above the T_g curve, the product becomes sticky and rubbery. To prevent the agglomeration of products, caking, and adhesion in the drying equipment during the drying process of

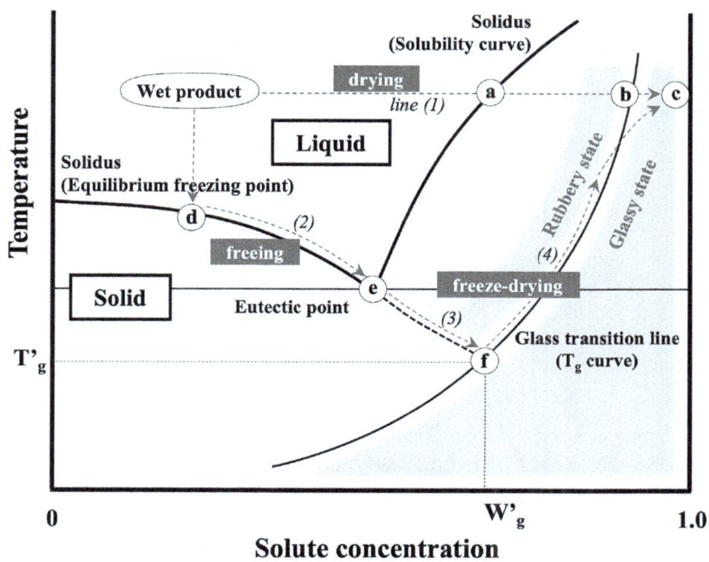

Figure 7.1 Schematic illustration of phase equilibrium.

powders, it is important to control the drying within the region below the T_g curve at the end of the drying process (point (c)). This can be achieved not only by controlling the water content, but also by controlling the temperature.

As shown in Fig. 7.2, the moisture sorption isotherms are redrawn as isobaric curves. The farther away from the glass transition line, the more the relative pressure of the water vapor becomes equivalent to that of bulk water. Near the glass transition line, it is observed to be significantly lower. Because the glass state is in relaxation, the water sorption isotherm does not strictly reflect equilibrium. Therefore, the isobaric line changes over a long timescale. As the water content of the product decreases, the apparent saturated vapor pressure decreases along these isobaric lines, as shown in Figs. 7.1 and 7.2. Thus, the drying rate decreases as the glass transition point approaches the later stages of drying.

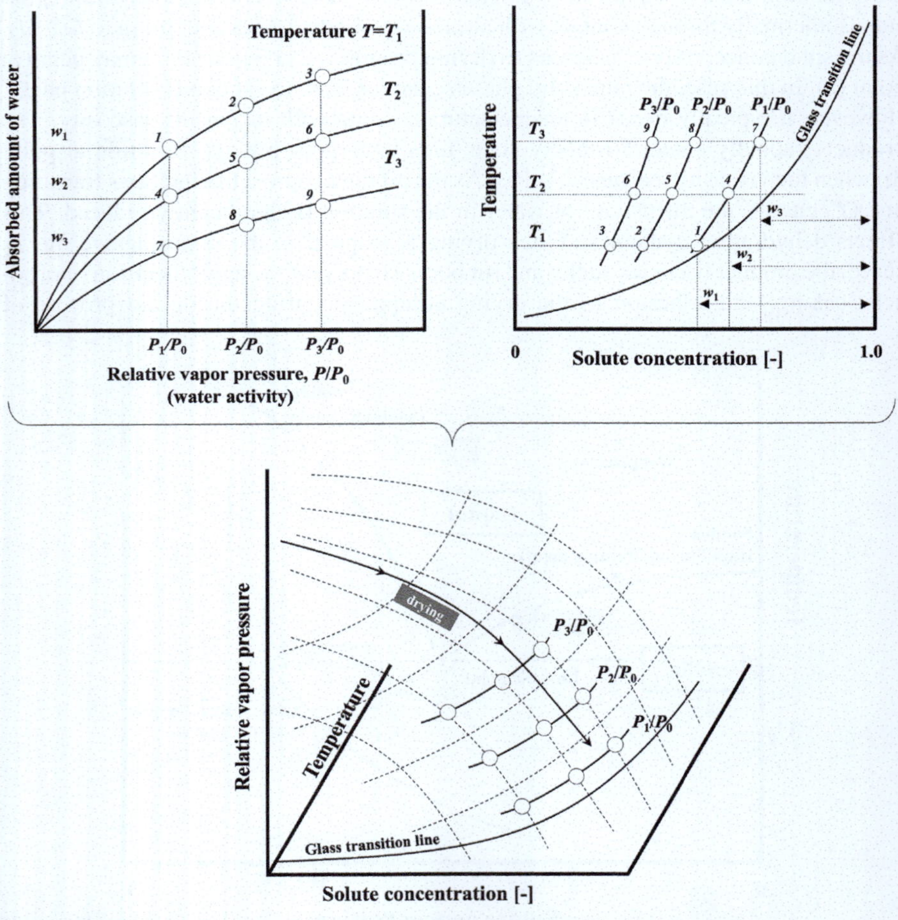

Figure 7.2 Schematic illustrations of sorption isotherms and isobaric curves.

Freeze drying is a special drying method that does not follow this path. The freezing process follows line (2). Freezing causes the formation of ice crystals in the product (point (d)) as well as the formation of a frozen concentrated phase consisting of water and other components. In most food freezing cases, the freeze-concentrated phase does not crystallize at the eutectic point (point (e)) and is not maximally freeze-concentrated (*line (3)*) until it forms a glass phase (point (f)). The glass transition temperature of the maximally freeze-concentrated phase is denoted as T'_g. During freeze drying, two dehydration mechanisms are involved: the sublimation of ice crystals and the evaporation of water in the freeze-concentrated phase (*line (4)*).

Considering food as a glass–rubber state matrix and tracing the drying process on a solid–liquid phase equilibrium diagram is a macroscopic approach. The microstructures of food, such as pores, are formed inside the food, and its water and heat transfer characteristics are highly dependent on them. In addition, the structure of cell tissue in natural food materials is a major factor affecting the transfer of water and heat. There are multiple possible pathways for water movement within the cell structure during thermal processing (Halder et al., 2011). The processing temperature affects the transcellular, symplastic, and apoplastic pathways (Halder et al., 2011). Therefore, when viewed at the mesoscopic scale, the movement phenomena cannot be predicted using the concept of glass transition (although it may be possible to approximate them). As shown in Fig. 7.3, the food drying process involves a multiscale transport process, including the transport of water and heat within the matrix and cellular tissue that make up the food (nanoscale), transport within the pore network structure that comprises these components (meso- and microscale), and transport from the product to the drying equipment. Mathematical models can contribute to comprehensively handling the dynamics of these interacting transport processes and controlling product quality.

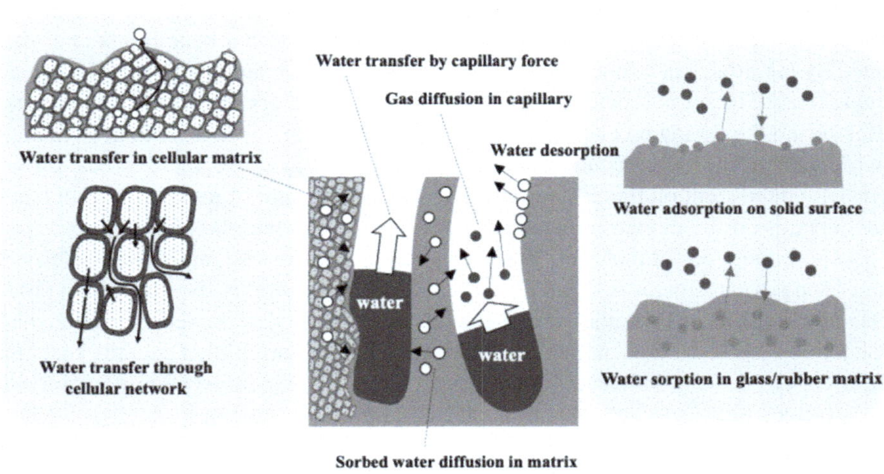

Figure 7.3 Schematic illustrations of water migration in food systems.

7.1.2 Needs of the food industry

The challenge in the food industry is to find the optimal point in the complex tradeoffs that arise from various conflicting relationships between process parameters and process goals (Banga et al., 2003; Banga and Paul Singh, 1994). Food drying is no exception. In food drying and controlling the drying rate, it is important to ensure quality, as described in the previous section, and there are often many complex quality items.

A typical example of the food drying process is that increasing the drying rate by increasing the temperature or air flow rate has a beneficial effect on the throughput, microbiological death rate, and shelf life. However, it can also reduce the quality of the food owing to decreased nutritional value, unfavorable shrinkage, cracking, and hardening; and can increase energy consumption, operating costs, food loss, and the energy contained in the product. Mathematical models are expected to realize the optimization of this multiobjective problem; thus, an integrated modeling approach is required (Van Boekel, 2008). As clearly stated in van Boekel's review, it is necessary to address all related processes simultaneously in an integrated manner, as well as the related quality and biochemical factors. The quality degradation processes involved in ingredient loss can be well explained by simple kinetic models such as the application of Arrhenius-type kinetic constants and first-order reaction rate equations (Devahastin and Niamnuy, 2010). Therefore, the priority issues are process modeling and integration.

Banga and Singh attempted to optimize the drying process by simulating the temperature of the product during the drying process and estimating the degree of quality deterioration that occurs in the product based on this history (Banga and Paul Singh, 1994). This approach considers the amount of residual ascorbic acid and enzyme activity in the model food as quality factors and seeks to determine drying conditions that minimize drying time (energy input) and maximize quality factors (Luyben et al., 1982; Mishkin et al., 1983; Villota and Karel, 1980). These studies successfully proposed operations such as gradually increasing the air temperature.

Nakagawa et al. considered the loss of betanin (i.e., due to hydrolysis) as a quality factor in a model drying system, and summarized the residual betanin content estimated by the model as a function of the drying temperature and humidity (Nakagawa et al., 2018). They found that betanin migrates toward the drying interface of the material during the drying process, and that the hydrolysis reaction proceeds at a reaction rate dependent on the moisture content. Therefore, the macroscopic degradation kinetics are dependent on the drying rate. The kinetic constant of the degradation reaction depends on the degree of drying, which is related to the glass transition.

Because the types of materials used in food products are vast and most are derived from living organisms, it is difficult to obtain material property data for many products, and in some cases, the properties have not been determined (McMinn and Magee, 1999). Therefore, it may be practical to apply simplified models that use empirical parameters. However, it should be noted that excessive simplification can mask various physical phenomena (as discussed in a later section).

The predictive capability inherent in process models allows for the design and optimization of virtual drying processes or dryers without the cost and effort of building pilot- or full-scale setups and the time required for validation (Datta, 2008). Modeling is expected to be a decisive role in advancing the optimization of conventional drying processes and be useful in conceptualizing, designing, and optimizing next-generation drying processes. In many of the food industry, batch-based production is performed using multiple pieces of equipment with different capabilities, and many factories face challenges in stabilizing quality. The ability to stabilize quality using a model-based approach would be advantageous for quantitatively demonstrating the cost-effectiveness of capital investment, such as the introduction of new equipment or repairs. To do this, it would be necessary for the physicochemical phenomena related to quality to be reflected in the operational variables or parameters that express the process model.

7.2 Mathematical modeling of drying

7.2.1 Sherwood's mechanistic model

The most important theoretical study on drying that describes the rate of drying in balance with the energy input is the early research of Sherwood and others (Sherwood, 1929a, 1929b, 1930, 1932; Sherwood and Comings, 1933). During the constant rate drying period (CRP), moisture evaporates from the surface of the material, where the driving force is the humidity difference in the gas film (Eq. 7.1). The heat on the material surface is attributed to heat transfer through the gas film and radiant heat from external heat sources (Eq. 7.2), and the mass transfer flux is calculated by dividing this value by the evaporation enthalpy.

$$\frac{\partial w}{\partial t} = RA(H_s - H_a) \tag{7.1}$$

$$\frac{\partial w}{\partial t} = \frac{1}{\Delta H_v}\left\{h_c A(T_a - T_s) + \varepsilon\sigma A\left(T_r^4 - T_a^4\right)\right\} \tag{7.2}$$

Where

A: wet surface area
h_c: heat transfer coefficient
H: absolute humidity (subscript, a: air; s: solid surface)
t: time
T: temperature (subscript, a: air; r: surrounding; s: solid surface)
R: constant
w: weight of water
ε: emissivity
σ: Stefan–Boltzmann constant
ΔH_v: enthalpy of evaporation

Because the material surface temperature can be calculated using this quasi-steady-state approximation, the corresponding drying rate can also be calculated. The concentration distribution inside a material differs depending on its characteristics. Sherwood regarded the stage where the rate of movement of this liquid water becomes dominant for the drying rate as the falling rate drying period (FRP) and expressed it as the movement of water vapor inside the material by diffusion.

$$\frac{\partial\theta}{\partial\tau} = \frac{\partial^2\theta}{\partial Y^2} \tag{7.3}$$

$$\theta = \frac{C - C_{eq}}{C_0 - C_{eq}}, Y = \frac{y}{L}, \tau = \frac{tD}{L^2} \tag{7.4}$$

Where

C: moisture concentration (subscript, 0: initial; eq: equilibrium)
D: moisture diffusion coefficient
L: thickness of material
y: distance to the drying surface

The dimensionless total quantity of moisture, Θ, during FRP can be expressed in the following form by solving the above differential equations.

$$\Theta(\tau) = \frac{8}{\pi^2}\sum_{m=1}^{\infty}(2m-1)^{-2}\exp\{-(2m-1)^2\pi^2\tau/4\} \tag{7.5}$$

Hougen et al. pointed out that this model, based on the diffusion equation, could not correctly predict the distribution of moisture content in the drying of various materials, especially granular materials, and explained the limitations of the Sherwood model (Hougen et al., 1940). However, although the Sherwood model had remained the most important mechanistic approach, the validity and limitations of diffusion-based models for drying have been discussed by many researchers since then (Puyate and Lawrence, 2006). In particular, the Sherwood model does not accurately predict the drying behavior when the drying rate is fast during FRP, and there are problems with applying a constant diffusion coefficient. The dependence of the diffusion coefficient on the moisture content is related to the material deformation and internal transport mechanism of water, which depend on the physical properties and structure of the material (Puyate and Lawrence, 2006).

Consequently, it is important to express the heat and mass exchange processes between the environment inside the drying equipment and the product, the macroscopic behavior of the water inside the product during the drying process, and the water transport mechanism that depends on the micro- and mesoscopic material structures that may change during drying.

7.2.2 Capillary flow and vapor diffusion modeling in porous media

Many dried products, including foodstuffs, are porous, and moisture movement within this pore structure is the most important mechanism involved in drying. Foodstuffs easily deform as drying progresses, and the mechanical effects acting within the material cannot be ignored. Krischer proposed a drying model that expresses the movement of moisture owing to capillary attraction within porous bodies (Krischer, 1938). The model assumes that porous materials are bundles of capillaries with different diameters arranged in parallel, and that water can move freely between different capillaries. As water evaporates from the surface of the material, the evaporation front gradually recedes. This regression occurs more rapidly in large capillaries and more slowly in small capillaries. Toei and Okazaki proposed an improved model based on the concept of Krischer (Toei and Okazaki, 1970). Water inside a capillary is divided into two categories: funicular water (i.e., water that can move freely) and pendular water (i.e., a combination of wedge water, which is held between particles by surface tension, and surface water, which is attached to the particle surface). Funicular water can move as water inside the material, but pendular water is thought to evaporate inside the material and diffuse as water vapor inside the material. Furthermore, by determining a distribution of capillary diameters, the progress of drying during the FRP of hydrophobic porous materials can be predicted with high accuracy.

In the drying of hydrophilic materials such as food, the effect of water adsorbed on the material (pendular water) cannot be ignored. Tamon et al. estimated the effective diffusion coefficient of water moving inside porous materials using a model that included adsorption theory. In addition to the movement of water during capillary condensation, the movement of water adsorbed on the pore surface is explained using the hopping model (Tamon et al., 1981; Toei et al., 1983). This theory considers that water in an adsorbed state (i.e., stabilized by the potential field at the surface) can hop from one adsorption site to another by receiving energy that is less than the desorption energy. By providing the thermal energy distribution of the molecules, it is possible to calculate the average residence time at an adsorption site, from which the surface diffusion coefficient can be calculated. The amount of water contributing to surface diffusion can be estimated from the adsorption isotherm. It has been estimated that the overall effective diffusion coefficient decreases as the moisture content decreases as drying progresses (Fig. 7.4).

The diffusion coefficient of water in a material, which is assumed to be constant in the Sherwood model, is strongly affected by the interaction between the material and water (e.g., capillary force, surface potential). This can be explained by the fact that the diffusion coefficient decreases as the water content decreases, and the meso- and microscopic internal structures of the material give rise to this interaction.

7.2.3 Luikov's model (irreversible thermodynamics approach)

Luikov proposed a drying model based on irreversible thermodynamics (Luikov, 1966, 1975). This is based on describing the transfer of heat, mass, and momentum

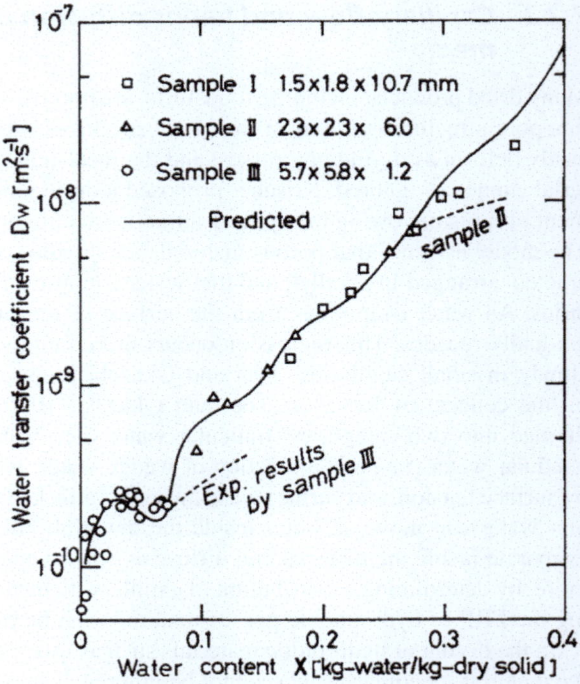

Figure 7.4 Water transfer coefficient as a function of water content: Comparison of experimentally obtained coefficient and that predicted by the hopping model (Toei et al., 1983).

in the capillaries using flux equations. This is viewed as a model that rigorously describes the transport processes in drying materials. The phenomenological coefficients introduced in the description of the flux equations depend on temperature and concentration. It is also possible to express the mass transfer induced by a temperature gradient (or vice versa). The problem with Luikov's approach is that it is difficult to determine the various phenomenological parameters.

Andrieu et al. have used this approach to modeling pasta (spaghetti) drying (Andrieu and Stamatopoulos, 1986; Andrieu et al., 1985; Andrieu, 1986).

$$\frac{\partial X}{\partial t}=\frac{\partial}{\partial x}\left\{D\left(\frac{\partial X}{\partial x}+\delta\frac{\partial T}{\partial x}\right)\right\} \tag{7.6}$$

$$\frac{\partial T}{\partial t}=\frac{1}{\rho C_p}\frac{\partial}{\partial x}\left(k\frac{\partial T}{\partial x}\right)+\frac{\varepsilon'\Delta H_v}{\rho C_p}\rho_s\frac{\partial X}{\partial t} \tag{7.7}$$

Where

X: dry basis moisture content
T: temperature
D: apparent water mass diffusivity
k: apparent thermal conductivity

ρC_p: volumetric heat capacity
ΔH_v: enthalpy of evaporation
ε': phase transition ratio
ρ_s: dry solid density

Some simplifications have been applied in determining the parameters. It is assumed that no new pore formation occurs inside the material as it dries, because the overall shrinkage rate is limited to 15%. In addition, experimentally obtained kinetic data are applied. The thermal diffusivity is constant at $1.45 \cdot 10^{-7}$ m^2/s, and the apparent mass diffusivity of water is given as a function of temperature (T) and moisture content (X [kg /kg]-dry basis) as follows:

$$D = \left(3.1 \times 10^{-7} X - 9.1 \times 10^{-9}\right) \exp\left(-\frac{21600}{RT}\right) \tag{7.8}$$

In addition, the boundary conditions are given as heat transfer and evaporation of moisture on the material surface, with the assumption that the material surface temperature is in equilibrium with ambient humidity. Calculations based on these assumptions were able to accurately predict the experimental results. When drying food products such as thin cylindrical pasta, heat transfer is very fast, and drying progresses isothermally. Therefore, the moisture transfer inside the product becomes dominant. This approach can be used to quantitatively derive the temperature, moisture distribution, and drying time during the drying process from the drying conditions at the time of manufacture.

7.2.4 Volume averaging theory

Whitaker's volume averaging theory (VAT) is recognized as important in drying of porous and/or deforming materials (Whitaker, 1977). VAT divides the interior of the material into regions composed of several continua (e.g., solid phase, liquid water, and gas phase) and introduces volume-averaged parameters for each region (Fig. 7.5). For example, the temperature (i.e., the volume-averaged temperature) is considered as follows.

The phase average of the temperature of the β-phase, $\langle T_\beta \rangle$, in the selected region v can be written as

$$\langle T_\beta \rangle = \frac{1}{v} \int_v T_\beta dV = \frac{1}{v} \int_{v_\beta} T_\beta dV \tag{7.9}$$

The average by the volume of the β-phase, which is denoted as the intrinsic phase average, $\langle T_\beta \rangle^\beta$, is

$$\langle T_\beta \rangle^\beta = \frac{1}{v_\beta} \int_{v_\beta} T_\beta dV \tag{7.10}$$

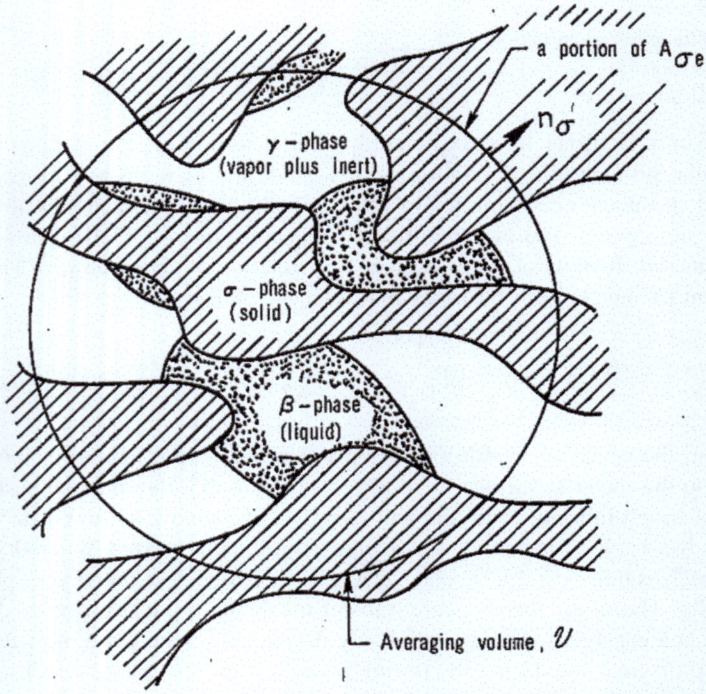

Figure 7.5 Averaging by a selected volume that contains different phases (Whitaker, 1977).

The volume fraction of the β-phase, ε_β, is

$$\varepsilon_\beta \langle T_\beta \rangle^\beta = \langle T_\beta \rangle \tag{7.11}$$

When drying a material, the change in ε_β as a function of time provides drying kinetics.

Based on these volume-averaged temperatures, for example, one can use Fourier's law in the following formula by using the contacting surface areas with the other phases (i.e., $A_{\beta\gamma}$, $A_{\beta\sigma}$):

$$\left\langle \boldsymbol{q}_\beta \right\rangle = -k_\beta \langle \nabla T_\beta \rangle = -k_\beta \left\{ \nabla \langle T_\beta \rangle + \frac{1}{v} \int_{A_{\beta\gamma}} T_\beta \boldsymbol{n}_{\beta\gamma} dA + \frac{1}{v} \int_{A_{\beta\sigma}} T_\beta \boldsymbol{n}_{\beta\sigma} dA \right\} \tag{7.12}$$

The following form is the heat flux balance equation.

$$\rho_\beta C_{p\beta} \frac{\partial \langle T_\beta \rangle}{\partial t} = -\left\langle \nabla \boldsymbol{q}_\beta \right\rangle + \langle \Phi_\beta \rangle$$

$$
= \nabla \left[k_\beta \left\{ \nabla \langle T_\beta \rangle + \frac{1}{v} \int_{A_{\beta\gamma}} T_\beta \boldsymbol{n}_{\beta\gamma} dA + \frac{1}{v} \int_{A_{\beta\sigma}} T_\beta \boldsymbol{n}_{\beta\sigma} dA \right\} \right] - \frac{1}{v} \int_{A_{\beta\gamma}} \boldsymbol{q}_\beta \boldsymbol{n}_{\beta\gamma} dA
$$

$$
+ \frac{1}{v} \int_{A_{\beta\sigma}} \boldsymbol{q}_\beta \boldsymbol{n}_{\beta\sigma} dA + \langle \Phi_\beta \rangle \tag{7.13}
$$

Where

$C_{p\beta}$: heat capacity of the β-phase
k_β: thermal conductivity of the β-phase
ρ_β: density of the β-phase
Φ_β: heat generation in the β-phase

Similarly, using volume-averaged variables, the fluxes and balance equations for the mass and momentum can be described.

The following restrictions can be adopted to derive the analytical solution (Whitaker, 1977).

- The solid phase is a rigid matrix fixed in an inertial frame.
- The enthalpy of the σ, β, and γ phases is independent of pressure.
- The enthalpy in each phase is a linear function of the temperature.
- The thermal conductivity of each phase is constant.
- The liquid phase contains only a single component.
- Compressional work and viscous dissipation are negligible in the liquid phase.
- There is no chemical reaction in the γ phase.
- Compressional work and viscous dissipation are negligible in the gas phase.
- Diffusional body force work and kinetic energy are negligible in the gas phase.
- Interfacial energy for each interface is negligible in the thermal sense.
- The density of the liquid phase is constant.
- The gas phase is ideal in the thermodynamic sense.

Additional assumptions such as local equilibrium, ignoring diffusive transport of thermal energy, gas phase continuity, hydrostatic pressure, and pore homogeneity are applied to derive the following formula:

$$
\frac{\partial S}{\partial t} = \nabla (D \nabla S) \tag{7.14}
$$

where

$$
D = K_\epsilon + \left\{ \left(\frac{\partial D}{\partial S} \right) \frac{D_{\text{eff}}}{\epsilon_\gamma} \right\} \tag{7.15}
$$

$$
K_\epsilon = \epsilon_\gamma \xi K_\beta k_\epsilon / \mu_\beta \tag{7.16}
$$

Where

D: drying diffusion coefficient
D_{eff}: gas phase total effective diffusivity tensor
k_{ϵ}: thermal conductivity tensor in the β-phase (liquid phase)
K_{β}: permeability tensor in the β-phase (liquid phase)
S: moisture content (fraction of moisture saturation)
ξ: thermal diffusion vector
ϵ_{γ}: volume fraction of the γ-phase (gas phase)
μ_{β}: shear viscosity

This formula shows an example of the dependence of the effective diffusion coefficient on moisture content (nonlinear diffusion) during the drying of porous media.

7.2.5 Modeling of drying stress and deformation

The drying models introduced in the previous sections are cases in which drying occurs in a nondeformable solid. In many cases, food drying involves deformation, such as shrinkage and cracking. There are uniform and nonuniform types of shrinkage, and the residual stress that occurs because of nonuniform shrinkage can increase the risk of cracking. The motivation for modeling the stress caused by drying is to prevent product damage by controlling drying. Lewis et al. estimated the stress field inside the material generated by drying using the finite element method (Lewis et al., 1977, 1979). The Luikov model was applied to calculate the drying progress of porous materials, and the stress field was derived from the distribution of the moisture potential to identify locations with a high risk of cracking. Research on the stress caused by drying has been summarized in a review by Hasatani and Itaya (1996). The nonuniformity of the temperature in the material during the drying process causes direct thermal stress. The temperature gradient that generated the stress occurred primarily during the preheating step. The moisture distribution within the material causes shrinkage stress, which is more severe than thermal stress. The stress can be reduced by increasing the absolute humidity of the drying medium. For the prediction and control of drying stress, the nonlinear diffusion of moisture transfer through the material is of great importance (Musielak, 1996, 2000).

Peczalski simulated the time-dependent evolution of the internal stress generated during the drying of spaghetti (Peczalski et al., 1996). This practical model consists of mass transfer, enthalpy balance, and constitutive equations. As an example of the model prediction, the stress (σ) along with the radial direction (r) of spaghetti progressed as a function of drying time (t); the stress increases more rapidly in the outer shell than in the inner shell. Ponsart et al. (2003) applied the incremental model (Mrani et al., 1995) to the drying of chocolate, and were able to simulate the time evolution of stress during the drying process and investigate the drying conditions to avoid cracks (Ponsart et al., 2003).

Many studies have reported the progress of drying accompanied by deformation, and mathematical approaches have been discussed (Hasatani and Itaya, 1996; Jan Kowalski, 1996; Katekawa and Silva, 2006; Kowalski, 2001). The basis for many of these studies was the model proposed by Whitaker, who applied VAT to the continuum model (Whitaker, 1986a, 1986b). This approach describes the conservation equations for each phase and then derives macroscopic partial differential equations by integrating these microscopic conservation equations at a representative volume. The shrinkage is modeled as being caused by water migration, and Darcy's law is applied to describe water migration instead of the conservation equation of water momentum. The problem lies in handling the average pressure of the water phase, and it is difficult to estimate the average pressure of the liquid phase in a multiphase material. Roque et al. developed a two-phase system model consisting of a solid phase and a liquid phase only, and showed that it is possible to estimate the average pressure of the aqueous phase from the mass transfer balance of the solid phase, even without applying a phenomenological law (Sarmento et al., 2004). This model can be applied to materials that are significantly deformed such as hydrogels. This model rigorously explains the relationship between stress and drying behavior as the progress of drying (Couture et al., 2007). The decrease in capillary diameter that accompanies shrinkage increases the capillary tension and changes the driving force for water migration. The migration phenomenon, which depends on the rheological properties of the material, is closely related to drying behavior.

The approaches described above are macroscopic modeling approaches. The pore network model is well known; it models meso- and microscale pore structures as a mutually connected lattice network and simulates water migration due to capillary forces (Fatt, 1956). This is useful for precisely understanding the movement of water at the meso- and microscale during the drying process, and it has been applied to model powder drying with agglomeration, such as fluidized bed drying and granulation processes (Carmeliet et al., 1999; Mezhericher et al., 2012; Mortier et al., 2011; Pham et al., 2022; Tsotsas, 2015).

7.2.6 Conjugate modeling approach

The mechanism of moisture transfer within materials is now better understood in detail than it was in the early 1900s based on several theoretical studies. The progress of drying is influenced by the interaction between the environment in the dryer and the product. The mass and heat transfer in the product is not only affected by the internal structures and structural changes accompanying drying but also interact with the environment in the dryer. When estimating the drying progress, it is reasonable to consider only the rate-limiting process. However, when the dynamics of the processes are comparable in terms of kinetics, models that describe the intrinsic features should be integrated. Conjugate modeling by Defraeye is an excellent example of an integrated modeling approach (Defraeye, 2011, 2014; Defraeye et al., 2012). This approach combines computational fluid dynamics (CFD) simulations of airflow in the dryer with finite element simulations of the drying behavior inside the material. The drying behavior model uses a continuum with a volume-of-fluid approach.

The greatest advantage of this simulation is that it does not require constant convective transfer coefficients (CTCs) at the boundary film. The transient mass and heat transfer are simulated using CFD, assuming continuity of the mass and heat flux at the interface. An interesting insight is the change in the CTC as drying progresses (although CTCs are not required for the simulation, Defraeye calculates mass and heat transfer coefficients from the results of the simulation). As the drying process approaches the transition from the CRP to FRP, the CTCs, which were constant up to that point, increase, reach a peak value, and then begin to decrease, and then remain constant at a higher value than that of the CRP. The fact that the peak shape differs between the mass transfer and heat transfer coefficients suggests that the Reynolds analogy is no longer valid. In other words, CTCs are not only related to the flow state within the boundary film but also depend on the temperature and moisture distribution around the boundary and at the interface of the material. In the FRP, in which the rate-limiting step in drying is the migration of water within the material, the effect of the CTCs is less important (Defraeye et al., 2012).

7.3 Practical use of mathematical modeling

Rigorous modeling clarifies the physical phenomena induced during the drying process and is useful not only for shortening the drying time but also for designing product quality and ensuring uniformity between products (Gong and Mujumdar, 2008). However, the use of advanced numerical methods is limited, particularly in the food industry. As explained in previous sections, many dried products are porous, and drying in porous media is a multiphase process that involves stress and deformation. This complexity is a major hurdle that must be overcome when performing simulations. Food manufacturers purchase industrial dryers from machinery companies that can satisfy product specifications (e.g., water activity, shelf life, shape, taste, texture, and nutritional value). The dryer company is responsible for equipment design, installation, drying procedures, and drying software that can realize the product specifications. Consequently, although the dryer company accumulates knowledge and technology related to drying a wide variety of products, it is not involved in the development of the product itself. In addition, in cases where there is a confidentiality obligation related to the product, significant meaningful knowledge is not disclosed to the public. If the only results of drying modeling and simulation are equipment design and scale-up, then manufacturers of industrial dryers that have a long lifespan (30–50 years) and produce custom-made products rather than mass-produced products would probably decide that it would be cheaper to solve problems using accumulated know-how and trial and error rather than performing advanced simulations.

The information provided by the model must satisfy user goals. The goals of dryer manufacturers and users differ (although there may be some shared goals). Dryer manufacturers must be able to predict the output of an operation (i.e., the performance of the equipment) based on the model system in order to design the equipment and set the operation protocol. On the other hand, food manufacturers need to control the quality of their products. If a defect occurs in a product, it is necessary to deal with it by changing the operation procedures or formulating new protocols; therefore, there is a need for rational guidelines. The robustness of a mathematical model is due to its flexibility of use, which allows both dryer companies and food manufacturers to perform simulations on their preferred platforms.

The modeling approaches introduced in the previous section target simplified systems (because the objective is to construct a theory). Van Boekel pointed out that we should be cautious about directly applying models to food manufacturing that are built on the basis of simplified systems (Van Boekel, 2008). Although it would be ideal to apply a model that fully reflects the complex characteristics of food, this is not possible with current knowledge. It is possible to analyze the dynamics directly using actual food; however, the parameters derived from this approach have little physical meaning and are difficult to interpret. The derived empirical or semi-empirical equations are highly case-dependent, but this does not cause any substantial problems, depending on the purpose of use. However, when using the model outside the parameter range on which the models are based, one should be very cautious (Van Boekel, 2008).

What is interesting about the research on modeling the drying of shrinking foods with cell structures (Crapiste et al., 1985, 1988a) is that the resulting mathematical expression is similar in form to a nonlinear diffusion equation. This has formed the basis for the development of simplified mathematical models of drying kinetics (Keey, 1994). The approach of Keey and Suzuki, who proposed the characteristic drying curve (CDC), is well known for its simplicity and good agreement with experiments (Keey and Suzuki, 1974).

In porous materials or drying systems where shrinkage is progressing, it has been found that the total effective diffusion coefficient of water movement inside the material depends on the moisture content. This is influenced by the pore structure of the material, interaction between water and the material, and structural changes due to mechanical stress. The derivation of the effective diffusion coefficient dependent on the water content from the experimental data indicates that the complex factors inherent in the material are reflected here. The regular regime approach was proposed as a method for deriving the effective diffusion coefficient, which depends on the water content, from experimental data (Schoeber W. J. A, 1976). This method estimates the effective diffusion coefficient of water in the form of a power law function from the drying curve obtained from isothermal drying experiments. A shortcut calculation method has also been proposed to simplify the calculation, based on which the calculation can be performed using two isothermal drying experiments; it is possible to estimate the effective diffusion coefficient regardless of the shape or shrinkage of the material (Thijssen and Coumans, 1985; Yoshida et al., 1991a, 1991b). Yamamoto has proposed a method that further simplifies the calculation (Yamamoto, 2001). The regular regime approach has been recognized as effective in food drying systems (Adhikari et al., 2002; Báez-González et al., 2004).

The main concern with empirical approaches is that the mechanisms become black boxes. The parameters involved in heat, mass, and momentum transfer are quantified in a complex manner; therefore, it is not possible to identify the physical parameters necessary for predicting and controlling the drying process. For example, if a product has not been sufficiently dried, even if the drying tendency is detected as a deviation from the model-predicted value, it is not possible to estimate the cause when the model does not contain physical parameters. The technical basis for solving various problems is a correct understanding of the phenomena. However, empirical parameters cannot be used to extract information related to specific phenomena. In recent years, an even greater simplification that assumes a constant diffusion coefficient (or fittings by power law equations without any physical background) has been adopted by many researchers, and there have been many reports on the fitting parameters (Onwude et al., 2016). These fitting parameters are case dependent; therefore, there is no room for further discussion. Some of these methods are based on an analytical solution of Fick's law for a specific sample shape. However, it is unacceptable to apply this method as a general rule to systems in which the diffusion coefficient may vary (McMinn and Magee, 1999). The author is also concerned that these approaches may be recognized as legitimate drying models (that adequately express the phenomenon) based on the number of search hits on the Internet. For information on the positioning and concepts of simplified models, the review by McMinn and Magee is helpful (McMinn and Magee, 1999).

7.4 Modeling of food drying

7.4.1 Agricultural products

The drying of agricultural and livestock products is largely dependent on the cell structure. Crapiste et al. modeled the drying of food products with cell structures based on VAT, as shown in Fig. 7.6, that underwent shrinkage when subjected to vacuum-assisted drying (Crapiste et al., 1985, 1988a, 1988b). The phases handled in the modeling (e.g., solid, liquid, and gas phases in the case of a porous body model)

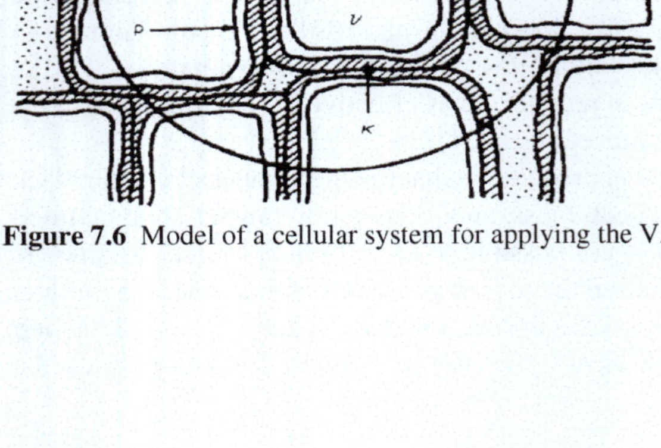

ν-phase: vacuole
η-phase: cytoplasm
γ-phase: intercellular space
κ-phase: cell wall
t : tonoplast
p : plasmalemma
υ : averaging volume

Figure 7.6 Model of a cellular system for applying the VAT approach (Crapiste et al., 1988a).

were divided into four phases: vacuole, cytoplasm (surrounded by the tonoplast and plasmalemma), cell wall, and intercellular space, and volume averaging was applied. Water flux is given by the chemical potential gradient. In addition, shrinkage is given as a function of water content. This shows that macroscopic changes in water content can be expressed by a formula similar to the diffusion equation, which applies a shrinkage coefficient that depends on the water content and a total effective diffusion coefficient.

Fanta et al. microscopically modeled water movement within fruit cell tissues (Fanta et al., 2013). They applied the effective water conductivity of the tissue, which is dependent on the water potential, by providing a water movement flux via the chemical potential gradient. This effective conductivity is considered to be the total mobility divided into mobility within the cell, through the cell membrane, and in the extracellular space. Furthermore, it is clear that the permeability of water through the cell membrane has a dominant effect on the macroscopic water conductivity of the entire tissue during dehydration, and it is suggested that the relationship between storage, drying conditions, and dehydration can be quantitatively obtained through a combination of macroscopic models (Fanta et al., 2014). A simulation that combines the microscopic modeling of water movement within cells with convective drying has been reported (Defraeye and Verboven, 2017). When the permeability of water within the cell tissue is low, the effect of the CTC at the interface where the food comes into contact with air will naturally be small. Interestingly, the magnitude of these effects can be quantitatively evaluated, and it is possible that the CTC may change depending on the level of dehydration. A review of recent research on fruit drying modeling was summarized by Castro et al. (2018).

7.4.2 Food powder

Spray drying is the most widely used method for producing food powders. The granulation process, in which a fine powder produced by spray drying is humidified and agglomerated in a fluidized bed, is also a drying technique that is widely used in the food industry. Spray droplets are very fine, and a modeling approach has been proposed that considers drying as a chemical reaction and expresses it using a reaction rate equation (Place et al., 1959). As detailed in Langrish's review, modeling the spray-drying process also requires a multiscale approach (Langrish, 2009; Langrish and Fletcher, 2001), which means considering the airflow in meter-scale equipment and the drying rate of the particles at the microscale. The flow of air within the device depends on the shape of the housing (i.e., short-form or tall-form) and the flow pattern, which differs depending on whether the particles and air are well mixed, whether it is a parallel flow (concurrent flow), or whether it is a more complex flow including turbulence (Truong et al., 2005a; Zbiciński, 1995). CFD is effective for modeling a system under flow, but it also has many issues, especially when turbulent flows are involved (Afshar et al., 2019; Langrish and Fletcher, 2001).

The simplified models introduced in the previous section can be effectively applied to predict the drying rates of powders such as models based on the CDC (Keey and Suzuki, 1974; Langrish and Kockel, 2001), nonlinear diffusion models that introduce an approximate concentration-dependent diffusion coefficient (Liou and Bruin, 1982), and the reaction engineering approach (REA), which treats the progress of drying as a chemical reaction (Chen, 2008; Chen and Lin, 2005; Patel and Chen, 2005). A comparison was made between the results of using both the CDC and REA for spray drying whole milk and skimmed milk, and it was found that the predictions by the REA model were closer to the experimental results than those by the CDC model, but the difference between the two models was less than 5%, confirming that both models had good accuracy (Chen and Lin, 2005).

The aggregation of particles during the drying process and their adhesion to the inside of the equipment are related to glass transition. Many spray-dried food powders are glass–rubber transition systems. By maintaining the temperature of the dried powder near or below its glass transition point, it is possible to prevent agglomeration and adhesion (Ozmen and Langrish, 2002). Truong et al. proposed a model-based search for a set of inlet variables that minimizes the difference between the air temperature at the dryer outlet and the glass transition temperature of the final product (Truong et al., 2005a, 2005b). In addition, there have been reports of approaches that use CFD to predict adhesion inside the dryer (Woo et al., 2008a, 2008b). It is also interesting to note that modeling of freeze drying of fine particles in a packed bed using the discrete element method (DEM) and CFD has been reported in recent years (Capozzi et al., 2019).

7.4.3 Freeze drying/lyophilization

Freeze drying, also called lyophilization, is a drying method that uses the sublimation of ice from frozen materials as the main dehydration principle. As drying progresses, a drying layer is formed, and the water vapor generated on the sublimation surface is released through the pores in the dried layer. This dehydration process by sublimation is called primary drying. Sorbed water (unfreezable water) is present in the dried layer and desorbed by evaporation. This process is called secondary drying. Primary drying determines the total drying time. The microstructure of the ice crystals inside the material forms a porous dried layer as the ice is removed by sublimation. The vapor generated during the drying process moves within this porous structure; thus, the freezing conditions are related to the drying kinetics. There have been reports of modeling approaches for ice crystal formation during the freezing process in foods (Woinet et al., 1998a, 1998b). There have also been reports of approaches that simulate ice crystal formation in solution and predict its effect on the freeze-drying rate (Nakagawa et al., 2007). Because freeze drying involves dehydration in a frozen state, there is no significant liquid water migration, and the degree of shrinkage is small. In vacuum freeze drying, drying progresses via conduction heat transfer from the shelf on which the product is loaded and radiation from an external heat source. The effect of radiant heat becomes relatively significant because of the lower contribution of convective heat transfer (Fig. 7.7).

The Liapis continuum model comprehensively explains the freeze-drying process (Liapis and Bruttini, 1994, 1995; Liapis and Litchfield, 1979). From a practical perspective, mechanistic models that simulate the primary drying stage are widely

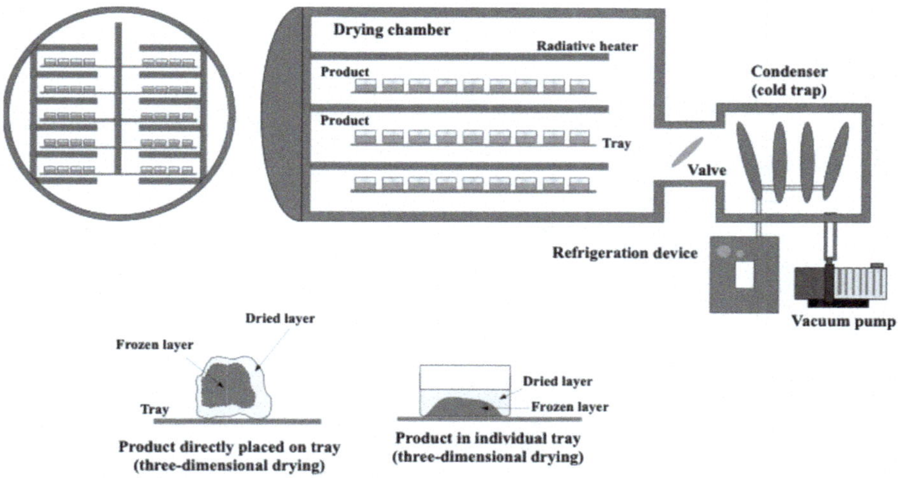

Figure 7.7 Freeze-dryer components for food products.

used in pharmaceutical freeze drying. They are designed to experimentally obtain heat and mass transfer parameters, and improve prediction accuracy (Giordano et al., 2011; Rambhatla et al., 2004; Velardi and Barresi, 2008). This is a two-zone model consisting of frozen and dried layers and only considers the movement of water vapor generated at the sublimation surface through the drying layer. The drying rate is used to express the velocity of the sublimation interface, and the sublimation enthalpy is simplified as being supplied by conductive heat through the frozen layer.

In the freeze-drying process of food products, the mass and heat transfer pathways are often three-dimensional, depending on the product layout and geometry (Nakagawa, 2024). Nakagawa et al. proposed a mechanistic model of freeze drying that progresses three-dimensionally with radiant heat as the main heat source (Nakagawa and Ochiai, 2015). The sublimation surface area, which changes as the drying progresses, was determined experimentally. The heat-receiving surface of the product (outer surface) and sublimation surface area were arranged in spherical coordinates, and the average thickness of the drying layer was determined by approximating it to the difference in the diameters of the spheres. This model introduces the overall effective conductance between the product exterior and condenser (i.e., cold trap). These features allow drying simulations to reflect radiant heat and equipment performance with low calculation effort (Nakagawa and Tamiya, 2019). Nakagawa et al. verified this using freeze-dried apple cubes and soup in a cup.

Modeling was also investigated for freeze drying under atmospheric pressure. Wolff and Gibert (1990) applied a two-layer model consisting of frozen and drying layers (Wolff and Gibert, 1990). As mentioned earlier, it can be assumed that shrinkage and liquid water migration do not occur; therefore, the diffusion coefficient of water vapor simply depends on the pore structures. Claussen et al. have verified their mechanistic model by using apples, cabbage, and fish meat (Claussen et al., 2007a, 2007b). Nakagawa et al. pointed out that in the process of freeze drying fruit

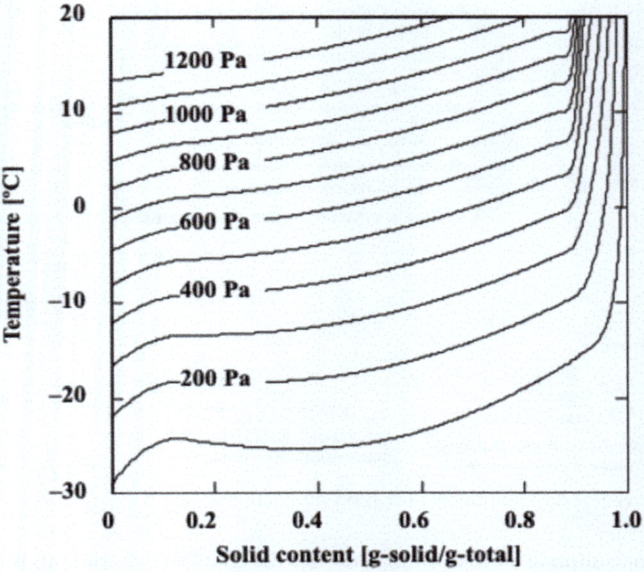

Figure 7.8 Experimentally obtained apparent vapor pressure of apple as a function of temperature and solid content (Nakagawa et al., 2020a).

at temperatures significantly above the glass transition point, the product of the drying process can be considered to be in a single rubbery state rather than a two-layered state. Therefore, a single-zone model is proposed, in which the experimentally obtained apparent vapor pressure of the glass–rubber phase (Fig. 7.8) is applied to a simple mechanistic model (Nakagawa et al., 2020a). This apparent vapor pressure can be considered a unique characteristic of food; therefore, simple and robust simulations can be performed using various models that are adapted to the dry form.

7.5 Model-based approach for quality management

One practical application of the modeling approach to food drying is quality stabilization. As mentioned earlier, being able to perform quality management in a model-based manner is not only an advantage for stabilizing quality, but also for knowing the cost-effectiveness of equipment improvements. The modeling approach discussed thus far relates various physicochemical phenomena to the operating variables of the process. Because many foods are non-equilibrium systems, the quality required for foods is often kinetic stability. Therefore, it is important to link the dynamics of quality changes to the processing variables. The mechanism of moisture transfer within materials has been precisely understood based on modeling studies. In addition, research on the structure of food itself is being undertaken by many food engineers with the aim of rationally designing processes to improve product quality (Aguilera, 2005). Here, we introduce examples of research conducted on the control or management of food quality using a model-based approach.

Perdana et al. conducted model-based research to determine the optimal drying route for maintaining the count of viable lactic acid bacteria during the spray-drying process of maltodextrin (Perdana et al., 2015). The bacterial inactivation kinetics was expressed by an empirical power law equation as a function of the temperature and water activity. Changes in cell viability were simulated based on changes in temperature, moisture content, and water activity as the progress of drying (Fig. 7.9). The simulation results predicted that viability decreased significantly when the temperature rose during the preheating stage, and a drying route that minimizes the reduction in viability can be found on a map of viability as a function of moisture content and temperature. They predicted that if the temperature during drying could be kept sufficiently low, a reduction in viability can be avoided, even if the residence time is long, and reported that reducing the droplet size is effective for preventing overheating in the early stages of drying (Perdana et al., 2015).

Tsotsas (2015) conducted various advanced modeling studies on drying, including pore network modeling. During granulation, water droplets on the surface of wet particles are involved in particle aggregation. Therefore, the agglomerate structure of the

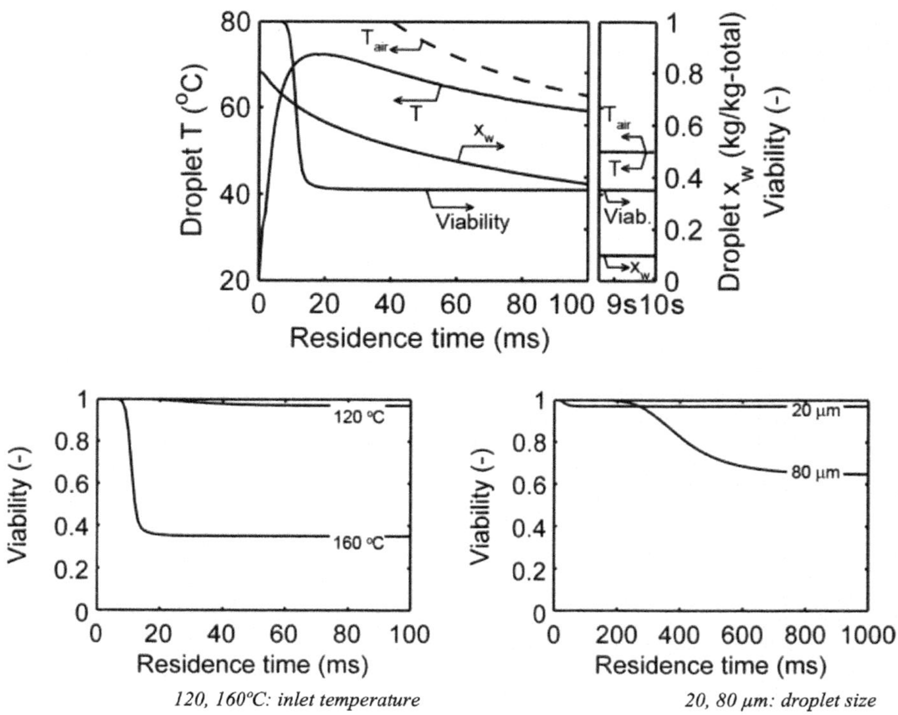

Figure 7.9 Processing route during spray drying and time-dependent trend of cell viability (Perdana et al., 2015).

particles depends on the drying rate (Fig. 7.10). They conducted particle observations using X-ray computed tomography (CT) (the granulation experiment was conducted with alumina powder) and examined the relationship between the particle structure and the simulation data for the inside of the dryer using CFD (Tsotsas, 2015). They found that the porosity and fractal dimension of the agglomerated particles formed were correlated with the drying conditions, and specifically reported that the higher the temperature of the inlet gas, the lower the porosity and the higher the fractal dimension of the structure. This suggests that the microstructure of dried fine particles can be controlled by varying the drying rate. However, for food products that undergo glass transition, this trend may be slightly more complex.

Nakagawa et al. calculated the design space for the pressure and temperature inside the drying chamber in the food freeze-drying process of instant soup and estimated the sublimation surface temperature as a function of control variables, that is, chamber pressure and heater temperature (Nakagawa and Tamiya, 2019; Nakagawa et al., 2020b). When the sublimation surface temperature exceeds the glass transition temperature of the freeze-concentrated phase, the risk of defects increases due to foaming and collapse. Therefore, it is desirable to distinguish the conditions that minimize drying time from the operating conditions that do not significantly exceed it. This can be realized through a model-based design space visualization, as shown in Fig. 7.11. In this example, the comparison results are based on the assumption of equipment with different performances, which are difficult to obtain experimentally. The overall effective conductance (described as the inner vapor transfer coefficient in this study) applied to the model can be considered as a parameter that represents the performance of the equipment. It is predicted that if this value decreases, the operating range in which foaming and collapse do not occur will become narrower. In the food industry, where batch operations are performed in parallel with multiple dryers, this is considered a promising approach for achieving operational equivalence between devices.

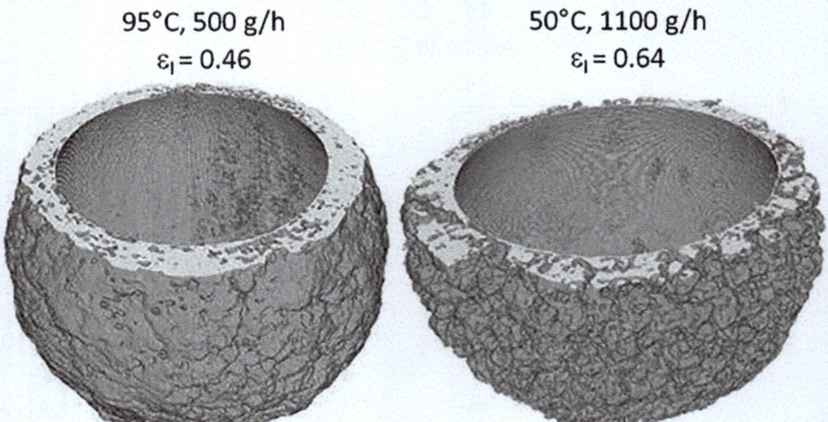

Figure 7.10 X-ray μ-CT of granulated particles (Tsotsas, 2015).

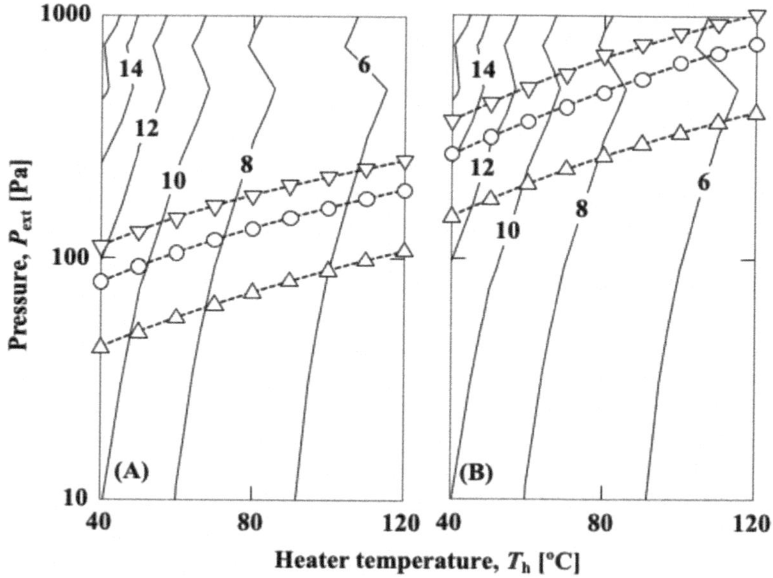

Figure 7.11 Contour line of the sublimation surface and the freeze-drying operations line; (A) larger conductance value, (B) smaller conductance value, (▽): Maximum vapor pressure values of at the product surface (P_{ext}), (○): Mean values of P_{ext}, (△): Minimum values of P_{ext} (Nakagawa and Tamiya, 2019).

7.6 Conclusion

Industrial drying has long been recognized as a unit operation with many issues, and the development of mathematical models has been ongoing for more than a century. Owing to the complexity of food drying, many issues remain unresolved. In reviewing the modeling of food drying in this chapter, the author has tried to show the history of research in chronological order from the earliest studies in 1929. This was because the author wanted to understand how the modeling issues were recognized and how the approaches were developed. The refinement of models to correctly understand the phenomenon and the simplification and securing of accuracy for practical use must be carried out in parallel. While paying the greatest respect to pioneering researchers, it would be a great pleasure for the author if this review contributes to future developments.

References

Adhikari, B., Howes, T., Bhandari, B.R., Yamamoto, S., Truong, V., 2002. Application of a simplified method based on regular regime approach to determine the effective moisture diffusivity of mixture of low molecular weight sugars and maltodextrin during desorption. J. Food Eng. 54, 157–165.

Afshar, S., Metzger, L., Patel, H., Selomulya, C., Woo, M.W., 2019. A practical CFD modeling approach to estimate outlet boundary conditions of industrial multistage spray dryers: inert particle flow field investigation. Dry. Technol. 37, 824–838.

Aguilera, J.M., 2005. Why food microstructure? J. Food Eng. 67, 3–11.

Andrieu, J., Stamatopoulos, A., 1986. Durum wheat pasta drying kinetics. Lebensm.-Wiss. Technol. 19, 448–456.

Andrieu, J., Stamatopoulos, A., Zafirpoulos, M., 1985. Equation for fitting desorption isotherms of durum wheat pasta. Int. J. Food Sci. Technol. 20, 651–657.

Andrieu, S., 1986. Moisture and heat transfer modelling during durum wheat pasta drying. In: Mujumdar, A.S. (Ed.), Drying'86. Hemisphere Publishing Corp, Washington, New York, London, pp. 492–498.

Báez-González, J.G., Pérez-Alonso, C., Beristain, C.I., Vernon-Carter, E.J., Vizcarra-Mendoza, M.G., 2004. Effective moisture diffusivity in biopolymer drops by regular regime theory. Food Hydrocoll. 18, 325–333.

Banga, J.R., Balsa-Canto, E., Moles, C.G., Alonso, A.A., 2003. Improving food processing using modern optimization methods. Trends Food Sci. Technol. 14, 131–144.

Banga, J.R., Paul Singh, R., 1994. Optimization of air drying of foods. J. Food Eng. 23, 189–211.

Capozzi, L.C., Barresi, A.A., Pisano, R., 2019. A multi-scale computational framework for modeling the freeze-drying of microparticles in packed-beds. Powder Technol. 343, 834–846.

Carmeliet, J., Descamps, F., Houvenaghel, G., 1999. A multiscale network model for simulating moisture transfer properties of porous media. Transport Porous Media 35, 67–88.

Castro, A.M., Mayorga, E.Y., Moreno, F.L., 2018. Mathematical modelling of convective drying of fruits: a review. J. Food Eng. 223, 152–167.

Chen, X.D., 2008. The basics of a reaction engineering approach to modeling air-drying of small droplets or thin-layer materials. Dry. Technol. 26, 627–639.

Chen, X.D., Lin, S.X.Q., 2005. Air drying of milk droplet under constant and time-dependent conditions. AIChE J. 51, 1790–1799.

Claussen, I.C., Andresen, T., Eikevik, T.M., Str⊘mmen, I., 2007a. Atmospheric freeze drying—modeling and simulation of a tunnel dryer. Dry. Technol. 25, 1959–1965.

Claussen, I.C., Ustad, T.S., Str⊘mmen, I., Walde, P.M., 2007b. Atmospheric freeze drying—a review. Dry. Technol. 25, 947–957.

Couture, F., Laurent, S., Roques, M.A., 2007. Drying of two-phase media: simulation with liquid pressure as driven force. AIChE J. 53, 1703–1717.

Crapiste, G.H., Whitaker, S., Rotstein, E., 1985. Fundamentals of Drying of Foodstuffs. Springer Berlin Heidelberg, Berlin, Heidelberg, pp. 297–302.

Crapiste, G.H., Whitaker, S., Rotstein, E., 1988a. Drying of cellular material—I. A mass transfer theory. Chem. Eng. Sci. 43, 2919–2928.

Crapiste, G.H., Whitaker, S., Rotstein, E., 1988b. Drying of cellular material—II. Experimental and numerical results. Chem. Eng. Sci. 43, 2929–2936.

Datta, A.K., 2008. Status of physics-based models in the design of food products, processes, and equipment. Compr. Rev. Food Sci. Food Saf. 7, 121–129.

Defraeye, T., 2011. Convective heat and mass transfer at exterior building surfaces. Katholieke Universiteit Leuven.

Defraeye, T., 2014. Advanced computational modelling for drying processes – a review. Appl. Energy 131, 323–344.

Defraeye, T., Blocken, B., Carmeliet, J., 2012. Analysis of convective heat and mass transfer coefficients for convective drying of a porous flat plate by conjugate modelling. Int. J. Heat Mass Tran. 55, 112–124.

Defraeye, T., Verboven, P., 2017. Convective drying of fruit: role and impact of moisture transport properties in modelling. J. Food Eng. 193, 95–107.

Dev, S.R.S., Raghavan, V.G.S., 2012. Advancements in drying techniques for food, fiber, and fuel. Dry. Technol. 30, 1147–1159.

Devahastin, S., Niamnuy, C., 2010. Invited review: modelling quality changes of fruits and vegetables during drying: a review. Int. J. Food Sci. Technol. 45, 1755–1767.

Fanta, S.W., Abera, M.K., Aregawi, W.A., Ho, Q.T., Verboven, P., Carmeliet, J., Nicolai, B.M., 2014. Microscale modeling of coupled water transport and mechanical deformation of fruit tissue during dehydration. J. Food Eng. 124, 86–96.

Fanta, S.W., Abera, M.K., Ho, Q.T., Verboven, P., Carmeliet, J., Nicolai, B.M., 2013. Microscale modeling of water transport in fruit tissue. J. Food Eng. 118, 229–237.

FAO, 2021. In: Agency, T.I.R.E. (Ed.), Renewable Energy for Agri-Food Systems.

Fatt, I., 1956. The network model of porous media. Transactions of the AIME 207, 144–181.

Giordano, A., Barresi, A.A., Fissore, D., 2011. On the use of mathematical models to build the design space for the primary drying phase of a pharmaceutical lyophilization process. J. Pharmaceut. Sci. 100, 311–324.

Gong, Z.-X., Mujumdar, A.S., 2008. Software for design and analysis of drying systems. Dry. Technol. 26, 884–894.

Halder, A., Datta, A.K., Spanswick, R.M., 2011. Water transport in cellular tissues during thermal processing. AIChE J. 57, 2574–2588.

Hasatani, M., Itaya, Y., 1996. Drying-induced strain and stress: a review. Dry. Technol. 14, 1011–1040.

Hougen, O., McCauley, H., Marshall, J.,W., 1940. Limitations of diffusion equations in drying. Trans. AIChE 36, 183–206.

Jan Kowalski, S., 1996. Mathematical modelling of shrinkage during drying. Dry. Technol. 14, 307–331.

Katekawa, M.E., Silva, M.A., 2006. A review of drying models including shrinkage effects. Dry. Technol. 24, 5–20.

Keey, R., 1994. Progress in understanding drying: 1964-1994. Drying 94, 75–88.

Keey, R.B., Suzuki, M., 1974. On the characteristic drying curve. Int. J. Heat Mass Tran. 17, 1455–1464.

Kowalski, S.J., 2001. Thermomechanical approach to shrinking and cracking phenomena in drying. Dry. Technol. 19, 731–765.

Krischer, O., 1938. The drying of solid substances as a problem in the movement of capillary moisture and the diffusion of vapor (In German) Z. Ver. deut. Ing. Verfahrenstechnik Beih. 4, 104.

Labuza, T.P., 1970. Properties of water as related to the keeping quality of foods. In: Proceedings 3rd International Conference Food Science and Technology, SOS70. Institute Food Technologists, Washington, DC, p. 618.

Langrish, T.A.G., 2009. Multi-scale mathematical modelling of spray dryers. J. Food Eng. 93, 218–228.

Langrish, T.A.G., Fletcher, D.F., 2001. Spray drying of food ingredients and applications of CFD in spray drying. Chem. Eng. Process. Process Intensif. 40, 345–354.

Langrish, T.A.G., Kockel, T.K., 2001. The assessment of a characteristic drying curve for milk powder for use in computational fluid dynamics modelling. Chem. Eng. J. 84, 69–74.

Lewis, R.W., Morgan, K., Thomas, H.R., Strada, M., 1979. Drying-induced stresses in porous bodies —an elastoviscoplastic model. Comput. Methods Appl. Mech. Eng. 20, 291–301.
Lewis, R.W., Strada, M., Comini, G., 1977. Drying-induced stresses in porous bodies. Int. J. Numer. Methods Eng. 11, 1175–1184.
Liapis, A.I., Bruttini, R., 1994. A theory for the primary and secondary drying stages of the freeze-drying of pharmaceutical crystalline and amorphous solutes: comparison between experimental data and theory. Sep. Technol. 4, 144–155.
Liapis, A.I., Bruttini, R., 1995. Freeze-drying of pharmaceutical crystalline and amorphous solutes in vials: dynamic multi-dimensional models of the primary and secondary drying stages and qualitative features of the moving interface. Dry. Technol. 13, 43–72.
Liapis, A.I., Litchfield, R.J., 1979. Numerical solution of moving boundary transport problems in finite media by orthogonal collocation. Comput. Chem. Eng. 3, 615–621.
Liou, J.K., Bruin, S., 1982. An approximate method for the nonlinear diffusion problem with a power relation between diffusion coefficient and concentration—I. Computation of desorption times. Int. J. Heat Mass Tran. 25, 1209–1220.
Luikov, A.V., 1966. Application of irreversible thermodynamics methods to investigation of heat and mass transfer. Int. J. Heat Mass Tran. 9, 139–152.
Luikov, A.V., 1975. Systems of differential equations of heat and mass transfer in capillary-porous bodies (review). Int. J. Heat Mass Tran. 18, 1–14.
Luyben, K.C.A.M., Liou, J.K., Bruin, S., 1982. Enzyme degradation during drying. Biotechnol. Bioeng. 24, 533–552.
McMinn, W.A.M., Magee, T.R.A., 1999. Principles, methods and applications of the convective drying of foodstuffs. Food Bioprod. Process. 77, 175–193.
Mezhericher, M., Naumann, M., Peglow, M., Levy, A., Tsotsas, E., Borde, I., 2012. Continuous species transport and population balance models for first drying stage of nanosuspension droplets. Chem. Eng. J. 210, 120–135.
Mishkin, M., Saguy, I., Karel, M., 1983. Mimimizing ascorbic acid loss during air drying with a constraint on enzyme inactivation for a hypothetical foodstuff. J. Food Process. Preserv. 7, 193–210.
Mortier, S.T.F.C., De Beer, T., Gernaey, K.V., Remon, J.P., Vervaet, C., Nopens, I., 2011. Mechanistic modelling of fluidized bed drying processes of wet porous granules: a review. Eur. J. Pharm. Biopharm. 79, 205–225.
Mrani, I., Fras, G., Benet, J.C., 1995. Numerical study of drying stresses in agar gel. Dry. Technol. 13, 551–570.
Mujumdar, A.S., 2006. Handbook of Industrial Drying. CRC press.
Musielak, G., 1996. Internal stresses caused by outflow of moisture and phase change inside dried material. Dry. Technol. 14, 289–306.
Musielak, G., 2000. Influence of the drying medium parameters on drying induced stresses. Dry. Technol. 18, 561–581.
Nakagawa, K., 2024. Mathematical modeling of freeze-drying process. Freeze Drying of Food Products 31–71.
Nakagawa, K., Horie, A., Nakabayashi, M., 2020a. Modeling atmospheric freeze-drying of food products operated above glass transition temperature. Chem. Eng. Res. Des. 163, 12–20.
Nakagawa, K., Hottot, A., Vessot, S., Andrieu, J., 2007. Modeling of freezing step during freeze-drying of drugs in vials. AIChE J. 53, 1362–1372.
Nakagawa, K., Ochiai, T., 2015. A mathematical model of multi-dimensional freeze-drying for food products. J. Food Eng. 161, 55–67.

Nakagawa, K., Tamiya, S., 2019. Influence of inner vapor transfer property of a freeze dryer on the design space of drying process. AIChE J. 65, e16571.
Nakagawa, K., Tamiya, S., Ochiai, T., 2020b. Evaluation of degree of collapse and its relationship with retention of organic volatiles in freeze-dried dextrin matrices. Ind. Eng. Chem. Res. 59, 18298–18306.
Nakagawa, K., Tamura, A., Adachi, S., 2018. Optimization of food dye (betanin) retention during hot air drying: design space calculation with consideration of reaction and substrate transfer kinetics. Dry. Technol. 36, 1920–1929.
Onwude, D.I., Hashim, N., Janius, R.B., Nawi, N.M., Abdan, K., 2016. Modeling the thin-layer drying of fruits and vegetables: a review. Compr. Rev. Food Sci. Food Saf. 15, 599–618.
Ozmen, L., Langrish, T.A.G., 2002. Comparison of glass transition temperature and stickey point temperature for skim milk powder. Dry. Technol. 20, 1177–1192.
Patel, K.C., Chen, X.D., 2005. Prediction of spray-dried product quality using two simple drying kinetics models. J. Food Process. Eng. 28, 567–594.
Peczalski, R., Laurent, P., Andrieu, J., Boyer, J., Boivin, M., 1996. Drying-induced stress build-up within spaghetti. Drying 96, 805–816.
Perdana, J., Fox, M.B., Boom, R.M., Schutyser, M.A.I., 2015. Establishing guidelines to retain viability of probiotics during spray drying. Dry. Technol. 33, 1560–1569.
Pham, S.T., Chareyre, B., Tsotsas, E., Kharaghani, A., 2022. Pore network modeling of phase distribution and capillary force evolution during slow drying of particle aggregates. Powder Technol. 407, 117627.
Place, G., Ridgway, K., Danckwerts, P., 1959. Investigation of air-flow in a spray-drier by tracer and model techniques. Trans. IChemE 37, 269–276.
Ponsart, G., Vasseur, J., Frias, J.M., Duquenoy, A., Méot, J.M., 2003. Modelling of stress due to shrinkage during drying of spaghetti. J. Food Eng. 57, 277–285.
Puyate, Y.T., Lawrence, C.J., 2006. Sherwood's models for the falling-rate period: a missing link at moderate drying intensity. Chem. Eng. Sci. 61, 7177–7183.
Rambhatla, S., Ramot, R., Bhugra, C., Pikal, M., 2004. Heat and mass transfer scale-up issues during freeze drying: II. Control and characterization of the degree of supercooling. AAPS PharmSciTech 5, 54–62.
Sarmento, A.L.S.C., Couture, F., Laurent, S., Roques, M.A., 2004. Modeling of heat and mass transport in two-phase media by considering liquid pressure. Dry. Technol. 22, 81–90.
Schoeber, W.J.A.,H., 1976. Regular Regimes in Sorption Processes. Ph.D. Thesis. Technical University, Eindhoven.
Sherwood, T., 1929a. The drying of solids—II. Ind. Eng. Chem. 21, 976–980.
Sherwood, T., 1930. The drying of solids—III. Mechanism of the drying of pulp and paper. Ind. Eng. Chem. 22, 132–136.
Sherwood, T., 1932. The drying of solids—IV. Application of diffusion equations. Ind. Eng. Chem. 24, 307–310.
Sherwood, T.K., 1929b. The drying of solids—I. Ind. Eng. Chem. 21, 12–16.
Sherwood, T.K., Comings, E., 1933. The drying of solids—V. Mechanism of drying of clays. Ind. Eng. Chem. 25, 311–316.
Slade, L., Levine, H., Reid, D.S., 1991. Beyond water activity: recent advances based on an alternative approach to the assessment of food quality and safety. Crit. Rev. Food Sci. Nutr. 30, 115–360.
Tamon, H., Okazaki, M., Toei, R., 1981. Flow mechanism of adsorbate through porous media in presence of capillary condensation. AIChE J. 27, 271–277.

Thijssen, H.A.C., Coumans, W.J., 1985. Short-Cut Calculation for Non-isothermal Drying of Shrinking and Non-shrinking Particles and of Hollow Spheres Containing an Expanding Central Gas Core. Springer Berlin Heidelberg, Berlin, Heidelberg, pp. 11–20.

Toei, R., Imakoma, H., Tamon, H., Okazaki, M., 1983. Water transfer coefficient in adsorptive porous body. J. Chem. Eng. Jpn. 16, 364–369.

Toei, R., Okazaki, M., 1970. Drying mechanism of capillary-porous bodies. J. Eng. Phys. 19, 1123–1131.

Truong, V., Bhandari, B.R., Howes, T., 2005a. Optimization of co-current spray drying process of sugar-rich foods. Part I—moisture and glass transition temperature profile during drying. J. Food Eng. 71, 55–65.

Truong, V., Bhandari, B.R., Howes, T., 2005b. Optimization of cocurrent spray drying process for sugar-rich foods. Part II—optimization of spray drying process based on glass transition concept. J. Food Eng. 71, 66–72.

Tsotsas, E., 2015. Multiscale approaches to processes that combine drying with particle formation. Dry. Technol. 33, 1859–1871.

Van Boekel, M.A.J.S., 2008. Kinetic modeling of food quality: a critical review. Compr. Rev. Food Sci. Food Saf. 7, 144–158.

Velardi, S.A., Barresi, A.A., 2008. Development of simplified models for the freeze-drying process and investigation of the optimal operating conditions. Chem. Eng. Res. Des. 86, 9–22.

Villota, R., Karel, M., 1980. Prediction of ascorbic acid retention during drying I. Moisture and temperature distribution in a model system. J. Food Process. Preserv. 4, 111–134.

Whitaker, S., 1977. Simultaneous heat, mass, and momentum transfer in porous media: a theory of drying. In: Hartnett, J.P., Irvine, T.F. (Eds.), Advances in Heat Transfer. Elsevier, pp. 119–203.

Whitaker, S., 1986a. Flow in porous media I: a theoretical derivation of Darcy's law. Transport Porous Media 1, 3–25.

Whitaker, S., 1986b. Flow in porous media III: deformable media. Transport Porous Media 1, 127–154.

Woinet, B., Andrieu, J., Laurent, M., 1998a. Experimental and theoretical study of model food freezing. Part I. Heat transfer modelling. J. Food Eng. 35, 381–393.

Woinet, B., Andrieu, J., Laurent, M., Min, S., 1998b. Experimental and theoretical study of model food freezing. Part II. Characterization and modelling of the ice crystal size. J. Food Eng. 35, 395–407.

Wolff, E., Gibert, H., 1990. Atmospheric freeze-drying part 2: modelling drying kinetics using adsorption isotherms. Dry. Technol. 8, 405–428.

Woo, M.W., Daud, W.R.W., Mujumdar, A.S., Wu, Z., Meor Talib, M.Z., Tasirin, S.M., 2008a. CFD evaluation of droplet drying models in a spray dryer fitted with a rotary atomizer. Dry. Technol. 26, 1180–1198.

Woo, M.W., Daud, W.R.W., Tasirin, S.M., Talib, M.Z.M., 2008b. Effect of wall surface properties at different drying kinetics on the deposition problem in spray drying. Dry. Technol. 26, 15–26.

Yamamoto, S., 2001. A short-cut method for determining concentration dependent diffusivity in liquid foods and polymer solutions from regular regime drying curves. Dry. Technol. 19, 1479–1490.

Yoshida, M., Imakoma, H., Okazaki, M., 1991a. Determination of diffusivity using the characteristic function for the regular regime. J. Chem. Eng. Jpn. 24, 720–726.

Yoshida, M., Imakoma, H., Okazaki, M., 1991b. Estimation of drying rate curves for nonhygroscopic porous slabs using the characteristic function for the regular regime. Chem. Eng. Process. Process Intensif. 30, 87–96.

Zbiciński, I., 1995. Development and experimental verification of momentum, heat and mass transfer model in spray drying. Chem. Eng. J. Biochem. Eng. J. 58, 123–133.

Modeling of food evaporation processes

8

Christian O. Díaz-Ovalle
Tecnológico Nacional de México/I.T. Roque, Departamento de Ingenierías, Celaya, Guanajuato, México

8.1 Introduction

The evaporation processes eliminate water from a liquid foodstuff by applying a heating medium. This action concentrates the containing solids into the liquid foodstuff, which is the main objective of the evaporation. This volume reduction benefits the storage, handling, and transport of concentrated foodstuff. The concentration of solids guarantees the properties of a new product and prepares the liquid for further processes, such as spray drying. During evaporation, the heating treatment reduces the microbiological risk and augments the shelf life. Even that, the produced vapor from concentrated juices contains flavoring products recovered by distillation. All these points have demonstrated the importance of evaporation processes in the food industry.

Energy consumption and thermal treatment demand attention to be mitigated or reduced. The multiple-effect design and the thermal recycling systems have set on the energy consumptions, and nowadays, this is a usual design in the evaporation processes. This design considers several evaporation units or effects and reuses the produced vapor of each effect to heat the others. Fig. 8.1 illustrates a multi-effect evaporator system. The diversity of geometrical design in the evaporation unit has focused on providing a stable process with low fouling risk and reduction of thermal damage. The literature presents the fundamentals and basis for understanding the evaporation process (Díaz-Ovalle and Jafari, 2023; Jafari et al., 2024). However, improving the design and operation of an evaporation unit is possible by applying modeling techniques to guarantee suitable approaches. Herein, some process system engineering principles apply to steer toward acceptable food evaporation processes.

8.2 Fundamentals of heat transfer in evaporators

The evaporation of a liquid foodstuff is possible in a non-direct contact unit, where a metal wall separates the liquid from the heating medium, in most cases motive steam or vapor for a multiple-effect evaporator system (MEE). The heat transfer corresponds to the transfer of energy from the heating medium to the liquid and experiences a convective phenomenon with boiling-condensing simultaneously, which forms heat

Modeling of Unit Operations in the Food Industry. https://doi.org/10.1016/B978-0-12-822180-8.00014-6

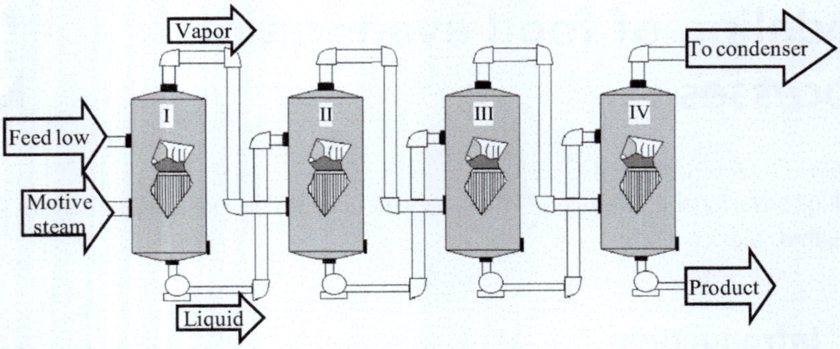

Figure 8.1 Example of forward multiple-effect evaporator.
Adapted from Díaz-Ovalle, C. O., & Jafari, S. M. (2023). Multiple-Effect Evaporators in the Food Industry: Fundamentals, Design, Simulation, Control, and Applications. Food Engineering Reviews, 15(4), 691-717. https://doi.org/10.1007/s12393-023-09350-6. With permission of Springer.

resistances. The heat flow rate, Q, quantifies the energy flow and involves the temperature difference between the fluids, $\Delta T = Ts - T$, Ts and T are the heating medium and liquid foodstuff temperatures, respectively. The heat resistances are the conductive and convective coefficients, and the overall heat transfer coefficient, U, relates them and includes the fouling resistance. Thus, the heat flow rate is $Q = UA\Delta T$, A is the heat transfer area, and $U = 1/(1/h_L + 1/h_V + \Psi)$, h_L and h_V are the boiling and condensing convective coefficients, respectively, and Ψ is the thermal fouling resistance.

The literature has reported values for the parameters U, h_L, and h_V and correspond to cases of either evaporators or liquid foodstuffs, such as whole milk, cane sugar, etc. The reader might consult the summarized values presented by Díaz-Ovalle and Jafari (2023). Nevertheless, theoretical correlations of h_L and h_V are suitable options for modeling MEE processes. For the h_V coefficient, Table 8.1 contains the reported correlations despite the low thermal resistance.

In contrast, the correlations for the h_L coefficient have required studies to understand the boiling phenomenon. Nukiyama (1934, 1966, 1984) has classified the boiling in theoretical regimes based on the value of ΔT, such as: a) ΔT<5°C, non-nucleate boiling regimen (free convection phenomenon), b) 5°C $\leq \Delta T \leq$30°C, nucleate boiling (presence of bubbles), and c) 30°C $< \Delta T$, transition boiling (unstable multiphase system). From this theoretical approach and the heat transfer principles, food engineering has cataloged a set of correlations for h_L, and Adib and Vasseur (2008) reviewed most of them. Tables 8.2 and 8.3 present the models for the nucleate pool and falling-film evaporators, respectively.

8.3 Modeling of food properties

The thermo-physical properties are essential for modeling heat transfer unit operations. For liquid foodstuffs, the obtention of properties limits most cases, however,

Table 8.1 Additional convective correlations for evaporation systems.

Correlation model and description	**Reference**
Condensing vapor as falling-film outside long vertical tube: $h_V = \kappa_L {\rho_L}^{2/3} g^{1/3} {\mu_L}^{-2/3} \left[{\mathrm{Re}_L}^{-0.44} + 5.28 \times 10^{-6} {\mathrm{Re}_L}^{-0.44} {\mathrm{Pr}_L}^{1/3}\right]^{1/2}$	Chen et al. (1987)
Condensing vapor outside a tube: $h_V = 1.13 \left[\frac{\rho_L g \lambda (\rho_L - \rho_V)(\kappa_L)^3}{\mu_L \mathrm{L} T (T^* - T)}\right]^{\frac{1}{4}}$	Presented by Pitts and Sissom (1998)
Boiling liquid in vertical tubes: $h_L = 0.002 \left(\frac{4L_i}{\pi \phi_{T,i} \mu_V}\right)^{0.6} \left[\frac{g \rho_V (\rho_L - \rho_V)(\kappa_L)^3}{(\mu_V)^2}\right]^{\frac{1}{3}}$	

Adapted from Díaz-Ovalle, C. O., & Jafari, S. M. (2023). Multiple-Effect Evaporators in the Food Industry: Fundamentals, Design, Simulation, Control, and Applications. *Food Engineering Reviews*, 15(4), 691-717. https://doi.org/10.1007/s12393-023-09350-6. With permission of Springer.

Choi and Okos (1986) proposed outstanding and practical correlations to determine the properties in function of temperature. The models are polynomial expressions of each component (water, protein, carbohydrate, fat, fiber, and ash), which relate the mass fraction of each one, x_i, such as:

$$\left.\begin{aligned} Cp &= \sum_j Cp_j x_j \\ \frac{1}{\rho} &= \sum_j \frac{x_j}{\rho_j} \\ \kappa &= \rho \sum_j \frac{\kappa_j x_j}{\rho_j} \end{aligned}\right\} \tag{8.1}$$

where Cp is the heat capacity, ρ is the density, κ is the thermal conductivity, and j is the index for the components.

For evaporation processes, Galván-Ángeles et al. (2015) and Díaz-Ovalle et al. (2017) demonstrated the difference between using constant and variable properties by applying the Choi-Okos correlations. Thus, the variable properties with the temperature are convenient for rigorous analysis of MEEs. Table 8.4 contains the values of the parameters for the Choi-Okos correlations. Based on this, the liquid enthalpy calculation might come from a variable heat capacity, $H_L = \int Cp(T)dT$ from T_{ref} to T, or constant heat capacity, $H_L = Cp(T - T_{ref})$. The 0°C is the recommended T_{ref} in liquid phase.

Table 8.2 Parameters for nucleate pool correlations.

Model	$h_L = Z \cdot (q)^a (\kappa_L)^b (Cp_L)^c (\sigma_L)^d (\lambda)^e (\rho_L - \rho_V)^f (\mu_L)^g (\rho_L)^h (\rho_V)^i \Phi$										
Author/Parameters	***Z***	***a***	***b***	***c***	***d***	***e***	***f***	***g***	***h***	***i***	**Φ**
McNelly	0.225	0.69	0.31	0.69	−0.31	−0.69	0.33	0	0	−0.33	$P^{0.31}$
Rohsenow	1.464	0.667	1.7	−0.7	−0.167	−0.667	0.167	−1.367	0	0	76.92
Glushenko	1112.5	0.7	0.3	−0.1	−0.15	0.1	0.15	−0.8	−0.1	0.1	$T^{*-0.8} \mu_V^{0.8}$
Gilmour	0.38	0.7	0.6	0.4	−0.425	−0.7	0	−0.3	0.7	−0.7	$1.35(10)^{-4} P^{0.85} \phi_T^{-0.3}$
Miyauchi-Yagui	1.33	0.74	0.37	0.63	0	−0.74	0	−0.11	0.74	−0.05	$\rho_{V,atm}^{-0.69}$
Kutateladze	$6.9(10)^{-4}$	0.7	0.65	0.35	−0.5	−0.7	−0.2	−0.35	0.7	−0.7	$P^{0.7}$

Adapted from Díaz-Ovalle, C. O., & Jafari, S. M. (2023). Multiple-Effect Evaporators in the Food Industry: Fundamentals, Design, Simulation, Control, and Applications. *Food Engineering Reviews*, 15(4), 691-717. https://doi.org/10.1007/s12393-023-09350-6. With permission of Springer.

Table 8.3 Parameters of vertical tube in falling-film evaporator.

Model		$h_L = Z \cdot (\rho_L)^a (\kappa_L)^b (\mu_L)^c (\Gamma)^d (Cp_L)^e (q)^f$						
Parameters		Z	a	b	c	d	e	f
Author	**Description**							
Nusselt	L-NN	1.48	0.66	1	−0.33	−0.33	0	0
Chun and Seban	W-NN	0.96	0.66	1	−0.44	−0.22	0	0
Prost et al.	W-NN-ss	6.93	0.66	1	−0.28	−0.38	0	0
McAdams	T-NN	0.034	0.67	0.33	−0.66	0.33	0.33	0
Ahmed and Kaparthi	T-NN	0.024	0.66	1	−0.61	0.35	0.4	0
Herbert and Stern	T-NN	0.0044	0.66	1	−1.31	0.65	0	0
Chun and Seban	T-NN	0.014	0.66	0.35	−0.41	0.4	0.65	0
Bouman et al.	N-sm	0.77	0	0	−0.41	0	0	0.69
Bouman et al.	N-wm	6.05	0	0	−0.44	0.26	0	0.47

L: laminar regime; W: wavy-laminar regime; T: turbulent regime; ss: solution of sucrose; sm: skim milk; wm: whole milk; NN: non-nucleate boiling; N: nucleate boiling; $\Gamma_i = L_i/(\pi\phi_{Ti}N_{Ti})$; *See nomenclature section.
Adapted from Díaz-Ovalle, C. O., & Jafari, S. M. (2023). Multiple-Effect Evaporators in the Food Industry: Fundamentals, Design, Simulation, Control, and Applications. *Food Engineering Reviews*, 15(4), 691-717. https://doi.org/10.1007/s12393-023-09350-6. With permission of Springer.

8.4 Models for mass and energy balances

Evaporation processes reclaim a considerable quantity of motive steam to reach the boiling condition for the liquid foodstuff, whereby the MEE design is widely used (Minton, 1986). The mass and energy balances aim to determine energy consumption (Maroulis and Saravacos, 2008) and are correspond to a system of either linear or nonlinear algebraic equations. The linear model features the search for mass flow rates, and the nonlinear model involves the temperature as an unknown variable. These models are systematic in MEE processes and depend on the sequential statement of the evaporation system. Díaz-Ovalle and Jafari (2023) have described remarkable works under several approaches. Herein, the mass and energy balances follow a model under the liquid flow sequence Galván-Ángeles et al. (2015), where the liquid pattern, into the evaporation system, numbers the effects of the vapor sequence instead of the vapor pattern that numbers the effects of the liquid sequence. Fig. 8.2 illustrates the renumbering of the liquid sequence III-IV-II-I in the vapor pattern to the vapor sequence IV-III-I-II in the liquid pattern. Díaz-Ovalle (2024b) has succinctly described this formulation that Table 8.5 summarizes.

The heat flow rate comes from either an effect, $\lambda_k V_k$, or the heating medium, $\lambda_S V_S$, where k is the index for the effect that heats the i-effect, S is the motive steam stream,

Table 8.4 Choi-Okos parameters.

Heat capacity, Cp[=]J/(kg°C): $Cp_j = a_{1,j} + b_{1,j}T - c_{1,j}T^2$ With $a_{1,j}$[=]J/(kg°C), $b_{1,j}$[=]J/(kg°C^2) and $c_{1,j}$[=]J/(kg°C^3).			
Component (*j*)	$a_{1,j}$x10^{-3}	$b_{1,j}$	$c_{1,j}$x10^3
Water	4.1762	−9.0864	−5.4731
Protein	2.0082	1.2089	1.3129
Carbohydrate	1.5488	1.9625	5.9399
Fat	1.9842	1.4733	4.8008
Fiber	1.8459	1.8306	4.6509
Ash	1.0926	1.8896	3.6817
Density, ρ[=]kg/m^3: $\rho_j = a_{2,j} - b_{2,j}T$ with $a_{2,j}$[=]kg/m^3, and $b_{2,j}$[=]kg/(m^{3}K).			
Componen (*j*)	$a_{2,j}$x10^{-3}		$b_{2,j}$
Water	0.9972		−0.0031439
Protein	1.3990		0.51484
Carbohydrate	1.5991		0.31046
Fat	0.9256		4.1757
Fiber	1.3115		3.6589
Ash	2.4238		0.28063
Thermal conductivity, κ[=]W/(mK): $\kappa_j = a_{3,j} + b_{3,j}T - c_{3,j}T^2$ with $a_{3,j}$[=]W/(mK), $b_{3,j}$[=]W/(mK^2) and $c_{3,j}$[=]W/(mK^3).			
Componen (*j*)	$a_{3,j}$	$b_{3,j}$x10^3	$c_{3,j}$x10^6
Water	0.57109	1.7625	6.7036
Protein	0.17881	1.1958	2.7178
Carbohydrate	0.20141	1.3874	4.3312
Fat	0.18071	−2.7604	0.1775
Fiber	0.18331	1.2497	3.1683
Ash	0.32962	1.4011	2.9069

Adapted from Becker, B. R., & Fricke, B. A. (2003). Freezing. In *Encyclopedia of Food Sciences and Nutrition*, B. Caballero (Ed.), (pp. 2706-2711). Oxford: Academic Press. With permission of Elsevier.

and λ: condensing specific enthalpy. An additional concept is the evaporation economy (*EE*), which represents a practical element to visualize the thermal efficiency of the system. The evaporation economy is the rate of the sum of produced vapors and the motive steam: $EE = \Sigma_i V_i / V_S$, high values of EE indicate an acceptable thermal efficiency.

Some industrial systems include derivations from the produced vapor in a *p*-effect to preheaters, and the heat flow rate is not at all disposed of to the effect. The correction is the fraction of vapor that flows to a preheater, δ_p, and the *i*-effect receives

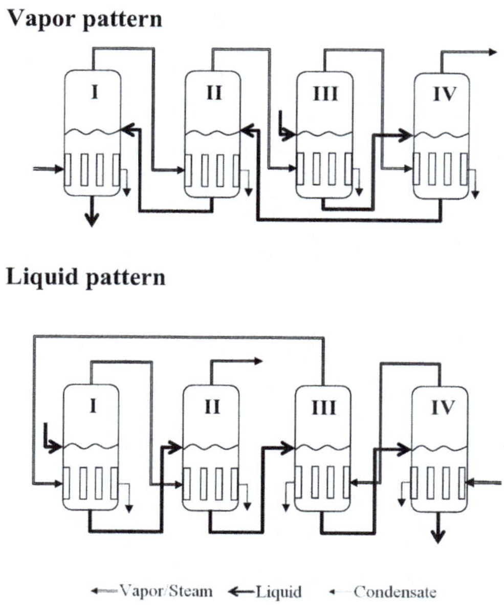

Figure 8.2 Comparison between the vapor pattern III-IV-II-I and liquid pattern IV-III-I-II evaporation sequences.
Adapted from Díaz-Ovalle, C. O. (2024). Design, model, and simulation of evaporators for the food industry. In S.M. Jafari, E. Çapanoglu & A.C. Karaca (Eds.), Evaporation Technology in Food Processing (pp. 269-313): Woodhead Publishing. With permission of Elsevier.

Table 8.5 Mass and energy balances for a MEE by following the liquid pattern.

Description	Equation
First effect, $i = 1$	$Q_i + V_i(H_{L,i} - H_{V,i}) = F(H_{L,i} - H_F)$
Last effect, $i = n$	$Q_i - H_{L,i-1}\sum_{m=1}^{m<i} V_m - V_i H_{V,i} = F\left(\frac{x_F}{x_i}H_{L,i} - H_{L,i-1}\right)$
Other effect: $1<i<n$	$Q_i + (H_{L,i} - H_{L,i-1})\sum_{m=1}^{m<i} V_m + V_i(H_{L,i} - H_{V,i}) = F(H_{L,i} - H_{L,i-1})$
Mass balances	$\sum_{m=1}^{n} V_m = F\left(1 - \frac{x_F}{x_n}\right)$ $L_i = F - \sum_{m=1}^{i} V_m$ $x_i = \frac{F x_F}{L_i}$
Nomenclature	i: effect index, n: total number of effects, F: mass flow rate of the feed stream, V: vapor stream, H specific enthalpy, L and V subscripts for the liquid and vapor phases, respectively, H_F: specific enthalpy at the feed stream, x_F: mass fraction of solids at the feed stream, x_i: mass fraction of solids, x_n: mass fraction of the product, Q_i: heat flow rate from the heating medium.

$Q_i(1-\delta_p) = \lambda_p V_p(1-\delta_p)$ with $k = p$. In addition, the MEEs recycle the produced vapor to increment the thermal efficiency and maintain the vacuum inside. This action is possible by using a thermo-compression unit, which mixes the vapor from a h-effect with the motive steam to heat the first effect of the sequence (Galván-Ángeles et al., 2015; Sharan and Bandyopadhyay, 2016b). Thus, this device modifies the mass flow rate of vapor upon an i-effect from an h-effect, such case the vapor mass flows are: a) heating vapor for the first effect: $V_S + V_h\theta_h$, with θ is the fraction of vapor to the thermo-compressor, b) heating vapor for an i-effect: $V_h(1-\theta_h)$ with $k = h$, and c) heating vapor for an i-effect and sending vapor to preheaters for two cases of vapor distribution, see Fig. 8.3a):

i) Distributive case: Both derivations from the mainstream yield a remaining stream of vapor: $V_h(1-\theta_h-\delta_h)$, θ_h and δ_h are fractions of the mainstream and $\theta_h+\delta_h<1$, with $k = h = p$ (Díaz-Ovalle, 2024b).

ii) Consecutive case: Derivations where the mainstream sends to preheater and later to thermo-compressor: $V_h(1-\theta_h)\ (1-\delta_h)$, θ_h and δ_h are fractions of the consecutive streams with $k = h = p$.

The thermo-compressor unit mixes the motive steam and the produced vapor and its modeling involves the pressure of the streams such as: P_S, P_M and P_T for the motive steam, mixture and produced vapor streams, respectively. Fig. 8.3b) schematizes these elements around a thermo-compressor unit. Due to the complex thermodynamic prediction, some empirical correlations model the thermo-compressor, for instance (Power, 1994):

$$4.6\ \ln\left(\frac{P_M}{P_T}\right) = \ln\left(\frac{P_S}{P_T}\right)\ln\left(\frac{V_S}{0.4V_h\theta_h}\right) \tag{8.2}$$

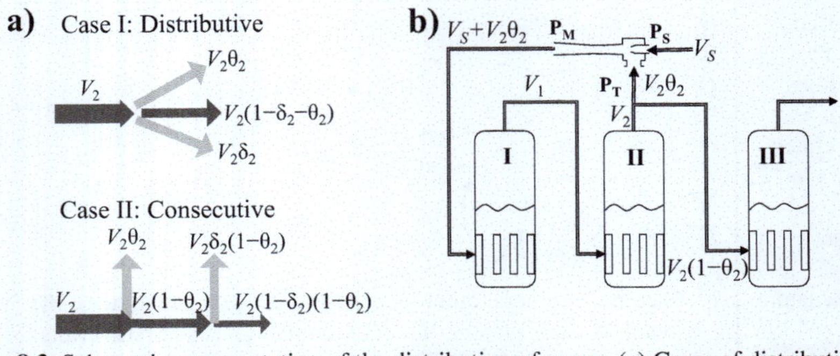

Figure 8.3 Schematic representation of the distribution of vapor: (a) Cases of distribution, (b) vapor mass flows around a thermo-compressor device ($h = 2$).
Adapted from Díaz-Ovalle, C.O. (2024). Design, model, and simulation of evaporators for the food industry. In S.M. Jafari, E. Çapanoglu & A.C. Karaca (Eds.), Evaporation Technology in Food Processing (pp. 269-313): Woodhead Publishing. With permission of Elsevier.

8.5 Exergetical modeling for evaporators

Exergy represents the maximum shaft work done by a system and the environment or the useful part of energy in a thermal process (Dincer and Rosen, 2013). In difference to energy, this thermodynamic property involves the environmental conditions. The food industry is a high-energy consumer (Briam et al., 2015; Ladha-Sabur et al., 2019), particularly, the dairy industry is the biggest energy consumer, and the evaporation process is representative (Moejes and van Boxtel, 2017; Zhang et al., 2018).

An exergetical analysis is fundamental for designing, incrementing energy efficiency, revealing the inefficiencies in the system, and achieving a sustainable process (Balkan et al., 2005). Zisopoulos et al. (2017) have reviewed the application of exergetical analysis in the food industry processes and observed the possibility of reducing energy waste streams and involving the nutritive aspects under thermal efficient processes. The food drying processes receive special attention by applying exergetical analysis (Aghbashlo et al., 2013), where the dairy industry is worth developing exergy and thermo-economic analysis (Singh et al., 2019). In addition, the evaporation process has reduced energy consumption by adapting devices and strategies, such as the multiple-effect evaporation, preheating systems, and fouling mitigation elements (Minton, 1986; Morison and Hartel, 2019). Nowadays, the exergetical analysis in evaporation processes is limited, and only a few works apply it in food processes, such as:

- Balkan et al. (2005) studied a triple-effect evaporator of orange juice under variable properties. They indicated a high thermal efficiency in low-temperature differences inside the effects and re-using the vapor.
- Sogut et al. (2010) determined the increment of exergetical efficiency by using insulation in a tomato paste concentrator. Their study revealed the lowest exergetical efficiency for the first and last effects.
- Bapat et al. (2013, 2016) analyzed a five-effect evaporator in the cane sugar production, where the boiling-point elevation demerits the exergetical efficiency.
- Mojarab Soufiyan et al. (2016) observed the highest exergy destruction in the last effect of a forced circulation evaporator in a tomato paste process.
- Singh et al. (2021) developed an energy-exergy-enviro economical (3E) analysis in a powder milk process and determined a high exergetical efficiency in the thermo-compressor in a three-effects evaporator.
- Díaz-Ovalle et al. (2022a) indicated a reduction of exergy destroyed for: a) incrementing of the number of effects, b) using constant properties, and c) applying thermo-compressor from the last effects.
- Piri et al. (2023) detected the heat losses in a sugar processing plant by comparing a 3E analysis to thermography studies where the final effect presented the largest exergy loss rate.

8.5.1 Fundamentals of exergy in evaporators

Exergetical analysis comes from the mass and energy balances in the system and relates to the environmental temperature, T_{∞}. Balkan et al. (2005) have described the principles for developing exergetical analysis in MEEs. From the Second law of Thermodynamics, the irreversibility of a process relates the generated entropy, S_G, and the exergy destroyed, ExD, by the expression $ExD = T_{\infty}S_G$. Conceptually, the exergy destroyed indicates the condition of the process based on: a) $ExD>0$: irreversible process, b) $ExD = 0$: equilibrium process, and c) $ExD<0$: impossible process. In addition, the concept of efficiency in exergetical analysis, η, relates the exergy of outputs and the exergy in inputs and is considered a "true" or "real" efficiency at a difference to energy efficiency called approximated efficiency (Dincer and Rosen, 2013). Table 8.6 contains all these definitions.

These concepts apply into the energy balances, from Table 8.5, and the thermo-compressor model (expression (8.2)) to yield a generalized exergetical model for MEEs, see Table 8.7. Thus, the total exergy destroyed for a MEE system is $ExD_T = \Sigma ExD_i + ExD_{Termo}$, where a desired evaporation process presents low values of ExD_T.

8.5.2 Case study of exergetical modeling

A four-effect evaporator is a prospect to produce commercially concentrated apple juice. The unit might receive 5100 kg/h of diluted apple juice with 11% of solids at 58°C. This proposal of design will send 7% of the produced vapor of each effect to a pre-heater (do not analyze herein). The third effect feeds the thermo-compressor in consecutive streams (Case II) at the same time as saturated motive steam at 2 kg_f/cm^2 in order to produce a mixed vapor at 90°C. The expected temperatures are (°C): 66, 62, 58 and 54. The components of the apple juice keep in proportion through the process, and the mass fraction for the concentrated apple juice are: carbohydrates, 65.9%, protein, 0.54%, fat, 0.16%, fiber, 0.2%, and water. For this desired process unit, develop an exergetical analysis that includes the effect of percent of sending vapor to pre-heater. Fig. 8.4 depicts this evaporation system. Considers the environmental temperature of 25°C.

Solution:

The exergetical analysis is after the mass and energy balances. Therefore, the initial step is the handling of the thermal properties. The heat capacity of the components depends on the mass fractions through the evaporation unit, but it is unknown. Thus, the heat capacity might be considered for water and components by $Cp = x_iCp_C+(1-x_i)Cp_W$, with x_i the mass fraction of the solids in the i-effect, and Cp_C the heat capacity of the components mixture. The value of Cp_C changes if the mass proportion of the components is modified; for this case, this proportion is constant and corresponds to: carbohydrates, 0.98653, protein, 0.00808, fat, 0.0024, fiber, 0.00299. Thus, the heat capacity for the solids and water comes from the parameters in Table 8.4, such as:

$$Cp_C = 1,554.45 + 1.9548T + 0.005896T^2$$

$$Cp_W = 4,176.2 - 9.0864T - 0.0054731T^2$$

Thus, the specific enthalpy for the apple juice is:

Table 8.6 Definitions of concepts related to exergy.

Description	Expression
Definition of exergy	$Ex(T) = H(T) - H(T_\infty) - T_\infty(S(T) - S(T_\infty)) = \int_{T_\infty}^{T} Cp dT - T_\infty \int_{T_\infty}^{T} \frac{Cp}{T} dT$
Exergy for constant heat capacity	$Ex(T) = Cp\left[T - T_\infty\left(1 + \ln\frac{T}{T_\infty}\right)\right]$
Exergy for a heat flow rate	$Ex(Q) = Q\left(1 - \frac{T_\infty}{T_Q}\right)$
Exergy for phase change	$Ex_\lambda(T) = Ex_V(T) - Ex_{L,Water}(T) = [H_V(T) - H_V(T_\infty) - T_\infty(S_V(T) - S_V(T_\infty))] - [H_L(T) - H_L(T_\infty) - T_\infty(S_L(T) - S_L(T_\infty))]_{Water}$
Exergetical balance	$ExD = \sum Ex_{inputs} - \sum Ex_{outputs}$
Exergy efficiency	$\eta = \frac{\sum Ex_{outputs}}{\sum Ex_{inputs}}$
Note	Ex_L: exergy for the concentrated liquid or product, and $Ex_{L,Water}$: exergy for water.

Table 8.7 Exergetical balances and exergy efficiency for a MEE.

Description	Equation
First effect, $i = 1$	$V_k Ex_\lambda(T_k) + V_1(Ex_L(T_1) - Ex_V(T_1)) - F\left(Ex_L(T_1) - Ex_{L,F}(T_F)\right) = ExD_1$ $\eta_1 = \dfrac{(F - V_1)Ex_L(T_1) + V_1 Ex_V(T_1) + V_k Ex_{L,Water}(T_k)}{V_k Ex_V(T_k) + F \cdot Ex_{L,F}(T_F)}$
Last effect: $i = n$	$V_k Ex_\lambda(T_k) - Ex_L(T_{n-1})\sum_{m=1}^{n} V_m - V_n Ex_V(T_n) - F\left(\dfrac{x_F}{x_n} Ex_L(T_n) - Ex_L(T_{n-1})\right) = ExD_n$ $\eta_n = \dfrac{\dfrac{x_F}{x_n} F \cdot Ex_L(T_n) + V_n Ex_V(T_n) + V_k Ex_{L,Water}(T_k)}{V_k Ex_V(T_k) + \left(F - \sum_{m=1}^{n} V_m\right) Ex_L(T_{n-1})}$
Other effect: $1 < i < n$	$V_k Ex_\lambda(T_k) + (Ex_L(T_i) - Ex_L(T_{i-1}))\sum_{m=1}^{m<i} V_m +$ $V_i(Ex_L(T_i) - Ex_V(T_i)) - F(Ex_L(T_i) - Ex_L(T_{i-1})) = ExD_i$ $\eta_i = \dfrac{\left(F - \sum_{m=1}^{i} V_m\right) Ex_L(T_i) + V_i Ex_V(T_i) + V_k Ex_{L,Water}(T_k)}{V_k Ex_V(T_k) + \left(F - \sum_{m=1}^{m<i} V_m\right) Ex_L(T_{i-1})}$
Thermo-compressor	$ExD_{Thermo} = V_S Ex_V(T_S) + V_h\theta_h Ex_V(T_h) - (V_S + V_h\theta_h)Ex_V(T_M)$ $\eta_{Thermo} = \dfrac{(V_S + V_h\theta_h)Ex_V(T_M)}{V_S Ex_V(T_S) + V_h\theta_h Ex_V(T_h)}$

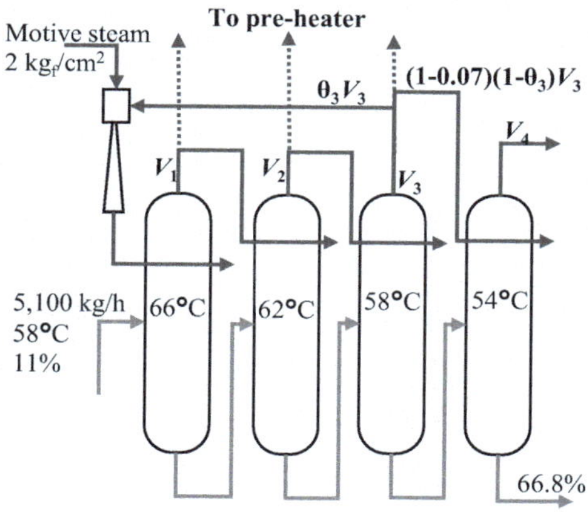

Figure 8.4 Evaporation system for the case study 8.5.2.

$$H_L = (x_i) \int_{T=0}^{T} Cp_C dT + (1 - x_i) \int_{T=0}^{T} Cp_W dT$$

The mass and energy balances must consider the sending of vapor to pre-heaters (7% of the vapor stream), whereby Effects II and III receive the mass flow rates of $0.93V_1$ and $0.93V_2$, respectively. In addition, the unknowledge of the percent of recycled vapor to thermo-compressor, θ_3 ($h = 3$), adds the bilinear term $V_3\theta_3$, which comes from the expression (8.2) with the pressures: $P_S = 2$ kgf/cm^2, $P_T(58°C) = 0.1861$ kgf/cm^2, and $P_M(90°C) = 0.7149$ kgf/cm^2, such as:

$$4.6 \ln\left(\frac{0.7149}{0.1861}\right) = \ln\left(\frac{2}{0.1861}\right)\ln\left(\frac{V_S}{0.4V_3\theta_3}\right)$$

$$V_3\theta_3 = 0.1843V_S$$

The mass and energy balances come from Table 8.5, whereas the enthalpies of vapor are in Table 8.8. The term $0.1843V_S$ substitutes $V_3\theta_3$ in order to reduce the nonlinearities in the equations. For the consecutive case: $V_3(1–0.07)$ $(1–\theta_3)$, and $0.93V_3–0.93V_3\theta_3 = 0.93V_3–0.1714V_S$. The liquid enthalpies are unknown due to

Table 8.8 Vapor enthalpies for case study 8.5.2.

Effect	Temperature (°C)	H_V (kcal/kg)	λ (kcal/kg)
1	66	625.62	559.69
2	62	624	562.06
3	58	622.32	564.38
4	54	620.6	566.66
Motive steam	119.62	645.8	525.93

the dependency of the mass fraction of solids, but for the feed stream and the product (kcal/kg): $H_F = 50.844$, and $H_{L,4} = 30.703$. Thus, the mass and energy balances with the vapor enthalpies:

$$V_1 + V_2 + V_3 + V_4 = 4,260.18\frac{kg}{h}$$

$$i = 1: 1.1843V_S\left(525.93\frac{kcal}{kg}\right) + V_1\left(H_{L,1} - 625.62\frac{kcal}{kg}\right)$$
$$= 5,100\frac{kg}{h}\left(H_{L,1} - 50.844\frac{kcal}{kg}\right)$$

$$i = 2: 0.93V_1\left(559.69\frac{kcal}{kg}\right) + V_1(H_{L,2} - H_{L,1}) + V_2\left(H_{L,2} - 624\frac{kcal}{kg}\right)$$
$$= 5,100\frac{kg}{h}(H_{L,2} - H_{L,1})$$

$$i = 3: 0.93V_2\left(562.06\frac{kcal}{kg}\right) + (V_1 + V_2)(H_{L,3} - H_{L,2}) + V_3\left(H_{L,3} - 622.32\frac{kcal}{kg}\right)$$
$$= 5,100\frac{kg}{h}(H_{L,3} - H_{L,2})$$

$$i = 4: (0.93V_3 - 0.17143V_S)\left(564.38\frac{kcal}{kg}\right) - (V_1 + V_2 + V_3)H_{L,3} - V_4\left(620.6\frac{kcal}{kg}\right)$$
$$= 5,100\frac{kg}{h}\left(\frac{0.11}{0.668}\left(30.703\frac{kcal}{kg}\right) - H_{L,3}\right)$$

The temperature per each effect is defined, but the mass fraction of solids is an unknown variable. This point yields a nonlinear problem that demands mathematical effort in providing the model and computing the derivatives for a Newton–Raphson model in a successive matrix solution. However, another option follows the steps: 1) provide guess values of mass fraction per each effect, 2) compute the enthalpies, 3) solve the mass and energy balances, and 4) compare the mass fraction of solids to decide to return to step 1. Hence, this strategy is adopted herein. The initial values of mass fraction composition correspond to: $x_i = x_{i-1} + (x_n - x_F)/n$. Table 8.9 contains the results of this iterative procedure, and the $\theta_3 = 0.2$, with the liquid streams (kg/h): $L_1 = 3869.75$, $L_2 = 2716.83$, and $L_3 = 1640.91$.

The exergy analysis applies the definitions in Table 8.6 and completes the expressions from Table 8.7. For Effect I (66°C), the destroyed exergy corresponds to:

$$(V_S + V_3\theta_3)Ex_\lambda(90°C) + V_1(Ex_L(66°C) - Ex_V(66°C)) - F(Ex_L(66°C) - Ex_{L,F}(58°C))$$
$$= ExD_1$$

$$\left(1,383.28\frac{kg}{h}\right)Ex_\lambda(90°C) + \left(1,230.25\frac{kg}{h}\right)(Ex_L(66°C) - Ex_V(66°C))$$
$$-\left(5,100\frac{kg}{h}\right)(Ex_L(66°C) - Ex_{L,F}(58°C)) = ExD_1$$

Table 8.9 Results of the mass and energy balances for case study 8.2.

Iteration	**1**	**2**	**3**	**4**
x_1	0.2495	0.1453	0.1450	0.1450
x_2	0.3890	0.2079	0.2065	0.2065
x_3	0.5285	0.3453	0.3419	0.3419
$H_{L,1}$	52.373	56.115	56.126	56.126
$H_{L,2}$	44.633	50.794	50.844	50.844
$H_{L,3}$	37.405	43.287	43.396	43.397
V_1	1238.09	1230.16	1230.25	1230.25
V_2	1163.91	1153.00	1152.92	1152.92
V_3	1073.48	1076.08	1075.93	1075.92
V_4	784.70	800.94	801.09	801.09
V_S	1151.99	1167.95	1168.09	1168.09
Economy	3.6981	3.6476	3.6471	3.6471

The exergy values use the entropies and enthalpies values from Steam Tables, therefore:

a) Phase change at 90°C:

$$Ex_\lambda(90°\mathrm{C}) = Ex_V(90°C) - Ex_{L,Water}(90°C) =$$
$$Ex_\lambda(90°\mathrm{C}) = [H_V(90°C) - H_V(25°C) - (298K)(S_V(90°C) - S_V(25°C))] -$$
$$[H_L(90°C) - H_L(25°C) - (298K)(S_L(90°C) - S_L(25°C))]_{Water} =$$
$$Ex_\lambda(90°\mathrm{C}) = \left[635.1\frac{kcal}{kg} - 608.2\frac{kcal}{kg} - (298K)\left(1.7858\frac{kcal}{kgK} - 2.0436\frac{kcal}{kgK}\right)\right] -$$
$$\left[89.98\frac{kcal}{kg} - 25.02\frac{kcal}{kg} - (298K)\left(0.2848\frac{kcal}{kgK} - 0.0876\frac{kcal}{kgK}\right)\right]_{Water} =$$
$$Ex_\lambda(90°\mathrm{C}) = 103.7244\frac{kcal}{kg} - 6.1944\frac{kcal}{kg} = 97.53\frac{kcal}{kg}$$

b) Vapor at 66°C:

$$Ex_V(66°\mathrm{C}) = H_V(66°C) - H_V(25°C) - (298K)(S_V(66°C) - S_V(25°C))$$
$$Ex_V(66°\mathrm{C}) = 625.62\frac{kcal}{kg} - 608.2\frac{kcal}{kg} - (298K)\left(1.866\frac{kcal}{kgK} - 2.0436\frac{kcal}{kgK}\right) = 70.3448\frac{kcal}{kg}$$

c) Liquid at 66°C and $x_l = 0.145$ (for entropy consider to convert Cp's to K or reduce to $Cp = A_1 + B_1$T):

$$Ex_L(66^\circ C) = \int_{25^\circ C}^{66^\circ C} CpdT - (298K) \int_{298K}^{339K} \frac{Cp_C x_1 + (1 - x_1)Cp_W}{T} dT$$

$$= 33.97 \frac{kcal}{kg} - (298K)\left(0.0436 \frac{kcal}{kgK}\right) = 20.977 \frac{kcal}{kg}$$

d) Feed stream: $Ex_{L,F}(58°C, x_F = 0.11) = 17.5845$ kcal/kg.
The exergy destroyed for Effect I:

$$\left(1,383.28 \frac{kg}{h}\right)\left(97.53 \frac{kcal}{kg}\right) + \left(1,230.25 \frac{kg}{h}\right)\left(20.977 \frac{kcal}{kg} - 70.166 \frac{kcal}{kg}\right) - \left(5,100 \frac{kg}{h}\right)\left(20.977 \frac{kcal}{kg} - 17.5845 \frac{kcal}{kg}\right) = 57,094.78 \frac{kcal}{h} = ExD_1$$

The efficiency of Effect I is:

$$\eta_1 = \frac{(F - V_1)Ex_L(66^\circ C) + V_1 Ex_V(66^\circ C) + (V_S + V_3\theta_3)Ex_L(90^\circ C)}{(V_S + V_3\theta_3)Ex_V(90^\circ C) + F \cdot Ex_{L,F}(58^\circ C)}$$

$$\eta_1 = \frac{\left(3,869.75 \frac{kg}{h}\right)\left(20.977 \frac{kcal}{kg}\right) + \left(1,230.25 \frac{kg}{h}\right)\left(70.166 \frac{kcal}{kg}\right) + \left(1,383.28 \frac{kg}{h}\right)\left(6.1944 \frac{kcal}{kg}\right)}{\left(1,383.28 \frac{kg}{h}\right)\left(103.724 \frac{kcal}{kg}\right) + \left(5,100 \frac{kg}{h}\right)\left(17.5845 \frac{kcal}{kg}\right)}$$

$$\eta_1 = \frac{176,066.06 \frac{kcal}{h}}{233,160.28 \frac{kcal}{h}} = 0.7551$$

For the thermo-compressor:

$$ExD_{Thermo} = V_S Ex_V(120^\circ C) + V_3\theta_3 Ex_V(58^\circ C) - (V_S + V_3\theta_3)Ex_V(90^\circ C)$$

$$ExD_{Thermo} = \left(1168.09 \frac{kg}{h}\right)\left(126.752 \frac{kcal}{kg}\right) + \left(215.185 \frac{kg}{h}\right)\left(57.849 \frac{kcal}{kg}\right) - \left(1383.28 \frac{kg}{h}\right)\left(103.724 \frac{kcal}{kg}\right)$$

$$ExD_{Thermo} = 17,026.646 \frac{kcal}{h}$$

$$\eta_{Thermo} = \frac{(V_S + V_3\theta_3)Ex_V(90^\circ C)}{V_S Ex_V(120^\circ C) + V_3\theta_3 Ex_V(58^\circ C)} = \frac{\left(1,383.28 \frac{kg}{h}\right)\left(103.724 \frac{kcal}{kg}\right)}{\left(1,168.09 \frac{kg}{h}\right)\left(126.752 \frac{kcal}{kg}\right) + \left(215.185 \frac{kg}{h}\right)\left(57.849 \frac{kcal}{kg}\right)}$$

$$\eta_{Thermo} = \frac{143,479.335 \frac{kcal}{h}}{160,505.98 \frac{kcal}{h}} = 0.8939$$

This procedure applies to the further Effects. Table 8.10 summarizes the values of exergies and efficiency for the case study.

Table 8.10 Values of the exergetical analysis of case study 8.5.2.

Effect	Ex_{Li}	Ex_{Vi}	$Ex_{L,Water,i}$	$Ex_{\lambda i}$	ExD_i	η_i
I	20.971	70.166	2.575	67.591	57,094.78	0.755
II	17.235	64.094	2.122	61.972	33,458.68	0.901
III	12.115	57.849	1.702	56.147	25,643.22	0.870
IV	3.772	51.456	1.321	50.135	10,723.30	0.810

Ex_{Li}, Ex_{Vi}, $Ex_{L,Water,i}$, $Ex_{\lambda i}$, ExD_i [=]kcal/h.

Table 8.11 Efficiency values for the exergy analysis for case study 8.5.2.

	Values of δ_i,				
Effect	**0.05**	**0.07**	**0.1**	**0.15**	**0.2**
I	0.761	0.755	0.747	0.733	0.720
II	0.902	0.901	0.900	0.899	0.897
III	0.871	0.870	0.868	0.865	0.860
IV	0.815	0.810	0.801	0.784	0.760

The problem indicates an exergetical analysis of the sending of vapor to the preheater. Thus, all effects might consider an equal value of δ_i, which takes the arbitrary values of 0.05, 0.1, 0.15, and 0.2 in order to complete the study. Table 8.11 contains the results for the efficiency per each effect. The results indicate a reduction of the efficiency that suggests preserving the produced vapor in the evaporation system instead of losses to surroundings. However, this analysis did not include the exergy analysis for the pre-heater, in such a case, the analysis might provide a global conclusion.

8.6 Optimization and design in evaporators

Process optimization is a complex and extended topic that applies to the design and synthesis of processes. The reader might consult the theory and fundamentals in mandatory texts, such as: McCormick (1982), Floudas (1995), Edgar et al. (2001), Avriel (2003), Floudas and Pardalos (2009), Diwekar (2008), Biegler (2010) and Bazaraa et al. (2013). Herein, the mention of optimization motives for modeling evaporation processes, where the minimum value of capital cost and energy are the scope of the problems. Díaz-Ovalle and Jafari (2023) have indicated some references that developed models, for instance, Nishitani and Kunugita (1979) (optimal evaporation sequence), Smith (2008) (optimal number of effects), Sharan and Bandyopadhyay (2016a) (energy integration of MEEs), etc.

8.6.1 Fundamentals of optimization: brief description

Optimization is a mathematical technique to solve decision-making problems based on determining either the minimum or maximum value of a function. This function (objective function) represents the desired element to optimize, mainly cost, which involves energy consumption, total profit, capabilities, risk, pollution, equipment size, etc. The formulation of this function is decisive to reach acceptable results. Moreover, the diversity of cases demonstrates the high applicability level of the optimization, and the technique of solution is particular for the class of problem. The nature of the model and the variables classify the problem, Fig. 8.5 presents a generalized classification that corresponds to:

a) mathematical structure: unconstrained model (only the objective function is the equation) or constrained model (additional equations to the objective function).
b) mathematical operation: linear model or nonlinear model.
c) type of variables: continuous and discrete (integer) variables.
d) number of variables: univariable or multivariable model.
e) nature of variable: deterministic or stochastic model.

Linear programming (LP) applies to models without complex mathematical operators and formulates additions and subtractions. Nevertheless, this apparent simplicity involves a high quantity of variables and equations that increment the complexity of the solving. Dantzig (1963) proposed a practical solution of LP under the Simplex method, which has evolved to sophisticated algorithms, for instance, the included algorithms into the CPLEX optimizer software. On the other hand, nonlinear programming (NLP) applies to complex mathematical operators. For rigorous solving procedures, the solution determines the points with derivates equal to zero and demands numerical methods. Nowadays, the NLP employs advanced numerical methods in several optimizers, such as NLPQL, MINOS, and BARON. Even, metaheuristics methods apply in solving NLP problems by searching strategies, such as simulated annealing, genetic and evolutionary algorithms, etc. In addition, mixed

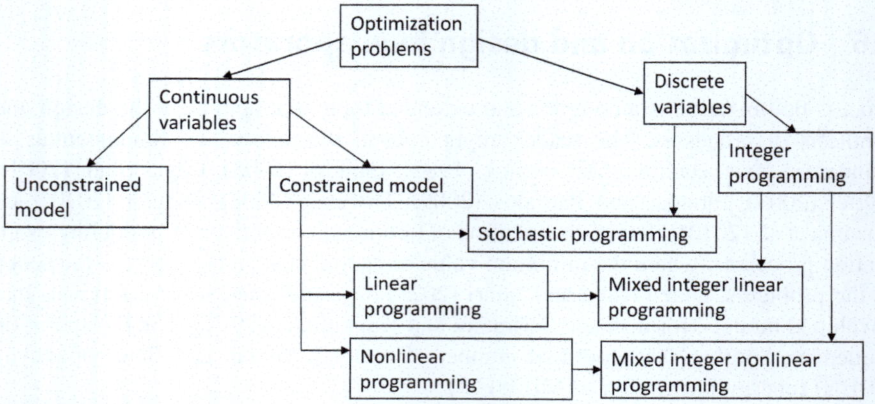

Figure 8.5 Generalized classification of optimization problems.

integer linear programming (MILP) and mixed integer nonlinear programming (MINLP) are fundamental for designing processes based on the optimal selection for an element from a group of candidates. The discrete variables select the elements by binary statements (1 select and 0 otherwise).

The disjunctive formulation is very common to select one element from a group and contains particular constraints and values. This formulation requires an arrangement to present a conventional formulation via the big-M method or convex-hull method (Floudas, 1995; Sawaya and Grossmann, 2005; Vecchietti and Grossmann, 2000). Fig. 8.6 illustrates with an example the convex-hull method for a linear problem, where the variable x is disaggregated into $x = \Sigma_c x_c$, x_c is a variable per each disjunction element, $c \epsilon C$, Y_c is a Boolean variable to indicate if the c-element is selected, B_c is a binary variable for c-element, and $B_c = 1$ indicates the selection. The binary variables guarantee that only one c-element is selected by $\Sigma_c B_c = 1$ and the bound $B_c Lx \leq x_k \leq B_c Ux$, Lx and Ux are the lower and upper bounds of x, respectively. The linear functions, $f_c(x)$ and $g_c(x)$, modify to the product of the constants u_c and v_c with B_c. The nonlinear problems require a modified procedure (Lee and Grossmann, 2000, 2003).

8.6.2 Elemental model for optimizing evaporators

Process engineering develops mathematical formulations for optimization based on mass and energy balances, design equations, and cost (Biegler et al., 1997; Smith, 2008). The optimization of MEEs has determined the optimal flow patterns, optimal heat transfer area, and optimal temperatures to minimize the utilities under diverse approaches such as energy integration (pinch analysis and grand composite curve) and mass and energy balances. Díaz-Ovalle and Jafari (2023) have summarized the remarkable works on optimization of MEEs, for instance, the design of optimal flow patterns (Nishitani and Kunugita, 1979), and the superstructure models to reduce energy consumption (Ahmetović et al., 2018; Jiang et al., 2018).

An elemental model considers the mass and energy balances from Table 8.5 and the design equations from the convective heat transfer coefficients from Tables 8.1 to 8.3. Furthermore, the foodstuff properties might come from the Choi-Okos model, but this consideration might increase the nonlinearities into the model. The

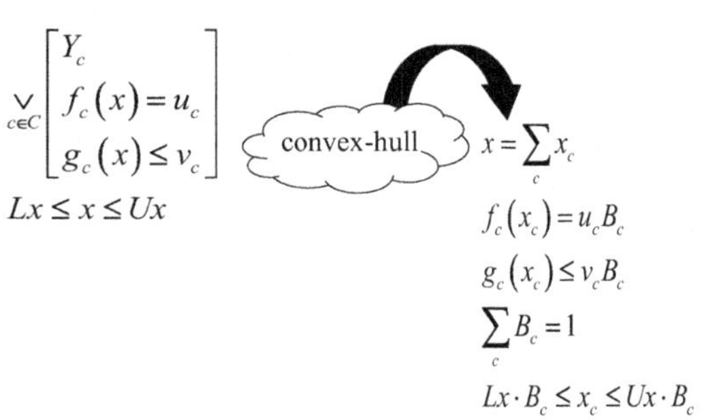

Figure 8.6 Application of the convex-hull method in a disjunctive problem.

geometrical features are fundamental to computing the heat transfer area and the final cost of design. Díaz-Ovalle et al. (2013) developed a simplified methodology to optimize MEEs which is taken into account herein. Table 8.12 concentrates on the model constraints, where ϕ_{Ti} is the selected diameter from a set of candidates, $\Phi_{i,d}$, in a disjunction such as $d \epsilon D$, where D and d are a set and index of diameters, respectively. The objective function is the capital cost and includes the tube length cost, $C_{L,d}$ (\$/m·tube), and the shell cost, $C_{D,d}$ (\$/tube). Both costs depend on the number of tubes per effect and the selected diameter. After the convex-hull procedure, the objective function is:

$$\min C_T = \sum_{i \in I} NT_i \left(\sum_{d \in D} C_{D,d} Y_{i,d} + LT_i \sum_{d \in D} C_{L,d} Y_{i,d} \right) \tag{8.3}$$

where C_T must be minimized.

Solving programming problems depends on specialized computational packages. The commercial computational packages are GAMS, LINGO, etc., and particular solvers apply per each mathematical model. For instance, the GAMS package uses the DICOPT solver for MINLP problems (this solver applies the outer-approximation strategy developing two sub-problems: NLP and MIP), CONOPT solver for NLP problems and CLPEX for MIP problems. Also, the reader might apply the SOLVER Tool from MS Excel, which demands an effort to allocate the equation and variables into the cell system. The upcoming case study describes the optimization of a MEE.

Table 8.12 Constraints to formulate an optimization model for optimizing MEEs.

Description	**Equation**
Expression for the heat transfer area.	$Q_i = U_i A_i^{Ave}(Ts_i - T_i), \forall i \in I$
Overall heat transfer coefficient by involving the inner and outer area.	$U_i = \left(\frac{1}{h_i^I} + \frac{A_i^I}{A_i^{II} h_i^{II}} \right)^{-1}, \forall i \in I$
Inner heat transfer area of the tube.	$A_i^I = \pi \phi_{Ti} LT_i NT_i, \forall i \in I$
Outer heat transfer area of the tube.	$A_i^{II} = \pi(\phi_{Ti} + 2\Delta r) LT_i NT_i, \forall i \in I$
Disaggregated expression for selecting a tube diameter from the set of diameters.	$\phi_{Ti} = \sum_{d \in D} \Phi_{i,d} Y_{i,d}, \forall i \in I$ $0 \le \phi_{Ti} \le \max_{d \in D}(\Phi_{i,d}), \forall i \in I$
Treatment of the binary variables in the model.	$\sum_{d \in D} Y_{i,d} = 1, \forall i \in I$ $Y_{i,d} = \{0, 1\}, \forall i \in I, d \in D$
Bounding expressions.	$0 < V_i < F, \forall i \in I$ $x_F < x_i < x_n, \forall i \in I$ $0 < h_i^I, h_i^{II}, A_i^{LM}, Q_i, \forall i \in I$

8.6.3 Case study of optimization of evaporators

Pearce (1992) has presented a MEE to process whey from 6.5% of solids to 60% of solids. Díaz-Ovalle (2024b) adapted the process for a six-effect evaporator under the sequence I-II-III-VI-V-IV at the temperatures of (°C): 68, 65, 61, 45, 50, and 57. The process operates 22,430 kg/h of whey at 80°C and reclaims steam from a thermo-compressor at 72.55°C. Fig. 8.7 conceptualizes the evaporation process with 20% of recycling vapor. Due to the unknowledge of the composition per effect, the whey properties come from Table 8.4 by considering an average concentrated whey based on the mass composition: water (66%), carbohydrate (23.8%), and protein (10.2%), whereby the whey properties depend on the temperature in exception for the viscosity, which is a constant value of 59 mPa s (Morison and Mackay, 2001). Table 8.13 contains the values of the whey and steam properties per temperature in the system.

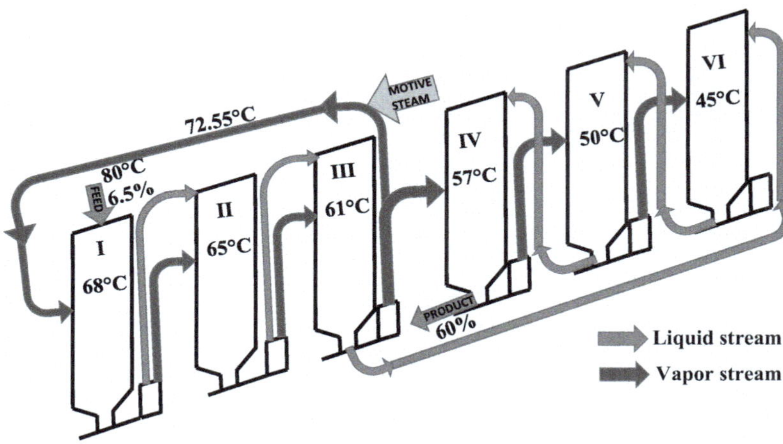

Figure 8.7 Multiple-effect evaporator for case study 8.6.3.

Table 8.13 Whey and vapor properties for case study 8.6.3.

Effect	Temperature (°C)	Whey			Vapor		
		ρ_L	Cp_L	κ_L	ρ_V	κ_V	μ_V
I	68	1100.27	3.0354	0.5498	0.6487	0.021686	11.29
II	65	1100.57	3.0504	0.5473	0.6535	0.021461	11.18
III	61	1100.97	3.0705	0.5438	0.6599	0.021161	11.04
IV	57	1101.37	3.0907	0.5402	0.6663	0.020861	10.89
V	50	1102.06	3.1264	0.5333	0.6775	0.020336	10.64
VI	45	1102.56	3.1521	0.5280	0.6856	0.019961	10.45

Cp_L[=]kJ/(kg°C), ρ_L, ρ_V [=]kg/m^3, κ_L, κ_V[=]W/(m°C), μ_V[=]μPa·s.

This system is a falling-film evaporator; hence, the overall heat transfer coefficient comes from the expressions: a) condensing vapor outside a tube, Table 8.1b) a non-nucleate boiling liquid in a falling-film evaporator by Prost et al. (2006), see Table 8.3, recommended by Díaz-Ovalle (2024b). Thus, the design considers the tube diameter candidates (in): 1, 1½, 2. At the same sort, the tube length cost is per candidate ($/(m**NT*)): 160, 148, 157; the shell cost per candidate ($/m): 940, 920, 830. Determine the optimum number of tubes, the tube length, the evaporation economy, and the minimum cost.

Solution:

The mathematical model involves the mass and energy equations of a six-effect evaporator without thermo-compression, which were developed previously by Díaz-Ovalle (2024b) and modified herein based on the properties and the expression $V_{\mathrm{Mix}} = V_S + 0.2V_3$, such as:

$$\begin{cases} V_1 + V_2 + V_3 + V_4 + V_5 + V_6 = 20,000\dfrac{kg}{h} \\ V_{\mathrm{Mix}}\left(555.75\dfrac{kcal}{kg}\right) - V_1\left(577.1\dfrac{kcal}{kg}\right) = -170,013\dfrac{kcal}{h} \\ V_1\left(560.42\dfrac{kcal}{kg}\right) - V_2\left(577.77\dfrac{kcal}{kg}\right) = -43,632.25\dfrac{kcal}{h} \\ V_2\left(557.62\dfrac{kcal}{kg}\right) - V_1\left(2.63\dfrac{kcal}{kg}\right) - V_3\left(578.8\dfrac{kcal}{kg}\right) = -58,894.84\dfrac{kcal}{h} \\ V_3\left(454.84\dfrac{kcal}{kg}\right) + (V_1 + V_2 + V_5 + V_6)\left(4.75\dfrac{kcal}{kg}\right) - V_4\left(579.79\dfrac{kcal}{kg}\right) = 106,515.13\dfrac{kcal}{h} \\ V_4\left(568.4\dfrac{kcal}{kg}\right) + (V_1 + V_2 + V_3 + V_6)\left(3.46\dfrac{kcal}{kg}\right) - V_5\left(581.6\dfrac{kcal}{kg}\right) = 77,675.95\dfrac{kcal}{h} \\ V_5\left(564.94\dfrac{kcal}{kg}\right) - (V_1 + V_2 + V_3 + V_4)\left(4.18\dfrac{kcal}{kg}\right) - V_5\left(577.2\dfrac{kcal}{kg}\right) = 93,757.4\dfrac{kcal}{h} \end{cases}$$

The selection of the candidate tubes per effect is:

$$\phi_{\mathrm{T}i} = (1\ \mathrm{in})Y_{i,1\mathrm{in}} + \left(1\,{}^{1}\!/_{2}\ \mathrm{in}\right)Y_{i,1\,{}^{1}\!/_{2}\,\mathrm{in}} + (2\ \mathrm{in})Y_{i,2\mathrm{in}}, \forall i \in I$$

$$Y_{i,1\mathrm{in}} + Y_{i,1\,{}^{1}\!/_{2}\,\mathrm{in}} + Y_{i,2\mathrm{in}} = 1, \forall i \in I$$

$$0 \le \phi_{\mathrm{T}i} \le 2\ \mathrm{in}, \forall i \in I$$

The FEVAT software is a suitable tool to solve this problem for any number of effects (Galván-Ángeles et al., 2015). FEVAT creates the model inside the GAMS interface and remotely executes the solution by linking the code with the solvers. For this case study, the solvers were DICOPT for the MINLP problem, CONOPT for the NLP problem, and CPLEX for the LP problem. The model statistics indicated 80 equations,

98 single variables, and 24 discrete variables. The author appreciates using of the license from the Technological Institute of Celaya, México. Finally, the solution indicated a total cost of $2,935,656.56, the motive steam of 2974.48 kg/h, and an evaporation economy of 6.724. Table 8.14 summarizes the results of this case study and Table 8.15 presents an extract of the GAMS code.

Table 8.14 Results for case study 8.6.3.

Effect	T (°C)	ϕ_{Ti}	*NTi*	*LT,i*	x_i	V_i	U_i
I	68	2	163	13.14	0.0786	3883.3	1255.36
II	65	2	272	13.17	0.0991	3816.2	1442.67
III	61	2	174	13.20	0.1335	3760.7	1331.72
IV	57	2	168	13.24	0.6000	2875.9	1368.04
V	50	1½	105	15.83	0.2738	2762.6	1317.22
VI	45	2	123	13.30	0.1802	2901.3	1377.20

ϕ_{Ti}[=]in, LT_i[=]m, V_i[=]kg/h, U_i[=]W/(m^2°C).

Table 8.15 Extract of the GAMS code for case study 8.6.3.

```
integer variable
NN(s);
binary variables Y(s,i);
equation obj Objective function expression (12.3);
obj.. cost =e= sum(s, NN(s)*(CostP_(s)+L(s)*Long(s)))+S_S*1000;
equation A1Y(s) sum of binary variables;
A1Y(s).. sum(i,y(s,i)) =e= 1;
equation A2D(s) selection of tube diameter;
A2D(s).. sum(i,y(s,i)*DD(i)) =e= D(s);
equation A3D(s) cost of tube;
A3D(s).. sum(i,y(s,i)*CostP(i)) =e= CostP_(s);
equation A4(s) length of tube;
A4(s).. sum(i,y(s,i)*Long_(i)) =e= Long(s);
equation Q_A(s) Heat transfer;
Q_A(S)$(Prev(s) = 0)..  QQ1(S) =e= S_S*HH*(1-Deriv(s));
equation Mat mass balance;
Mat.. F*(1-xf/xs) =e= sum(s, V(s));
equation Mat1(s) calculation of liquid streams;
Mat1(s).. Liq(s) =e= F- sum(ss$(ord(ss) <= ord(s)), V(ss));
equation Mat2(s) calculation of mass fraction of solids;
Mat2(s).. x(s) =e= (F*xf/Liq(s))$(ord(s) = 1) + (Liq(s-1)*x(s-1)/Liq(s))$(ord(s) <> 1);
equation BofH1(s) energy balance for effect I;
BofH1(s)$(ord(s) = 1).. QQ1(s) +  V(s)*(CP(s)*T(s)-Hv(s)) =e= (T(s)*CP(s)-TempF*CP_F)*F;
equation U1(s) overall heat transfer coefficient;
U1(s).. U(s) =e= 1/(1/ho(s)+(pi*(D(s)+2*DX)*L(s)*NN(s))/((pi*D(s)*L(s)*NN(s))*hi(s)));
```

8.7 Dynamic models with fouling formation

The food process demands the prediction of operative conditions before making decisions and solving problems. The dynamic analysis of unit operations is a valuable tool that requires a level of detail in proportion to the level of precision. The reader might consult fundamental and advanced information from Himmelblau and Bischoff (1976), Luyben (1996), Ingham et al. (2000), Hangos and Cameron (2001), and Stephanopoulos (2017).

Evaporators seem to be simple equipment with one feed flow and two outlets, evaporation phenomenon, and heating medium but the existent phenomena limit the use of a high level of detail. This point restricts MEEs that employ the successive formulation from only one effect. Nowadays, the dynamic analysis of evaporators improves performance and guarantees a controllable process. Unfortunately, the food processes suffer from fouling formation that demerits thermal efficiency, operative cost, and product quality. This section presents the typical dynamic models for a MEE and adds the fouling formation to provide a better perception during operation.

8.7.1 Development of dynamic models in evaporators

The literature contains a diversity of strategies for dynamic modeling for evaporators. Díaz-Ovalle and Jafari (2023) have provided some remarkable approaches on this topic, and some of them are considered rigorous models for the properties and the overall heat transfer coefficients. In Díaz-Ovalle (2024a), the basics of developing dynamic models for evaporators comply with the elemental points:

a) An evaporator is a hermetic unit with an accumulative liquid mass under heating and boiling.
b) The geometrical aspect of an evaporator is similar to a stirred tank, where the liquid is a constant mass and covers the heating medium.
c) The phase equilibrium predicts the vapor production for a short approach.
d) A MEE process systematically uses the basic model.
e) The basic model requires a Runge-Kutta method for solving.

Evaporators are continuous processes that follow the mass and energy balances with an accumulative term. Fig. 8.8 schematizes an evaporation unit with one inlet, F, and two outlets, V and L, similar to a flash separator, $F = L + V$, and heat transfer. The accumulative term is the change in time of the mass and energy quantities into the process unit that yields a time derivative. For this case, the accumulative mass might be either constant or variable with time in response to geometrical changes. The constant accumulative mass is a valid assumption for reducing the complexity.

For a MEE, Table 8.15 contains the dynamic model with constant mass, which Andre and Ritter (1968), Newell and Fisher (1972), Tonelli et al. (1990), Runyon et al. (1991) and Miranda and Simpson (2005) declared previously. On the other hand, the vapor stream is the response of the heat transfer in the liquid and the phase equilibrium during the boiling phenomenon. Tonelli et al. (1990) declared the correspondence to heat transfer, and Díaz-Ovalle et al. (2017) applied the phase

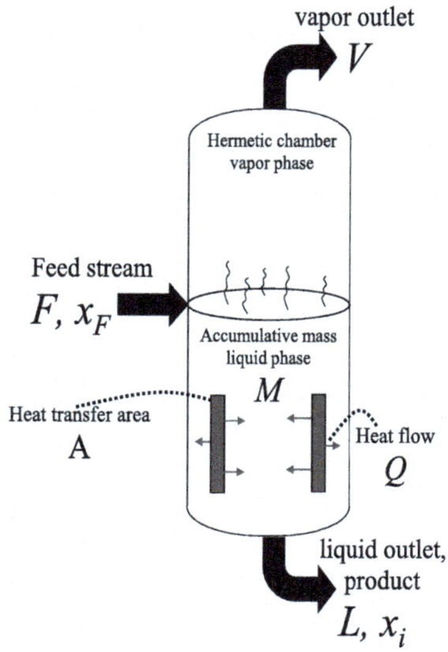

Variable accumulative mass:

$$F - V - L = \frac{dM}{dt}$$

$$Fx_F - Lx_i = \frac{d(Mx_i)}{dt}$$

$$FH_{L,F} + Q - LH_{L,i} - VH_{V,i} = \frac{d(MH_i)}{dt} = \frac{d(MCpT_i)}{dt}$$

Constant accumulative mass:

$$F - V - L = 0$$

$$\frac{Fx_F - Lx_i}{M} = \frac{dx_i}{dt}$$

$$\frac{FH_{L,F} + Q - LH_{L,i} - VH_{V,i}}{M} = \frac{d(CpT_i)}{dt}$$

Figure 8.8 Representation of an evaporation unit with the dynamic models for constant and variable accumulative mass.

Table 8.16 Dynamic model for MEE.

Description	Equation
Mass balance of solids	$\frac{dx_i}{dt} = \frac{F_i x_{i-1} - (F_i - V_i)x_i}{M_i}$
Energy balance	$\frac{dT_i}{dt} = \frac{F_i H_{L,i-1} + Q_i - (F_i - V_i)H_{L,i} - V_i H_{V,i}}{M_i Cp(T_i)}, H_{L,i} = \int_0^{T_i} Cp(T)dT$
Considerations for the first effect.	$\begin{cases} i = 1, F_{i-1} = F, T_{i-1} = T_F, x_{i-1} = x_F, Q_1 = U_1 A_1 (Ts - T_1) \\ i \neq 1, F_i = F_{i-1} - V_{i-1}, Q_i = U_i A_i (T_{i-1} - T_i) \end{cases}$
Vapor stream in correspondence to heat transfer	$V_i = \frac{Q_i}{\lambda_i} = \frac{U_i A_i (T_{i-1} - T_i)}{\lambda_i}$
Vapor stream in correspondence to phase equilibrium,	$V_i = \frac{P*(T_i)}{P_{atm}} F_i$ $P*(T_i) = 218.1428 \, \mathrm{Exp}\left(6.537 - \frac{7,173.79}{421.4747 + 1.8T_i}\right), \; P*[=]atm, \; T_i[=]°C$

equilibrium from the change of density, and late reduced by Díaz-Ovalle (2024b). Both proposals are in Table 8.16.

The motive steam, V_S, is the heating medium for the first effect and generates disturbances that influence the evaporation performance. The inclusion of the motive steam flow rate into the evaporation is through the overall heat transfer coefficient, U. Some convective correlations include the heat flux, q, that relates the heat flow rate, Q, and the heat transfer area, A, such as: $q = Q/A = \lambda_S V_S/A$. Thus, the diversity of convective correlations in Tables 8.2 and 8.3 offer suitable mathematical approaches. For instance.

- Plate evaporators: the experimental data by Hoffman (2004) are correlated with the logarithm function:

$$U_{0,i} = 2,441.25\ \ln\left(\frac{\lambda_{i-1}V_{i-1}}{A_i}\right) - 17,340.6107, \text{for } i = 1, \lambda_{i-1} = \lambda_S, V_{i-1} = V_S$$
$$R^2 = 0.9741 \tag{8.4}$$

where the specific condensation enthalpy might follow: $\lambda = 605.4e^{(-0.0012437)}$, $\lambda[=]$kcal/kg and $T[=]$°C (Díaz-Ovalle, 2024b).

- Vertical tubes: the correlations reported by Adib and Vasseur (2008) in Table 8.2 contain the heat flux, q, for computing the boiling convective coefficient, h_L. For instance, McNelly's correlation presented vapor properties with dependence on the motive steam and liquid temperatures in the product: $\lambda(T_i)^{-0.69}P(T_i)^{0.31}\sigma_L^{-0.31}\kappa_L(T_i)^{0.31}$, for water liquid properties this expression correlates to: 9.101414exp($8.26 \times 10^{-4}T_i$), $R^2 = 0.999$, and for $U_{0,i} \approx h_{L,i}$:

$$U_{0,i} = 0.225\left(\frac{\lambda_{i-1}V_{i-1}}{A_i}\right)^{0.69} Cp_L^{\ 0.69}\left(9.101414e^{0.000826T_i}\right), \text{for } i = 1, \lambda_{i-1} = \lambda_S, V_{i-1} = V_S \tag{8.5}$$

On the other hand, the literature also presents the modeling of falling-film evaporators, which suffer complex phenomena, such as bubble formation, break-up of the film, and non-Newtonian flow. Díaz-Ovalle and Jafari (2023) have summarized the outstanding approaches to model the falling-film evaporation, which reclaimed elaborated solving techniques. Table 8.17 contains the existent models for falling-film evaporators. The frustum shape approach has been reduced and applied to the milk process concentration (Díaz-Ovalle, 2024b), and extended in the liquid stream recycling into the effects (Díaz-Ovalle et al., 2022b).

8.7.2 Fouling formation modeling

Fouling is an undesirable phenomenon for the food industry that affects the quality of the foodstuffs and increases energy consumption by reducing the heat transfer capacity (Fryer, 1989; Jeurnink and Brinkman, 1994). The compounds, temperature, and pH of the foodstuffs, and geometrical aspects of the medium remarkably influence the fouling formation. The fouling phenomenon modifies the value of the overall heat transfer coefficient by adding a thermal resistance. The modeling of this stage is (Fryer, 1989):

Table 8.17 Some models of falling-film MEEs by involving mass and energy balances.

Description	Model
Winchester model 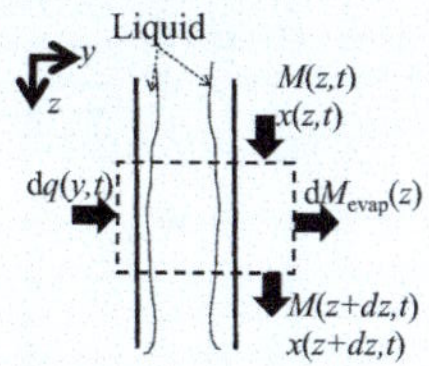	Mass balance: $\frac{1}{v}\frac{\partial M(z,t)}{\partial t}+\frac{\partial M(z,t)}{\partial z}=\frac{\partial M_{Evap}(z,t)}{\partial z}$ Mass fraction of solids balance: $\frac{1}{v}\frac{\partial[M(z,t)x(z,t)]}{\partial t}+\frac{\partial[M(z,t)x(z,t)]}{\partial z}=0$ Energy balance: $\frac{1}{v}\frac{\partial[M(z,t)Cp(x(z,t))T_{film}(z,t)]}{\partial t}+\frac{\partial[M(z,t)Cp(x(z,t))T_{film}(z,t)]}{\partial z}=U(x(z,t))\pi\phi_T\cdot N_T\cdot\left(T_{wall}(z,t)-T_{film}(z,t)\right)$ $-\left(\lambda+Cp_{Water}T_{film}(z,t)\right)\frac{\partial M_{evap}(z,t)}{\partial z}$
Stefanov-Hoo model 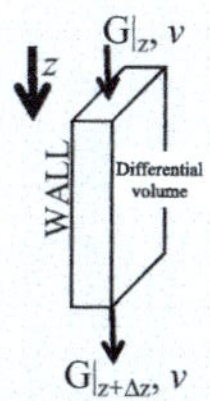	Dimensionless expression of mass and energy conservation: Mass of liquid balance: $\frac{\partial G}{\partial t}+v\frac{\partial G}{\partial z}=-W\cdot v\cdot\eta$ Mass of condensing vapor balance: $\frac{\partial G_C}{\partial t}+v\frac{\partial G_C}{\partial z}=W_C\cdot v\cdot\eta_C$ Mass fraction of solids balance: $\frac{\partial X}{\partial t}+v\frac{\partial X}{\partial z}=W\cdot v\cdot\eta\cdot\frac{X}{G}$ Energy balance: $\frac{\partial T}{\partial t}+v\frac{\partial T}{\partial z}=\eta_T\left(T_{wall}-T_{film}\right)$
Díaz-Ovalle Model 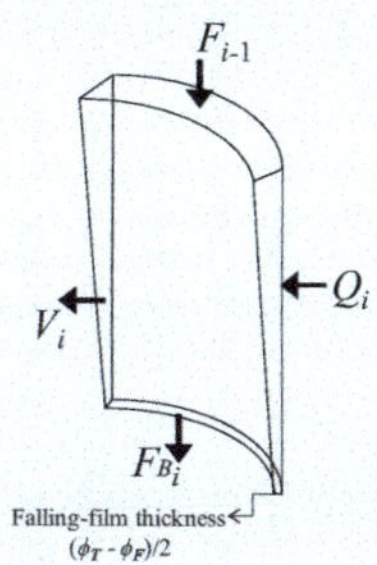	Reduced form of the model for a frustum Mass of falling-film: $M_F=\frac{\pi LT\rho_L}{4}\left(\phi_T^2-\frac{\phi_F^2+\phi_U^2+\phi_U\phi_F}{3}\right)$ Mass fraction of solids balance: $\frac{dx_i}{dt}=\frac{F_{i-1}(x_{i-1}-x_i)+V_ix_i}{M_{Fi}}$ Energy balance: $\frac{dT_i}{dt}=\frac{F_{i-1}Cp(T_{i-1}-T_i)+Q_i-V_i\lambda_i}{M_{Fi}Cp}$ Falling-film thickness expression (from mass balance): $\frac{d\phi_{F,i}}{dt}=\frac{F_{i-1}-V_i-FB_i}{-\frac{\pi LT_i\rho_L}{12}(2\phi_{F,i}+\phi_{U,i})}$ where $FB_i=\frac{\pi g\rho_L^2\phi_{T,i}^2(\phi_{T,i}^2-\phi_{F,i}^2)}{16\mu_L(1-\psi_i^2)}\left(\frac{1}{8}-\frac{\psi_i^2}{2}+\frac{3\psi_i^4}{8}+\frac{\psi_i^4}{2}\ln\frac{1}{\psi_i}\right)\quad\psi_i=\frac{\phi_{F,i}}{\phi_{T,i}}$

Adapted from Díaz-Ovalle, C. O., & Jafari, S. M. (2023). Multiple-Effect Evaporators in the Food Industry: Fundamentals, Design, Simulation, Control, and Applications. *Food Engineering Reviews*, 15(4), 691-717. https://doi.org/10.1007/s12393-023-09350-6. With permission of Springer.

$$\frac{1}{U} = \frac{1}{U_0} + \Psi \tag{8.6}$$

where U_0 is the overall heat transfer coefficient in the absence of fouling and Ψ is the fouling resistance. An initial intent to compute the fouling resistance through time is the balance of deposition and removal material on the surface:

$$\frac{d\Psi}{dt} = M_D - M_R \tag{8.7}$$

with M_D and M_R are the mass fluxes of deposited and removed material from the surface, respectively.

The heat transfer resistance has been treated as a kinetic model in order to represent the fouling formation by including the Biot number, Bi (Kern and Seaton, 1959), such as:

$$\frac{d\mathrm{Bi}}{dt} = \frac{k_d}{\mathrm{Re}} \, exp\left[\frac{-Ea_F}{\mathrm{R}} \cdot \frac{1 + \mathrm{Bi}}{T_W + \mathrm{Bi}T}\right] - k_r\mathrm{Bi} \tag{8.8}$$

where R is the universal gas constant, 8.314 J/molK, Ea_F is the activation energy for the fouling reaction, 150 kJ/mol, k_d, and k_r are parameters of the fouling reaction, $4.85 \times 10^{13}\ \mathrm{s}^{-1}$ and $1.3\mathrm{x}10^{-3}\ \mathrm{s}^{-1}$ respectively (Fryer, 1989), Re is the Reynolds number of the fluid, T_W and T are the wall and fluid temperature, respectively. In principle, the Biot number relates the internal and external thermal resistances such as: ΨU_0 or $\delta^F U_0/k^F$, with δ^F and k^F are the fouling thickness and the thermal conductivity of fouling ($k^F = 0.27$ W/m·K), respectively. Thus, an algebraic substitution of the Biot number into the expression (8.8) yields in terms of the fouling resistance:

$$\frac{d\Psi}{dt} = \frac{k_d}{\mathrm{Re}} \, exp\left[\frac{-Ea_F}{\mathrm{R}} \cdot \frac{1 + U_0\Psi}{T_W + U_0\Psi T}\right] - k_r\Psi - \frac{\Psi}{U_0}\left(\frac{dU_0}{dT}\right)\frac{dT}{dt} \tag{8.9}$$

and in terms of the fouling thickness (Molina-Pérez et al., 2015):

$$\frac{d\delta^F}{dt} = \frac{k_d k^F}{U_0\mathrm{Re}} \, exp\left[\frac{-Ea_F}{\mathrm{R}} \cdot \frac{k^F + \delta^F U_0}{k^F T_W + \delta^F T U_0}\right] - k_r\delta^F - \frac{\delta^F}{U_0}\left(\frac{dU_0}{dT}\right)\frac{dT}{dt} \tag{8.10}$$

The derivative of U_0 with respect to T is approached by finite difference to $[U_0(T+\Delta T) - U_0(T-\Delta T)]/(2\Delta T)$ with ΔT as a low value. This term is negligible for the U_0 constant.

The dairy industry suffers from the fouling phenomenon due to the denaturalization of the protein β-lactoglobulin (Fryer and Slatter, 1985; Lalande et al., 1984). The literature has indicated the benefit of modeling this phenomenon in dairy thermal processes to optimize the design of the equipment (Alhuthali et al., 2022; Georgiadis and Macchietto, 2000; Georgiadis et al., 1998a, 1998b; Loveday, 2016; Mahdi et al., 2009; Sharma and Macchietto, 2021). Jun and Puri (2005b) classified the existing fouling models in the literature, such as: a) one-phase model: based on temperature changes one example is the Delplace and Leuliet (1995) model, b) two-phase model: dynamic expression of the fouling rate that represents deposited solids and heat transfer resistance, such as the Schreier and Fryer (1995) model proposed for industrial processes, and c) four-phase model: protein denaturalization process where the native (N) β-lg becomes in the states unfolded (U), aggregated (A) and deposited (D) in surface and bulk reactions, where the mathematical model proposed by de Jong et al. (1993) describes this principle. Fig. 8.9 depicts the four-phase fouling case.

The four-phase model describes the possible source of the fouling formation on a hot surface during milk evaporation. Thus, the literature has determined the whole kinetic expressions, which correspond to (de Jong et al., 1993):

$$\begin{aligned} -r_N &= k_U C_N \\ r_U &= k_U C_N - k_A C_U^2 - k_D C_U \\ r_A &= k_A C_U^2 \\ r_D &= k_D C_U \end{aligned} \tag{8.11}$$

where r is the rate reaction, k is the reaction constant, C is the molar concentration, and the subscripts N, U, A, and D correspond to the β-lg states: native, unfolded, aggregated, and deposited, respectively. The k parameters have been determined at a diversity of conditions and fit an Arrhenius model (de Wit and Klarenbeek, 1989):

$$k = k_0 \exp\left(-\frac{Ea}{RT}\right) \tag{8.12}$$

where k_0 is the pre-exponential factor, Ea is the activation energy, R is the universal gas constant. Table 8.18 contains the values of these parameters for β-lg denaturation.

The four-phase fouling model has been applied in CFD simulation with suitable results that explain the fouling phenomena, for instance, the works by Jun and Puri (2005a), Choi et al. (2013), Pan et al. (2019) and Ramírez-Rivera et al. (2024).

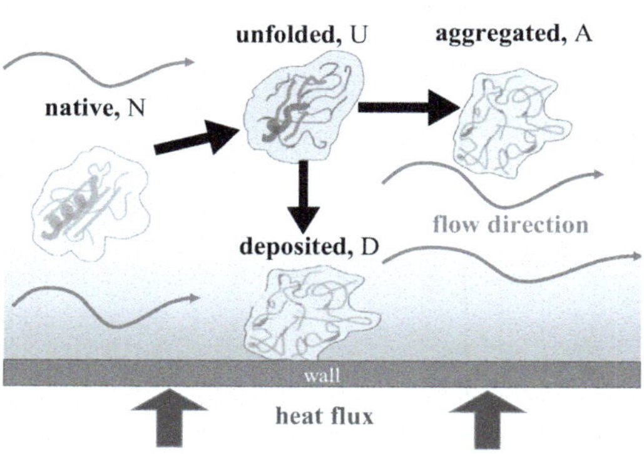

Figure 8.9 Schematic representation of the four-phases fouling. Adapted from Ramírez-Rivera, M. J., Díaz-Ovalle, C. O., Ramos-Ojeda, E., & Castrejón-González, E. O. (2024). CFD simulation analysis of fouling formation in a milk falling-film evaporator. Food and Bioproducts Processing, 143, 242-254. https://doi.org/10.1016/j.fbp.2023.12.003. With permission of Elsevier.

Table 8.18 Values of the kinetic parameters of the Arrhenius equation for the four-phase β-lg denaturalization.

β-lg state	k_0[=]1/s	*Ea*[=] kJ/kgmol	References
Unfolding, U	3.37x10^{37}	261,000	de Wit and Klarenbeek (1989) reported
Aggregation, A	1.36x10^{43}	312,000	by de Jong et al. (1992).
Deposition, D	0.51	45,100	de Jong et al. (1993)

Adapted from Ramírez-Rivera, M. J., Díaz-Ovalle, C. O., Ramos-Ojeda, E., & Castrejón-González, E. O. (2024). CFD simulation analysis of fouling formation in a milk falling-film evaporator. *Food and Bioproducts Processing, 143*, 242-254. https://doi.org/10.1016/j.fbp.2023.12.003. With permission of Elsevier.

Expression (8.11) is not available for continuous milk evaporators. Thus, its recommended use follows the conventional reactor design in a dynamic model (Ingham et al., 2000; Tijskens et al., 2001).

The model requires the inclusion of the inlet molar concentration of β-lg and the formation of β-lg denaturation products. For whole milk, the mass compositions correspond to: solids, 13.7%, proteins, 3.4%, and β-lg, 0.32% with a molecular weight of 18,283 $kg_{lg}/kgmol_{lg}$ (Mehta, 2015). These compositions vary for each kind of whole milk and herein are mentioned to illustrate their applicability in evaporation modeling. A convenient approach is the using mass fraction of β-lg, $x_{L,i}$, and solids free of β-lg, $x_{F,i}$, to compute the consumption of β-lg during the evaporation process without affecting the mass fraction of solids. However, the molar concentration of β-lg (N), $C_{N,i}$, is related to $x_{L,i}$, such as: $C_{N,i} = x_{L,i} \cdot \rho_{milk}/(18{,}283\ kg_{lg}/kgmol_{lg})$. The equation per each case and the denaturation products are (t[=]s):

$$\frac{dx_{F,i}}{dt} = \frac{F_i x_{F,i-1} - (F_i - V_i)x_{F,i}}{M_i}$$

$$\frac{dx_{L,i}}{dt} = \frac{F_i x_{L,i-1} - (F_i - V_i)x_{L,i} - \left(3.37 \times 10^{37}\frac{1}{s}\right)e^{-\frac{261{,}000\frac{kJ}{kgmol}}{RT}} M_i x_{L,i}}{M_i}$$

$$\frac{dC_{U,i}}{dt} = \left(3.37 \times 10^{37}\frac{1}{s}\right)e^{-\frac{261{,}000\frac{kJ}{kgmol}}{RT_i}} \frac{\rho_{milk}\left(x_{L,i}\right)}{18,283\frac{kg_{lg}}{kgmol}} - \left(1.36 \times 10^{43}\frac{1}{s}\right)$$

$$e^{-\frac{312{,}100\frac{kJ}{kgmol}}{RT_i}} C_{U,i}^2 - \left(0.51\frac{1}{s}\right)e^{-\frac{45{,}100\frac{kJ}{kgmol}}{RT_i}} C_{U,i} \frac{dC_{D,i}}{dt} = \left(0.51\frac{1}{s}\right)e^{-\frac{45{,}100\frac{kJ}{kgmol}}{RT_i}} C_{U,i} \tag{8.13}$$

with $x_i = x_{F,i} + x_{L,i}$ and the quantity of deposited protein per area corresponds to: $C_{D,i} M_i (18{,}283\ kg_{lg}/kgmol_{lg})/A \cdot \rho_{milk}$.

The modeling of MEEs by considering fouling formation and β-lg denaturation employs the energy balance from Table 8.16, the fouling formation equation, expression (8.9), and the protein denaturation process, expression (8.13). An n-effect evaporator presents a total of $6n$ ordinary differential equations that must be solved simultaneously.

8.7.3 Runge-Kutta method for evaporation process

The Runge-Kutta method of 4 th order (RK method) is a recommended fixed-time step method for dynamic simulation. The RK method computes the function per each derivative, whereby the derivatives must be clear without mathematical operators, such as: $\mathbf{y}' = \mathbf{f}(t,\mathbf{y})$, where $\mathbf{y}$ is a vector of dynamic variables, and $\mathbf{f}$ is a set of derivatives. The start point is the proposing of initial values by the physical conditions of the system: $\mathbf{y}(0) = \mathbf{y_0}$, $\mathbf{y_0}$ is a vector of initial values. The solution procedure is a successive calculation of the RK method parameters, such as:

$$\left.\begin{aligned}
\mathbf{y}_{i,j+1} &= \mathbf{y}_{i,j} + \left(\mathbf{k}_{1,i,j} + 2\cdot\mathbf{k}_{2,i,j} + 2\cdot\mathbf{k}_{3,i,j} + \mathbf{k}_{4,i,j}\right)\cdot\frac{\Delta t}{6} \\
\mathbf{k}_{1,i,j} &= \mathbf{f}_i\left(t_j, \mathbf{y}_{i,j}\right) \\
\mathbf{k}_{2,i,j} &= \mathbf{f}_i\left(t_j + \frac{\Delta t}{2}, \mathbf{y}_{i,j} + \frac{\Delta t}{2}\cdot\mathbf{k}_{1,i,j}\right) \\
\mathbf{k}_{3,i,j} &= \mathbf{f}_i\left(t_j + \frac{\Delta t}{2}, \mathbf{y}_{i,j} + \frac{\Delta t}{2}\cdot\mathbf{k}_{2,i,j}\right) \\
\mathbf{k}_{4,i,j} &= \mathbf{f}_i\left(t_j + \Delta t, \mathbf{y}_{i,j} + \Delta t\cdot\mathbf{k}_{3,i,j}\right)
\end{aligned}\right\} \tag{8.13}$$

where Δt is the time step, $\mathbf{k}_{1,i,j}$, $\mathbf{k}_{2,i,j}$, $\mathbf{k}_{3,i,j}$ and $\mathbf{k}_{4,i,j}$ are the parameters of the RK method in vector form indexed by w ($\mathbf{k}_{w,i,j}$) and the subscripts i and j refer to the evaporator effect and the iteration counting, respectively. The number of iterations depended on the time step and total time ($iterations = t_{\mathrm{Total}}/\Delta t$). Table 8.19 represents the VBasic code under a MACRO in MSExcel for solving a MEE by considering fouling formation and β-lg denaturation model. The upcoming case study illustrates the dynamic modeling of a conventional multi-effect evaporator system.

8.7.4 Case study of dynamic evaporators

A four-effect plate evaporator concentrates fresh milk with the compositions: 86.3% of water, 4.9% of fat, 3.4% of protein, 4.1% of carbohydrates, and 0.7% of ash. This

Table 8.19 VBasic code for solving a MEE by considering fouling formation and β-lg denaturation model by applying the RK method.

Runge-Kutta structure code:

```
For i=1 To Neff
    Select Case m
      Case 1
        x_(i)=x(i)
        T_(i)=T(i)
        R_(i)=R(i)
        xF_(i)=xF(i)
        CD_(i)=CD(i)
        CU_(i)=CU(i)
      Case 2
        x_(i)=x(i)+dt*kx(m-1,i)*0.5
        T_(i)=T(i)+dt*kT(m-1,i)*0.5
        R_(i)=R(i)+dt*kR(m-1,i)*0.5
        xF_(i)=xF(i)+dt*kxF(m-1,i)*0.5
        CD_(i)=CD(i)+dt*kCD(m-1,i)*0.5
        CU_(i)=CU(i)+dt*kCU(m-1,i)*0.5
      Case 3
        x_(i)=x(i)+dt*kx(m-1,i)*0.5
        T_(i)=T(i)+dt*kT(m-1,i)*0.5
        R_(i)=R(i)+dt*kR(m-1,i)*0.5
        xF_(i)=xF(i)+dt*kxF(m-1,i)*0.5
        CD_(i)=CD(i)+dt*kCD(m-1,i)*0.5
        CU_(i)=CU(i)+dt*kCU(m-1,i)*0.5
      Case 4
        x_(i)=x(i)+dt*kx(m-1,i)
        T_(i)=T(i)+dt*kT(m-1,i)
        R_(i)=R(i)+dt*kR(m-1,i)
        xF_(i)=xF(i)+dt*kxF(m-1,i)
        CD_(i)=CD(i)+dt*kCD(m-1,i)
        CU_(i)=CU(i)+dt*kCU(m-1,i)
    End Select
Next i

xn(i)=x(i)+dt*(kx(1,i)+2*kx(2,i)+2*kx(3,i)+kx(4,i))/6
Tn(i)=T(i)+dt*(kT(1,i)+2*kT(2,i)+2*kT(3,i)+kT(4,i))/6
Rn(i)=R(i)+dt*(kR(1,i)+2*kR(2,i)+2*kR(3,i)+kR(4,i))/6
xFn(i)=xF(i)+dt*(kxF(1,i)+2*kxF(2,i)+2*kxF(3,i)+kxF(4,i))/6
CDn(i)=CD(i)+dt*(kCD(1,i)+2*kCD(2,i)+2*kCD(3,i)+kCD(4,i))/6
CUn(i)=CU(i)+dt*(kCU(1,i)+2*kCU(2,i)+2*kCU(3,i)+kCU(4,i))/6
```

Differential equations:

```
Definitions:
*i=1, T_(i-1)=TS, V(i-1)=V_S
Tw(i)=(T_(i-1)+T_(i))/2
Lambda(i)=605.46*Exp(-0.001243*T_(i-1))
Uo(i)=2441.2499*Log(V(i-1))*Lambda(i)/Area(i))-17340.6107
U(i)=1/(1/Uo(i)+R_(i))
Q(i)=U(i)*Area(i)*(T_(i-1)-T_(i))
*V(i)=Q(i)/lambda(TS), i=1
*V(i)=Psat(i)*F(i)/14.7, i>1
Evaporation: Table 12.15 (Eqs. 1 and 2 from case study 12.7.4)
kx(m,i)=(F(i)*xo(time)-(F(i)-V(i))*x_(i))/Mass(i)
kT(m,i)=(F(i)*HL_(To,xo)-(F(i)-V(i))*HL_(T_(i),x_(i))+Q(i)-V(i)*(lambda(i)+T_(i)))/
          (Mass(i)*Cp_(T_(i),x_(i)))
Fouling-Reaction: Expression 12.12 (Eqs. 3 to 6 from case study 12.7.4)
kR(m,i)=(kd/Re)*Exp(-Ea*(1+Uo(i)*R_(i))/(8.314*(Tw(i)+273.15+Uo(i)*R_(i)*(T_(i)+273.15))))-k_r*R_(i)
        RCN=3.37E+37*Exp(-261000/(8.314*(T_(i)+273.15)))*3600
        RCU=1.36E+43*Exp(-312000/(8.314*(T_(i)+273.15)))*CU_(i)*CU_(i)*3600
        RCD=0.51*Exp(-45100/(8.314*(T_(i)+273.15)))*CU_(i)*3600
kxF(m,i)=(F(i)*0.0032-RCN*Mass(i)*xF_(i)-(F(i)-V(i))*xF_(i))/Mass(i)
kCU(m,i)=(RCN*1017*xF_(i)/18283-RCU-RCD)
kCD(m,i)=RCD
Cp (kcal/kg°C) and Enthalpy (kcal/kg): Properties for case study 12.7.4.
Cp_=(x*(413.253+0.36144*T+0.00095*T*T)+(1-x)*(1000-2.1738*T-0.00130935*T*T))/1000
        HL1=x*(413.253*T+0.36144*T*T/2+0.00095*T*T*T/3)
        HL2=(1-x)*(1000*T-2.1738*T*T/2-0.00130935*T*T*T/3)
HL_=(HL1+HL2)/1000
```

Continued

Table 8.19 VBasic code for solving a MEE by considering fouling formation and β-lg denaturation model by applying the RK method.—cont'd

Nomenclature:

Code	model	code	model	Code	model	code	model
m	**k**$_1$, **k**2, **k**3 or **k**4.	Lambda(i)	λ_i	T_(i), Tn(i)	T_i	kT(m,i)	$(T_i)' = \mathrm{d}T_i/\mathrm{d}t$
i	*i* index for effect	Uo(i)	Uo_i	x_(i), xn(i)	$x_{F,i}$	kx(m,i)	$(x_{F,i})'$
TS	*Ts*	U(i)	U_i	xF_(i), xFn(i)	$x_{L,i}$	kxF(m,i)	$(x_{L,i})'$
V_S	*Vs*	Q(i)	Q_i	R_(i), Rn(i)	Ψ_i	kR(m,i)	$(\Psi_i)'$
Tw(i)	T_W	Mass(i)	M_i	CU_(i), CUn(i)	$C_{U,i}$	kCD(m,i)	$(C_{U,i})'$
Psat(i)	$P^*(T_i)$	F(i)	F_i	CD_(i), CDn(i)	$C_{D,i}$	kCD(m,i)	$(C_{D,i})'$
Area(i)	A_i	V(i)	V_i	HL_(T,x)	$H_L(T_i,x_i)$	Cp_(T,x)	$Cp(T_i,x_i)$
dt	Δt	k_r	k_r	kd	k_d	Neff	n

composition corresponds to 13.7% of solids and 0.33% of β-lg (Mehta, 2015). The inlet flow comes from a preheater at 80°C and 8,700 kg/h. The first effect receives a thermo-compressed vapor at 90°C and 450 kg/h. The charged mass and heat transfer area per effect are (kg, m^2): I: 1,100, 15, II: 1,500, 25, III: 1,500, 22, and IV: 1,500, 25. The Reynolds number for the plate heat exchanger corresponds to the average experimental value of 10 (Zhong et al., 2020). Simulate the evaporation chamber (10 h and $\Delta t = 0.1$ h) per each effect by quantifying: a) mass fraction of solids, b) temperature, c) deposited β-lg (D) per area (kg/m^2), and d) fouling resistance.

Solution:

The plate exchanger is modeled as an intrinsic element of the evaporation system that corresponds to $Q_i = U_i A_i (T_{i-1} - T_i)$, and the value of the overall heat transfer coefficient comes from the expression (8.4). To involve the flow rate of heating fluid in effect I, the vapor stream follows the heat transfer approach for effect I, and the phase equilibrium approach for the remaining effects, and Table 8.15 indicates both approaches.

The heat capacity, $Cp(T_i,x_i)$, follows the Choi-Okos model from Table 8.4, where the composition varies through the concentration process. The case study 8.5.2 illustrates the calculation of $Cp(T_i,x_i)$ for maintaining the composition proportion of solids. Similarly, this case study uses ($Cp[=]$J/kgK):

$$Cp_S = 1,727.4 + 1.5108T + 0.004T^2$$

$$Cp_W = 4,176.2 - 9.0864T - 0.00547T^2$$

Thus, heat capacity is $Cp(T_i,x_i) = x_i Cp_S(T_i) + (1 - x_i) Cp_W(T_i)$ and the specific enthalpy for the processed milk is:

$$H_{L,i} = (x_i) \int_{T=0}^{T_i} Cp_S dT + (1 - x_i) \int_{T=0}^{T_i} Cp_W dT$$

The solution of the equations from Table 8.16 answers the points a) and b) of this case study. The expression (8.9) solves the point c), fouling resistance, by neglecting the last term. Finally, the solving of Eq. (8.13) determines the points d) and e) by computing $C_{D,i} M_i (18{,}360\ \text{kg/kgmol}) / A_i \rho$. To simplify the numerical solution, the density is assumed constant to the experimental value of 1,017 kg/m^3. In conclusion, the case study requires the simultaneous solution of six ordinary differential equations per effect (24 ordinary differential equations in total):

1) $$\frac{dT_i}{dt} = \frac{F_i H_{L,i-1} + U_i \frac{A_i (T_{i-1} - T_i)}{\lambda_i} - (F_i - V_i) H_{L,i} - V_i H_{V,i}}{M_i Cp(T_i)}$$

2) $$\frac{dx_{F,i}}{dt} = \frac{F_i x_{F,i-1} - (F_i - V_i) x_{F,i}}{M_i}$$

3) $$\frac{d\Psi_i}{dt} = \frac{\left(4.85 \times 10^{13} \frac{1}{s}\right)}{Re} exp\left[\frac{-150,000 \frac{kJ}{kgmol}}{R} \cdot \frac{1 + U_{0,i}\Psi_i}{T_{i-1} + U_{0,i}\Psi_i T_i}\right] - \left(0.0013 \frac{1}{s}\right)\Psi_i$$

4) $$\frac{dC_{D,i}}{dt} = \left(0.51 \frac{1}{s}\right) e^{-\frac{45,100 \frac{kJ}{kgmol}}{RT_i}} C_{U,i}$$

5) $$\frac{dx_{L,i}}{dt} = \frac{F_i x_{L,i-1} - (F_i - V_i) x_{L,i} - \left(3.37 \times 10^{37} \frac{1}{s}\right) e^{-\frac{261,000 \frac{kJ}{kgmol}}{RT}} M_i x_{L,i}}{M_i}$$

6) $$\frac{dC_{U,i}}{dt} = \left(3.37 \times 10^{37} \frac{1}{s}\right) e^{-\frac{261,000 \frac{kJ}{kgmol}}{RT_i}} \frac{\rho_{milk}\left(x_{L,i}\right)}{18,283 \frac{kg_{lg}}{kgmol}} - \left(1.36 \times 10^{43} \frac{1}{s}\right) e^{-\frac{312,100 \frac{kJ}{kgmol}}{RT_i}} C_{U,i}^2 - \left(0.51 \frac{1}{s}\right) e^{-\frac{45,100 \frac{kJ}{kgmol}}{RT_i}} C_{U,i}$$

by using expression (8.4):

$$\frac{1}{U_i} = \frac{1}{2,441.25 \ln\left(\frac{\lambda_{i-1} V_{i-1}}{A_i}\right) - 17,340.6107} + \Psi_i.$$

The application of the VBasic code from Table 8.19 solved the problem. In detail, Table 8.19 explains the code lines in correspondence with Eq. A MEE requires a loop to count the number of effects involved in the simulation. Fig. 8.10 depicts the results to answer all the points of the case study in particular, the point e) corresponds to $C_{D,i} M_i (18{,}283\ kg_{lg}/kgmol_{lg})/A \cdot \rho_{milk}$. The results demonstrate the high fouling level during operation for the effect I although the operative conditions are proportional to the remaining effects. The fouling thermal resistance reached values close to experimental data for the plate evaporator: $5.8 \times 10^{-8}\ m^2h/kcal$ (Hoffman, 2004).

8.8 Control of evaporation processes

The quality requirements for foodstuffs push to elaborate healthy and stable products. The food industry has maintained a high rank in quality with the help of automatic process control. Nowadays, energy consumption is included in high production efficiency by reducing the waste of time, raw material, fouling production, and consumption of services. Thus, a well-controlled food process is fundamental to guarantee high-quality standards and acceptable profit.

8.8.1 Basic elements of process control

In principle, the process control maintains desirable variables (controlled variables) under a range of values or desired values (set-point, SP) that do not engage the product quality. To achieve the control targets, the manipulation of mass flow rates are typical

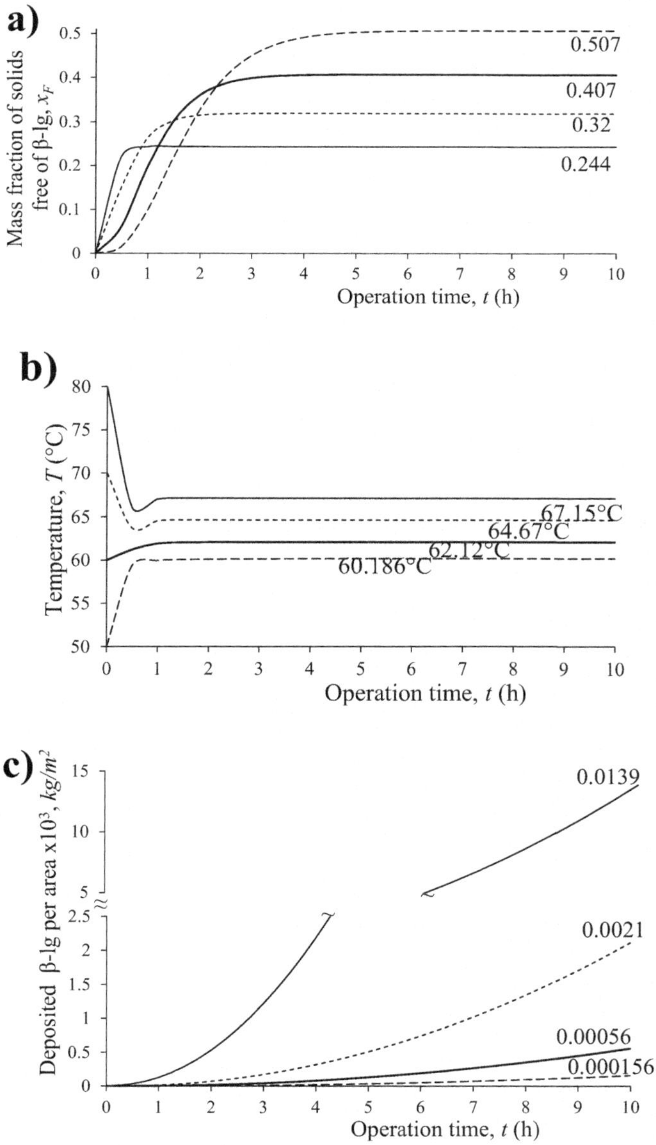

Figure 8.10 Results for the case study 8.7.4: (a) Mass fraction of solids free of β-lg, $x_{F,i}$, (b) temperature, Ti, (c) deposited β-lg in kg/m^2, and (d) thermal resistance, Ψ_i.

changes (manipulated variables). A control valve is the element for manipulating the mass flow in process control. In most cases, the SP values come from observed operative conditions in industrial process units with acceptable results for the product. In a simple control system, one manipulated variable responds to one controlled variable,

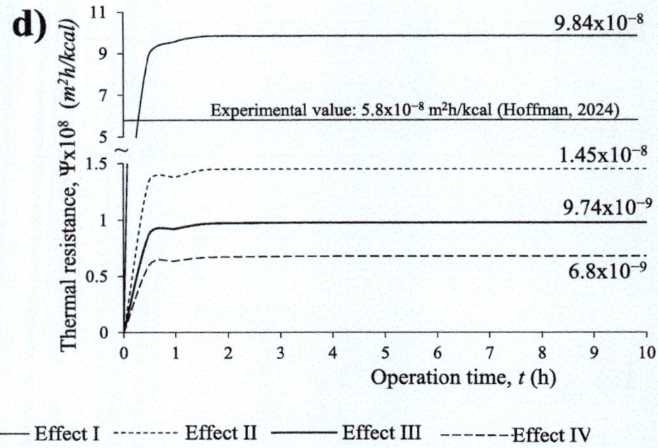

Figure 8.10 cont'd.

which is a SISO (Single-Input Single-Output) system; nevertheless, the process might require a MIMO (Multiple-Input Multiple-Output) system. The control of SISO and MIMO systems requires observing a low level of fluctuations in response to either input disturbances or changes in SP. Therefore, the control theory describes several techniques to determine the correct control in response to the process nature, such as (Díaz-Ovalle and Jafari, 2023):

- Feedback control is a conventional process control technique that evaluates the difference between the values of SP and the controlled variable, ε, and corrects it by modifying the manipulated variable. The PID (Proportional-Integral-Derivative) is an outstanding technique that aids process control by requiring control parameters from particular tuning procedures. The PID elements correspond to:
  - Proportional action provides the value of the manipulated variable in response to Δ, and, in some cases, does not eliminate the offset error (existence of ε).
  - Integral action reduces the offset error by trying to achieve the SP ($\varepsilon = 0$) through the sum of ε over time.
  - Derivative action corrects the rate of change for ε.

The food industry has employed feedback control since the applicability of automatic control systems. Similarly, to other process industries, the use of PID control is not overridden by another control technique. Fig. 8.11 represents a systematic PID control loop with the PID elements.

- Feedforward control predicts the level of the system response to disturbances and corrects them either before or during their occurrence. This control requires a tested mathematical model and historical data.
- Cascade control applies to MIMO systems based on a multi-loop control where an elemental loop is a primary controller, and a secondary loop reduces disturbances.
- Artificial Intelligence systems apply to complex systems with a high demand for adaptability and overcoming the limits of traditional control techniques. The usual techniques are: fuzzy logic technique (similar to human response), neural networks (fundaments of experience learning), and generic control algorithms (global optimum solution).

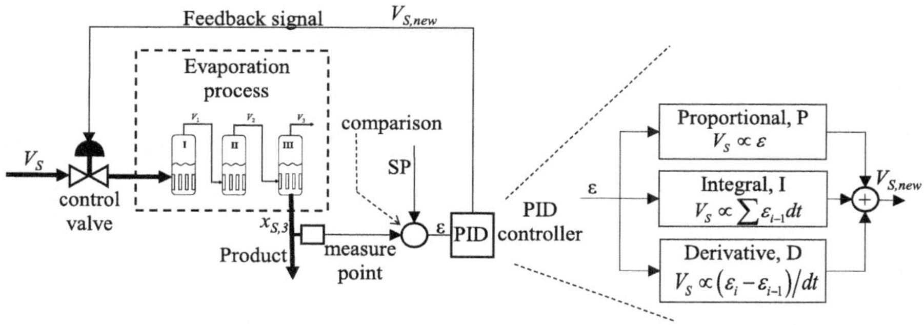

Figure 8.11 Representation of a PID feedback control in an evaporation process.

The diverse control techniques suit MEEs (Minton, 1986). Díaz-Ovalle and Jafari (2023) have summarized the application of these strategies for diverse evaporators. The feedforward control presents an option for the evaporation process, for instance, the falling-film evaporation process, but it reclaims historical data that limits an extension for the industry. Nevertheless, the PID control strategy acceptably adjusts to industrial processes (Haasbroek et al., 2013b), and its application to falling-film evaporators demonstrated a suitable performance (Díaz-Ovalle et al., 2022b). This chapter adopted the PID control strategy in a MEE based on the models from Section 8.7.

8.8.2 Elements of process control in evaporation

The evaporation process reclaims adequate process control to avoid undesirable changes in the product and inconveniences during performance. Nevertheless, an evaporation unit undergoes phenomena that complicate the well-prediction during the evaporation performance and limit the process control, such as boiling point rise, fouling formation, non-Newtonian fluid behavior, etc. Particularly, the fouling phenomenon requires special attention for determining cleaning time, and the process control only retards this undesirable phenomenon but not eliminates it (Díaz-Ovalle et al., 2022b). Thus, the evaporation process under control must be stable for changes in SP and unpredictable delays and disturbances.

In principle, for control of evaporators, the controlled variables are the mass fraction of solids, liquid inner level, and temperature of the effects, and the manipulated variables are the motive steam and feed mass flows. A SISO system might apply for the controlled variables in one of the effects by manipulating either V_S or F. Thus, the selection of the variables to control and manipulate creates different control configurations, similar to Runyon et al. (1991) and Díaz-Ovalle et al. (2022b). Fig. 8.12 schematizes a P&ID proposal for global PID control configurations in a MEE. The conventional control configurations are (Bloore and O'Callaghan, 2009): a) manipulating V_S to control any value of x_i, b) manipulating V_S to control any value of T_i, and c) manipulating F to control any value of x_i. The food evaporation processes follow

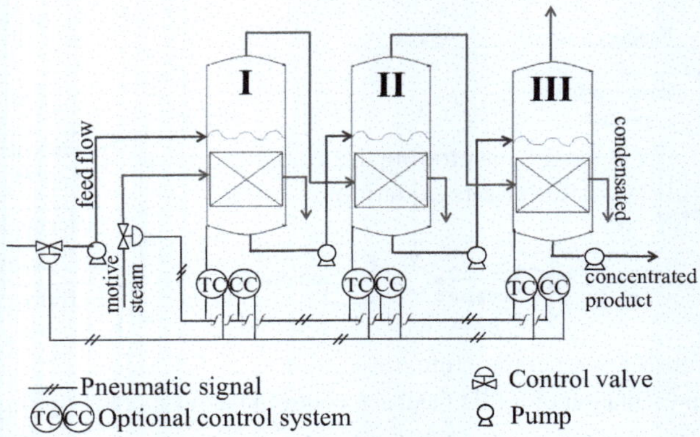

Figure 8.12 PID controller of a MEE with optional control systems.
Adapted from Díaz-Ovalle, C. O., & Jafari, S. M. (2023). Multiple-Effect Evaporators in the Food Industry: Fundamentals, Design, Simulation, Control, and Applications. Food Engineering Reviews, 15(4), 691-717. https://doi.org/10.1007/s12393-023-09350-6. With permission of Springer.

the control configurations: a) the product quality based on x_N, and b) the highest temperature into the system to avoid thermal damage, T_1.

A sensitivity analysis benefits the understanding of a MEE performance, which aids in fitting the process control. Díaz-Ovalle et al. (2017) simulated a falling-film MEE. They observed: a) the increment of feed flow reduces the mass fraction of solids, b) the motive steam conditions are directly proportional to the mass fraction of solids through the system, c) the temperature of the feed stream directly affects the temperature of the system, and b) the increment of the mass fraction of solids in the feed stream is proportional to the mass fraction of solids in the system. These points are valuable for reaching an acceptable tuning of a PID controller.

8.8.3 PID control in evaporators based on the time-domain

The PID control strategy is adequate for the MEE performance and has guaranteed the stability of the operating variables due to the disturbance rejection and acceptable set point tracking (Haasbroek et al., 2013a). The PID control in the time domain is a valid option for nonlinear systems, such as the case of a MEE with rigorous heat transfer correlations and variable properties (Coughanowr, 1991; Luyben, 1996). In the time domain, the conventional model for a PID system is (Ingham et al., 2000):

$$V_m = V_{m,0} + Kc\left(\varepsilon(t) + \frac{1}{\tau_I}\int \varepsilon(t)dt + \tau_D\frac{d(\varepsilon(t) - \varepsilon(t-1))}{dt}\right) \quad (8.14)$$

where V_m and $V_{m,0}$ are the manipulated variables for the PID response and $\varepsilon = 0$ (BIAS), respectively, and Kc, τ_I, and τ_D are the PID parameters, controller gain, integral time, and derivative time, respectively. Thus, these parameters replace the proportion sings from Fig. 8.11. Luyben (1996) presented the adequate location of each PID action into an RK code for dynamic systems:

a) Proportional action: a PID function reads the value of the controlled variable from expression (12.14): `PID=V0+Kc*(SP-C1)`.
b) Integral action: an accumulative variable (`sum`) initializes to zero (`sum=0`) before solving the RK method, computes the integral argument (`(SP-C1)*dt`), and adds the current value to the previous after the "For" loop header (`sum=sum+(SP-C1)*dt`).
c) Derivative action: the current and previous values (`C2` and `C1`) of the controlled variable determine the derivative (`deriv`) after the calculation of the current value (`deriv=(C2-C1)/dt`).

The simulation of a PID control in a MEE is possible by modifying the code of the RK for a dynamic model. Therefore, Table 8.20 contains the PID controller code into the VBasic code from Table 8.19.

The PID feedback control requires a tuning strategy to determine the control parameters from the expression (8.14). This point has demanded special attention for the control theory to reduce the time response and the error, ε (Johnson and Moradi, 2005; O'Dwyer, 2009). The tuning strategies contemplate the controlled variable response to a disturbance of the manipulated variable, whereby there are several approaches to determine the control parameters and there is no general tuning method (Aström and Hägglund, 2005). For instance, in the trial-and-error method, the more simple strategy, the value of the control parameters is changed by observation, such as (Luyben and Luyben, 1997):

1. Indicate a small value of *Kc* and increase it until to observe an oscillation in the response and reduce one-half.
2. Reduce the value of τ_I until an oscillation in the response and duplicate.
3. Increase the value of τ_D until noise and reduce one-half.
4. Make short modifications of *Kc* until the best results.

However, the trial-and-error method does not guarantee acceptable control results. Therefore, the literature presents methods based on the process-reaction curve (graph of the controlled variable in response to a disturbance in the manipulated variable), such as the Ziegler et al. (1942) and the Cohen et al. (1953). Both methods are established in the graphical parameters (Ingham et al., 2000): a) time for the disturbance, t^*, b) lag time, T_L, c) the magnitude of the controlled variable before and after the disturbance, A_1 and A_2, respectively, d) slope at the inflection point after the disturbance, s^*, and e) times of the slope crosses A_1 and A_2, t_1 and t_2, respectively. Fig. 8.13 depicts all these elements in a visual example and Table 8.21 summarizes the calculation of the control parameters. The next case study applies this principle to controlling a MEE.

8.8.4 Case study of controlling evaporation processes

The tomato concentrate is a relevant product for the food industry and highly consumed worldwide. Hayes et al. (1998) have declared its main process aspects, where the evaporation process is one of the highest production cost steps. The USDA (U.S. Department of Agriculture) indicates the composition of tomato products in the Food Data Central (2019), in solid proportion: 1.53% of fat, 14.12% of protein, 61.78% of carbohydrates, 13.4% of fiber, and 9.15% of ash. Runyon et al. (1991) established a well-known case study for controlling short vertical tube evaporators in

Table 8.20 VBasic code for including a PID controller into a MEE.

Declarations:

```
Public variable sum, deriv, SP, ti, td As Double
```

Location of the PID code:

```
Variable declaration here
Variable initialization
sum=0
deriv=0
For k=1 to num_iteration 'For loop
     sum=sum+(SP-C1)*dt
     ******
     RK main code here
     'Example of calling the PID function (substitution of F(i) from
     Table 12.18):
     kx(m,i)=(PID(C1)*xo(time)-(PID(C1)-V(i))*x_(i))/Mass(i)
     *****
     deriv=(C2-C1)/dt
     C1=C2
Next k
```

PID function:

```
Function PID(C As Double)
     PID=V0+Kc*((SP-C)+sum/ti+deriv*td)
End Function
```

Nomenclature:

Kc	controller gain
ti	integral time constant
td	derivative time constant
dt	RK time step
SP	set-point value at time
C	reading value of the controlled variable
num_iteration	number of iterations for the RK method
k	counter for the For loop

a counterflow process and Yadav and Jana (2010) analyzed the same case under other techniques of PI feedback control. Herein, the system presents a cocurrent process with a feed flow of 26,082 kg/h at 85°C with 8.7% of solids and a motive steam flow of 4000 kg/h at 115.6°C. The mass charge and the heat transfer area for each effect are: I: 3800 kg, 102 m^2, and II: 3800 kg, 412 m^2.

Apply the tuning methods from Table 8.21 and compare them to control the mass fraction of solids in Effect II (SP of $x_2 = 26.5\%$). Report ($t_{Total} = 10$ h and $\Delta t = 0.01$h): a) values of the control parameters per each method, and b) a graph

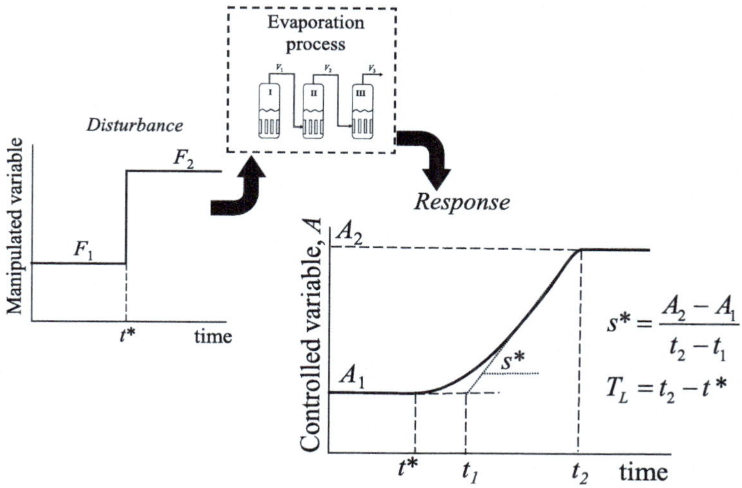

Figure 8.13 Illustrative example of the process-reaction curve in a MEE.

Table 8.21 Controlled settings for a PID configuration.

Method	***Kc***	τ_I	τ_D
Ziegler-Nichols	$\frac{1.2}{T_L s^*}$	$2T_L$	$0.5T_L$
Cohen-Coon	$\frac{\tau^*}{T_L K^*}\left(\frac{4}{3}+\frac{T_L}{12\tau^*}\right)$	$T_L\left(\frac{32\tau^*+6T_L}{13\tau^*+8T_L}\right)$	$T_L\left(\frac{4\tau^*}{12\tau^*+2T_L}\right)$
Nomenclature	$K^* = A_2/A_1$ $\quad \tau^* = A_2/s^*$		

Adapted from Díaz-Ovalle, C. O., Rico-Ramírez, V., & Ramos-Ojeda, E. (2022). Control of falling-film evaporators with modeling of fouling. Food and Bioproducts Processing, 132, 68-82. https://doi.org/10.1016/j.fbp.2021.12.010. With permission of Elsevier.

for variables (x_2, T_i, Ψ_2, and V_S) for the control test: disturbance of $Ts = 115°\text{C} + (5.75°\text{C})\cdot\sin(35t)$ and SP change from $x_i = 0.24$ to $x_i = 0.26$ at 4 h.

Solution:

The case study solution contains the dynamic model, the numerical simulation without control, and the application of control. The dynamic model follows the strategy from case study 8.7.4 without β-lg denaturalization reactions. Thus, the first step specifies the heat capacity, $Cp(T_i,x_i)$, from the Choi-Okos model from Table 8.4. Similarly to previous case studies, the mathematical expression for $Cp(T_i,x_i)$ considers the solid proportions of the tomato paste:

$$Cp_S = 1,618.43 + 1.8242T + 0.0048T^2$$

$$Cp_W = 4,176.2 - 9.0864T - 0.00547T^2$$

and the specific enthalpy of the tomato during evaporation is:

$$H_{L,i} = (x_i)\int_{T=0}^{T_i} Cp_S dT + (1 - x_i)\int_{T=0}^{T_i} Cp_W dT$$

The dynamic model follows Table 8.16 and includes the fouling formation (expression 8.8):

$$1)\ \frac{dT_i}{dt} = \frac{F_i H_{L,i-1} + U_i \frac{A_i (T_{i-1} - T_i)}{\lambda_i} - (F_i - V_i) H_{L,i} - V_i H_{V,i}}{M_i Cp(T_i)}$$

$$2)\ \frac{dx_{F,i}}{dt} = \frac{F_i x_{F,i-1} - (F_i - V_i) x_{F,i}}{M_i}$$

$$3)\ \frac{d\Psi_i}{dt} = \frac{\left(4.85 \times 10^{13} \frac{1}{s}\right)}{\text{Re}}\, exp\left[\frac{-150,000 \frac{kJ}{kgmol}}{\text{R}} \cdot \frac{1 + U_{0,i}\Psi_i}{T_{i-1} + U_{0,i}\Psi_i T_i}\right] - \left(0.0013 \frac{1}{s}\right)\Psi_i$$

by using Re = 20, $\lambda_i = 605.4e^{(-0.001243T)}$ and the expression (8.5):

$$\frac{1}{U_i} = \frac{1}{0.225\left(\frac{\lambda_{i-1} V_{i-1}}{A_i}\right)^{0.69} Cp_L{}^{0.69}(9.101414e^{0.000826T})} + \Psi_i.$$

The dynamic simulation without control is in Fig. 8.14. Each graph contains the steady state value for the dynamic variables. These results are similar to those presented by the literature ($x_2 = 0.2637$, $T_I = 74.7$). The mass fraction of solids reached the steady state at 1.5 h for Effect I and 2.5 h for Effect II. The Effect I presented more fouling formation during the evaporator performance.

The value $x_2 = 0.2411$ was considered the initial SP value for applying the tuning methods. The motive steam flow rate suffered an increment of 20% (from 4000 kg/h to 4800 kg/h). This step disturbance activated after 6 h of reaching the steady state, this instruction was programmed with a conditional function. Fig. 8.15 contains the process-reaction curve for x_2 with the required values to compute the control parameters, similar to Fig. 8.13. Hence, the control parameters are: Ziegler Nichols ($Kc = 46.6926$ kg/h, $\tau_I = 1.7$ h, $\tau_D = 0.425$) and Cohen-Coon ($Kc = 12.5837$ kg/h, $\tau_I = 2.011$ h, $\tau_D = 0.2789$ h). However, for both methods, the values of the control parameters required extra tuning to reach the SP during a change of SP ($Kc = 12{,}000$ kg/h, $\tau_I = 0.475$ h, $\tau_D = 0.425$ h). There is no comparison between the tuning methods due to the similarity of the results. This description solves Point a) from the case study.

The evaporation control testing solves Point b) for a disturbance of Ts, which represents the typical steam oscillation under a sinusoidal function. The SP change is an expected manual variation during the evaporation performance. Fig. 8.16 depicts the

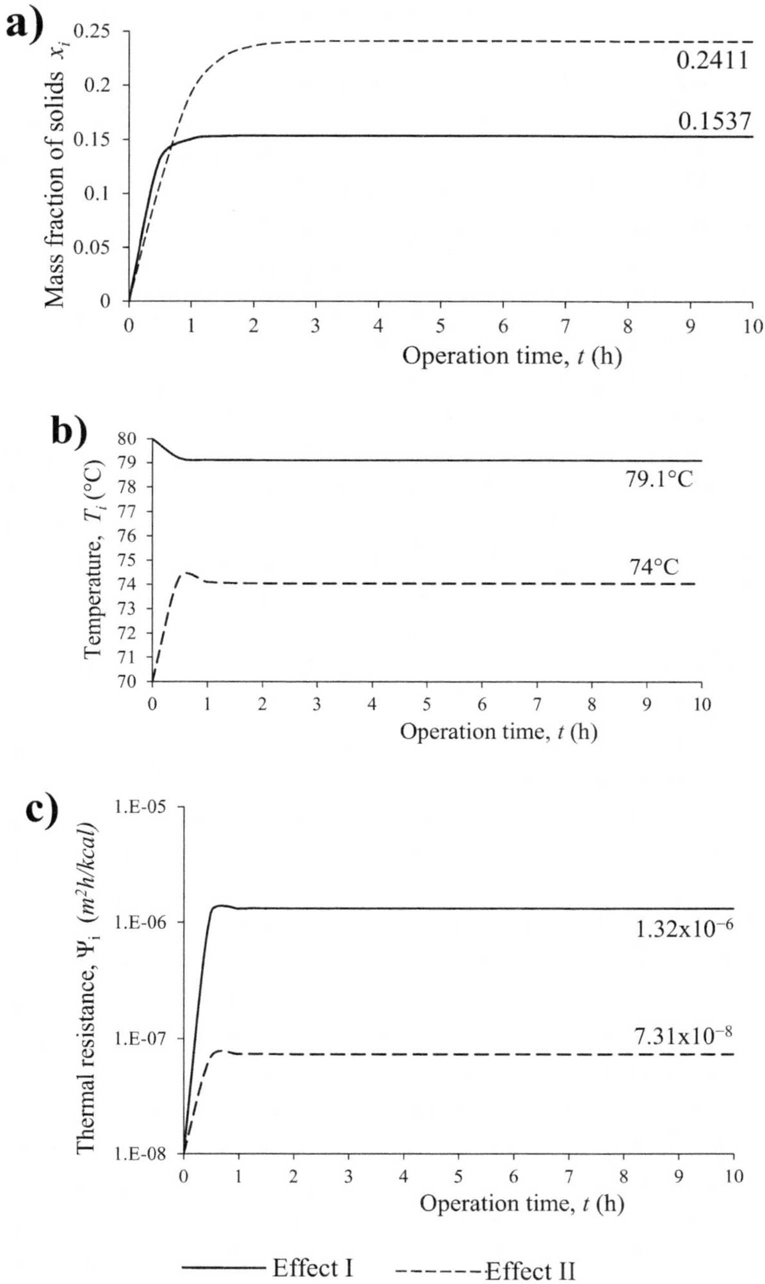

Figure 8.14 Dynamic simulation without control of case study 8.8.4 for the dynamic variables: (a) Mass fraction of solids, (b) temperature, and (c) fouling thermal resistance.

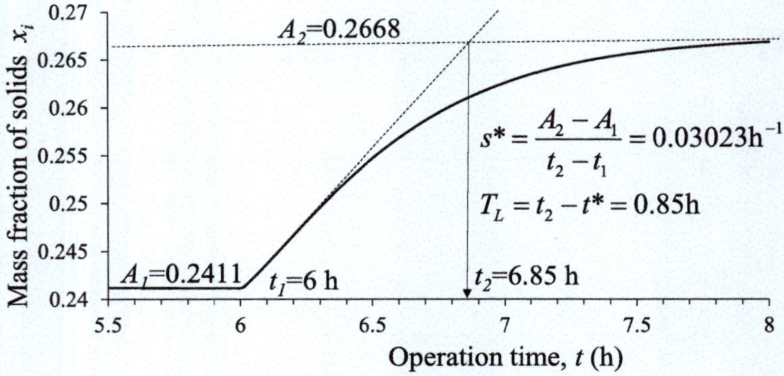

Figure 8.15 Process-reaction curve for the case study 8.8.4.

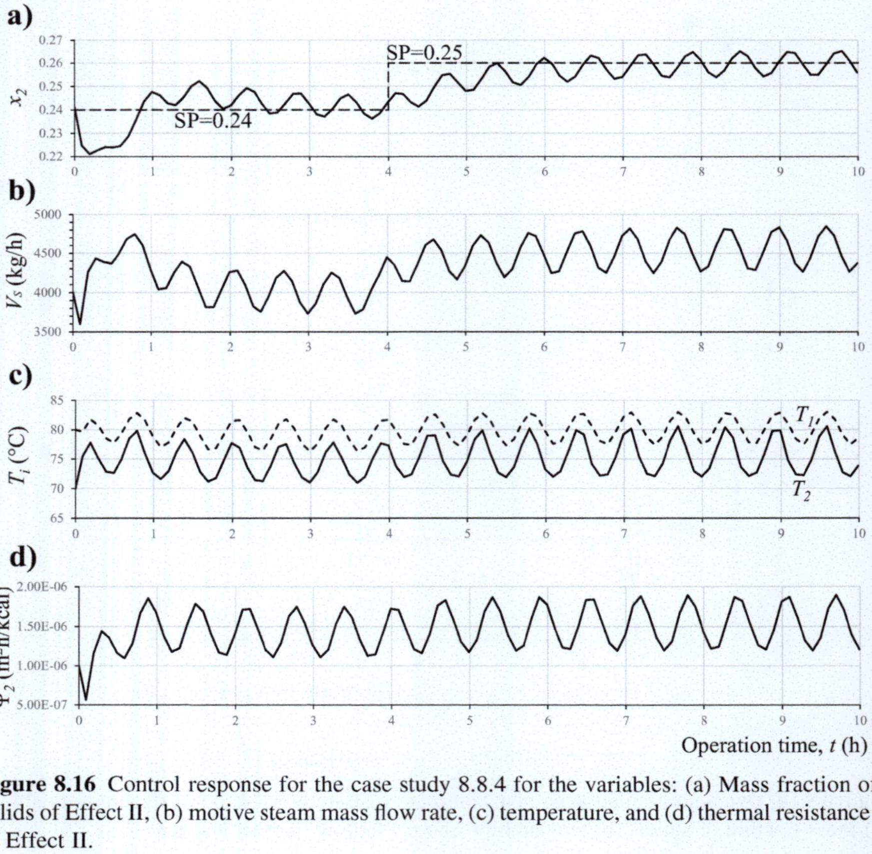

Figure 8.16 Control response for the case study 8.8.4 for the variables: (a) Mass fraction of solids of Effect II, (b) motive steam mass flow rate, (c) temperature, and (d) thermal resistance of Effect II.

results where the sinusoidal behavior of the disturbance dominated all the variables. This behavior increased the difficulty of controlling the mass fraction, x_2. Besides, the proposed control system generates a suitable response.

8.9 Conclusion

Mathematical modeling of unit operations is fundamental to understanding the transformation of foodstuffs and involves the principles of thermodynamics, rheology, fluid properties, heat transfer, etc. The scope of this technique is to predict the unit operation performance, approximate foodstuff changes, and influence decision-making. In particular, the evaporation processes receive special attention due to thermal risk during operation and the energy savings. Also, the literature has presented an evaporation process with a high mathematical demand to complete a design or performance enhancement. In this chapter, some elements of mathematical modeling are applied to the evaporation process: exergetical modeling, optimization, dynamic simulation with fouling formation, and time domain control. The chapter presents the theoretical elements of each topic and their application in the food industry. Indeed, each topic contains a suggested numerical solution with the computer aid. Therefore, all these topics are guidelines for improving the evaporation processes to reach quality requirements and sustainability restrictions.

Nomenclature

A	heat transfer area
$\boldsymbol{a, b, c}$	parameters for the Choi-Okos model
$\boldsymbol{A_1, A_2}$	magnitude of the controlled variable in the process-reaction curve
$\boldsymbol{A^{ave}}$	average heat transfer area
$\boldsymbol{A^I, A^{II}}$	inner and outer heat transfer areas
Bi	Biot number
$\boldsymbol{C_A}$	molar concentration of the aggregated β-lg
$\boldsymbol{C_D}$	shell cost
$\boldsymbol{C_L}$	tube length cost
$\boldsymbol{C_N}$	molar concentration of the native β-lg
$\boldsymbol{Cp}$	specific heat capacity
$\boldsymbol{C_T}$	total cost
$\boldsymbol{C_U}$	molar concentration of the unfolded β-lg
D	set of candidate diameters
$\boldsymbol{Ea}$	activation energy for the Arrhenius model
$\boldsymbol{Ea_F}$	activation energy for the fouling reaction
$\boldsymbol{Ex}$	specific exergy
$\boldsymbol{ExD}$	destroyed exergy/*i*-effect
$\boldsymbol{ExD}$**T**	total destroyed exergy
$\boldsymbol{ExD_{Thermo}}$	total exergy for thermocompressor
$\boldsymbol{Ex_L}$	exergy for the concentrated liquid or product

$Ex_{L,Water}$	exergy for water
Ex_{λ}	specific exergy for phase change
F	mass flow rate of the feed
$\mathbf{f}$	set of derivatives
$\mathbf{g}$	gravity acceleration
H	specific enthalpy
h_{L}	boiling convective coefficient
h_{V}	condensing convective coefficient
k	reaction constant for the β-lg states
$\mathbf{k}_{w,i,j}$	parameters of the RK method in vector form
K^*	control parameter, A_2/A_1
k_0	pre-exponential factor for the Arrhenius model
Kc	controller gain
k_d, k_r	parameters of the fouling reaction
k^{F}	thermal conductivity of fouling
L	mass of liquid that leaves an *i*-effect
L_{T}	length of the tube
$\bar{M}$	mass charge into the evaporator
M_D, M_R	mass fluxes of deposited and removed material from the surface
n	number of effects in a MEE
N_{T}	number of tubes per effect
P	pressure of vapor
P^*	pressure of vapor at saturation
$\mathbf{P_M}$	pressure of discharge vapor
Pr	Prandtl number
$\mathbf{P_S}$	pressure of motive steam
$\mathbf{P_T}$	pressure of input vapor
Q	heat flow rate from the heating medium
q	heat flux
$\mathbf{R}$	universal gas constant
R	Pearson correlation coefficient
r	rate reaction
Re	Reynolds number
S	specific entropy
s^*	slope at the inflection point in the process-reaction curve
S_G	generated entropy
T	temperature of effect
T^*	temperature at saturation
T_L	lag time
T_M	temperature of the thermo-compressor mix
T_Q	temperature from the heat flow rate
Ts	temperature of heating medium
t_{Total}	total time of simulation
T_w	temperature of the wall
$\mathbf{T}_{\infty}$	temperature of surroundings
U	overall heat transfer coefficient
U_0	overall heat transfer coefficient in the absence of fouling
V	mass flow rate of vapor in an *i*-effect
V_S	mass flow rate of motive steam

x	mass fraction of solids/*j*-component
x_F	mass fraction of a solid free foodstuff
x_L	mass fraction of β-lg
Z	parameter of convective coefficient correlations

Indexes

d	index for candidate diameters
h	index of effect that sends vapor to thermo-compressor
i	index of effect in a multiple-effect sequence
j	index for components
k	index of effect that heats an *i*-effect
m	index of effect in a multiple-effect sequence
p	index of effect that sends vapor to a preheater
w	index for the parameters of the RK method

Subscripts

F	condition of feed stream
L	liquid phase
S	heating medium
V	vapor phase
W	water liquid

Greek letters

Γ	wetting rate
δ	fraction of vapor that flows to a preheater
ε	difference between the SP and the controlled variable
θ	fraction of vapor to the thermo-compressor
η	exergy efficiency
κ	thermal conductivity
λ	specific condensation enthalpy
μ	dynamic viscosity
ρ	density
σ	surface tension
Φ	diameter candidates
Ψ	thermal fouling resistance
Ω	function of nucleate pool correlations
τ^*	control parameter, A_2/s^*
τ_D	derivative time
δ^F	fouling thickness
τ_I	integral time
Δr	thickness of the metal wall
ΔT	difference of temperature
Δt	time step
ϕ_T	diameter of the tube

Acronyms

LP	linear programming
MEE	multiple-effect evaporator

MILP	mixed integer linear programming
MIMO	multiple-input multiple-output system
MINLP	mixed integer nonlinear programming
NLP	nonlinear programming
P&ID	Piping and instrumentation diagram
PID	Proportional-Integral-Derivative feedback control
RK	Runge-Kutta method
SISO	single-input single-output system
SP	set-point value

References

Adib, T.A., Vasseur, J., 2008. Bibliographic analysis of predicting heat transfer coefficients in boiling for applications in designing liquid food evaporators. J. Food Eng. 87, 149–161. https://doi.org/10.1016/j.jfoodeng.2007.12.013.

Aghbashlo, M., Mobli, H., Rafiee, S., Madadlou, A., 2013. A review on exergy analysis of drying processes and systems. Renew. Sustain. Energy Rev. 22, 1–22. https://doi.org/10.1016/j.rser.2013.01.015.

Ahmetović, E., Ibrić, N., Kravanja, Z., Grossmann, I.E., Maréchal, F., Čuček, L., et al., 2018. Simultaneous optimisation and heat integration of evaporation systems including mechanical vapour recompression and background process. Energy 158, 1160–1191. https://doi.org/10.1016/j.energy.2018.06.046.

Alhuthali, S., Delaplace, G., Macchietto, S., Bouvier, L., 2022. Whey protein fouling prediction in plate heat exchanger by combining dynamic modelling, dimensional analysis, and symbolic regression. Food Bioprod. Process. 134, 163–180. https://doi.org/10.1016/j.fbp.2022.05.009.

Andre, H., Ritter, R.A., 1968. Dynamic response of a double effect evaporator. Can. J. Chem. Eng. 46 (4), 259–264. https://doi.org/10.1002/cjce.5450460409.

Aström, K. J., & Hägglund, T. (2005). Advanced PID control: ISA.

Avriel, M., 2003. Nonlinear Programming: Analysis and Methods. Dover Publications.

Balkan, F., Colak, N., Hepbasli, A., 2005. Performance evaluation of a triple-effect evaporator with forward feed using exergy analysis. Int. J. Energy Res. 29 (5), 455–470. https://doi.org/10.1002/er.1074.

Bapat, S.M., Majali, V.S., Ravindranath, G., 2013. Exergetic evaluation and comparison of quintuple effect evaporation units in Indian sugar industries. Int. J. Energy Res. 37 (12), 1415–1427. https://doi.org/10.1002/er.2940.

Bapat, S.M., Majali, V.S., Ravindranath, G., 2016. Exergy and sustainability analysis of quintuple effect evaporation unit in a sugar industry - a case study. Int. J. Renew. Energy Technol. 7 (1), 46–68. https://doi.org/10.1504/ijret.2016.073401.

Bazaraa, M.S., Sherali, H.D., Shetty, C.M., 2013. Non-Linear Programming: Theory and Algorithms. Wiley and Sons.

Biegler, L.T., 2010. Nonlinear Programming: Concepts, Algorithms, and Applications to Chemical Processes. SIAM.

Biegler, L.T., Grossmann, I.E., Westerberg, A., 1997. Systematic Methods for Chemical Process Design. Prentice Hall, Englewood Cliffs, NJ.

Bloore, C.G., O'Callaghan, D.J., 2009. Process control in evaporation and drying. In: Tamime, A.Y. (Ed.), Dairy Powders and Concentrated Products. Wiley-Blackwell, pp. 332–350.

Briam, R., Walker, M.E., Masanet, E., 2015. A comparison of product-based energy intensity metrics for cheese and whey processing. J. Food Eng. 151, 25–33. https://doi.org/10.1016/j.jfoodeng.2014.11.011.

Chen, S.L., Gerner, F.M., Tien, C.L., 1987. General film condensation correlations. Exp. Heat Transf. 1 (2), 93–107. https://doi.org/10.1080/08916158708946334.

Choi, C.Y., Okos, M.R., 1986. Effects of temperature and composition on the thermal properties of foods. In: Maguer, M.L., Jelen, P. (Eds.), Food Engineering and Process Aplications: Transport Phenomena. Elsevier Applied Science Publishers, London.

Choi, W., Jun, S., Nguyen, L.T., Rungraeng, N., Yi, H., Balasubramanian, S., et al., 2013. 3-D milk fouling modeling of plate heat exchangers with different surface finishes using computational fluid dynamics codes. J. Food Process. Eng. 36 (4), 439–449. https://doi.org/10.1111/j.1745-4530.2012.00684.x.

Cohen, G.H., Coon, G.A., Rochester, N.Y., 1953. Theoretical consideration of retarded control. Trans ASME 75, 827–834.

Coughanowr, D., 1991. Process Systyems Analysis and Control. McGraw-Hill.

Dantzig, G.B., 1963. Linear Programming and Extensions. The Rand Co.

de Jong, P., Bouman, S., van der Linder, H.J., 1992. Fouling of heat treatment equipment in relation to the denaturation of B-lactoglobulin. J. Soc. Dairy Technol. 45, 3–8. https://doi.org/10.1111/j.1471-0307.1992.tb01715.x.

de Jong, P., Waalewijn, R., van Der Linder, H.J.L.J., 1993. Validity of a kinetic fouling model for heat-treatment of whole milk. Lait 73, 293–302. https://doi.org/10.1051/lait:1993327.

de Wit, J.N., Klarenbeek, G., 1989. Technological and functional aspects of milk proteins. In: Barth, C.A., Schlimme, E. (Eds.), Milk Proteins: Nutritional, Clinical, Functional and Technological Aspects. Steinkopff, Heidelberg, pp. 211–222.

Delplace, F., Leuliet, J.C., 1995. Modelling fouling of a plate heat exchanger with different flow arrangements by whey protein solutions. Food Bioprod. Process. 73, 112–120.

Díaz-Ovalle, C.O., 2024a. 11 - design, model, and simulation of evaporators for the food industry. In: Jafari, S.M., Çapanoglu, E., Karaca, A.C. (Eds.), Evaporation Technology in Food Processing. Woodhead Publishing, pp. 269–313.

Díaz-Ovalle, C.O., 2024b. Design, model, and simulation of evaporators for the food industry. In: Jafari, S.M., Çapanoglu, E., Karaca, A.C. (Eds.), Evaporation Technology in Food Processing. Woodhead Publishing, pp. 269–313.

Díaz-Ovalle, C.O., Galván-Ángeles, E., Ramos-Ojeda, E., Castillo-Borja, F., 2013. Metodología para el diseño óptimo de evaporadores de película descendente [An approach for optimal design in falling-film evaporators]. Avances en Ciencias e Ingeniería 4 (3), 49–61.

Díaz-Ovalle, C.O., González-Alatorre, G., Alvarado, J.F.J., 2017. Analysis of the dynamic response of falling-film evaporators considering fouling. Food Bioprod. Process. 104, 124–136. https://doi.org/10.1016/j.fbp.2017.05.007.

Díaz-Ovalle, C.O., Jafari, S.M., 2023. Multiple-effect evaporators in the food industry: fundamentals, design, simulation, control, and applications. Food Eng. Rev. 15 (4), 691–717. https://doi.org/10.1007/s12393-023-09350-6.

Díaz-Ovalle, C.O., Ramos-Ojeda, E., Herrera-Enciso, F., Castillo-Borja, F., 2022a. Análisis exergético para un evaporador multiefecto de alimentos con termocompresión [Exergetical analysis for a food multi-effect evaporator with thermocompression]. Avances en Ciencias e Ingeniería 13 (4), 1–8.

Díaz-Ovalle, C.O., Rico-Ramírez, V., Ramos-Ojeda, E., 2022b. Control of falling-film evaporators with modeling of fouling. Food Bioprod. Process. 132, 68–82. https://doi.org/10.1016/j.fbp.2021.12.010.

Dincer, I., Rosen, M.A., 2013. Exergy. Elsevier.

Diwekar, U., 2008. Introduction to Applied Optimization. Springer Verlag.

Edgar, T.F., Himmelblau, D.M., Lasdon, L.S., 2001. Optimization of Chemical Processes. McGraw-Hill, Inc.

Floudas, C. A. (1995). Nonlinear and mixed-integer optimization. fundamentals and applications.

Floudas, C.A., Pardalos, P.M., 2009. Encyclopedia of Optimization. Springer.

Fryer, P.J., 1989. The uses of fouling models in the design of food process plant. Int. J. Dairy Technol. 42 (1), 23–29. https://doi.org/10.1111/j.1471-0307.1989.tb01703.x.

Fryer, P.J., Slatter, N.K.H., 1985. A direct simulation procedure for chemical reaction fouling in heat exchangers. Chem. Eng. J. 31, 97–107. https://doi.org/10.1016/0300-9467(85)80048-4.

Galván-Ángeles, E., Díaz-Ovalle, C.O., González-Alatorre, G., Castrejón-González, E.O., Vázquez-Román, R., 2015. Effect of thermo-compression on the design and performance of falling-film multi-effect evaporator. Food Bioprod. Process. 96, 65–77. https://doi.org/10.1016/j.fbp.2015.07.004.

Georgiadis, M.C., Macchietto, S., 2000. Dynamic modelling and simulation of plate heat exchnagers under milk fouling. Chem. Eng. Sci. 55, 1605–1619. https://doi.org/10.1016/S0098-1354(99)80050-7.

Georgiadis, M.C., Rotstein, G.E., Macchietto, S., 1998a. Modeling and simulation of shell and tube heat exchangers under milk fouling. AIChE J. 44, 959–971. https://doi.org/10.1002/aic.690440917.

Georgiadis, M.C., Rotstein, G.E., Macchietto, S., 1998b. Optimal design and operation of heat exchangers under milk fouling. AIChE J. 44 (9), 2099–2111. https://doi.org/10.1002/aic.690440917.

Haasbroek, A., Auret, L., Steyn, W.H., 2013a. A comparison of control techniques for dairy falling film evaporators. IFAC Proc. Vol. 46 (32), 253–258. https://doi.org/10.3182/20131218-3-IN-2045.00132.

Haasbroek, A., Steyn, W.H., Auret, L., 2013b. Advanced control with fundamental and data-based modeling for falling film evaporators. In: Paper Presented at the 2013 IEEE International Conference on Industrial Technology (ICIT).

Hangos, K.M., Cameron, I.T., 2001. Process Modelling and Model Analysis. Academic Press.

Hayes, W.A., Smith, P.G., Morris, A.E.J., 1998. The production and quality of tomato concentrates. Crit. Rev. Food Sci. Nutr. 38 (7), 537–564. https://doi.org/10.1080/10408699891274309.

Himmelblau, D.M., Bischoff, K.B., 1976. Análisis y Simulación de Procesos. Editorial Reverté, S. A.

Hoffman, P., 2004. Plate evaporators in food industry–theory and practice. J. Food Eng. 61 (4), 515–520. https://doi.org/10.1016/S0260-8774(03)00296-6.

Ingham, J., Dunn, I.J., Heinzle, E., Přenosil, J.E., 2000. Chemical Engineering Dynamics: An Introduction to Modelling and Computer Simulation. Wiley-VCH Verlag.

Jafari, S.M., Çapanoğlu, E., Karaca, A.C., 2024. Evaporation Technology in Food Processing. Elsevier-Woodhead Publishing.

Jeurnink, T.J.M., Brinkman, D.W., 1994. The cleaning of heat exchangers and evaporators after processing millk or whey. Int. Dairy J. 4, 347–368.

Jiang, Y., Kang, L., Liu, Y., 2018. Simultaneous synthesis of a multiple-effect evaporation system with background process. Chem. Eng. Res. Des. 133, 79–89. https://doi.org/10.1016/j.cherd.2018.02.037.

Johnson, M.A., Moradi, M.H., 2005. PID Control: New Identification and Design Methods. Springer-Verlag, London.

Jun, S., Puri, V.M., 2005a. 3D milk-fouling model of plate heat exchangers using computational fluid dynamics. Int. J. Dairy Technol. 58 (4), 214–224. https://doi.org/10.1111/j.1471-0307.2005.00213.x.

Jun, S., Puri, V.M., 2005b. Fouling models for heat exchangers in dairy processing: a review. J. Food Process. Eng. 28 (1), 1–34. https://doi.org/10.1111/j.1745-4530.2005.00473.x.

Kern, D.Q., Seaton, R.E., 1959. A theoretical analysis of thermal surface fouling. Br. Chem. Eng. 4 (5), 258–262.

Ladha-Sabur, A., Bakalis, S., Fryer, P.J., Lopez-Quiroga, E., 2019. Mapping energy consumption in food manufacturing. Trends Food Sci. Technol. 86, 270–280. https://doi.org/10.1016/j.tifs.2019.02.034.

Lalande, M., Tissier, J.P., Corrieu, G., 1984. Fouling of a plate heat exchanger used in ultra-high-temperature sterilization of milk. J. Dairy Res. 51, 557–568.

Lee, S., Grossmann, I.E., 2000. New algorithms for nonlinear generalized disjunctive programming. Comput. Chem. Eng. 24 (9–10), 2125–2141.

Lee, S., Grossmann, I.E., 2003. Global optimization of nonlinear generalized disjunctive programming with bilinear equality constraints: applications to process networks. Comput. Chem. Eng. 27 (11), 1557–1575.

Loveday, S.M., 2016. β-Lactoglobulin heat denaturation: a critical assessment of kinetic modelling. Int. Dairy J. 52, 92–100. https://doi.org/10.1016/j.idairyj.2015.08.001.

Luyben, M.L., Luyben, W.L., 1997. Essentials of Process Control. McGraw-Hill.

Luyben, W., 1996. Process Modeling, Simulation and Control for Chemical Engineers. McGraw-Hill.

Mahdi, Y., Mouheb, A., Oufer, L., 2009. A dynamic model for milk fouling in a plate heat exchanger. Appl. Math. Model. 33 (2), 648–662. https://doi.org/10.1016/j.apm.2007.11.030.

Maroulis, Z.B., Saravacos, G.D., 2008. Food Plant Economics. Taylor & Francis Group, Boca Raton, FL.

McCormick, G.P., 1982. Nonlinear Programming, Theory, Algorithms and Applications. Wiley.

Mehta, B.M., 2015. Chemical composition of milk and milk products. In: Cheung, P.C.K., Mehta, B.M. (Eds.), Handbook of Food Chemistry. Springer Berlin Heidelberg, Berlin, Heidelberg, pp. 511–553.

Minton, P.E., 1986. Handbook of Evaporation Technology. Noyes Publications.

Miranda, V., Simpson, R., 2005. Modelling and simulation of an industrial multiple effect evaporator: tomato concentrate. J. Food Eng. 66 (2), 203–210. https://doi.org/10.1016/j.jfoodeng.2004.03.007.

Moejes, S.N., van Boxtel, A.J.B., 2017. Energy saving potential of emerging technologies in milk powder production. Trends Food Sci. Technol. 60, 31–42. https://doi.org/10.1016/j.tifs.2016.10.023.

Mojarab Soufiyan, M., Dadak, A., Hosseini, S.S., Nasiri, F., Dowlati, M., Tahmasebi, M., et al., 2016. Comprehensive exergy analysis of a commercial tomato paste plant with a double-effect evaporator. Energy 111, 910–922. https://doi.org/10.1016/j.energy.2016.06.030.

Molina-Pérez, H., Cano-Gómez, J.J., Díaz-Ovalle, C.O., Castillo-Borja, F., 2015. Equivalencia del espesor de ensuciamiento con b-lg desnaturalizada en el calentamiento de leche

[Equivalency of fouling thickness with denatured b-lg in heating of milk]. Avances en Ciencias e Ingeniería 6 (1), 49–62.

Morison, K.R., Hartel, R.W., 2019. Evaporation and freeze concentration. In: Heldman, D.R., Lund, D.B., Sabliov, C.M. (Eds.), Handbook of Food Engineering. Taylor & Francis Group, Boca Raton.

Morison, K.R., Mackay, F.M., 2001. Viscosity of lactose and whey protein solutions. Int. J. Food Prop. 4 (3), 441–454.

Newell, R.B., Fisher, D.G., 1972. Model development, reduction, and experimental evaluation for an evaporator. Ind. Eng. Chem. Process Des. Dev. 11 (2), 213–221. https://doi.org/10.1021/i260042a011.

Nishitani, H., Kunugita, E., 1979. The optimal flow pattern of mutieffect evaporator systems. Comput. Chem. Eng. 3, 261–268. https://doi.org/10.1016/0098-1354(79)80044-7.

Nukiyama, S., 1934. The maximum and minimum values of the heat Q transmitted from metal to boiling water under atmospheric pressure. J. Japan Soc. Mech. Engrs. 37 (207), 367.

Nukiyama, S., 1966. The maximum and minimum values of the heat Q transmitted from metal to boiling water under atmospheric pressure. Int. J. Heat Mass Tran. 9 (12), 1419–1433. https://doi.org/10.1016/0017-9310(66)90138-4.

Nukiyama, S., 1984. Memories of my research on boiling. Int. J. Heat Mass Tran. 27 (12), 955–957. https://doi.org/10.1016/0017-9310(84)90111-X.

O'Dwyer, A., 2009. Handbook of PI and PID Controller Tunning Rules. Imperial Colleges Press, London, UK.

Pan, F., Chen, X.D., Mercadé-Prieto, R., Xiao, J., 2019. Numerical simulation of milk fouling: taking fouling layer domain and localized surface reaction kinetics into account. Chem. Eng. Sci. 197, 306–316. https://doi.org/10.1016/j.ces.2018.12.021.

Pearce, R.J., 1992. Whey processing. In: Zadow, J.G. (Ed.), Whey and Lactose Processing. Springer Verlag, Dordrecht, The Netherlands.

Piri, A., Nikbakht, A.M., Hazervazifeh, A., Sarnavi, H.J., Fanaei, A.R., 2023. Role of 3E analysis in detection of thermodynamic losses in the evaporation line and steam and power production unit in the sugar processing plant. Biomass Convers. Biorefinery 13 (4), 3049–3069. https://doi.org/10.1007/s13399-021-01348-6.

Pitts, D., Sissom, L., 1998. Heat Transfer. McGraw-Hill.

Power, R.B., 1994. Steam Jet Ejectors for the Process Industries. McGraw-Hill Inc, New York.

Prost, J.S., González, M.T., Urbicain, M.J., 2006. Determination and correlation of heat transfer coefficients in a falling film evaporator. J. Food Eng. 73 (4), 320–326. https://doi.org/10.1016/j.jfoodeng.2005.01.032.

Ramírez-Rivera, M.J., Díaz-Ovalle, C.O., Ramos-Ojeda, E., Castrejón-González, E.O., 2024. CFD simulation analysis of fouling formation in a milk falling-film evaporator. Food Bioprod. Process. 143, 242–254. https://doi.org/10.1016/j.fbp.2023.12.003.

Runyon, C.H., Rumsey, T.R., McCarthy, K.L., 1991. Dynamic simulation of a nonlinear model of a double effect evaporator. J. Food Eng. 14 (3), 185–201. https://doi.org/10.1016/0260-8774(91)90007-F.

Sawaya, N.W., Grossmann, I.E., 2005. A cutting plane method for solving linear generalized disjunctive programming problems. Comput. Chem. Eng. 29 (9), 1891–1913.

Schreier, P.J.R., Fryer, P.J., 1995. Heat exchanger fouling: a model study of the scaleup of laboratory data. Chem. Eng. Sci. 50 (8), 1311–1321. https://doi.org/10.1016/0009-2509(95)98843-4.

Sharan, P., Bandyopadhyay, S., 2016a. Energy integration of multiple effect evaporators with background process and appropriate temperature selection. Ind. Eng. Chem. Res. 55 (6), 1630–1641. https://doi.org/10.1021/acs.iecr.5b03516.

Sharan, P., Bandyopadhyay, S., 2016b. Integration of thermo-vapor compressor with multiple-effect evaporator. Appl. Energy 184, 560–573. https://doi.org/10.1016/j.apenergy.2016.10.037.

Sharma, A., Macchietto, S., 2021. Fouling and cleaning of plate heat exchangers: dairy application. Food Bioprod. Process. 126, 32–41. https://doi.org/10.1016/j.fbp.2020.12.005.

Singh, G., Chopra, K., Tyagi, V.V., Pandey, A.K., Ma, Z., Ren, H., 2021. A comprehensive energy, exergy and enviroeconomic (3-E) analysis with carbon mitigation for multistage evaporation assisted milk powder production unit. Sustain. Energy Technol. Assessments 43, 100925. https://doi.org/10.1016/j.seta.2020.100925.

Singh, G., Singh, P.J., Tyagi, V.V., Barnwal, P., Pandey, A.K., 2019. Exergy and thermo-economic analysis of ghee production plant in dairy industry. Energy 167, 602–618. https://doi.org/10.1016/j.energy.2018.10.138.

Smith, R., 2008. Chemical Process Design and Integration. Wiley & Sons, Ltd.

Sogut, Z., Ilten, N., Oktay, Z., 2010. Energetic and exergetic performance evaluation of the quadruple-effect evaporator unit in tomato paste production. Energy 35 (9), 3821–3826. https://doi.org/10.1016/j.energy.2010.05.035.

Stephanopoulos, G., 2017. Chemical Process Control. An Introduction to Theory and Practice. pEARSON.

Tijskens, L.M.M., Hertog, M.L.A.T.M., Nicolaï, B.M., 2001. Food Process Modelling. Woodhead Publishing Limited and CRC Press LLC.

Tonelli, S.M., Romagnoli, J., Porras, J., 1990. Computer package for transient analysis of industrial multiple-effect evaporators. J. Food Eng. 12 (4), 267–281. https://doi.org/10.1016/0260-8774(90)90002-P.

Vecchietti, A., Grossmann, I.E., 2000. Modeling issues and implementation of language for disjunctive programming. Comput. Chem. Eng. 24 (9–10), 2143–2155.

Yadav, P., Jana, A.K., 2010. Simulation and control of a commercial double effect evaporator: tomato juice. Chem. Prod. Process Model. 5 (1). https://doi.org/10.2202/1934-2659.1443. Art. 6.1-Art. 6.20.

Zhang, Y., Munir, M.T., Udugama, I., Yu, W., Young, B.R., 2018. Modelling of a milk powder falling film evaporator for predicting process trends and comparison of energy consumption. J. Food Eng. 225, 26–33. https://doi.org/10.1016/j.jfoodeng.2018.01.016.

Zhong, Y., Deng, K., Zhao, S., Hu, J., Zhong, Y., Li, Q., et al., 2020. Experimental and numerical study on hydraulic performance of chevron brazed plate heat exchanger at low Reynolds number. Processes 8 (9), 1076.

Ziegler, J.G., Nichols, N.B., Rochester, N.Y., 1942. Optimum settings for automatic controllers. Trans ASME 64, 759–765. https://doi.org/10.1115/1.2899060.

Zisopoulos, F.K., Rossier-Miranda, F.J., van der Goot, A.J., Boom, R.M., 2017. The use of exergetic indicators in the food industry – a review. Crit. Rev. Food Sci. Nutr. 57 (1), 197–211. https://doi.org/10.1080/10408398.2014.975335.

Modeling of food extraction processes

9

A. Perez-Vazquez [1,3], P. Barciela [1,3], F. Chamorro [1,3], S. Seyyedi-Mansour [1,3], J. Echave [1,2,3], A.G. Pereira [1,3], Aurora Silva [1,4], P. Donn [1,3], M.A. Prieto [1,3] and M. Carpena [1,3]
[1]Instituto de Agroecoloxía e Alimentación (IAA), Universidade de Vigo, Nutrition and Food Group (NuFoG), Campus Auga, Ourense, Spain; [2]CIMO, LA SusTEC, Instituto Politécnico de Bragança, Campus de Santa Apolonia, Bragança, Portugal; [3]Nutrition and Food Group (NuFoG), Galicia Sur Health Research Institute (IIS Galicia Sur), Vigo, Spain; [4]REQUIMTE/LAQV, Instituto Superior de Engenharia do Porto, Instituto Politécnico do Porto, Porto, Portugal

9.1 Introduction

9.1.1 Overview of food extraction processes

In food industry, extraction processes involve the separation and concentration of valuable components including macronutrients (lipids, proteins, carbohydrates, fibers), micronutrients (minerals and vitamins), phytochemical compounds (phenolic compounds, terpenoids, carotenoids, alkaloids), pigments, and flavors (among others) from raw food materials, food waste or by-products. The goal of food extraction processes is to recover these compounds, intended to be used as foods, nutraceuticals, or in food products as supplements or additives, enhancing thereby their nutritional and functional properties. In the production of natural food products, food ingredients and nutraceuticals, this step is crucial, with various methods being employed to achieve effective extraction. Thus, extraction methodologies can be classified according to: (1) the state of sample, liquid or solid, involving solid-liquid extraction (SLE) and liquid-liquid extraction (LLE); (2) the extraction technique, with maceration, Soxhlet, infusion, heat, ultrasound or microwave assisted extractions (UAE and MAE), supercritical fluid extraction (SFE), and pressurized liquid extraction (PLE); (3) the physical action (mechanical processes) or could be enzymatic processes (Sinha et al., 2022; Sridhar et al., 2022). In this book chapter, an emphasis will be placed on SLE, LLE and SFE.

LLE methods rely on the differential solubility of an analyte between an aqueous phase and an immiscible organic solvent (Pena-Pereira et al., 2009). LLE has been widely used for extracting volatile compounds from alcoholic beverages, although newer techniques have emerged (Marín-San Román et al., 2020). Despite its simplicity, LLE remains popular in analytical chemistry due to its ability to extract compounds of varying volatilities. However, it has limitations, such as high solvent

Modeling of Unit Operations in the Food Industry. https://doi.org/10.1016/B978-0-12-822180-8.00002-X

use, potential chromatographic interferences, and time-consuming procedures (Pena-Pereira et al., 2009).

SLE is a common analytical method for preparing solid samples by partitioning analytes between the solid matrix and a liquid extractant. The efficiency of the process depends on three factors: the extractant's penetration into the matrix, the analyte diffusion, and the analyte solubility in the extractant. Traditionally, SLE had low efficiency, but improvements like high-temperature and pressure methods, along with microwave, ultrasound and pressurized liquid assistance, have enhanced its performance. These advancements have reduce extraction time, automate the process, and minimize the use of organic solvents (Chanioti et al., 2014).

SFE is an innovative extraction method that used fluids like carbon dioxide at their critical points, allowing for the manipulation of key properties like viscosity, diffusivity, and density to selectively extract compounds (Braga et al., 2023). By adjusting pressure and temperature, SFE enables fine-tuned selectivity and extraction efficiency in the recovering of both polar and non-polar compounds. Additionally, the use of co-solvents can modify the solvent's polarity, further enhancing extraction capabilities (T. Gong et al., 2021). This green extraction technique offers advantages such as high selectivity and versatility due to the ability to control extraction conditions (Perez-Vazquez et al., 2023a).

These different extractions methods intended to recover food ingredients or compounds, involve unit operations such as diffusion, solubilization, centrifugation, hydrodistillation (HD), filtration, and pressing, (Vannier et al., 2022), which could be modeled and optimized.

9.1.2 Importance of modeling in food extraction

Modeling is the process of simulating real-world systems to analyze, predict, and optimize their behavior. It involves many approaches, such as mechanistic models (which include process-based and rate-based/kinetic models), empiric models (including statistical models and Response Surface Methodology (RSM)), and hybrid models (combining multiple methods) (Herrero et al., 2010; Sanket et al., 2024; Tiwari et al., 2013). Modeling plays a crucial role in food extraction processes, providing insights that optimize efficiency, reducing costs, and enhancing product quality. Mathematical models help predict and control these processes, offering a scientific basis for optimizing extraction conditions, including temperature, solvent concentration, time, and pressure (R. F. da Silva et al., 2022). By simulating different extraction scenarios, models reduce the need for trial-and-error approaches, leading to better energy utilization and resource conservation.

Process optimization is one of the key benefits of modeling in food extraction. For example, SFE is highly dependent on factors like pressure and temperature. Modeling allows the prediction of the most efficient conditions, minimizing solvent use while maximizing yield and selectivity for targeted compounds (Perez-Vazquez et al., 2023a). Similarly, in enzyme assisted extraction (EAE), models simulate enzyme-substrate interactions, enabling more efficient breakdown of cell walls to enhance the physicochemical properties of the matrix (Hashemi et al., 2022). Modeling also facilitates scaling-up from laboratory to industrial-scale extraction. Many extraction processes behave differently when scaled, making it difficult to predict performance. However, reliable models account for variables that affect process dynamics at larger scales, reducing uncertainty and the risk of inefficiencies (Lloyd and van Wyk, 2011). Furthermore, modeling supports the development of green extraction technologies, where environmental impact is considered. Models can help minimize solvent consumption, energy use, and waste production, aligning with sustainability goals in food processing (R. F. da Silva et al., 2022).

9.2 Fundamental concepts in extraction

9.2.1 Principles of extraction

Extraction in food processing is fundamentally concerned with separating desired compounds, such as bioactive compounds or flavors, from a source material (feed) using a solvent. The process relies on the transfer of solutes from a solid or liquid matrix into a fluid phase, which involves the principles of solubility and mass transfer (R. F. da Silva et al., 2022). Solubility is defined as the process in which the solute distributes between the source material and the solvent until equilibrium is achieved. It is determined by the chemical potential, which is influenced by both concentration and activity of the solute, dictating its distribution. Mass transfer is another critical principle, involving the movement of solutes from the feed material into the solvent. Macroscopically, this process is governed by concentration gradients and the resistance encountered at the interface between the feed and solvent, which is in fact promoted by chemical potential. Mass transfer is hindered by three primary barriers: the layer where diffusion occurs on the side of the feed material; the interface between the feed and solvent; and the layer where diffusion happens on the solvent side, which depends on the chemical potential gradient. In food processing, where the feed material often has a complex structure (e.g., semi-solid or gel-like), minimizing these resistances is essential to improve extraction efficiency (Lloyd and van Wyk, 2011). Therefore, enhancing mass transfer rates through reducing particle size, increasing surface area, or applying energy-intensive techniques like agitation, heating, or ultrasound is often employed to optimize extraction. These methods aim to increase the contact between the solvent and the feed material, facilitating faster and more efficient solute transfer (Hashemi et al., 2022; Lloyd and van Wyk, 2011).

In food processing, solvent selection is also a crucial step to achieve optimal solubility. Ideally, the solvent should dissolve the desired solute effectively while minimizing the dissolution of unwanted materials. Moreover, it should be stable, not chemically reactive with the solute, and easily separable from both the extracted solute and the remaining feed material. Factors like polarity, temperature, and pressure must be considered to ensure that the target compounds are efficiently dissolved. The solvent's ability to be reused or recovered after extraction is also important, affecting both economic and environmental sustainability (Hashemi et al., 2022; Lloyd and van Wyk, 2011).

9.2.2 Types of extraction methods

The choice of extraction technique is shaped by multiple considerations, such as the attributes of the target compounds, the nature of the material being processed, and the quality requirements of the final product. Below is a concise summary of key points regarding several extraction methods and their recent updates.

9.2.2.1 Solid-liquid extraction

SLE is a widely utilized method in food processing for isolating valuable bioactive compounds, such as antioxidants, flavors, and pigments, from solid matrices like fruits, seeds, and plant materials. The process involves the partitioning of analytes between a solid matrix and a liquid extractant, governed by three primary mechanisms: the penetration of the solvent into the solid matrix, the diffusion of analytes from the interior to the surface, and their subsequent solubilization in the extractant (Chuo et al., 2022; Priego-Capote, 2021). SLE offers advantages due to its simplicity, low cost, and adaptability across various applications. However, its traditional form often suffers from lower extraction efficiencies, leading to the development of advanced techniques to improve performance. Strategies like using high temperatures and pressures and incorporating auxiliary energies such as microwaves and ultrasound have significantly enhanced the solvent's penetration and mass transfer rates, resulting in higher extraction yields. These modifications have led to techniques like MAE and UAE, which not only shorten extraction times but also reduce solvent use, aligning with principles of green chemistry (Patrice Didion et al., 2023; Priego-Capote, 2021).

An essential factor in optimizing SLE is solvent, which must be carefully selected based on the properties of target compounds and the intended analytical methods. Solvents with varying polarities, from intermediate (e.g., methyl tert-butyl ether, acetonitrile) to more polar options (e.g., methanol, ethanol), are employed depending on the nature of the analytes. However, in food processing, the selection of the solvent is also limited by its toxicity (Celeiro et al., 2021). Recent advancements in SLE, such as the use of superheated solvents and sustainable alternatives like natural deep eutectic solvents (NADES), have improved efficiency (Ferri et al., 2023; Patrice Didion et al., 2023). Modeling approaches are also increasingly applied to optimize extraction conditions, enabling better control over variables such as solvent type, temperature, and time to maximize yield and minimize environmental impact. These advancements ensure SLE remains a versatile and effective method for food extraction processes, allowing efficient recovery of high-quality compounds.

9.2.2.2 Liquid-liquid extraction

LLE is a commonly employed technique for separating compounds based on their differential solubility in two non-mixing liquid phases. This method is grounded in principles of phase equilibria, allowing solutes to distribute selectively between the phases. It is particularly valuable in the food and bioprocessing industries for isolating bioactive compounds, flavors, and essential oils (EOs). The effectiveness of LLE depends on the careful selection of a solvent system, where the target compounds preferentially partition into the solvent phase while impurities remain in the feed phase. Key factors influencing this process include the distribution coefficient, solvent density, interfacial tension, and solute properties. The choice of solvents is critical, since they must offer high selectivity and low toxicity, be easily recoverable, and pose minimal environmental risks. Thus, although traditional solvents like hexane, chloroform or ethyl acetate have been widely used in LLE due to their strong solvating power and ability to effectively separate a wide range of compounds, their toxicity, volatility, and environmental concerns make them unsuitable for food processing (Almohasin et al., 2023; Antony et al., 2021; Kannouma et al., 2022). In this regard, the use of green solvents in LLE has already been studied. For instance, ionic liquids (ILs), known for their low volatility and high thermal stability, are effective in extracting a broad range of compounds but present challenges due to their cost and potential environmental impact; while NADES provide a more economical and sustainable alternative, with high extraction efficiencies, particularly for bioactive compounds from natural sources and agricultural by-products (Rodríguez-Llorente et al., 2020).

Advanced LLE techniques have emerged to improve efficiency, such as reactive extraction, which incorporates a reversible chemical reaction to enhance selectivity and recovery, and membrane-based solvent extraction, which reduces solvent loss and prevents emulsion formation. Aqueous biphasic systems (ABS), formed by water-soluble polymers or salts, offer a low-toxicity environment ideal for extracting sensitive biomolecules like proteins and enzymes. ABS has also proven effective in recovering rare earth elements and other metals, providing a green alternative to conventional solvent methods (Pereira Neves et al., 2022). Despite its benefits, LLE faces challenges, including the cost of solvent recovery and the need for optimization to suit various applications. Future developments will likely focus on the creation of novel bio-based solvents and the integration of LLE with other techniques, such as SFE and membrane processes, to enhance both efficiency and sustainability. With ongoing innovation, LLE continues to be a flexible and effective method for diverse industrial uses, combining process optimization with environmental consciousness (Almohasin et al., 2023).

9.2.2.3 Supercritical fluid extraction

SFE uses supercritical fluids, primarily carbon dioxide (CO_2), to efficiently extract bioactive compounds from various materials. A fluid reaches a supercritical state when it is subjected to conditions above its critical temperature and pressure, where it exhibits both liquid-like solvency and gas-like diffusivity. This unique state enables the fluid to penetrate solid matrices deeply and dissolve target compounds effectively. CO_2 is the preferred supercritical fluid because of its low critical point (31 °C, 73.8 bar), non-toxic and non-flammable properties, and suitability for extracting heat-sensitive substances. The use of co-solvents like ethanol can further enhance its ability to dissolve polar compounds (Chemat et al., 2017; Echave et al., 2024). The performance of SFE is shaped by several critical factors, including the temperature and pressure conditions, the rate of CO_2 flow, and the particle size of the input material. Adjusting these parameters allows for selective extraction of both polar and non-polar compounds. High pressures increase CO_2 density, improving its capacity to dissolve compounds, while moderate temperatures (35 –50 °C) protect sensitive substances from thermal degradation. Optimal extraction conditions are achieved by carefully balancing these factors to maximize yield and maintain product integrity. For example, higher temperatures can decrease CO_2 density, reducing its solvating power; however, they also increase solute vapor pressure, which can enhance solubility under certain conditions (Dashtian et al., 2024).

SFE is widely regarded as a green extraction technique due to its minimal use of harmful solvents, lower energy consumption, and efficient extraction capabilities. It has been successfully applied in decaffeinating coffee and tea, extracting EOs, and recovering antioxidants and polyphenols from various food sources. The method has demonstrated effectiveness in extracting complex bioactive compounds, such as omega fatty acids from fruits and flavonoids from plant leaves (Boateng, 2024; Perez-Vazquez et al., 2023a). To further enhance its efficiency, SFE is often combined with different green techniques. Microfluidic supercritical extraction is an innovative method that operates with small fluid volumes (μL to pL) enabling real-time monitoring and precise control of extraction conditions, reducing energy consumption and optimizing yields (Dashtian et al., 2024). Despite its advantages, SFE does face challenges, including high initial costs for equipment and the need for specialized knowledge to manage high-pressure systems. Additionally, its scalability and economic feasibility for large-scale applications remain major challenges. While SFE is effective on a laboratory scale, further research is needed to establish its cost-effectiveness in industrial settings. Future developments in SFE are likely to focus on integrating it with other technologies and optimizing operational parameters to enhance scalability and reduce costs. Advances in co-solvent use, equipment design, and process configurations will help expand its applications in the food industry. As these innovations progress, SFE is expected to become a more viable option for producing high-quality extracts sustainably (Boateng, 2024; Perez-Vazquez et al., 2023a).

9.3 Mechanistic models of extraction

Mechanistic modeling is a powerful tool for understanding and predicting the behavior of complex systems by applying fundamental physical, chemical, and biological principles. It is typically based on systems of differential equations, as it is represented in Fig. 9.1A,B, and is particularly effective when its parameters are well-defined (Zhang, 2010). In the food industry, mechanistic models are employed to optimize processes and improve product quality.

Mechanistic models are often employed to simulate food heating using microwaves. A one-dimensional model, for example, integrates conductive heat transfer with electromagnetic heating to mimic the heating of spherical food products (Yang and Chen, 2021). Another application of mechanistic modeling in food processing includes simulating volume changes and deformation in food items, such as

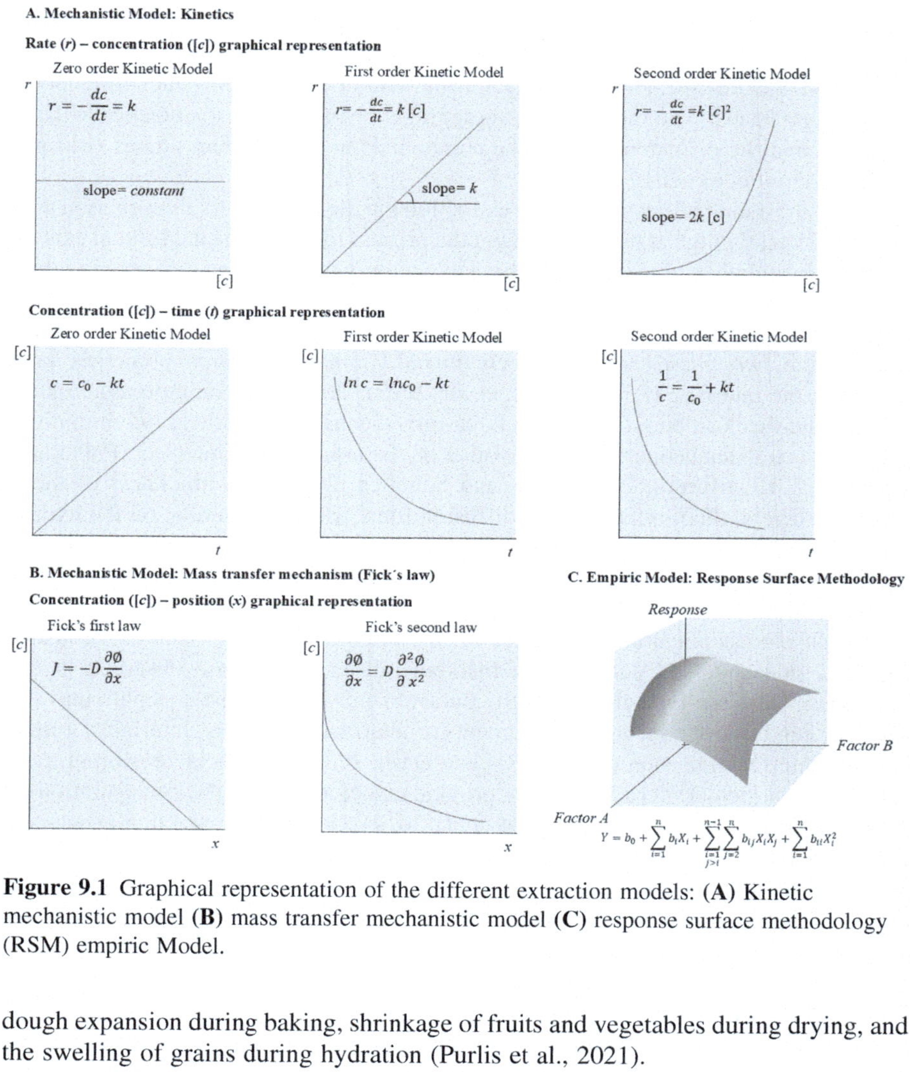

Figure 9.1 Graphical representation of the different extraction models: **(A)** Kinetic mechanistic model **(B)** mass transfer mechanistic model **(C)** response surface methodology (RSM) empiric Model.

dough expansion during baking, shrinkage of fruits and vegetables during drying, and the swelling of grains during hydration (Purlis et al., 2021).

9.3.1 *Kinetic models of extraction*

The extraction of bioactive compounds, flavors, EOs, colors, and nutrients from food raw materials is a fundamental operation in the food industry (Usman et al., 2022). Since extraction processes involve the movement of solutes from one phase to another (Lebovka et al., 2016), understanding the kinetics of these processes is crucial for the optimization of its efficiency, predicting extraction yields, and therefore, for the efficient design and optimization of unit operations (Belwal et al., 2020). This is because kinetic models provide a quantitative description of the rate at which compounds are transferred from one phase to another, offering valuable information about the mechanisms that govern these processes (Capuano et al., 2018). The extraction kinetics depend on the chemical reactions occurring within the system (including bond breaking and formation or aggregation-disaggregation processes), the kinetics of these reactions, and the diffusion rates of the compounds in the different phases (Bazant, 2013; Mircioiu et al., 2019).

One of the main factors influencing extraction kinetics is the activation energy, as it can be sufficiently high at the beginning of the process to determine the solvent extraction rate (Piskulich et al., 2019). Another key factor is the diffusion rate, which refers to amount of moles diffusing per unit area per unit time, governed by Fick's law (Varzakas, 2014). However, in most practical extraction processes (analytical or technological), the two phases are effectively mixed, limiting diffusion processes to a zone near the interface (Lainez-Cerón et al., 2022). There is a common assumption that the interface can be modeled as a three-dimensional region where two immobile layers, one on the aqueous side and the other on the organic side, intersect (Fuller and Vermant, 2012; Murray, 2002; Sagis and Scholten, 2014). The thickness of these immobile layers, often referred to as diffusion films, varies depending on the hydrodynamic conditions within the extraction apparatus. When these conditions allow for negligible thickness, the contribution of diffusion to the extraction rate diminishes (Assis et al., 2016; Ho et al., 2013).

The physicochemical characteristics of this interfacial region, such as viscosity, dielectric constant, and density, can markedly differ from those observed in the bulk phases and are typically not well characterized. Nonetheless, insights into the interface can be derived from measurements of interfacial viscosity, interfacial potential, and interfacial tension (Bhagya Raj and Dash, 2022). Interfacial tension investigations are particularly crucial as they provide essential data on the concentration of the extractant at the interface (Chemat et al., 2019). Generally, extraction agents are characterized by their dual hydrophilic and hydrophobic nature, depending on the specific compounds being extracted (polar compounds with hydrophilic solvents and nonpolar compounds with hydrophobic solvents) (Ortega-Zamora et al., 2021). The adsorption of extraction reagents at the water-organic interface can confer surfactant properties, increasing the interfacial concentration and promoting interfacial reactions. These reactions can create interfacial resistance if they are slow. However, if they are fast, the extraction process depends solely on diffusion within the two diffusion films, referred to as a "diffusion regime". When diffusion is instantaneous compared to chemical reactions, the process is in a "kinetic regime" where the rate of chemical reactions is determinant (Servis and Clark, 2018). In practice, diffusion and kinetic regimes often coexist, which gives rise to a combined mass transfer regime characterized by slow chemical reactions, where both diffusion and chemical reactions are important. This field is interdisciplinary and requires understanding the influence of hydrodynamics on solvent extraction kinetics (Kockmann and Agar, 2022). In designing equipment and processes for food processing and storage, the mass transfer coefficient at the interface is crucial, affecting drying, separation, and storage processes (Li et al., 2023).

Several kinetic models have been developed to describe and quantify the rate at which solutes are extracted under different conditions. The integration of these kinetic models enables food engineers and scientists to better understand and control the complex interactions and limiting steps in extraction processes, leading to improved process design and superior product quality in the food industry (Giannakourou et al., 2021). This subsection explores some of the fundamental kinetic models used in extraction studies.

9.3.1.1 Zero-order kinetic models

Reaction order is the sum of the exponents of reactant concentration terms in the rate equation (Toledo, 2019). Zero-order kinetic models play a crucial role in food science for the assessment and prediction of food product shelf life and quality under different storage conditions (Cui et al., 2023). This model assumes that the rate of reaction remains constant regardless of reactant concentration (Savva, 2019), being especially pertinent for phenomena such as nutrient degradation, enzymatic browning, and microbial proliferation, where environmental factors like temperature and pH are the primary determinants of the reaction rate (Ahmed et al., 2012; Amodio et al., 2017).

In food systems, zero-order reactions are often observed in situations where the concentration of the reactant is sufficiently high so that its level remains virtually unchanged over time. This leads to a reaction rate that does not depend on the concentration of the reactant (Santacesaria and Tesser, 2018; Toledo, 2019). A classic example of zero-order kinetics in food science is the Maillard reaction, where the rate of color development in foods can be described by zero-order kinetics (Van Boekel, 2008). The mathematical description of a zero-order reaction is given by the differential equation (Eq. 9.1):

$$-\frac{dc}{dt} = k \tag{9.1}$$

Integrating this equation results in the concentration of the reactant as a function of time (Eq. 9.2):

$$c = c_0 - kt \tag{9.2}$$

The integration of zero-order kinetics facilitates the prediction of both the quality and shelf life of different fruits such as kiwifruits. This integrated approach supports distributors in refining their inventory and sales strategies, while also helping consumers identify the optimal time for consumption (Zhang et al., 2021a).

9.3.1.2 First-order kinetic models

First-order kinetic models are among the simplest and most widely used to describe extraction processes (Toledo, 2019). These models serve as a cornerstone in food science for analyzing and predicting the dynamics of various chemical and biological reactions that affect food products (Van Boekel, 2008). They are based on the principle that the reaction rate (r) is directly proportional to the concentration of a single reactant, expressed mathematically as Eq. (9.3), where k is the rate constant and c represents the concentration of the reactant.

$$r = -\frac{dc}{dt} = kc \tag{9.3}$$

The integration of this equation leads to Eq. (9.4):

$$c = c_0 \exp(-kt) \tag{9.4}$$

where c_0 represents the initial concentration. As first-order reaction is defined by a logarithmic alteration in the concentration of a reactant over time, the previous nonlinear equations are frequently transformed into the linear form reported in Eq. (9.5):

$$ln\ c = ln\ c_0 - kt \tag{9.5}$$

This linear relationship facilitates the analysis of numerous processes in food systems, including the degradation of vitamins, the loss of flavor compounds, the evolution of off-flavors, and the activity of preservatives (Kong and Singh, 2016). For example, first-order kinetic models explain the thermal degradation of betanin, a natural pigment derived from red beets (Van Boekel, 2008) or the thermal degradation of color in peach puree (Ávila and Silva, 1999). In the degradation of vitamin D_2 in mushroom powder, first-order kinetics helps estimate the rate at which the vitamin degrades over during storage under different environmental conditions, which is crucial for determining the shelf life of products fortified with this nutrient (Pedrali et al., 2020). Commonly employed first-order kinetics models include single first-order kinetics, logarithm of the slope transformed kinetics, parallel first-order kinetics, and a combination of parallel and sequential kinetics models (Yu et al., 2021). By incorporating temperature, pH, or other variables like environmental conditions into these models, researchers can simulate real-world conditions and make predictions about how these factors influence reaction rates (Arcus and Mulholland, 2020; Peleg, 2022).

9.3.1.3 Second-order kinetic models

Second-order kinetic models provide a comprehensive description for extraction processes involving interactions between solute molecules and a solid matrix. In these cases, the extraction rate is proportional to the square of the concentration of the extractable compound. Therefore, a second-order bimolecular reaction exhibits a similar concentration-versus-time plot as a first-order unimolecular reaction, but with a rate constant that varies with the concentration of the second reactant (Fernández-Ramos et al., 2006). The second-order kinetic equation ($n = 2$) is typically expressed as:

$$r = -\frac{dc}{dt} = kc^2 \tag{9.6}$$

Upon integration, this equation yields:

$$c = \frac{c_0}{1 + c_0 kt} = kc^2 \tag{9.7}$$

which can be linearized as:

$$\frac{1}{c} = \frac{1}{c_0} + kt \tag{9.8}$$

This linear form is useful for plotting 1/c against time. Second-order bimolecular reactions also follow a similar rate equation:

$$r = -\frac{dc}{dt} = kcb \tag{9.9}$$

where *c* and *b* represent reactants. Integrating this equation with b held constant gives:

$$ln\frac{c}{c_0} = -K't \tag{9.10}$$

where *K'* is the pseudo-first-order rate constant ($K' = k \cdot b$). Such reactions exhibit concentration-dependent rate constants for different levels of the second reactant, typically oxygen.

Second-order kinetics find application in diverse scenarios, including the degradation of amino acids in the Maillard reaction, such as the loss of lysine in sterilized milk, which serves as an indicator of quality degradation (Van Boekel, 2008), or in processes like the aerobic degradation of ascorbic acid (Villota and Hawkes, 2006). These models effectively describe situations where the extraction rate decreases notably over time due to the depletion of easily accessible solute molecules and increased resistance to mass transfer within the solid matrix (Götz, 1995; Zhang et al., 2021b). Despite their utility, second-order kinetics are less prevalent in food science, possibly due to the intricate nature of reactions such as fat oxidation and the Maillard reaction, which involve complex networks of reactions rather than simple equations (Ahmed et al., 2012; Van Boekel, 2008). Multi-response mechanistic modeling techniques are valuable for unraveling such networks, providing detailed scientific insights, albeit with limited direct practical applicability. Simplified models for color formation in the Maillard reaction focus on critical reaction steps (De Vleeschouwer et al., 2009; Selim and Zhang, 2007). In practical applications, deterministic models like those discussed above yield consistent outputs under fixed conditions, but may oversimplify the inherent variability and uncertainty in biological systems. To address this challenge, stochastic models are introduced, accommodating multiplicative errors or employing logarithmic transformations to enhance accuracy (Mundi et al., 2019).

As the number of parameters in a model increases, the fit generally improves to a dataset. However, this improvement comes at a cost: the uncertainty in estimating these parameters also increases, which is undesirable. Importantly, increased parameter uncertainty leads to significantly higher uncertainty in the predictions of the model due to error propagation (Van Boekel, 1996). Therefore, it is advisable to seek a model with the fewest parameters that still provides an acceptable fit. This principle is known as parsimony or Occam's razor (Guthery et al., 2003), that favors simpler models with fewer parameters to minimize uncertainty in predictions. When multiple models appear to fit reasonably well, a useful approach is model discrimination. One effective tool for this purpose is the Akaike criterion, which evaluates the residual sums-of-squares of different models while penalizing those with more parameters (Guthery et al., 2003). These refinements aim to improve the clarity and coherence of the scientific text while preserving the essential references and logical flow of information.

Moreover, there is frequently no clear theoretical basis for selecting a specific reaction order. Often, the order is merely determined as a parameter that fits experimental data, as observed in numerous studies. Consequently, alternative empirical models have been suggested recently (Corradini and Peleg, 2006a, 2006b). These models should not be seen as superior or inferior to traditional models with fixed orders; rather, they offer additional options that should be considered. In certain scenarios, these alternative models may offer a more accurate fit to experimental data.

9.3.2 Mass transfer mechanism

Mass transfer mechanisms in the food industry are critical to processes such as drying, osmotic dehydration, and fermentation. Mechanistic modeling provides valuable insights into these processes by describing the underlying physical phenomena. Models based on principles like Fick's laws of diffusion are commonly used to predict mass transfer rates.

According to Fick's First law of diffusion (Eq. 9.11), the rate at which a substance diffuses through a medium is directly proportional to the concentration gradient, or the rate of change in concentration relative to position within the medium. This law is particularly applicable in steady-state conditions, where J is the diffusion gradient, D the diffusivity constant, and $\frac{\partial \varnothing}{\partial x}$ the concentration gradient.

Fick's Second Law (Eq. 9.12), on the other hand, describes how concentration changes over time due to diffusion, making it useful for non-steady-state conditions (Xing et al., 2023). Thus, diffusion-based kinetic models are essential for understanding the internal mass transfer mechanisms that govern the extraction of compounds from solid food matrices (Whitehead, 2021).

$$J = -D\frac{\partial \varnothing}{\partial x} \tag{9.11}$$

$$\frac{\partial \varnothing}{\partial x} = D\frac{\partial^2 \varnothing}{\partial x^2} \tag{9.12}$$

In practical extraction applications, this partial differential equation is often simplified by assuming a uniform initial distribution of the solute and constant boundary conditions. Solutions derived from this equation offer valuable insights into the temporal evolution of concentration profiles within both the solid matrix and the extraction medium.

Within the realm of food industry applications, diffusion-based models serve as indispensable tools for the design and optimization of various processes such as solvent extraction, SFE, and EAE (Wu et al., 2017). These models facilitate the determination of optimal parameters including extraction duration, temperature, and solvent-to-solid ratios, thereby maximizing yield and efficiency while preserving the quality and bioactivity of the extracted compounds (Martinez Fernandez, 2020; Pinela et al., 2019).

There are several examples of mechanistic modeling applications in the food industry. For instance, in osmotic dehydration of fruits and vegetables. Since water and solute transfer between food and the osmotic solution are driven by concentration gradients, mechanistic models, such as Fick's laws of diffusion, are often used to predict these mass transfer rates (González-Pérez et al., 2021). It is also important to consider that, in drying processes, heat and mass transfer occur simultaneously. Mechanistic models help in understanding the kinetics of moisture removal and the impact of variables like temperature and air velocity. These models can be enhanced with machine learning (ML) techniques to improve accuracy and efficiency (Khan et al., 2022).

As for volume change and deformation phenomena that occur in foods during processes like drying, cooking, and baking, mechanics-based modeling offers significant benefits in terms of its ability to predict and interpret the phenomena involved. It also offers helpful tools and insights to enhance comprehension of the connections between raw material composition and structure, processing parameters, and final product properties (Purlis et al., 2021).

9.3.3 Thermodynamic considerations

Thermodynamic properties play a fundamental role in the behavior of food systems, which consist of diverse molecular classes such as volatile compounds, proteins, carbohydrates, lipids, organic acids, and minerals. In these complex systems, overall behavior is influenced more by the interactions between components than by the properties of individual compounds in their pure form. While the thermodynamic properties of individual molecules are well understood, predicting the behavior of complex food matrices under varying conditions remains challenging. This is largely because molecular interactions can lead to non-linear or unpredictable behaviors that are difficult to model based solely on isolated components (Polachini et al., 2023).

When it comes to extraction processes, thermodynamic considerations are crucial for understanding mass transfer, phase equilibria, and energy changes, all of which influence the efficiency and dynamics of extracting target compounds from a food matrix. Whether dealing with solids, liquids, or gases, key thermodynamic factors include Gibbs free energy, enthalpy (ΔH), entropy (ΔS), partition coefficient (K), temperature, chemical potential, phase equilibrium, solvent-solute interactions, mass transfer, diffusion, and activity coefficients. These principles provide a framework for optimizing extraction processes, making them more predictable and effective (Albarri and Şahin, 2023; Haghighi and Khajenoori, 2013). For this reason, several studies have conducted thermodynamic analysis for the optimization of extraction processes performed in food matrices. For instance, the thermodynamic analysis of UAE process was performed in *Moringa oleifera* leaves. The parameters analyzed were ΔH, ΔS, and Gibbs free energy, and results showed that the process is endothermic, random and spontaneous (Albarri and Şahin, 2023). Changes in ΔH and ΔS were also studied in garlic pennycress seed oil recovered by maceration, showing an endothermic, irreversible, and spontaneous process (Kostić et al., 2024).

9.3.4 Computational fluid dynamics (CFD) in extraction

Computational Fluid Dynamics (CFD) is a scientific field that uses digital technology to generate quantitative predictions of fluid flow behavior. These predictions are based on the fundamental conservation laws (mass, momentum, and energy) that govern fluid motion. The simulations typically depend on specific conditions, such as flow geometry, fluid properties, and boundary and initial conditions of the flow field. Results often include flow variable values sets, like velocity, pressure, or temperature, at different points within the domain and over time. Additionally, CFD can evaluate overall flow characteristics, such as flow rate or hydrodynamic forces acting on objects within the flow (Kajishima, 2012).

In the context of the food industry, including traditional and modern extraction processes, multiple complex and interacting phenomena must be considered. These include heat and mass transfer, various mechanisms of heat generation, phase changes, chemical reactions, and operational conditions (Roohi and Hashemi, 2019). CFD is recognized as an essential, cost-effective, and valuable tool for analyzing and predicting the behavior of food systems during extraction. It aids in modeling solvent flow through food matrices, optimizing parameters such as flow rate, pressure, and temperature to enhance extraction efficiency. By simulating heat and mass transfer dynamics, CFD helps ensure uniform solvent distribution, improves the design of extraction equipment, and reduces the need for trial-and-error experimentation (Xia and Sun, 2002). This makes CFD highly valuable for understanding and optimizing extraction processes in the food industry. For instance, a study employed transient CFD simulations to analyze the effects of salinity concentration and applied voltage on the extraction rate of EO during the ohmic heating process. This method provided a comprehensive understanding of how these factors influence mass transfer and optimize extraction efficiency over time, and experimental data was in agreement with the numerical results obtained (Roohi and Hashemi, 2019). In another study, CFD modeling was used to determine solid-liquid extraction applied to *Robinia pseudocacia*. The optimization as performed is based on two factors: the maceration time and the flow rate. CFD results were correlated with the data obtain after the experiment was run, showing the efficiency of this method to predict qualitative and quantitative data in an accurate manner (Caré et al., 2023).

9.4 Empiric model of extraction: response surface methodology (RSM)

Response Surface Methodology (RSM) is a set of mathematical and statistical techniques used to model and analyze problems where a response (output variable) is influenced by several independent factors (input variables), as it is represented in Fig. 9.1C. It is widely applied in experimental design to optimize workflows and identify the set of parameters that produce the best possible outcomes (Barton, 2013). RSM combines design of experiments (DoE), regression analysis, and optimization methods to improve or optimize the performance or response of a process or product (Barton, 2013). It was first introduced by George E.P. Box and K.B. Wilson in 1951 (Box and Wilson, 1951) and has since been applied successfully in various fields, including chemical process design, engineering, and manufacturing. Since the early 2000s, RSM has been widely used in many food processing applications, including the optimization of yeast conditions for the production of enzymes, the determination of ideal spray and fluid bed drying conditions, the optimization of osmotic dehydration conditions of various products, and the optimization of baking, pasteurization, and extrusion process conditions, among others (Worku Kidane, 2021).

As an example, the following second-order mathematical model with interactive factors (Eq. 9.13) was applied recently to the optimization of an antioxidant rich -extract from an edible algae specie (Cassani et al., 2024).

$$Y = b_0 + \sum_{i=1}^{n} b_i X_i + \sum_{\substack{i=1 \\ j>i}}^{n-1} \sum_{j=2}^{n} b_{ij} X_i X_j + \sum_{i=1}^{n} b_{ii} X_i^2 \tag{9.13}$$

The response variable, or dependent variable, in the equation is called Y. The independent variables are X_i and X_j. The constant coefficient is denoted by b_0. The coefficients of linear effect, interaction effect, and quadratic effect are denoted by b_i, b_{ij} and b_{ii}, respectively, and the number of variables considered in the analysis is indicated by n.

9.4.1 Design of experiments (DoE)

DoE is a branch of applied statistics that focuses on planning, conducting, analyzing, and interpreting controlled tests to evaluate the factors that control the value of a parameter or group of parameters. It allows for multiple input factors to be manipulated simultaneously to determine their effect on a desired output (response) (Del Vecchio, 2007). This method helps identify important interactions that might be missed when experimenting with one factor at a time. DoE is used in various fields, including engineering, manufacturing, and social sciences, to optimize processes, improve quality, and increase efficiency. Key concepts in DoE include blocking, randomization, and replication, which help ensure the validity, reliability, and replicability of the experiments (Lamidi et al., 2024).

9.4.1.1 *Factorial designs*

Classical designs are a fundamental starting point for constructing test designs. These include full factorial and fractional factorial designs. A notable example of a full factorial design is the 3^k factorial design, where k represents the number of factors, each with two levels. A full factorial design encompasses all possible combinations of factors in a test, varying the factors simultaneously rather than one at a time. This approach allows to examine both main effects (the impact of independent variables on the dependent variable) and interactions (the combined effect of independent variables on the dependent variable) for both categorical and continuous factors (Box and Wilson, 1951).

9.4.1.2 *Central composite designs*

This design is particularly useful for estimating curvature and interactions between factors, making it a powerful tool for optimization and process improvement (Bhattacharya, 2021). Advantages of Central Composite Design (CCD) include efficiency, as it requires fewer experimental runs compared to a full factorial design; flexibility, as it can accommodate different types of factors and be easily adapted to various experimental conditions; excellent estimation of curvature, which helps in identifying optimal conditions; and robustness to missing data (Stamenković et al., 2018). However, CCD also has limitations such as complexity in design and analysis, and is only suitable to continuous factors (Gilman et al., 2021).

9.4.1.3 *Box-Behnken designs*

The Box-Behnken Design (BBD) is an experimental design used in RSM to fit quadratic models while avoiding extreme treatment combinations, such as corner points. It assigns each factor to one of three equally spaced levels, typically coded as -1, 0, and $+1$, and is structured to prevent overly extreme combinations (Gilman et al., 2021). This design is often favored for its efficiency, requiring fewer experimental runs compared to alternatives like the central composite design (Nguyen et al., 2023). One of its key advantages is that it provides sufficient data for quadratic models with fewer runs, making it both cost-effective and time-efficient (Stamenković et al., 2018). Additionally, by avoiding extreme points, it reduces the risk of issues with experimental boundaries (Ferreira et al., 2007). However, its limitations include poor coverage of the corners of the design space, which can hinder exploration of the entire response surface (Gilman et al., 2021). It is also less suitable for experiments requiring more than three levels per factor or involving discontinuous variables (Ferreira et al., 2007; Gilman et al., 2021). Moreover, comparative studies involving BBD, Face Central Composite (FCCD) and Full Factorial Designs (FFD) for optimizing solid-liquid extraction processes indicate that BBD and FCCD combined with RSM is recommended, since they have better statistical performances and are economically advantageous (Stamenković et al., 2018).

9.4.2 Model fitting and validation

Validation in RSM studies is a crucial step to ensure the reliability and accuracy of the developed models. First, it is essential to assess the confidence levels of each parameter during model construction, by perform an analysis of variance (ANOVA) to verify their significance using Fisher's F-test and the probability (*P*-value) (Peng et al., 2020). After determining the model's coefficients, validation typically involves comparing the predicted responses from the RSM models with actual experimental data (from independent runs) to evaluate the model's performance. Statistical validation metrics include the coefficient of determination (R^2), which indicates how well the model explains the variability in the response data, and the root mean square error (RMSE), which measures the differences between the predicted and observed values (Loan et al., 2024; Peng et al., 2020).

9.4.3 Optimization of extraction parameters

Optimizing extraction parameters is essential for enhancing efficiency, yield, and quality in various extraction processes, particularly in the food industry. While optimization requires significant time, energy, and resources, it addresses specific challenges related to process intensification and ensures optimal results. During the optimization phase, key factors identified through screening or prior experience are thoroughly analyzed to determine ideal conditions. The outcome of any extraction process is influenced by numerous factors and their interactions, such as mass transfer and extraction kinetics, all of which must be considered to achieve the best possible result (A. K. Das and Dewanjee, 2018).

Among these parameters, the most relevant in the extraction process are solvent, temperature, time, and particle size of the matrix. Additionally, the solvent's polarity and solvent-to-solid ratio are critical factors that influence mass transfer and extraction efficiency, since they directly affect the ability to dissolve the target compounds and the concentration gradient driving mass transfer (Y. Gong et al., 2012; Zakaria et al., 2023). Other important parameters include pressure in techniques like pressurized liquid extraction and electrical parameters in techniques such as ohmic heating (Al-Hilphy et al., 2023; Perez-Vazquez et al., 2023b). The combined effects of these variables must be carefully balanced, as their interactions can significantly impact on the overall efficiency of the extraction process.

9.5 Hybrid models in extraction

9.5.1 Combining mechanistic and empiric dynamic models

Hybrid models combine the descriptive character of empirical models with the predictive capability of mechanistic modeling to provide a more comprehensive representation of extraction processes. Dynamical models are essential for understanding how extraction variables evolve over time and for optimizing process efficiency (I. Das and Arora, 2021). For instance, Peleg's model, usually applied in processes such as adsorption, offers useful parameters describing the initial extraction rates and the maximum amount extracted, while first- and second-order kinetic models best describe the exponential decay or concentration-dependent behavior of solute transfer, respectively (Kaderides et al., 2019).

Mechanistic models largely implement ordinary differential equations (ODEs) to recreate the dynamic biological phenomena, metamorphosing biological systems with emphasis on chemical and physical core components. They explain how variables changes through time and space incorporating reaction rates, diffusion and equilibrium for example (Noordijk et al., 2024). For example, in the case of subcritical water extraction (SCWE) and SFE-CO_2 modeling, mechanistic models focus on how diffusion coefficients and desorption rates affect the extraction rate. On the other hand, empirical models complement this by fitting experimental data to estimate parameters that are difficult to measure directly. Coming back to the previous example, integrating both approaches enables more accurate and practical modeling of complex processes, combining detailed theoretical insights with experimental validation and parameterization (B. G. da Silva et al., 2023).

For example, in a study of *Marrubium vulgare* L., HD and volatile solvent extraction were modified and optimized using dynamic models. The extraction temperature, particle size, and solvent were the selected variables. The optimal conditions were found to be 3% of mass ratio, 180 min extraction time, particle size range between 0.1–0.63 mm, and an extraction temperature of 120 °C (Rezazi et al., 2017). Using a sigmoid model successfully explained and predicted the relation between temperature and yield, with the efficiency of the extraction being improved at higher temperatures. This methodology allowed combining data on rate-based kinetics and data on the mechanism of diffusion in order to optimize the extraction process (Rezazi et al., 2017).

In another example focusing on optimizing SFE-CO_2, researchers used a hybrid model: the precursor-dependent synthesis and first order independent degradation. The mechanistic model described extraction dynamics with a differential equation for solute concentration, where K_{solute} and K_{decay} are the solute extraction and decay rate constants (Eq. 9.14):

$$\frac{dC_{solute}(t)}{dt} = K_{solute} \times C_{Co2}(t) - K_{decay} \times C_{solute}(t) \quad (9.14)$$

This model used constraint-based methods to optimize parameters such as pressure and temperature. By refining fluxes with the kinetic model and analyzing equilibrium behavior with Essential Flux Modes (EFMs), the study achieved a significant improvement in the extraction efficiency and yield (Moulin et al., 2021).

9.5.2 Hybrid–machine learning approaches

To overcome obstacles in extraction approaches, hybrid approaches that combine multiple technologies are increasingly employed. By integrating mechanistic or empiric models with predictive machine learning (ML), these hybrid systems achieve superior performance than using the systems individually (Adeib Idris and Markom, 2019; Baker et al., 2018; Sansana et al., 2021). ML methods such as data classification, data clustering, regressive predictive modeling, ensemble methods, clustering, transfer learning, reinforcement learning, natural language processing, and deep learning are regularly used in engineering with minimal human supervision (Adeib Idris and Markom, 2019). Mechanistic models offer deep biological insights but struggle with scalability and computational complexity. ML, while handling large datasets and making accurate predictions, often lacks mechanistic clarity. To address these challenges, hybrid modeling integrates both methodologies (Noordijk et al., 2024).

In this scenario, first-principles modeling is employed in conjunction with ML to allow consideration of multiple factors. The objective is to separate the state space of the complex system (CSS) into different operating regimes, which is done by extracting low-dimensional deep features from the data. Sub-models are then constructed for each regime, and the overall dynamic model is expressed as an interpolation of all these sub-models. The associated degrees, which indicate the probability that the current operating point belongs to a certain condition, are not constant, but depend on the displacements of the system operating point (Sun et al., 2020). This level of dynamization improves predictive and comprehensibility performances, making the processes such as extraction very efficient by using both the mechanistic and ML type of modeling (Noordijk et al., 2024). For instance, researchers have combined mechanistic modeling with ML to optimize the thickness of frozen microwaveable meals, enhancing heating uniformity and efficiency, which leads to higher-quality products (Yang et al., 2021). In other study, mechanistic modeling combined with ML was applied to optimize container thickness for better heating and refreezing uniformity in microwaveable food products. The authors employed Bayesian optimization techniques, demonstrating that this combined approach is highly effective for the design and optimization of complex food items (Yang et al., 2021).

Focusing on the isolation of different food components, an ML based hybrid model was suggested to increase the performance of the extraction. Food images were captured, then processed and segmented with the RCNN Mask Region based model. After that, the segmented images were processed in a discriminatory deep learning model, to extract features from them. These features were then used in different ML classifiers including support vector machine (SVM) and random forest (RF) for the estimation of the concentration of analytes in food. This hybrid strategy significantly enhanced extraction efficiencies, hence leading to higher accuracy and consistency in quantification of analytes (Agarwal et al., 2023).

Table 9.1 Comparison of predicted and experimental extraction results of compounds present in food matrices using Response Surface Methodology (RSM).

Sample	Target compound	Extraction Technique	Independent variables	Levels	Experi-mental design	N runs	Predicted results	Experimental results	% Deviation	Validation of the method			Ref.
										R^2	*p*	*LoF*	
Plants													
Olea europaea (L)	Oleuropein	UAE	*MUAE, L–S, t*	3	BBD	16	Y: 12.898 mg/g dw	Y: 13.386 mg/g dw	3.784	0.500	<0.005	14.426	Giacometti et al. (2018)
	Verbascoside						Y: 0.324 mg/g dw	Y: 0.363 mg/g dw	12.037	0.840	<0.005	0.005	
	Luteolin-4′-O-glucoside						Y: 0.412 mg/g dw	Y: 0.527 mg/g dw	27.913	0.856	<0.005	0.009	
Vernonia amygdalina (L)	Flavonoid	MAE	*L–S, P, t, T*	3	FCCCD	30	TFC: 86.5051 mg QE/g dw	TFC: 87.05 mg QE/g dw	0.630	0.9889	0.0992	–	Alara et al. (2018)
							DPPH: 94.3254%	DPPH: 95.93%	1.701	0.9886	0.1153	–	
							ABTS: 95.8915%	ABTS: 97.22%	1.385	0.9891	0.1241	–	
Arnica montana (F)	Polyphenols	HAE	*t, T, % EtOH*	5	CCC	28	Y: 1.877 g E/g dw	Y:2.147 g E/g dw	14.385	0.766	<0.0001	0.126	(Garcia-Oliveira et al., 2024a, 2024b)
							TPC: 64.20 mg/g dw	TPC: 83.61 mg/g dw	30.234	0.846	<0.0001	0.21	
							TPC: 440.73 mg/g E	TPC: 389.98 mg/g E	11.515	0.837	<0.0001	0.103	
		UAE	*t, T, P, % EtOH*				Y: 1.098 g E/g dw	Y: 1.242 g E/g dw	13.115	0.767	<0.0001	0.376	
							TPC: 50.82 mg/g dw	TPC: 55.16 mg/g dw	8.540	0.821	0.0004	0.0565	
							TPC: 499.79 mg/g E	TPC: 447.94 mg/g E	10.374	0.983	<0.0001	0.112	
Arnica montana (F)	Polyphenols	PLE	*t, T, % EtOH*	5	CCC	28	Y: 0.1910 g/g dw	Y: 0.1955 g/g dw	2.356	0.948	<0.0001	0.126	(Garcia-Oliveira et al., 2024a, 2024b)
							TPC: 78.61 mg/g dw	TPC: 71.76 mg/g dw	8.714	0.834	<0.0001	0.567	
		MAE	*t, T, % EtOH*				TPC: 388.88 mg/g E	TPC: 368.28 mg/g E	5.297	0.851	<0.0001	0.687	
							Y: 0.3033 g/g dw	Y: 0.3019 g/g dw	0.462	0.952	<0.0001	0.469	
							TPC: 245.73 mg/g E	TPC mg/g E: 286.36 mg/g E	16.534	0.771	0.0002	0.070	
Rosmarinus officinalis	Polyphenols	UAE	*t, T, WC, MD*	3	FCCD	30	TAC: 0.975 mmol TE/g dw	TAC: 0.498 mmol TE/g dw	48.923	0.9044	<0.0001	0.0105	Erdoğan et al. (2024)

Essential oils												
Rosmarinus officinalis (L)	Essential oil	HD	*t, T*	3	CCC	13	Y: 0.0270–0.0278 g/g dw	Y: 0.0285 g/g dw	5.556–2.518	0.937	<0.0001	0.005	Yeddes et al. (2022)
Citrus aurantium (P)	Essential oil	HD UAE	*L–S, t UAE, t HD*	–	CCRD	20	Y: 0.0099 mL/g	Y: 0.0090 mL/g	9.091	0.8445	<0.0001	0.047	Heydari et al. (2021)
							TPC: 1.0833 mg GAE/mL	TPC: 1.1003 mg GAE/mL	1.569	0.9169	<0.0001	0.8007	
							IC_{50}: 251.56 μg/mL	IC_{50}: 263.28 μg/mL	4.659	0.9457	<0.0001	0.5174	
Coriandrum sativum (L)	Essential oil	HD	*L–S, T, t, % H_2O*	3	CD	27	Y: 0.0110 g/g	Y: 0.0109 g/g	0.909	0.97	<0.001	0.0009	Scandar et al. (2024)
Thymus capitatus	–	NE	*X_1, X_2, X_3, X_4*	3	–	11	dsd: 451.88 nm	dsd: 463.0 nm	2.461	0.981	<0.001	–	BenJemaa et al. (2022)
Food waste												
Juglans regia (L)	Polyphenols	HAE	*L–S, T, t*	3	BBD	17	TPC: 880.39 mg GAE/L	TPC: 866.01 mg GAE/L	1.633	0.988	<0.0001	0.005	Simsek and Süfer (2021)
							TTC: 75.89 mg TAE/L	TTC: 53.23 mg TAE/L	29.859	0.901	<0.0001	0.005	
Actinidia deliciosa (B)	Polyphenols	HAE	*t, T, pH*	3	BBD	17	TPC: 79.5374 mg GAE/g dw	TPC: 76.6832 mg GAE/g dw	3.722	0.9945	<0.0001		Aires and Carvalho (2020)
							TFC: 636.61 mg CAE/g dw	TFC: 635.65 mg CAE/g dw	0.151	0.9999	<0.0001		
							DPPH: 79.74%	DPPH: 79.22%	0.652	0.9970	<0.0001		
Actinidia deliciosa (P)	Flavonoids	UAE	*t, P, % EtOH*	5	CCRD	20	Y: 46% extract	Y: 46% extract	0	0.8999	<0.0001	ns	Giordano et al. (2021)
							Y1: 0.426 mg/g dw	Y1: 0.432 mg/g dw	1.408	0.9816	<0.0001	ns	
							Y2: 0.78 mg/g dw	Y2: 0.78 mg/g dw	0	0.9718	<0.0001	ns	
							Y3: 0.148 mg/g dw	Y3: 0.150 mg/g dw	1.351	0.9899	<0.0001	ns	
							Y4: 0.145 mg/g dw	Y4: 0.1468 mg/g dw	1.241	0.8879	<0.0001	ns	
							TFC: 1.49 mg/g dw	TFC: 1.51 mg/g dw	1.342	0.9728	<0.0001	ns	
Solanum lycopersicum (SD)	Phenolic compounds	MAE	*L–S, P, t*	3	CCD		TPC: 2.6847 mg GAE/g dw	TPC: 2.6531 mg GAE/g dw	1.177	0.9699	<0.0001	0.8109	Ouatmani et al. (2022)
							AA: 84.27%	AA: 82.97%	1.543	0.9515	<0.0001	0.4308	
Vitis vinifera (PM)	Phenolic compounds	–	*L–S, t, % solv, T*	–	CCD	52	TPC: 43.6 mg GAE/g dw	TPC: 40.4 mg GAE/g dw	7.339	–	–	–	Rodrigues et al. (2023)
Musa paradisiaca and *Carica papaya* (S)	Pectin	MAE	*t, T, pH*	3	BBD	17	Y: 23.74%	Y: 23.78%	0.168	0.9926	<0.0001	0.5271	Mada et al. (2022)
							Anhydrouronic: 67.06%	Anhydrouronic: 67.27%	0.313	0.9875	<0.0001	0.0547	

Continued

Table 9.1 Continued

Sample	Target compound	Extraction Technique	Independent variables	Levels	Experi-mental design	N runs	Predicted results	Experimental results	% Deviation	Validation of the method			Ref.
										R^2	*p*	*LoF*	
Algae													
Nannochlor-opsis sp.	FAMEs	MAE	*C–C, % MeOH, t*	3	Three factorials	28	FAMEs: 68.59%	FAMEs: 70.17%	2.304	0.972	0.061	0.005	Patil et al. (2011)
Fucus vesiculosus	Phlorotannins	MAE	*t, % solv, T*	3	BBD	15	TPhC: 3.01 mg PGE/g dw	TPhC: 3.16 mg PGE/g dw	4.983	0.99	<0.001	–	Amarante et al. (2020)
Ulva pertusa	Polysaccharides	MAE	*t, P, Ph, L–S*	3	BBD	27	Y: 41.91%	Y: 42.12%	0.501	0.9830	<0.0001	–	Le et al. (2019)
Sargassum muticum	Phenolic compounds	MAE	*t, % EtOH, p*	5	CCCD	28	Y: 483.87 mg/g dw	Y: 469.4 mg/g dw	2.99	0.9298	<0.05	–	A. Silva et al. (2024)
							TPC: 66.10 mg/g dw	TPC: 44.28 mg/g dw	33.011	0.8140			
							DPPH: 58.98 nM R˙/g dw	DPPH: 50.15 nM R˙/g dw	14.97	0.9042			
							ABTS: 85.88 nM R˙/g dw	ABTS: 83.20 nM R˙/g dw	3.121	0.8289			
							BCM: 0.112 µM βC/g dw	BCM: 0.101 µM βC/g dw	9.821	0.8487			
Undaria pinnatifida	Fucoxanthin	UAE	*t, P, % EtOH*	5	CCCD	28	Y: 55.29% extract	Y: 54.13% extract	2.098	0.9395	>0.05	–	Lourenço-Lopes et al. (2023)
							Y: 20.89 mg/g algae	Y: 20.91 mg/g algae	0.096	0.9034	>0.05	–	
							Y: 124.07 mg/g E	Y: 124.39 mg/g E	0.258	0.9225	>0.05	–	
		MAE	*t, p, % EtOH*				Y: 58.43% extract	Y: 60.25% extract	3.115	0.9426	>0.05	–	
							Y: 10.18 mg/g algae	Y: 10.01 mg/g algae	1.67	0.8752	>0.05	–	
							Y: 70.04 mg/g E	Y: 58.83 mg/g E	16.005	0.8695	>0.05	–	

(L), leaf; (F), flower; (P), peel; (PM), pomace; (B), bagasse; (S), skin; (SD), seeds; HAE, hat assisted extraction; MAE, microwave assisted extraction; UAE, ultrasound assisted extraction; HD, hydrodistillation; NE, nanoemulsion; L–S, liquid–solid ratio; t, time; T, temperature; % solv, solvent; % EtOH, Ethanol; MeOH, methanol; MUAE, mode of ultrasound operation (pulsed and continuous); TPC, total phenolic content; TTC, total tannin content; TAC, total alkaloid content; TE, Trolox equivalent; TAE, tannin acid equivalents; CAE, catechin acid equivalents; PGE, phloroglucinol equivalents; C–C, catalyst concentration; FAMEs content, fatty acid methyl ester conversion; P, power; Y1, epicatechin dimer type B content; mg/g, Y2, epicatechin content; mg/g, Y3, quercetin-3-O-glucoside content; mg/g, Y4, quercetin-3-O-rhamnoside content; mg/g, AA, antioxidant activity; FCCD, face-centered composite design; WC, water content; MD, molar ratio; TPhC, Total Phlorotannins content; FCCCD, face-centered central composite design; TFC, Total flavonoid content determination; DPPH, 2; 2-diphenyl-1-picrylhydrazyl radical, ABTS, 2; 2-azino-bis (ethylbenzothiazoline-6-sulfonic acid (ABTS+•) radical cation, p, pressure; pr, percentage of recovery; BCM, β-Carotene bleaching method; CCRD, central composite rotatable design; IC_{50}, the half maximal inhibitory concentration; CD, custom design; X_1, percentage of the emulsifier in the continuous phase varied in the range of 1%–7%; X_2, stirring time ranged from 0 to 6 min; X_3, the sonication time varied from 0 to 30 min; and X_4 the percentage of the dispersed phase was 5%–15%, dsd, drop size diameter (nm); Pi, polydispersity index.

The authors are currently pioneering the use of hybrid models that combine predictive tools and intelligent systems to simulate and predict the extraction methods performance. Artificial neural networks (ANNs) are often used for these predictions because of their ability to learn from historical data. However, hybrid models that combine ANNs with adaptive neuro-fuzzy inference systems (ANFIS) address problems such as limited extrapolation capability and the black-box nature of ANNs, increasing the prediction performance. Another approach is to fuse pure mathematics and neural networks to solve black-box problems and improve accuracy. These models bring notable advantages such as matching efficiency and reduced dosage of larger tests (Adeib Idris and Markom, 2019). In one study, a hybrid model of UAE and ML was used to enhance the extraction of phytocompounds from *Solanum torvum* Sw. The FCCD in RSM was used to optimize key extraction parameters which were further validated through ANFIS and RF algorithms. Optimal conditions for the extraction were found to be 65% MeOH at an ultrasound intensity of 60 W/cm^2, 45 °C for 15 min with a particle size of 0.5 mm, which resulted in the highest concentrations of total polyphenols, total flavonoids, and free radical scavenging activity. Bioactive compounds were identified by gas chromatography-mass spectroscopy (GCMS) and liquid chromatography-mass spectroscopy (LCMS) techniques which contributed to the success of the hybrid-ML model (Petchimuthu et al., 2023).

In another study, to illustrate the fucoxanthin extraction process, a hybrid model was developed using ANNs, SVMs and ANFIS. ANNs are optimal for analyzing complex patterns even in noisy data because it can perform nonlinear processing. SVMs are useful because of their efficient application in both classification and regression problems, as well as the fact that it minimizes the risk of overfitting by accommodating both linear and nonlinear data. ANFIS achieves this by taking the best of ANNs learning capabilities and combining it with fuzzy logic for effective and efficient rule-based decision making. In this way, it manages to design a model that consists of: (1) information gathering, (2) raw data cleaning, (3) extraction of important aspects and (4) testing with the available options. This ensures a good level of accuracy in maintaining the model's ability to focus ($R^2 = 99.60\%$) (Chong et al., 2023).

9.6 Case studies in food extraction processes

RSM is widely used in food science and the food industry to optimize processes and product formulations, improving factors like texture, flavor, and nutritional content. This technique helps researchers identifying optimal conditions efficiently by analyzing interactions between variables (Yolmeh and Jafari, 2017). In Table 9.1, a selection of recent applications of RSM studies in food related research is presented.

9.6.1 Extraction of bioactive compounds from plants

As mentioned, the choice of the extraction method and solvent is highly dependent on the target matrix and its intrinsic composition that is, lignocellulosic content, as well as the target compounds (e.g., polyphenols, tocopherols, chlorophylls, carotenoids, terpenoids, polysaccharides, etc.). Indeed, depending on their structure, polarity and subsequent hydrophobicity, various groups of bioactive compounds can be co-extracted and obtained together in crude extracts, especially if no purification steps are added to the process (Carbone et al., 2020). Undoubtedly, among plant-derived bioactive compounds, polyphenols have been significantly highlighted in recent years and a great number of extraction studies have been performed to obtain them from agri-food by-products as well as traditional medicinal plants (Jimenez-Lopez et al., 2020). However, optimization studies are crucial to make extraction processes feasible and cost-effective. Taking traditional medicinal plants as an example, *Arnica montana* (chamomile) flowers subjected to an extraction optimization study using heat-assisted extraction (HAE) and UAE following a CCCD, achieved significant adjustment to the obtained models. This was achieved by implementing three independent variables in the model (*T*, *t*, *S*) at five levels, resulting in 28 experimental points to be performed to validate the model. The study revealed that HAE outperformed the polyphenol yield of UAE by 26% (yield and total phenolic content (TPC)/g extract) by comparing the optimal points of both methods. It was also found that the most determinant parameters were *t* and *S*, and despite the better results provided by HAE, the total extraction time by UAE was much shorter, going from 60 to 20 min (Garcia-Oliveira et al., 2024a, 2024b). Likewise, a study on rosemary (*Rosmarinus officinalis*) employing NADES, established optimal extraction conditions for UAE in this matrix. Using 1,2-propylene glycol combined with choline chloride as NADES and comparing it to conventional solvents (MeOH, EtOH), the authors established a model using four variables at three levels, while also testing six rosemary genotypes from different Mediterranean countries. The responses considered included antioxidant capacity, rosmarinic acid content, and pigment content, resulting in an appropriate model fit ($R^2 = 0.9044$). Nonetheless, the authors did not find any significant difference between the use of NADES and conventional solvents on rosmarinic acid yield, but did improve the yield of pigments and antioxidant activity (Erdoğan et al., 2024).

9.6.2 Extraction of essential oils

EOs are a complex mixture obtained from different plant fractions such as flowers, leaves or fruits peels among others, that are generally comprised of various plant secondary metabolites, with a high concentration of sesquiterpenes, but also fatty acids, coumarins, alkaloids and several polyphenols of low molecular weight (Ríos, 2015). Extraction from plants and food waste to obtain EOs is usually performed by water or steam distillation with possible varying pressure. Other possibilities include maceration, HAE or supercritical CO_2 extraction. Some significant examples of EOs optimization include aromatic herbs, spices, and *Citrus* peels, as they are generally rich in them. For instance, HD of rosemary was optimized by studying the effect of just two parameters, *t* & *T*, and establishing three levels for each of them, taking only extraction yield as response (Yeddes et al., 2022). Extraction optimization of EOs from *Citrus* peels by combining hydrodistillation and UAE in sequential steps has also proven to be an effective strategy to improve its yield, concentration of polyphenols and antioxidant activity (Heydari et al., 2021). By using a central composite rotatable design (CCRD), the effect of volume/mass ratio (X_1), UAE time (X_2) and HD time (X_3) were optimized based on TPC, antioxidant capacity, and yield. The resulting optimal conditions were then evaluated for antimicrobial activity to determine the overall best performance. Indeed, the developed model had a great accuracy in predicting the responses ($R^2 = 0.9457$ for antioxidant activity) (Heydari et al., 2021). Similar processes reported with conventional Soxhlet extraction in ginger (*Zingiber officinale*), with the adjustment of three independent variables at three levels (*t*, *S%*, and *weight of sample*) (Mohammed et al., 2022). This allowed the authors to perform only 15 experiments and obtain an optimum point of 15% regarding yield with a high adjustment to the model ($R^2 = 0.978$).

9.6.3 Extraction in food waste valorization

Food waste comprises various discarded fractions of foods such as peels, seeds, stems, leaves or exhausted pulps that although not feasible for the food market, can often be a very rich source of bioactive compounds, such as polyphenols, terpenoids, EOs, etc. This approach of using various food waste as source of phytochemicals has been a major focus of academic and industrial research in the last decade as it is a sustainable alternative to their valorization, while also posing an affordable raw material for extraction. For instance, peanut (*Juglans regia*) shells were studied for optimization by water infusion and establishing *t*, *T* and *S/L* ratio as variables, while color and TPC were chosen as response (Simsek and Süfer, 2021). By encoding 5 center points with a BB design, the study could be completed with 17 experiments, and the experiments showed great adjustment to the developed model, with $R^2 = 0.988$ and $R^2 = 0.901$ for yield and TPC, respectively. Likewise, studies with kiwifruit (*Actinidia deliciosa*) peel extracted by UAE under a CCRD, allowed for the optimization of polyphenol recovery from this fruit waste (Giordano et al., 2021). The independent variables considered were *t*, *ultrasonic power* and *%S*, coded at five levels and yield and individual flavonoids' content considered as the model's responses. By following the CCRD model, the optimized conditions were obtained with just 20 experiments, obtaining a model with very good adjustment ($R^2 > 0.88$ for all responses) that allowed to maximize yield while reducing extraction time (Giordano et al., 2021). On the other hand, tomato (*Solanum lycopersicum*) seeds' extraction was optimized by MAE, considering oil extraction yield and antioxidant activity as responses (Ouatmani et al., 2022). In the study, the extraction process was optimized by establishing time, microwave power and solvent type and establishing them at three levels following a CCD modeling. The authors did not only find that the optimum conditions were at 7 s of MAE time, all while yielding double TPC levels in the extracts as well as higher antioxidant activity in comparison to a 24 h maceration (Ouatmani et al., 2022). These cases illustrate not only the usefulness and feasibility of optimization studies for different target compounds and components in food and food waste, but also the practical applications in making the extraction process more cost-effective and understanding the influence of the various process-intrinsic parameters in the results.

9.7 Conclusion and future trends

This chapter presented an overview of the physico-chemical phenomena that govern the process of compound extraction, highlighting solubility, diffusion, and mass transfer, which are affected by experimental variables such as temperature, pressure, solvent, and solid-liquid ratio. This approach was carried out by focusing on three extraction techniques (SLE, LLE, and SFE), in which the factors that most influence the efficiency of the process are solvent, temperature, and time. Moreover, the chapter examined the mechanistic and empirical models used to optimize the efficiency of the extraction process, focusing on the mathematical equations that describe the behavior of each model and their main applications within industrial processes in the food industry. Hybrid models were also assessed as an approach in which the combination of different mathematical models seems to be a suitable strategy to minimize the limitations present when applied separately. Overall, this chapter provides an analysis on the importance of modeling to enhance extraction efficiency, as well as the relevance of predictive tools to minimize experimental runs while predicting the system.

Acknowledgements

This research leading to these results was supported by Proyectos de Generación de Conocimiento 2023 (PID2023-138814OA-C22) supporting the contract of F. Chamorro and by MICIU/AEI/10.13039/501100011033 supporting predoctoral industrial grant for A. Perez-Vazquez (DIN2024-013416) in collaboration with Mercantia Desarrollos Alimentarios S.L.; by Xunta de Galicia for supporting the post-doctoral grant of A.G. Pereira (IN606B-2024/011), and the predoctoral grant of P. Barciela (ED481A-2024-230). The authors thank the EU-FORA Fellowship Program (EUBA-EFSA-2023-ENREL-01) that supports the work of F. Chamorro (INNOV2-SAFETY-GA13) and S. Seyyedi-Mansour and P. Donn (ALGASAFE-GA14). The authors are grateful to the National funding by FCT, Foundation for Science and Technology, through the individual research grants of J. Echave (2023.04987.BD) with the DOI https://doi.org/10.54499/2023.04987.BD. Funding for open access charge: Universidade de Vigo/CISUG. Scientific evaluations presented in this article are independent and are not intended to represent the views/any official position of EFSA.

Abbreviations

ABS	Aqueous Biphasic System
ANFIS	Adaptive Neuro-Fuzzy Inference Systems
ANN	Artificial Neural Networks
ANOVA	Analysis of Variance
BBD	Box Behnken Design

CCD	Central Composite Design
CCRD	Central Composite Totable Design
CFD	Computational Fluid Dynamics
EAE	Enzyme-Assisted Extraction
EFM	Essentia Flux Models
EO	Essential Oil
FCCD	Face Central Composite Design
FFD	Full Factorial Desing
GC-MS	Gas Chromatography-Mass Spectroscopy
GCCK	General Catalytic Cycle Kinetics
HAE	Heat-Assisted Extraction
HD	Hydrodistillation
IL	Ionic Liquids
K	Partition coefficient
LLE	Liquid-Liquid Extraction
MAE	Microwave-Assisted Extraction
ML	Machine Learning
NADES	Natural Deep Eutectic Solvents
ODE	Ordinary Differential Equations
PLE	Pressurized Liquid Extraction
RMSE	Root Mean Square Error
RSM	Response Surface Methodology
SFE	Supercritical Fluid Extraction
SLE	Solid-Liquid Extraction
SPE	Solid Phase Extraction
TPC	Total Phenolic Content
UAE	Ultrasound-Assisted Extraction
ΔH	Enthalpy
ΔS	Entropy

References

Adeib Idris, S., Markom, M., 2019. Application of hybrid predictive tools for prediction and simulation in supercritical fluid extraction - an overview. IOP Conf. Ser. Mater. Sci. Eng. 551 (1), 3–8. https://doi.org/10.1088/1757-899X/551/1/012051.

Agarwal, R., Choudhury, T., Ahuja, N.J., Sarkar, T., 2023. Hybrid deep learning algorithm-based food Recognition and Calorie estimation. J. Food Process. Preserv. 15. https://doi.org/10.1155/2023/6612302.

Ahmed, J., Dolan, K., Mishra, D., 2012. Chemical reaction kinetics Pertaining to foods. In: Handbook of Food Process Design, pp. 113–166. https://doi.org/10.1002/9781444398274.ch6.

Aires, A., Carvalho, R., 2020. Kiwi fruit residues from industry processing: study for a maximum phenolic recovery yield. J. Food Sci. Technol. https://doi.org/10.1007/s13197-020-04466-7.

Al-Hilphy, A.R., Al-Behadli, T.K.M., Al-Mtury, A.A., Abd Al-Razzaq, A.A., Shaish, A.S., Liao, L., Zeng, X.A., Manzoor, M.F., 2023. Innovative date syrup processing with ohmic heating technology: Physiochemical characteristics, yield optimization, and sensory attributes. Heliyon 9 (9), e19583. https://doi.org/10.1016/j.heliyon.2023.e19583.

Alara, O.R., Abdurahman, N.H., Olalere, O.A., 2018. Optimization of microwave-assisted extraction of flavonoids and antioxidants from *Vernonia amygdalina* leaf using response surface methodology. Food Bioprod. Process. 107, 36–48. https://doi.org/10.1016/j.fbp.2017.10.007.

Albarri, R., Şahin, S., 2023. Kinetics, thermodynamics, and mass transfer mechanism of the ultrasound-assisted extraction of bioactive molecules from Moringa oleifera leaves. Biomass Convers. Biorefinery 13 (9), 7919–7926. https://doi.org/10.1007/s13399-021-01686-5.

Almohasin, J.A., Balag, J., Miral, V.G., Moreno, R.V., Tongco, L.J., Lopez, E.C.R., 2023. Green solvents for liquid–liquid extraction: recent advances and future trends. Eng. Proc. 56 (1). https://doi.org/10.3390/ASEC2023-16278.

Amarante, S.J., Catarino, M.D., Marçal, C., Silva, A.M.S., Ferreira, R., Cardoso, S.M., 2020. Microwave-assisted extraction of Phlorotannins from Fucus vesiculosus. Mar. Drugs 18 (11). https://doi.org/10.3390/md18110559.

Amodio, M.L., Derossi, A., Colelli, G., 2017. Modeling shelf life of packaged, ready-to-eat fruits and vegetables with reference to the fate of nutritional compounds. In: Fruit and Vegetable Phytochemicals: Chemistry and Human Health, second ed. John Wiley & Sons, Ltd, pp. 833–856. https://doi.org/10.1002/9781119158042.ch38.

Antony, F.M., Pal, D., Wasewar, K., 2021. Separation of bio-products by liquid-liquid extraction. Phys. Sci. Rev. 6 (Issue 4). https://doi.org/10.1515/psr-2018-0065.

Arcus, V.L., Mulholland, A.J., 2020. Temperature, dynamics, and enzyme-Catalyzed reaction rates. Annu. Rev. Biophys. 49, 163–180. https://doi.org/10.1146/annurev-biophys-121219-081520.

Assis, F.R., Morais, R.M.S.C., Morais, A.M.M.B., 2016. Mass transfer in osmotic dehydration of food products: comparison between mathematical models. Food Eng. Rev. 8 (2), 116–133. https://doi.org/10.1007/s12393-015-9123-1.

Ávila, I.M.L.B., Silva, C.L.M., 1999. Modelling kinetics of thermal degradation of colour in peach puree. J. Food Eng. 39 (2), 161–166. https://doi.org/10.1016/S0260-8774(98)00157-5.

Baker, R.E., Peña, J.M., Jayamohan, J., Jérusalem, A., 2018. Mechanistic models versus machine learning, a fight worth fighting for the biological community? Biol. Lett. 14 (5), 1–4. https://doi.org/10.1098/rsbl.2017.0660.

Barton, R.R., 2013. Response surface methodology. In: Gass, M.C.F.S.I. (Ed.), Encyclopedia of Operations Research and Management Science. Springer US, pp. 1307–1313. https://doi.org/10.1007/978-1-4419-1153-7_1143.

Bazant, M.Z., 2013. Theory of chemical kinetics and charge transfer based on nonequilibrium thermodynamics. Accounts of Chemical Research 46 (5), 1144–1160. https://doi.org/10.1021/ar300145c.

Belwal, T., Chemat, F., Venskutonis, P.R., Cravotto, G., Jaiswal, D.K., Bhatt, I.D., Devkota, H. P., Luo, Z., 2020. Recent advances in scaling-up of non-conventional extraction techniques: learning from successes and failures. TrAC, Trends Anal. Chem. 127, 115895. https://doi.org/10.1016/j.trac.2020.115895.

BenJemaa, M., Rahali, F.Z., Falleh, H., Beji-Serairi, R., Megdiche-Ksouri, W., Hamrouni, I., Hammami, M., Nehme, R., Pereira, R.B., Andrés, S., Bouhallab, S., Pereira, D.M., Ceciliani, F., Abdennebi-Najar, L., Ksouri, R., 2022. Essential oil Stabilisation by response surface methodology (RSM): nanoemulsion formulation, physicochemical,

Microbiological, and sensory investigations. Molecules 27 (21), 7330. https://doi.org/10.3390/molecules27217330.

Bhagya Raj, G.V.S., Dash, K.K., 2022. Comprehensive study on applications of artificial neural network in food process modeling. Crit. Rev. Food Sci. Nutr. 62 (10), 2756–2783. https://doi.org/10.1080/10408398.2020.1858398.

Bhattacharya, S., 2021. Central composite design for response surface methodology and its application in Pharmacy. In: Response Surface Methodology in Engineering Science. IntechOpen. https://doi.org/10.5772/intechopen.95835.

Boateng, I.D., 2024. Mechanisms, capabilities, limitations, and economic stability Outlook for extracting phenolics from Agro-byproducts using emerging thermal extraction technologies and their combinative effects. Food Bioprocess Technol. 17 (Issue 5). https://doi.org/10.1007/s11947-023-03171-5.

Box, G.E.P., Wilson, K.B., 1951. On the experimental Attainment of optimum conditions. Journal of the Royal Statistical Society Series B: Statistical Methodology 13 (1), 1–38. https://doi.org/10.1111/j.2517-6161.1951.tb00067.x.

Braga, M.E.M., Gaspar, M.C., de Sousa, H.C., 2023. Supercritical fluid technology for agrifood materials processing. Curr. Opin. Food Sci. 50. https://doi.org/10.1016/j.cofs.2022.100983.

Capuano, E., Oliviero, T., van Boekel, M.A.J.S., 2018. Modeling food matrix effects on chemical reactivity: challenges and perspectives. Crit. Rev. Food Sci. Nutr. 58 (16), 2814–2828. https://doi.org/10.1080/10408398.2017.1342595.

Carbone, K., Amoriello, T., Iadecola, R., 2020. Exploitation of kiwi juice pomace for the recovery of natural antioxidants through microwave-assisted extraction. Agriculture (Switzerland) 10 (10), 1–14. https://doi.org/10.3390/agriculture10100435.

Caré, F., Sangaré, D., Bostyn, S., Atwi-Ghaddar, S., Lafite, P., Buron, F., 2023. Computational Fluid Dynamics (CFD) modeling of static maceration in view to optimize continuous flow extractions of robinetin and dihydrorobinetin from Robinia pseudoacacia wood. Food Bioprod. Process. 141, 185–198. https://doi.org/10.1016/j.fbp.2023.07.007.

Cassani, L., Silva, A., Carpena, M., Pellegrini, M.C., García-Pérez, P., Grosso, C., Barroso, M. F., Simal-Gandara, J., Gómez-Zavaglia, A., Prieto, M.A., 2024. Phytochemical compounds with promising biological activities from Ascophyllum nodosum extracts using microwave-assisted extraction. Food Chem. 438, 138037. https://doi.org/10.1016/j.foodchem.2023.138037.

Celeiro, M., Garcia-Jares, C., Llompart, M., Lores, M., 2021. Recent advances in sample preparation for cosmetics and personal care products analysis. Molecules 26 (Issue 16). https://doi.org/10.3390/molecules26164900.

Chanioti, S., Liadakis, G., & Tzia, C. (2014). Solid–liquid extraction. November 2014, 253–286. https://doi.org/10.1201/b17803-7.

Chemat, F., Rombaut, N., Meullemiestre, A., Turk, M., Perino, S., Fabiano-Tixier, A.S., Abert-Vian, M., 2017. Review of green food processing techniques. Preservation, transformation, and extraction. Innov. Food Sci. Emerg. Technol. 41 (May), 357–377. https://doi.org/10.1016/j.ifset.2017.04.016.

Chemat, F., Vian, M.A., Ravi, H.K., Khadhraoui, B., Hilali, S., Perino, S., Tixier, A.S.F., 2019. Review of alternative solvents for green extraction of food and natural products: Panorama, principles, applications and prospects. Molecules 24 (16). https://doi.org/10.3390/molecules24163007.

Chong, J.W.R., Tang, D.Y.Y., Leong, H.Y., Khoo, K.S., Show, P.L., Chew, K.W., 2023. Bridging artificial intelligence and fucoxanthin for the recovery and quantification from microalgae. Bioengineered 14 (1), 2244232. https://doi.org/10.1080/21655979.2023.2244232.

Chuo, S.C., Nasir, H.M., Mohd-Setapar, S.H., Mohamed, S.F., Ahmad, A., Wani, W.A., Muddassir, M., Alarifi, A., 2022. A Glimpse into the extraction methods of active compounds from plants. Crit. Rev. Anal. Chem. 52 (Issue 4). https://doi.org/10.1080/10408347.2020.1820851.

Corradini, M.G., Peleg, M., 2006a. Linear and non-linear kinetics in the synthesis and degradation of acrylamide in foods and model systems. Crit. Rev. Food Sci. Nutr. 46 (6), 489–517. https://doi.org/10.1080/10408390600758280.

Corradini, M.G., Peleg, M., 2006b. Prediction of vitamins loss during non-isothermal heat processes and storage with non-linear kinetic models. Trends Food Sci. Technol. 17 (1), 24–34. https://doi.org/10.1016/j.tifs.2005.09.004.

Cui, F., Zheng, S., Wang, D., Tan, X., Li, Q., Li, J., Li, T., 2023. Recent advances in shelf life prediction models for monitoring food quality. Compr. Rev. Food Sci. Food Saf. 22 (2), 1257–1284. https://doi.org/10.1111/1541-4337.13110.

da Silva, B.G., Martin do Prado, J., Fileti, A.M.F., Foglio, M.A., Vieira e Rosa, P. de T., 2023. Kinetic models for extraction with supercritical carbon dioxide from pink pepper: theoretical, empirical, and semi-empirical models and artificial neural network approach. Chem. Eng. J. Adv. 15, 8. https://doi.org/10.1016/j.ceja.2023.100514.

da Silva, R.F., Carneiro, C.N., Cheila, C.B., Gomez, F., Espino, M., Boiteux, J., de los, Á., Fernández, M., Silva, M.F., de S. Dias, F., 2022. Sustainable extraction bioactive compounds procedures in medicinal plants based on the principles of green analytical chemistry: a review. Microchem. J. 175. https://doi.org/10.1016/j.microc.2022.107184.

Das, A.K., Dewanjee, S., 2018. Optimization of extraction using mathematical models and computation. Computational Phytochemistry 75–106. https://doi.org/10.1016/B978-0-12-812364-5.00003-1.

Das, I., Arora, A., 2021. Kinetics and mechanistic models of solid-liquid extraction of pectin using advance green techniques- a review. Food Hydrocoll. 120, 15. https://doi.org/10.1016/j.foodhyd.2021.106931.

Dashtian, K., Kamalabadi, M., Ghoorchian, A., Ganjali, M.R., Rahimi-Nasrabadi, M., 2024. Integrated supercritical fluid extraction of essential oils. J. Chromatogr. A 1733, 465240. https://doi.org/10.1016/J.CHROMA.2024.465240.

De Vleeschouwer, K., Van der Plancken, I., Van Loey, A., Hendrickx, M.E., 2009. Modelling acrylamide changes in foods: from single-response empirical to multiresponse mechanistic approaches. Trends Food Sci. Technol. 20 (3–4), 155–167. https://doi.org/10.1016/j.tifs.2009.01.060.

Del Vecchio, R.J., 2007. Design of experiments. In: Saul, M.C.F., Gass, I. (Eds.), Handbook of Vinyl Formulating, second ed. Springer, New York, pp. 515–527. https://doi.org/10.1002/9780470253595.ch22.

Echave, J., Fraga-Corral, M., Seyyedi-Mansour, S., Cassani, L., Garcia-Oliveira, P., Cao, H., Simal-Gandara, J., Prieto, M.A., Xiao, J., 2024. Developing polyphenols from agricultural wastes. https://doi.org/10.19103/as.2023.0120.21.

Erdoğan, Ü., Han Öztürk, T., Önder, S., Tonguç, M., 2024. Eco-friendly extraction of six different rosemary (Rosmarinus officinalis L.) genotypes-using natural deep eutectic solvents: optimization and modeling via response surface methodology (RSM). J. Mol. Liq. 407 (May). https://doi.org/10.1016/j.molliq.2024.125167.

Fernández-Ramos, A., Miller, J.A., Klippenstein, S.J., Truhlar, D.G., 2006. Modeling the kinetics of bimolecular reactions. Chem. Rev. 106 (11), 4518–4584. https://doi.org/10.1021/cr050205w.

Ferreira, S.L.C., Bruns, R.E., Ferreira, H.S., Matos, G.D., David, J.M., Brandão, G.C., da Silva, E.G.P., Portugal, L.A., dos Reis, P.S., Souza, A.S., dos Santos, W.N.L., 2007. Box-Behnken design: an alternative for the optimization of analytical methods. Anal. Chim. Acta 597 (2), 179–186. https://doi.org/10.1016/j.aca.2007.07.011.

Ferri, M., Rodríguez, Ó., Tassoni, A., 2023. Editorial: New green extraction methods for the sustainable recovery of functional plant secondary metabolites. Front. Plant Sci. 14. https://doi.org/10.3389/fpls.2023.1186180.

Fuller, G.G., Vermant, J., 2012. Complex fluid-fluid interfaces: rheology and structure. Annu. Rev. Chem. Biomol. Eng. 3, 519–543. https://doi.org/10.1146/annurev-chembioeng-061010-114202.

Garcia-oliveira, P., Chamorro, F., Simal-gandara, J., Prieto, M.A., Cassani, L., 2024a. Enhancing phenolic compounds recovery from *Arnica montana* L . flowers through optimized green extraction protocols. J. Appl. Res. Med. Aromat. Plants 43 (October), 100600. https://doi.org/10.1016/j.jarmap.2024.100600.

Garcia-Oliveira, P., Chamorro, F., Simal-Gandara, J., Prieto, M.A., Cassani, L., 2024b. Improving phenolic compound extraction from Arnica Montana flowers through multivariate optimization of heat and ultrasound-assisted methods. Sustain. Chem. Pharm. 41, 101722.

Giacometti, J., Žauhar, G., Žuvić, M., 2018. Optimization of ultrasonic-assisted extraction of major phenolic compounds from olive leaves (Olea europaea L.) using response surface methodology. Foods 7 (9), 1–14. https://doi.org/10.3390/foods7090149.

Giannakourou, M.C., Gogou, E., Taoukis, P.S., 2021. Reaction kinetics in food-processing engineering. In: Jafari, S.M. (Ed.), Engineering Principles of Unit Operations in Food Processing: Unit Operations and Processing Equipment in the Food Industry. Woodhead Publishing, pp. 443–470. https://doi.org/10.1016/B978-0-12-818473-8.00010-4.

Gilman, J., Walls, L., Bandiera, L., Menolascina, F., 2021. Statistical design of experiments for Synthetic biology. ACS Synth. Biol. 10 (1), 1–18. https://doi.org/10.1021/acssynbio.0c00385.

Giordano, M., Pinela, J., Dias, M.I., Calhelha, R.C., Stojković, D., Soković, M., Tavares, D., Cánepa, A.L., Ferreira, I.C.F.R., Caleja, C., Barros, L., 2021. Ultrasound-assisted extraction of flavonoids from kiwi peel: process optimization and bioactivity assessment. Applied Sciences (Switzerland) 11 (14). https://doi.org/10.3390/app11146416.

Gong, T., Liu, S., Wang, H., Zhang, M., 2021. Supercritical CO2 fluid extraction, physicochemical properties, antioxidant activities and hypoglycemic activity of polysaccharides derived from fallen Ginkgo leaves. Food Biosci. 42. https://doi.org/10.1016/j.fbio.2021.101153.

Gong, Y., Hou, Z., Gao, Y., Xue, Y., Liu, X., Liu, G., 2012. Optimization of extraction parameters of bioactive components from defatted marigold (Tagetes erecta L.) residue using response surface methodology. Food Bioprod. Process. 90 (1), 9–16. https://doi.org/10.1016/j.fbp.2010.12.004.

González-Pérez, J.E., Ramírez-Corona, N., López-Malo, A., 2021. Mass transfer during osmotic dehydration of fruits and vegetables: process factors and non-thermal methods. Food Eng. Rev. 13 (2), 344–374. https://doi.org/10.1007/s12393-020-09276-3.

Götz, P., 1995. Solvent extraction in biotechnology: recovery of primary and secondary metabolites. Chemie Ingenieur Technik. Springer Science & Business Media. https://doi.org/10.1002/cite.330670826.

Guthery, F.S., Burnham, K.P., Anderson, D.R., 2003. Model selection and Multimodel inference: a practical information-Theoretic approach. J. Wildl. Manag. 67 (Issue 3). https://doi.org/10.2307/3802723.

Haghighi, A., Khajenoori, M., 2013. Subcritical water extraction. Mass Transfer - Advances in Sustainable Energy and Environment Oriented Numerical Modeling. https://doi.org/10.5772/54993.

Hashemi, B., Shiri, F., Švec, F., Nováková, L., 2022. Green solvents and approaches recently applied for extraction of natural bioactive compounds. TrAC, Trends Anal. Chem. 157. https://doi.org/10.1016/j.trac.2022.116732.

Herrero, M., Mendiola, J.A., Cifuentes, A., Ibáñez, E., 2010. Supercritical fluid extraction: recent advances and applications. J. Chromatogr. A 1217 (Issue 16). https://doi.org/10.1016/j.chroma.2009.12.019.

Heydari, M., Rostami, O., Mohammadi, R., Banavi, P., Farhoodi, M., Sarlak, Z., Rouhi, M., 2021. Hydrodistillation ultrasound-assisted green extraction of essential oil from bitter orange peel wastes: optimization for quantitative, phenolic, and antioxidant properties. J. Food Process. Preserv. 45 (7), 1–15. https://doi.org/10.1111/jfpp.15585.

Ho, Q.T., Carmeliet, J., Datta, A.K., Defraeye, T., Delele, M.A., Herremans, E., Opara, L., Ramon, H., Tijskens, E., Van Der Sman, R., Van Liedekerke, P., Verboven, P., Nicolaï, B. M., 2013. Multiscale modeling in food engineering. J. Food Eng. 114 (3), 279–291. https://doi.org/10.1016/j.jfoodeng.2012.08.019.

Jimenez-Lopez, C., Fraga-Corral, M., Carpena, M., García-Oliveira, P., Echave, J., Pereira, A. G., Lourenço-Lopes, C., Prieto, M.A., Simal-Gandara, J., 2020. Agriculture waste valorisation as a source of antioxidant phenolic compounds within a circular and sustainable bioeconomy. Food Funct. 11 (6), 4853–4877. https://doi.org/10.1039/D0FO00937G.

Kaderides, K., Papaoikonomou, L., Serafim, M., Goula, A.M., 2019. Microwave-assisted extraction of phenolics from pomegranate peels: optimization, kinetics, and comparison with ultrasounds extraction. Chem. Eng. Process. Process Intensif. 137 (January), 1–11. https://doi.org/10.1016/j.cep.2019.01.006.

Kajishima, 2012. Computational fluid dynamics, applied computational fluid dynamics. Mechanical Engineering Series 1–19. https://doi.org/10.1016/B978-0-12-382100-3.10010-1.

Kannouma, R.E., Hammad, M.A., Kamal, A.H., Mansour, F.R., 2022. Miniaturization of Liquid-Liquid extraction; the barriers and the enablers. Microchem. J. 182. https://doi.org/10.1016/j.microc.2022.107863.

Khan, M.I.H., Batuwatta-Gamage, C.P., Karim, M.A., Gu, Y.T., 2022. Fundamental understanding of heat and mass transfer processes for Physics-Informed machine learning-based drying modelling. Energies 15 (24), 9347. https://doi.org/10.3390/en15249347.

Kockmann, N., Agar, D.W., 2022. Liquid-liquid processes: mass transfer processes and chemical reactions. Multiphase Flows for Process Industries: Fundamentals and Applications. John Wiley & Sons, Ltd, pp. 243–288. https://doi.org/10.1002/9783527812066.ch5.

Kong, F., Singh, R.P., 2016. Chemical deterioration and physical instability of foods and beverages. In: Subramaniam, P. (Ed.), The Stability and Shelf Life of Food, second ed. Woodhead Publishing, pp. 43–76. https://doi.org/10.1016/B978-0-08-100435-7.00002-2.

Kostić, M.D., Đorđević, B.S., Miladinović, M.R., Stamenković, O.S., 2024. Process kinetics, thermodynamics and optimization of oil extraction from the garlic pennycress (Thlaspi alliaceum L.) seeds. J. Food Meas. Char. 18 (8), 6494–6509. https://doi.org/10.1007/s11694-024-02665-w.

Lainez-Cerón, E., Ramírez-Corona, N., López-Malo, A., Franco-Vega, A., 2022. An overview of mathematical modeling for conventional and intensified processes for extracting essential oils. Chem. Eng. Process. Process Intensif. 178, 109032. https://doi.org/10.1016/j.cep.2022.109032.

Lamidi, S., Olalere, R., Yekinni, A., Adesina, K., 2024. Design of experiments (DOE): applications and benefits in quality control and Assurance. In: Qamar, S.Z., Al-Hinai, N. (Eds.), Quality Control and Quality Assurance - Techniques and Applications. IntechOpen. https://doi.org/10.5772/intechopen.113987.

Le, B., Golokhvast, K.S., Yang, S.H., Sun, S., 2019. Optimization of Microwave-Assisted Extraction of Polysaccharides from Ulva Pertusa and Evaluation of Their Antioxidant Activity.

Lebovka, N., Vorobiev, E., Chemat, F., 2016. Enhancing extraction processes in the food industry. In: Enhancing Extraction Processes in the Food Industry. https://doi.org/10.1201/b11241.

Li, J., Deng, Y., Xu, W., Zhao, R., Chen, T., Wang, M., Xu, E., Zhou, J., Wang, W., Liu, D., 2023. Multiscale modeling of food thermal processing for insight, comprehension, and utilization of heat and mass transfer: a state-of-the-art review. Trends Food Sci. Technol. 131, 31–45. https://doi.org/10.1016/j.tifs.2022.11.018.

Lloyd, P., van Wyk, J., 2011. Introduction to extraction in food processing. https://doi.org/10.1201/b11241-2.

Loan, L.T.K., Tat, T.Q., Minh, P.D.T., Thao, V.T.T., Hoang, P.T.M., Nhi, T.T.Y., Giang, B.L., Phat, D.T., Tai, N. Van, 2024. Prediction of the germination rate and antioxidant properties of VD20 Rice by utilizing Artificial neural network-coupled response surface methodology and product characterization. J. Food Meas. Char. https://doi.org/10.1007/s11694-024-02835-w.

Lourenço-Lopes, C., Carreira-Casais, A., Carperna, M., Barral-Martinez, M., Chamorro, F., Jiménez-López, C., Cassani, L., Simal-Gandara, J., Prieto, M.A., 2023. Emerging technologies to extract fucoxanthin from Undaria pinnatifida: microwave vs. Ultrasound assisted extractions. Mar. Drugs 21 (5). https://doi.org/10.3390/md21050282.

Mada, T., Duraisamy, R., Guesh, F., 2022. Optimization and characterization of pectin extracted from banana and papaya mixed peels using response surface methodology. Food Sci. Nutr. 10 (4), 1222–1238. https://doi.org/10.1002/fsn3.2754.

Marín-San Román, S., Rubio-Bretón, P., Pérez-Álvarez, E.P., Garde-Cerdán, T., 2020. Advancement in analytical techniques for the extraction of grape and wine volatile compounds. Food Res. Int. 137, 109712. https://doi.org/10.1016/j.foodres.2020.109712.

Martinez Fernandez, J.S., 2020. Sequential Hydrothermal extraction for the recovery of valuable bioactives from Potato peels. In: ProQuest Dissertations and Theses. Washington State University, United States – Washington.

Mircioiu, C., Voicu, V., Anuta, V., Tudose, A., Celia, C., Paolino, D., Fresta, M., Sandulovici, R., Mircioiu, I., 2019. Mathematical modeling of release kinetics from supramolecular drug delivery systems. Pharmaceutics 11 (3). https://doi.org/10.3390/pharmaceutics11030140.

Mohammed, H.H., Laftah, W.A., Noel Ibrahim, A., Che Yunus, M.A., 2022. Extraction of essential oil from Zingiber officinale and statistical optimization of process parameters. RSC Adv. 12 (8), 4843–4851. https://doi.org/10.1039/D1RA06711G.

Moulin, C., Tournier, L., Peres, S., 2021. Combining kinetic and constraint-based modelling to better understand metabolism dynamics. Processes 9 (10), 1–23. https://doi.org/10.3390/pr9101701.

Mundi, I., Alemany, M.M.E., Poler, R., Fuertes-Miquel, V.S., 2019. Review of mathematical models for production planning under uncertainty due to lack of homogeneity: proposal of a conceptual model. Int. J. Prod. Res. 57 (15–16), 5239–5283. https://doi.org/10.1080/00207543.2019.1566665.

Murray, B.S., 2002. Interfacial rheology of food emulsifiers and proteins. Curr. Opin. Colloid Interface Sci. 7 (5–6), 426–431. https://doi.org/10.1016/S1359-0294(02)00077-8.

Nguyen, N.-K., Borkowski J., J., Phuong Vuong, M., 2023. Perspective chapter: Cyclic generation of box-behnken designs and New second-order designs. In: Response Surface Methodology - Research Advances and Applications. IntechOpen. https://doi.org/10.5772/intechopen.107178.

Noordijk, B., Garcia Gomez, M.L., ten Tusscher, K.H.W.J., de Ridder, D., van Dijk, A.D.J., Smith, R.W., 2024. The rise of scientific machine learning: a perspective on combining mechanistic modelling with machine learning for systems biology. Frontiers in Systems Biology 4, 1–13. https://doi.org/10.3389/fsysb.2024.1407994.

Ortega-Zamora, C., González-Sálamo, J., Hernández-Borges, J., 2021. Deep eutectic solvents application in food analysis. Molecules 26 (22). https://doi.org/10.3390/molecules26226846.

Ouatmani, T., Haddadi-Guemghar, H., Boulekbache-Makhlouf, L., Mehidi-Terki, D., Maouche, A., Madani, K., 2022. A sustainable valorization of industrial tomato seeds (cv Rio Grande): sequential recovery of a valuable oil and optimized extraction of antioxidants by microwaves. J. Food Process. Preserv. 46 (1), 1–13. https://doi.org/10.1111/jfpp.16123.

Patil, P.D., Gude, V.G., Mannarswamy, A., Cooke, P., Munson-McGee, S., Nirmalakhandan, N., Lammers, P., Deng, S., 2011. Optimization of microwave-assisted transesterification of dry algal biomass using response surface methodology. Bioresour. Technol. 102 (2), 1399–1405. https://doi.org/10.1016/j.biortech.2010.09.046.

Patrice Didion, Y., Gijsbert Tjalsma, T., Su, Z., Malankowska, M., Pinelo, M., 2023. What is next? the greener future of solid liquid extraction of biobased compounds: novel techniques and solvents overpower traditional ones. Separ. Purif. Technol. 320. https://doi.org/10.1016/j.seppur.2023.124147.

Pedrali, D., Gallotti, F., Proserpio, C., Pagliarini, E., Lavelli, V., 2020. Kinetic study of vitamin D2 degradation in mushroom powder to improve its applications in fortified foods. Lwt 125, 109248. https://doi.org/10.1016/j.lwt.2020.109248.

Peleg, M., 2022. A New Look at models of the combined effect of temperature, pH, water activity, or other factors on microbial Growth rate. Food Eng. Rev. 14 (1), 31–44. https://doi.org/10.1007/s12393-021-09292-x.

Pena-Pereira, F., Lavilla, I., Bendicho, C., 2009. Miniaturized preconcentration methods based on liquid-liquid extraction and their application in inorganic ultratrace analysis and speciation: a review. Spectrochim. Acta B Atom Spectrosc. 64 (1). https://doi.org/10.1016/j.sab.2008.10.042.

Peng, Y., Khaled, U., Al-Rashed, A.A.A.A., Meer, R., Goodarzi, M., Sarafraz, M.M., 2020. Potential application of Response Surface Methodology (RSM) for the prediction and optimization of thermal conductivity of aqueous CuO (II) nanofluid: a statistical approach and experimental validation. Phys. Stat. Mech. Appl. 554, 124353. https://doi.org/10.1016/j.physa.2020.124353.

Pereira Neves, H., Max Dias Ferreira, G., Max Dias Ferreira, G., Rodrigues de Lemos, L., Dias Rodrigues, G., Albis Leão, V., Barbosa Mageste, A., 2022. Liquid-liquid extraction of rare earth elements using systems that are more environmentally friendly: advances, challenges and perspectives. Separ. Purif. Technol. 282. https://doi.org/10.1016/j.seppur.2021.120064.

Perez-Vazquez, A., Barciela, P., Carpena, M., Donn, P., Seyyedi-Mansour, S., Cao, H., Fraga-Corral, M., Otero, P., Simal-Gandara, J., Prieto, M.A., 2023a. Supercritical fluid extraction as a potential extraction technique for the food industry. Eng. Proc. 37 (1), 115.

Perez-Vazquez, A., Carpena, M., Barciela, P., Cassani, L., Simal-Gandara, J., Prieto, M.A., 2023b. Pressurized liquid extraction for the recovery of bioactive compounds from seaweeds for food industry application: a review. Antioxidants 12 (3), 612.

Petchimuthu, P., Sumanth, G.B., Kunjiappan, S., Kannan, S., Pandian, S.R.K., Sundar, K., 2023. Green extraction and optimization of bioactive compounds from Solanum torvum Swartz. using ultrasound-aided solvent extraction method through RSM, ANFIS and machine learning algorithm. Sustain. Chem. Pharm. 36, 20. https://doi.org/10.1016/j.scp.2023.101323.

Pinela, J., Prieto, M.A., Pereira, E., Jabeur, I., Barreiro, M.F., Barros, L., Ferreira, I.C.F.R., 2019. Optimization of heat- and ultrasound-assisted extraction of anthocyanins from *Hibiscus sabdariffa* calyces for natural food colorants. Food Chem. 275, 309–321. https://doi.org/10.1016/j.foodchem.2018.09.118.

Piskulich, Z.A., Mesele, O.O., Thompson, W.H., 2019. Activation energies and beyond. J. Phys. Chem. A 123 (33), 7185–7194. https://doi.org/10.1021/acs.jpca.9b03967.

Polachini, T.C., Morales, S.A.V., Filho, L.R.P., Ribeiro, E.F., Saraiva, L.S., Basso, R.C., 2023. Physical properties and molecular interactions applied to food processing and formulation. Processes 11 (7), 2181. https://doi.org/10.3390/pr11072181.

Priego-Capote, F., 2021. Solid–liquid extraction techniques. Analytical Sample Preparation With Nano- and Other High-Performance Materials 111–130. https://doi.org/10.1016/B978-0-12-822139-6.00002-X.

Purlis, E., Cevoli, C., Fabbri, A., 2021. Modelling volume change and deformation in food products/processes: an overview. Foods 10 (4), 778. https://doi.org/10.3390/foods10040778.

Rezazi, S., Abdelmalek, S., Hanini, S., 2017. Kinetic study and optimization of extraction process conditions. Energy Proc. 139, 98–104. https://doi.org/10.1016/j.egypro.2017.11.180.

Ríos, J.L., 2015. Essential oils. In: Essential Oils in Food Preservation, Flavor and Safety. Elsevier Inc. https://doi.org/10.1016/B978-0-12-416641-7.00001-8.

Rodrigues, R.P., Sousa, A.M., Gando-Ferreira, L.M., Quina, M.J., 2023. Grape pomace as a natural source of phenolic compounds: solvent screening and extraction optimization. Molecules 28 (6), 2715. https://doi.org/10.3390/molecules28062715.

Rodríguez-Llorente, D., Cañada-Barcala, A., Álvarez-Torrellas, S., Águeda, V.I., García, J., Larriba, M., 2020. A review of the use of eutectic solvents, terpenes and terpenoids in liquid–liquid extraction processes. Processes 8 (Issue 10). https://doi.org/10.3390/pr8101220.

Roohi, R., Hashemi, S.M.B., 2019. Experimental and computational fluid dynamics modeling of Satureja khuzestanica essential oil extraction during ohmic-hydrodistillation. J. Food Process. Eng. 42 (5), e13083. https://doi.org/10.1111/jfpe.13083.

Sagis, L.M.C., Scholten, E., 2014. Complex interfaces in food: structure and mechanical properties. Trends Food Sci. Technol. 37 (1), 59–71. https://doi.org/10.1016/j.tifs.2014.02.009.

Sansana, J., Joswiak, M.N., Castillo, I., Wang, Z., Rendall, R., Chiang, L.H., Reis, M.S., 2021. Recent trends on hybrid modeling for Industry 4.0. Comput. Chem. Eng. 151, 107365. https://doi.org/10.1016/j.compchemeng.2021.107365.

Santacesaria, E., Tesser, R., 2018. Kinetics of Homogeneous reactions and related mechanisms. In: The Chemical Reactor from Laboratory to Industrial Plant. Springer International Publishing, pp. 191–290. https://doi.org/10.1007/978-3-319-97439-2_4.

Savva, M., 2019. Rates and orders of reactions. In: Pharmaceutical Calculations. Springer International Publishing, pp. 269–299. https://doi.org/10.1007/978-3-030-20335-1_12.

Sanket, S., Sharma, P.K., Mani, I., Nain, L., Satheesh, N., 2024. Optimization of ohmic parameters in enzyme assisted aqueous extraction for better physico-chemical properties of the black cumin seed oil. Ind. Crop. Prod. 208. https://doi.org/10.1016/j.indcrop.2023.117892.

Scandar, S., Zadra, C., Lanari, D., Marcotullio, M.C., 2024. Boosting the essential oil yield of Coriandrum sativum L. using choline chloride-based NADES: an optimized and systematic study employing response surface methodology (RSM). Ind. Crop. Prod. 208 (December 2023), 117920. https://doi.org/10.1016/j.indcrop.2023.117920.

Selim, H.M., Zhang, H., 2007. Arsenic adsorption in soils: second-order and multireaction models. Soil Sci. 172 (6), 444–458. https://doi.org/10.1097/ss.0b013e31804fa2a1.

Servis, M.J., Clark, A.E., 2018. Interfacial heterogeneity is essential to water extraction into organic solvents. ChemRxiv 2, 1–24.

Silva, A., Cassani, L., Carpena, M., Lourenço-Lopes, C., Grosso, C., Chamorro, F., Garc\'\ia-Pérez, P., Carvalho, A., Domingues, V.F., Barroso, M.F., others, 2024. Exploring the potential of Invasive species Sargassum muticum: microwave-assisted extraction optimization and bioactivity profiling. Mar. Drugs 22 (8), 352.

Simsek, M., Süfer, Ö., 2021. Infusion of walnut (Juglans regia L.) shell tea: multi response optimization and antioxidant potential. J. Appl. Res. Med. Aromat. Plants 20 (November 2020), 100278. https://doi.org/10.1016/j.jarmap.2020.100278.

Sinha, D., Mukherjee, S., & Chowdhury, S. (2022). Methods of extraction of phytochemicals - isolation, Characterization, and Therapeutic Applications of Natural Products (pp. 250–279).

Sridhar, A., Vaishampayan, V., Senthil Kumar, P., Ponnuchamy, M., Kapoor, A., 2022. Extraction techniques in food industry: insights into process parameters and their optimization. Food Chem. Toxicol. 166. https://doi.org/10.1016/j.fct.2022.113207.

Stamenković, O.S., Kostić, M.D., Radosavljević, D.B., Veljković, V.B., 2018. Comparison of box-behnken, face central composite and full factorial designs in optimization of hempseed oil extraction by n-hexane: a case study. Period. Polytech. - Chem. Eng. 62 (3), 359–367. https://doi.org/10.3311/PPch.11448.

Sun, B., Yang, C., Wang, Y., Gui, W., Craig, I., Olivier, L., 2020. A comprehensive hybrid first principles/machine learning modeling framework for complex industrial processes. J. Process Control 86, 30–43. https://doi.org/10.1016/j.jprocont.2019.11.012.

Tiwari, B.K., Brunton, N.P., Brennan, C.S., 2013. Handbook of plant food phytochemicals: sources, stability and extraction. In: Handbook of Plant Food Phytochemicals: Sources, Stability and Extraction. https://doi.org/10.1002/9781118464717.

Toledo, R.T., 2019. Kinetics of chemical reactions. In: Kinetics of Chemical Reactions, pp. 285–299. https://doi.org/10.1002/9783527808397.

Usman, I., Hussain, M., Imran, A., Afzaal, M., Saeed, F., Javed, M., Afzal, A., Ashfaq, I., Al Jbawi, E., A. Saewan, S., 2022. Traditional and innovative approaches for the extraction of bioactive compounds. Int. J. Food Prop. 25 (1), 1215–1233. https://doi.org/10.1080/10942912.2022.2074030.

Van Boekel, M.A.J.S., 1996. Statistical aspects of kinetic modeling for food science problems. J. Food Sci. 61 (3), 477–486. https://doi.org/10.1111/j.1365-2621.1996.tb13138.x.

Van Boekel, M.A.J.S., 2008. Kinetic modeling of food quality: a critical review. Compr. Rev. Food Sci. Food Saf. 7 (1), 144–158. https://doi.org/10.1111/j.1541-4337.2007.00036.x.

Vannier, C., Cochrane, T.A., Reza, P.Z., Bellamy, L., 2022. An analysis of agricultural systems modelling approaches and examples to support future Policy development under Disruptive changes in New Zealand. Applied Sciences (Switzerland) 12 (5). https://doi.org/10.3390/app12052746.

Varzakas, T., 2014. Centrifugation - filtration. In: Food Engineering Handbook: Food Process Engineering. CRC Press, pp. 61–123.

Villota, R., Hawkes, J.G., 2006. Reaction kinetics in food systems. In: Handbook of Food Engineering, second ed. CRC Press, pp. 125–286. https://doi.org/10.1201/9780429449734-3.

Whitehead, F.A., 2021. *Diffusion coefficient Modelling of Natural Bioactives from Crosslinked Hydrocolloid Matrices under Moving Boundary Conditions* (Issue February). RMIT University. Doctoral dissertation.

Worku Kidane, S., 2021. Application of response surface methodology in food process modeling and optimization. In: Response Surface Methodology in Engineering Science. IntechOpen. https://doi.org/10.5772/intechopen.100113.

Wu, K., Ju, T., Deng, Y., Xi, J., 2017. Mechanochemical assisted extraction: a novel, efficient, eco-friendly technology. Trends Food Sci. Technol. 66, 166–175. https://doi.org/10.1016/j.tifs.2017.06.011.

Xia, B., Sun, D.W., 2002. Applications of computational fluid dynamics (CFD) in the food industry: a review. Comput. Electron. Agric. 34 (1–3), 5–24. https://doi.org/10.1016/S0168-1699(01)00177-6.

Xing, C., Li, M., Liu, L., Lu, R., Liu, N., Wu, W., Yuan, D., 2023. A comprehensive review on the blending condition between virgin and RAP asphalt binders in hot recycled asphalt mixtures: mechanisms, evaluation methods, and influencing factors. J. Clean. Prod. 398, 136515. https://doi.org/10.1016/j.jclepro.2023.136515.

Yang, R., Chen, J., 2021. Mechanistic and machine learning modeling of microwave heating process in domestic ovens: a review. Foods 10 (9), 2029. https://doi.org/10.3390/foods10092029.

Yang, R., Wang, Z., Chen, J., 2021. An integrated approach of mechanistic-modeling and machine-learning for thickness optimization of frozen microwaveable foods. Foods 10 (4), 763. https://doi.org/10.3390/foods10040763.

Yeddes, W., Ouerghemmi, I., Hammami, M., Gadhoumi, H., Affes, T.G., Mohamed, S.N., Aidi-Wannes, W., Witrowa-Rajchert, D., Saidani-Tounsi, M., Nowacka, M., 2022. Optimizing the method of rosemary essential oils extraction by using response surface methodology (RSM)-Characterization and Toxicological assessment. Sustainability 14 (7). https://doi.org/10.3390/su14073927.

Yolmeh, M., Jafari, S.M., 2017. Applications of response surface methodology in the food industry processes. Food Bioprocess Technol. 10 (3), 413–433. https://doi.org/10.1007/s11947-016-1855-2.

Yu, W., Zhou, X., Li, C., 2021. Application of first-order kinetics modeling to reveal the nature of starch digestion characteristics. Food Funct. 12 (15), 6652–6663. https://doi.org/10.1039/d1fo00450f.

Zakaria, N.H., Whanmek, K., Thangsiri, S., Chathiran, W., Srichamnong, W., Suttisansanee, U., Santivarangkna, C., 2023. Optimization of Cold Brew coffee using central composite design and its properties compared with hot Brew coffee. Foods 12 (12), 2412. https://doi.org/10.3390/foods12122412.

Zhang, P., 2010. Industrial control system simulation routines. In: Advanced Industrial Control Technology. Elsevier, pp. 781–810. https://doi.org/10.1016/b978-1-4377-7807-6.10019-1.

Zhang, W., Luo, Z., Wang, A., Gu, X., Lv, Z., 2021a. Kinetic models applied to quality change and shelf life prediction of kiwifruits. Lwt 138, 110610. https://doi.org/10.1016/j.lwt.2020.110610.

Zhang, X., Ma, F., Yin, S., Wallace, C.D., Soltanian, M.R., Dai, Z., Ritzi, R.W., Ma, Z., Zhan, C., Lu, X., 2021b. Application of upscaling methods for fluid flow and mass transport in multi-scale heterogeneous media: a critical review. Appl. Energy 303, 117603. https://doi.org/10.1016/j.apenergy.2021.117603.

Modeling of membrane separation processes

10

Ozgur Tarhan [1], *Ozlem Caglar Duvarci* [2] *and Osvaldo H. Campanella* [3]
[1]Department of Food Engineering, Faculty of Engineering and Life Sciences, Uşak University, Uşak, Türkiye; [2]Department of Chemical Engineering, Facultly of Engineering, Izmir Institute of Technology, Izmir, Türkiye; [3]Department of Food Science and Technology, the Ohio State University, Columbus, OH, United States

10.1 Introduction

Membrane technologies have become one of the major techniques in food processing and are widely used for separation processes such as clarification and/or increasing concentrations of components in liquid form of foods, as well as filtration, debacterization, and fractionation. The growing requirements for water and energy sources, driven by population growth, have further amplified the significance of membrane technologies, owing to their broad applicability across diverse fields, including food production, pharmaceuticals, water treatment, and biotechnology. The advantages of membrane operations have positioned them as attractive alternatives to conventional separation methods like evaporation and centrifugation, thereby garnering substantial research interest.

The studies on membrane technologies date back to the 18th century. Jean-Antoine Nollet described the permeation of water in bladders by osmosis in 1748. Then, the first artificial semipermeable membrane was prepared by Traube and Pfeffer in 1867 (Nambi Krishnan et al., 2022). A rapid progress in membrane preparation was seen, such as nitrocellulose membrane (Bechhold, 1907) and microcollodion membranes (Ferry, 1936; Zsigmondy and Bachmann, 1918). Another major step was the clinically successful hemodialysis achieved by William Johan Kolff in 1945 (Nambi Krishnan et al., 2022). Water-salt solution was separated by cellulose acetate reverse osmosis membranes in the 1950s (Reid and Breton, 1959). The membrane had very low flux due to its thickness. This is followed by the use of electrodialysis and ion-exchange membranes for saline sources (1954), the preparation of anisotropic reverse osmosis and asymmetric cellulose acetate membranes (1963), and a polyamide hollow fiber permeator (1970). The anisotropic reverse osmosis reduced the saline content as much as 99.5%. In the 80 and 90s, membranes started to be used in gas separation with developments in composite membrane preparation by interfacial polymerization (Cadotte et al., 1981; Cadotte, 1981). Membranes were placed in a bioreactor or in a separate tank for filtration, and membrane fouling is prevented by air bubbles (Yamamoto et al., 1988).

Modeling of Unit Operations in the Food Industry. https://doi.org/10.1016/B978-0-12-822180-8.00007-9

Within the last 50 years, numerous different membrane preparation methods and techniques have been investigated for certain types of food and chemical applications. Basically, membranes act as selective barriers that separate substances based on molecular size, with pressure, concentration, and electrical potential gradients serving as the primary driving forces (Kadival and Ghosh, 2018; Rajendran et al., 2021; Teodoro et al., 2023). The key membrane operations are included Microfiltration (MF), Ultrafiltration (UF), Nanofiltration (NF), Reverse Osmosis (RO), Osmotic Evaporation, Osmotic Distillation, Membrane Distillation (MD), Direct Contact Membrane Distillation (DCMD), Vacuum Membrane Distillation (VMD), Enzymatic Membrane Reactor (EMR), Electrodialysis (ED), Membrane Bioreactors (MBRs), Membrane Contactors (MCs), and Pervaporation (PV). Separation processes by membrane operations are based on a driving force depending on pressure difference (i.e., MF, UF, NF, RO), concentration difference (i.e., dialysis), electrical potential difference (i.e., electrodialysis), or vapor pressure difference (i.e., PV). By selectively allowing the passage of specific molecules or ions while blocking others, membranes exhibit semipermeable behavior determined by properties such as molecular size, electrical charge, and other physicochemical factors. Membranes are further classified based on several distinguishing features, including whether they are organic or inorganic, homogeneous or heterogeneous, charged or neutral, solid or liquid, and symmetric or asymmetric (Kadival and Ghosh, 2018).

Membrane separation processes are based on the transportation of solutes and solvents across a membrane system under the influence of a driving force (concentration, pressure, and electrical potential gradient). The type of membrane separation employed, the design and characteristics of membrane modules, the properties of the feed solution, and the operational conditions are critical parameters that govern system performance and overall process efficiency.

Mathematical models have great significance in designing and optimization of membrane processes. Theoretical modeling and computer simulations revealed attractive results, providing a comprehensive understanding of the integrated membrane processes. The membrane is a selective porous barrier separating molecules due to their molecular size. The main driving force, depending on the membrane process type, refers to a gradient or a factor forcing the substances to move through the membrane. Modeling of pressure-driven processes like MF, UF, and RO; concentration gradient-effective processes like forward osmosis (FO) and dialysis; electrical potential gradient-driven electrodialysis and temperature gradient-effective membrane distillation processes involve key parameters and specific considerations of individual systems (Y. Lee and Clark, 1998; Teodoro et al., 2023). The pore-flow and solution-diffusion mechanisms are the two main types of water transport mechanisms in membranes, which are driven by pressure gradient and concentration gradient, respectively (Heiranian et al., 2023). Mechanistic models are based on Navier–Stokes and continuity equations to estimate solute and solvent fluxes, diffusion and rejection coefficients, and fouling profiles. Numerical solution methods facilitate solvation of complex equations derived in these models using special boundary conditions of membrane systems. This chapter presents types of membrane processes applied to foods, mass transfer basics in membranes, and modeling of transport phenomena for various membrane processes.

10.2 Membrane-based separation processes

Membranes are thin layers of semipermeable and selective barriers that are used to separate dissolved substances or fine particles from solutions, depending on molecular or physical size, electrical charge, and other properties (Herrera-Valdez, 2012). From a materials perspective, membranes are broadly categorized into organic and inorganic types (Kadival and Ghosh, 2018; Younas and Rezakazemi, 2022). Organic membranes are generally more flexible, easier to handle, and exhibit good chemical stability, although they are sensitive to extreme pH conditions and certain solvents; they also offer thermal stability at moderately elevated temperatures, low production costs, and enhanced resistance to fouling due to their smooth surface morphology. In contrast, inorganic membranes are more rigid and display excellent chemical resistance (tolerant to high pH, solvents, and harsh environments), high thermal stability (withstanding elevated temperatures), prolonged operational lifetimes, microbial resistance, and suitability for high-pressure applications. Membranes are further classified as symmetric or asymmetric depending on pore structure. Symmetric membranes possess a uniform pore distribution throughout the cross-section, whereas asymmetric membranes exhibit a pore size gradient, with smaller pores at the surface that progressively increase in size toward the interior. Various membrane materials are used for Microfiltration, Ultrafiltration, Nanofiltration, and Reverse Osmosis depending on their pore size and structure (Daneluz et al., 2023; El-Okazy et al., 2024; S. Nath et al., 2024; Younas and Rezakazemi, 2022; Yun et al., 2024).

In membrane separation processes, the feed stream is fractionated into a permeate—comprising components that pass through the membrane—and a retentate, containing retained species (Younas and Rezakazemi, 2022). The permeate typically carries particles that are small enough to traverse the membrane pores, while the retentate retains those that cannot pass through. Membrane systems can operate in two primary modes: crossflow filtration and dead-end (perpendicular) filtration (Fig. 10.1a and b). In crossflow filtration, the feed stream flows parallel to the membrane surface (Rajendran et al., 2021), permitting a fraction of the fluid to permeate through the membrane while the remaining portion continues along the surface as retentate. This parallel flow configuration reduces fouling by promoting the continuous removal of deposited particulates from the membrane surface (Fig. 10.1a). In contrast, dead-end filtration operates with the flow that is perpendicular to the membrane, causing retained particles to accumulate directly on the surface. Over time, this leads to the formation of a fouling layer, or "cake," which becomes progressively thicker and results in a steady decline in permeate flux (Fig. 10.1b). Crossflow filtration, by minimizing surface fouling, extends membrane lifespan and maintains more consistent separation performance.

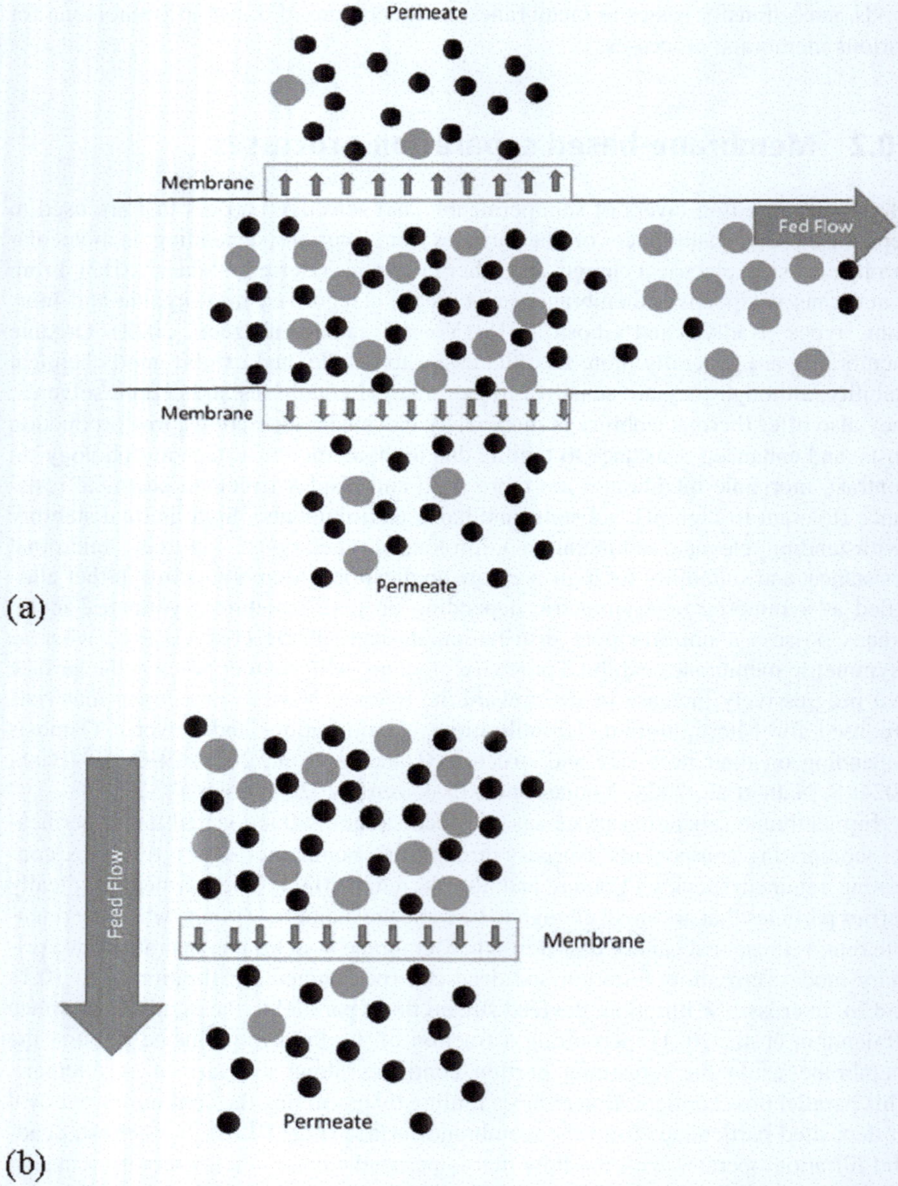

Figure 10.1 Schematic diagrams of membrane separation depending on the flow direction of feed: (a) Crossflow filtration and (b) dead-end filtration.

Hence, the materials used in the fabrication of membrane (inorganic or organic), thickness of membrane, its pore size&structure, porosity, and solvent permeability are important characteristics of membranes which directly affect both membrane selection for a specific application and determination of operating conditions. The operating conditions, such as flow rate of permeate, resistance to mechanical stresses, presence of chemicals, and application of heat, are also important conditions in membrane material selection (Rajendran et al., 2021; Soni et al., 2009; Teodoro et al., 2023; Tijing et al., 2015; Vinh-Thang and Kaliaguine, 2013; Younas and Rezakazemi, 2022).

Membrane technologies have many advantages compared to traditional technologies, and mostly operate at ambient temperatures, which offer:

- improvement in the desired properties of food by preservation of thermo-sensitive compounds,
- simultaneous increase in concentration without loss of bioactive compounds and nutritional value,
- enhanced process efficiency, high yield, low energy cost, water usage, reduced waste generation, and
- simple operating mechanisms which make membranes easy to scale up.

However, membranes also have some disadvantages such as;

- need for pretreatment of food materials,
- sensitivity to conditions of operation (pH, temperature, pressure, feed concentration, etc.),
- membrane dysfunction (membrane degradation over time or fouling problem during operation due to microorganisms and biofilm formation on the surface),
- damage risks of membranes due to exposure to mechanical stresses and chemicals during operation,
- improper handling during cleaning and replacement (need of experienced operators) due to complexity of system design and operation
- high cost of initial installation due to installation and instrument, etc.

Membrane processes play a vital role in a wide range of processes within the food industry, offering efficient and sustainable solutions for separation, concentration, and purification processes. Some membrane processes given below can serve critical applications in the food industry:

1. Filtration and Separation: Membrane processes like MF, UF, NF, and RO are extensively used for separating and purifying components in food processing. Applications include separating whey proteins in dairy processing, recovering bioactive compounds from fruit juices, clarification of juices, and extracting protein-based bioactive compounds from whey in the dairy industry (Q. Chen et al., 2018; Pouliot et al., 2014).
2. Increasing Concentration and Clarification: Membrane processes are employed for concentrating and clarifying fruit juices, recovery of sugar from foods, removal of yeast and microorganisms from alcoholic beverages such as beer and wine, and other liquid food products. Also, they are used in protein concentrate production from sources like whey and oil-seed mixtures.

3. Water Purification: Membrane processes are utilized for water purification in the food industry by removing bivalent ions like calcium and magnesium, ensuring the quality and safety of water used in various food processing operations.
4. De-alcoholization and Flavor Preservation: RO and PV are used to remove undesired molecules from beverages and to preserve flavorant molecules by selective permeation, which helps to retain flavor compounds under controlled temperature and pressure. PV-based membrane processes are applied for ethanol removal from alcoholic beverages to create nonalcoholic alternatives while preserving quality properties and aroma compounds.
5. Waste Valorization and Resource Recovery: Membrane processes are involved in waste valorization by recovering valuable compounds (selective separation of proteins, enzymes, sugars, organic acids, minerals, fats, oils, antioxidants, vitamins, aroma compounds, etc.) from food processing byproducts and waste streams, which promotes sustainability in the food industry. These membrane operations enable maximization of resource utilization and production of high-value products, otherwise discarded foods, and their reuse in various applications while reducing waste and contributing to sustainable resource management and environmental conservation.
6. Pre-treatment and Pre-filtration: MF is used for pretreatment processes such as removing bacteria from milk, pretreating cheese whey, and/or removing suspended particles, colloids, bacteria, and contaminants (such as oil or fat globules) from feed streams before further food processing operations. This pretreatment step in food processing ensures that downstream processes operate efficiently, acts as a physical barrier to separate particles

In membrane processing, the concentration gradient plays a crucial role by serving as a driving force for transport, influencing separation efficiency, impacting selectivity, affecting permeate flux, contributing to energy efficiency, and ensuring product quality and purity. The concentration gradient through a membrane induces the movement of substances by creating a difference in solute concentrations, facilitating transport from a higher concentration to a lower concentration. The relevance of concentration gradients to membrane function is summarized below and illustrated schematically in Fig. 10.2. Important features on concentration gradient for separation by membranes are listed below:

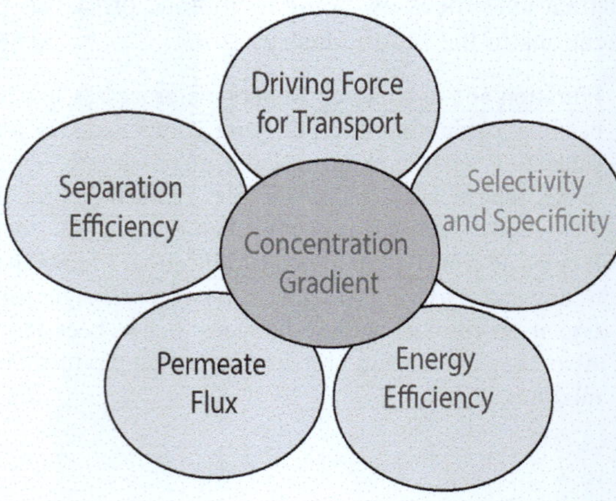

Figure 10.2 The crucial role of the concentration gradient in membrane processing.

Enhanced Separation Efficiency: A significant concentration gradient enhances the efficiency of separation processes by promoting higher flux rates of solutes across the membrane, leading to improved separation of components.

Selectivity and Specificity: The concentration gradient influences the selectivity and specificity of membrane processes by controlling the preferential transport of specific components based on their concentration differences.

Process Optimization: Maintaining an optimal concentration gradient is essential for optimizing membrane processes in terms of efficiency, productivity, and product quality, allowing operators to tailor the separation process to achieve desired outcomes.

Energy Efficiency: Proper control and management of the concentration gradient help minimize energy consumption in membrane processes, contributing to more sustainable and cost-effective operations.

The critical factor in membrane processes is permeability, which refers to the ability of a membrane to allow substances to pass through it. In the context of membrane separation technology, it is a measure of how easily a particular substance (such as water, ions, or organic molecules) can move through the membrane material. High permeability means that the membrane lets components of the feed stream pass through it more easily, while low permeability indicates that the membrane is more restrictive. Permeability is affected by factors such as the membrane's material, pore size, and the driving force applied (e.g., pressure, concentration gradient, charges, or electric potential) (Kadival and Ghosh, 2018). Different types of membrane processing used in food applications are listed below.

10.2.1 Pressure-driven membrane processes

Pressure-driven membrane separation encompasses a group of processes in which the pressure difference across the membrane serves as the primary driving force for separation. These techniques are widely used in the food industry and many other industries owing to their high efficiency and the ability to operate at lower temperatures, thereby minimizing thermal degradation of sensitive products. The most-used pressure-driven membrane processes include microfiltration, ultrafiltration, nanofiltration, and reverse osmosis. The main features are given below (Kadival and Ghosh, 2018).

1. Microfiltration (MF) includes the use of membranes with large pores (0.1–5 μm) to remove suspended particles, bacteria, and yeast from liquids such as milk, beer, and wine. Operates at low pressures (<2 bar).
2. Ultrafiltration (UF) utilizes membranes with smaller pores (1–100 nm) to lower the concentration of proteins and other macromolecules, to produce protein concentrates from whey, and to clarify fruit juices. Operates at pressures between 2–10 bars.
3. Nanofiltration (NF) employs membranes with even smaller pores (5–10 nm) to remove organic molecules, viruses, and divalent ions, for water softening, lactose desalting, protein concentrate production, and whey demineralization. Operates at pressures between 5–15 bars.

4. Reverse Osmosis (RO) uses membranes with the smallest pores (<0.5 nm) for water purification, increasing concentration or recovery of sugar from fruit/vegetable juices, and dealcoholization of beer and wine. Operates at high pressures (15–100 bars).

Pressure-driven membrane processes, including microfiltration (MF), ultrafiltration (UF), nanofiltration (NF), and reverse osmosis (RO), offer predominantly chemical-free operation and generate minimal waste, thereby promoting environmental sustainability and reducing the risk of chemical contamination in the treated product. These processes exhibit high separation efficiency, effectively removing a wide range of contaminants (suspended solids, bacteria, viruses, and dissolved substances) related to the pore size and membrane material. Pressure-driven membranes consistently produce high-quality permeate, which is essential for applications demanding stringent quality standards, such as drinking water production and pharmaceutical manufacturing. Their typical compact and modular design facilitates easy installation in space-constrained environments and enables integration into hybrid systems, enhancing overall process efficiency. Furthermore, their versatility allows them to accommodate a wide range of feed solutions, supporting diverse applications across water treatment, food processing, and industrial separation sectors. Operations at low temperatures and low-pressure differences (if possible) and minimum use of chemicals might be reasons for less energy consumption and low operational costs (Teodoro et al., 2023; Younas and Rezakazemi, 2022). These processes are effective for treating various feed mixtures like water, gas, and food, and are characterized by their ability to produce stable products with low energy consumption due to operations at low temperatures, which allows a better retention of volatile aromas and nutrients, without adding any chemicals such as solvents.

Organic membrane materials commonly include cellulose derivatives, polyamides, polyolefins, polycarbonates, and polysulfones, whereas inorganic membranes are primarily based on silicon, zirconium, titanium, and aluminum oxides. Table 10.1 summarizes recent research on microfiltration, ultrafiltration, nanofiltration, and reverse osmosis in food processing.

MF in one-step or multi-step operations is mostly applied to removing suspended particles, fat globules, proteins, yeast, bacteria, and other microorganisms under gentle operating conditions, such as dairy products, alcoholic beverages (i.e., beer, wine), fermented products (i.e., kombucha), oil-water mixtures, etc. A single-step crossflow MF process with a synthesized polysulfone (PS) membrane efficiently isolates fat globules from cow and buffalo milk, whereas multi-step MF systems, involving up to 10 stages, have been investigated for the recovery of tacrolimus from *Streptomyces tsukubaensis* fermentation broths (Bertan and Cremasco, 2024; Kapoor et al., 2024). The performance of membranes can be changed under the effect of acidic and basic conditions. For example, changing the acidic or basic conditions can minimize the adsorption of protein and the fouling of polyethersulfone (PES) microfiltration membranes enriched by polydopamine (PDA) coatings for sterile filtration of bovine serum albumin (BSA) solution (Yun et al., 2024). Another study was done on incorporating microfiltration after the first fermentation stage to enhance kombucha's stability (a fermented tea beverage) and visual appeal while maintaining

Table 10.1 Some recent studies on membrane separations used in food processing.

Process	Membrane material	Pore size of membrane	Application	References
Microfiltration	Polysulfone	0.8 mm	Removing fat globules from milk	Kapoor et al. (2024)
	Poly-ethersulfone	0.1–10 μm	Reducing turbidity and retaining suspended compounds, colloids, and microorganisms in kombucha.	Daneluz et al. (2023)
	Mixed cellulose esters	0.22 μm (with porosity of 0.75) and 3.0 μm (with porosity of 0.83)	Recovering tacrolimus by multi-step microfiltration	Bertan and Cremasco (2024)
	SiC ceramics	1.4 μm Highly hydrophilic	Separating oil from protein-rich rapeseed extracts	Chandrappa et al. (2024)

Continued

Table 10.1 Continued

Process	Membrane material	Pore size of membrane	Application	References
Ultrafiltration	Polysulfone Dialysis filters (commercial) - Composite polyamide (suez (GE)) - Polyethersulfone (PM UP010, microdyn nadir) - Graphene oxide on porous PVDF membrane (commercial)	Molecular weight ~15–20kDa –1 kDa –10kDa	Recovering microorganisms Concentrating the dairy proteins during lactose transmission by ultrafiltration	Kim et al. (2009) El-Okazy et al. (2024)
	Amicon ultrafiltration (UF) by membrane tubes	Molecular weight cut-off 10 kDa	Producing succinylated protein–iron complexes for food fortification purposes	Shilpashree et al. (2015)
	- MWCO PES membranes - Hydrophilized-PES membrane (MicrodynNadir, Germany)	–300 and 100 kDa –30 kDa	Concentrating alfalfa protein extract	Tanambell et al. (2024)
	Hollow-fiber ultrafiltration (polysulfone) Concentrating pipettes (80–100 kDa cut-off, cat. CC08003, innovaprep)	80–100 and ~600 kDa cutoff	Removing the mengo virus & hepatitis A virus from fresh and frozen fruit and vegetables	Trudel-Ferland et al. (2024)
	Regenerated cellulose (millipore, prep/scale spiral wound TFF)	~1 and ~10 kDa	Recovery of residual proteins (soluble) after mono- (*Kluyveromyces marxianus*) cultivation and mixed culture (*K. Marxianus* and *Saccharomyces cerevisiae*) on whey	Yadav et al. (2016)

Nanofiltration	Polypiperazineamide (PA) synthesized the interfacial polymerization (hydrophilic)	~200 kDa	Removing hazardous Congo red dye from water	Hairom et al. (2014)
	Fluoropolyamide and silica–fluoropolyamide on polyethersulfone membrane	Not mentioned	Optimizing preparation conditions (silica concentration, reaction time, monomer concentration, curing time, and oxidation) to enhance membrane performance and reduce high salt content in water	Hu et al. (2013)
	Alumina ceramic membrane	3.55 nm	Influence of thickness on water permeation	Liu et al. (2024)
	Polypiperazine, with a nonwoven polyester as support material	300–500	Investigating the impacts of solution pH and membrane pore size on the recovery percentages and concentrations of volatile fatty acids, and evaluating the performance of two commercial nanofiltration membranes under various conditions	Pervez et al. (2022)

Continued

Table 10.1 Continued

Process	Membrane material	Pore size of membrane	Application	References
Reverse osmosis	Tubular ceramic MF membranes (Isoflux™, tami industries, France) Spiral-wound polyamide composite RO membranes (dairy AF3838C30, general electrics)	Cut-off of 1.4 μm Pore size<0.1 nm	Improving water removal efficiency from skim milk by developing a cascade membrane process including both microfiltration (MF) and reverse osmosis (RO).	Blais et al. (2021)
	Two types of reverse osmosis membranes: The trisep X20 advanced composite membrane and the thin film composite alfa laval RO99 membrane.	Not mentioned	Evaluating membranes efficiency on concentration of valuable compounds from beetroot peel extracts	Zin et al. (2020)
	A polyamide thin-film composite membrane	Not mentioned	Investigating microbubble effect (novel swirl-flow microbubble generator-MBG) and two-phase flow in RO process performance for desalinization.	Y.-B. Kim et al. (2024)
	A multichannel, gradient ceramic MF membrane with a pore size of 1.4 mm (ISOFLUX) and DOW FILMTEC™ SW30-2540 membrane for RO.	Not mentioned	Assessing of the feasibility and limitations of the process for producing storable ESL RO whey concentrates.	Marx et al. (2018)

its probiotic benefits (Daneluz et al., 2023). Separation of oil-water mixtures or protein-rich extracts from oil can be successfully done by different membrane materials such as ceramic membrane or polymeric membrane (SnO_2/PVC) (Chandrappa et al., 2024; Sadek and Al-Jubouri, 2023).

UF effectively separates compounds with high molecular weight from low molecular weight, as evidenced by studies on alfalfa juice, sweet whey, and the detection of foodborne viruses (El-Okazy et al., 2024; Kim et al., 2009; Shilpashree et al., 2015; Tanambell et al., 2024; Trudel-Ferland et al., 2024; Yadav et al., 2016). Graphene UF membranes and polyethersulfone (PES) membranes show good performance in enhancing filtration efficiency and reducing fouling. Indeed, the performance of UF membranes can vary significantly depending on their material itself and also the molecular weight cut-off (MWCO). The importance of selecting appropriate membrane materials and pore sizes to optimize filtration performance and minimize fouling is important for maintaining efficiency and reducing operational costs. The optimization of UF processes involves fine-tuning operating parameters such as transmembrane pressure, temperature, and elution time to get the highest performance. e.g., the study on foodborne virus detection optimized elution time and buffer composition to improve virus recovery from produce. Similarly, the research on alfalfa juice demonstrated the importance of pretreatment steps and membrane selection in achieving effective protein concentration and fractionation. These optimizations are essential for scaling up UF applications and ensuring their practical viability in industrial settings.

NF membranes combine features of RO and UF systems, employing size exclusion, charge interactions, and diffusion to achieve selective separation. The NF separation mechanism is governed by a complex interplay of molecular sieving, electrostatic effects, and diffusive transport for effective separation of a wide range of solutes across diverse applications. Many NF membranes possess a negative surface charge, enhancing the rejection of anions and promoting preferential separation of multivalent ions while enabling partial passage of monovalent ions. During NF, solute particles can accumulate near the membrane surface, forming a concentration polarization layer that increases resistance to flow. Additionally, fouling arises from solute deposition on the membrane surface or within its pores, causing a decline in flux and reduced membrane performance. Strategies to optimize NF membrane design—such as eliminating interlayers, minimizing selective layer thickness, and tailoring surface properties—are critical for achieving high permeation rates and superior selectivity, making NF suitable for applications (i.e., textile wastewater treatment and desalination). Various fabrication and modification approaches, including nanoparticle incorporation, interfacial polymerization, and layer-by-layer assembly, have been investigated to enhance NF membrane performance (Hairom et al., 2014; Hu et al., 2013; Liu et al., 2024; K. Nath et al., 2018; Pervez et al., 2022).

RO is mostly combined with an MF in advance. Microfiltration reduces microbial load (i.e., microorganisms and spores) and particulate matter (i.e., fat globules and somatic cells), impurities, and other contaminants. By removing particulate matter and microorganisms, MF reduces the fouling and clogging potentials on the RO membranes. The removal of microorganisms by microfiltration (MF) enables reverse osmosis (RO) operations to be conducted at elevated temperatures (e.g., 50°C), thereby increasing permeate flux and reducing consumed energy per unit of water removed. Integrating MF as a pretreatment step enhances the overall efficiency of the concentration process by preconcentrating the feed solution and alleviating the operational load on the RO membranes (Blais et al., 2021; Kim et al., 2024; Marx et al., 2018; Zin et al., 2020).

Different combinations of RO were studied extensively to enhance and improve membrane efficiency. A combination of MF and RO was studied for different food applications, for example, for skim milk by using the cascade MF/RO hot process (Blais et al., 2021), for whey concentrates by using cold MF and a two-step preservation process for producing RO (Marx et al., 2018). Investigation on two RO membranes to concentrate valuable compounds from beetroot peel extracts or a combination of FO with membrane distillation for concentrated liquid food was also studied (Kim et al., 2024; Zin et al., 2020). Different attachments would be used to improve RO efficiency, such as adding a microbubbling unit to the RO for desalination (Kim et al., 2024).

10.2.2 Ion exchange membrane processes

Separation technologies such as ion exchange membrane processes utilize ion-selective membranes to facilitate ion movement due to an applied electric field while blocking other particles (Kadival and Ghosh, 2018; Rajendran et al., 2021; Tanaka, 2010; Teodoro et al., 2023). These processes are used in the food industry for the separation of bioactive compounds, purification of liquids such as removing salt and/or minerals from water (water desalination, fruit juice deacidification, and whey demineralization), etc.

Electrodialysis and its various types offer versatile and efficient methods for efficiently removing specific ions from solutions, require less energy compared to thermal processes, reducing the need for chemical additives, and minimizes waste generation. Electrodialysis is electic filed driven-an ion migration through ion-selective membranes. Here, hydrodynamic pressure has no importance on transportation of ions. The ion exchange membranes are divided into two categories: anion-exchange membranes (AEM) and cation-exchange membranes (CEM). Under the influence of the electric field, cations move toward the cathode and pass through CEMs, while anions move toward the anode and traverse AEMs, resulting in effective ion separation from the feed solution. Electrodialysis membranes are typically arranged in an alternating series of CEMs and AEMs, forming compartments that selectively transport ions. This arrangement concentrates ions in the concentrate stream while depleting them in the dilute stream, achieving efficient separation (Rajendran et al., 2021). The migration rate of ions can be tuned by changing the intensity of the electrical field. The electrodialytic separation is changed by pH and ionic strength.

Electrodialysis can be coupled with ultrafiltration (EDUF) by arranging arrays of ultrafiltration and/or ion-exchange membranes in an electrodialysis cell. This configuration enables the simultaneous separation of positively and negatively charged molecules, facilitating the isolation of bioactive molecular fractions from complex raw matrices. Electrodialysis can also be combined with nanofiltration (NF) to enhance separation efficiency. The integration of NF and electrodialysis has been successfully employed for monovalent ions removal from brine solutions, improving both selectivity and process performance (Al Kholif, Hermana, Bilad and Chao, 2024; Rajendran et al., 2021).

There are different approaches for modeling of ion-exchange membranes (Dammak et al., 2021; Mareev et al., 2022; Oren et al., 2024; Porozhnyy et al., 2017; Purpura et al., 2024). There are many research available in the literature about different operations on many different foods, as well as different modifications on models. In terms of ion and water passing through charged membranes, there are solution-diffusion models single phase and multiphase models: Teorell-Meyer-Sievers (TMS), Three-wire model, and microheterogeneous model and pore-flow models (Cell models and Space-charge models) (Mareev et al., 2022).

Solution and diffusion-based single-phase model is Teorell-Meyer-Sievers model in which the membrane is accepted as a single phase, meaning an aqueous polymer chain matrix solution with mobile ions and functional groups. This model uses the Nernst–Planck equation, assumes local electroneutrality (LEN), and applies the Donnan equilibrium at membrane/solution interfaces. Water transportation is not taken into account due to osmosis-electroosmosis counteraction.

Multi-phase models are microheterogeneous and three-wire models. The microheterogeneous model has the effective medium approach. It represents a quasi-homogeneous system. The membrane is considered an array of capillary pores, a cluster-channel network, or a uniformly distributed pore. The structure may have micro-, meso-, and macropores. The pores are assumed to be uniform and distinguished by the ratio of the pore radius r to several characteristics, such as the electrical double layer or the length of action of adsorption forces.

The three-wire model is based on the idea of the formation of three parallel conducting layers: The first layer is the passage of ions through the conductive spheres and the solution between them, which is connected in series, while the other two layers are connected in parallel. The second layer is a pure ion-exchange gel. The third layer is the pure electrolyte solution representing the conductance of macropores in a heterogeneous membrane, which is filled with solution.

The microheterogeneous model has the effective medium approach, considering a quasi-homogeneous system that includes a porous medium (an electrically neutral solution in uniformly distributed micro-, meso-, and macropores) and gel phase (a hydrated polymer matrix with fixed charged groups whose charge is compensated by the charge of mobile ions). The pores are distinguished by the ratio of radius of pores (r) to different characteristic lengths such as electrical double layer or length of action of adsorption forces (l): Micropores ($r < l$ (approximately, $r < 1.5$ nm)), mesopores ($r > l$, but r and l are in the same order) and macropores ($r >> l$ ($r > 50$ nm)). The remaining part is considered the gel phase, which is a microporous medium not containing an electroneutral solution (Mareev et al., 2022). Even though there is no distinguished phase separation between the first and second phases, the main idea of the model is to assign different physicochemical properties.

In membrane modeling, the Cell Model describes the concentrated disperse systems well (Filippov, 2022; Mareev et al., 2022). The parameters considered in the model are transport properties such as conductivity, hydraulic, and diffusion permeability of the units (grains, fibers, etc.) that form the media and their packing properties. The approach of the cell model is replacing a real set of chaotically organized ion exchanger grains with a periodic alignment of identical porous charged spheres enclosed by concentric spherical shells filled with an electrolyte. The common influence of neighboring particles is considered by creating certain boundary conditions on their surfaces. Since this is a porous medium, there are specific boundary conditions on the surface of the liquid shell. The effect of surface forces on the barriers of the micro- and mesopores is considered by stress jump at the fluid–porous interface in the Cell model.

Space-charge models are related to pore flow models describing the ion and water transport in a membrane pore (Mareev et al., 2022; Uzdenova & Urtenov, 2021). The starting point of this approach comes from the description of ion and water transport inside micropores with charged walls. Space-charge models focus on the formation and behavior of space charge regions near the surfaces of ion-exchange membranes. At the ion-selective surface, the concentration of counterion exceeds the concentration of co-ion, space charge regions are formed. Nernst–Planck–Poisson–Navier–Stokes equation is used to describe the transport of water and ions. The main parameters are surface charge and ion concentration. The wall charge can be controlled by an external voltage source.

The Manning condensation model describes the behavior of ions in polyelectrolyte solutions and ion-exchange membranes. In this model, it is assumed that counterions (ions of opposite charge to the fixed charges on the polymer) condense onto the charged polymer backbone when the linear charge density of the polymer exceeds a critical value. This condensation reduces the effective charge density of the polymer and affects the distribution and mobility of ions. Dimensionless charge density (ξ) describes the condensation of ions, which is defined as $\xi = \lambda\_B$ /b, where λ_B is the Bjerrum length (the distance at which the electrostatic interaction between two charges is comparable to thermal energy), and b is the average distance between charges on the polymer backbone. When $\xi > 1$, counterion condensation occurs. The condensation reduces the effective charge density and modifies the electrostatic potential around the polymer. The prediction of the model is the distribution of free and condensed counterions around the polymer. This model is used to describe ion partitioning and transport in ion-exchange membranes, polyelectrolyte solutions, and other systems with charged polymers. It helps in understanding ion selectivity, permeability, and other transport properties (Oren et al., 2024).

10.2.3 Pervaporation

Pervaporation is a membrane-based separation technique utilized for the fractionation of liquid mixtures via partial vaporization across a nonporous or semiporous membrane. The process involves the selective interpenetration of one or more components from the liquid mixture, which are subsequently collected as vapor on the filtrate side of the membrane. The driving force for the separation is partial vapor pressure differences between the feed and permeate sides. A partial vacuum is typically applied on the permeate side to maintain this pressure gradient. The vapor-phase permeate is collected downstream, while the remaining liquid is on the retentate side of the membrane (Castro-Muñoz, 2019; Kadival and Ghosh, 2018; Teodoro et al., 2023).

Pervaporation membranes are typically made from polymers or composites that have selective permeability for specific components of the mixture. Membrane material selection should be done carefully by considering the nature of the separation, such as hydrophilic membranes for water removal or organophilic membranes for organic solvent separation. Mostly, this technique is used for water removal from organic solvents like ethanol, isopropanol, acetone, bioethanol, and other biofuels (Castro-Muñoz, 2019; Kadival and Ghosh, 2018; Teodoro et al., 2023; Tijing et al., 2015). It is also used for the separation of azeotropes that cannot be separated by conventional distillation. Separating mixtures of organic compounds, such as separating aromatic compounds from aliphatic compounds, can be done by pervaporation. This method offers high selectivity for specific components and operates at relatively low temperatures and pressures, reducing thermal degradation of sensitive compounds. The selection of membrane materials is critical for achieving selectivity and permeability at a high level in the desired separation process.

The membrane material should be carefully selected by considering membrane selectivity depending on the type of application, feed composition, and operating temperature. The existence of impurities in the feed stream may affect membrane selectivity. Pressure difference increased at the permeate side can enhance the permeation rate as well as the feed pressure. The thickness of the membrane affects permeation rate; thinner membranes generally offer higher permeation rates but may be less durable and more prone to damage.

The pervaporation process can be described by different models (Castro-Muñoz, 2019; Farhan et al., 2023; Tijing et al., 2015). Solution-Diffusion (SD) model describes the mass transport mechanism through a dense polymeric membrane in the pervaporation process. It involves three steps: sorption of the permeating components from the liquid feed to the membrane surface, diffusion across the membrane, and desorption to the vapor phase on the permeate side. The SD model is a commonly approved model used to define the mass transport mechanism through compact polymeric membranes in the pervaporation process. It involves three major stages.

a. Sorption: The migrating components from the liquid feed preferentially sorb onto the membrane surface. This step is influenced by the chemical properties of the membrane material and the permeating molecules.

b. Diffusion: The sorbed components then penetrate across the membrane. The diffusion rate is estimated by the micromorphology of the membrane structure and the size and shape of the diffusing molecules.

c. Desorption: Finally, the diffusing components desorb from the membrane into the vapor phase on the permeate side. This step is typically very rapid via the employment of a vacuum or a sweeping gas on the permeate side.

The SD model assumes that a thermodynamic equilibrium exists at the two sides of the membrane interface (feed and permeate sides) and that the pressure across the membrane is constant and equal to the high-pressure value of the feed side. The driving force for mass transfer in this model is the chemical potential gradient for each component through the membrane, which is described by the concentration gradient between the feed and filtrate sides.

The Universal Quasi-Chemical (UNIQUAC) Model is used to predict the activity coefficient of nonideal alcohol-water systems in PVA membranes (Farhan et al., 2023). It is a thermodynamic model used to predict the activity coefficients of components in nonideal liquid mixtures. It is particularly useful for describing the phase behavior of complex mixtures, including those involving polymers and solvents. The model calculates the activity coefficient of each component in a mixture, which is essential for predicting phase equilibria. The activity coefficient is divided into two parts: (a) Combinatorial Part: Accounts for the entropy of mixing due to differences in the size and shape of the molecules. (b) Residual Part: Accounts for the enthalpy of mixing due to interaction energies between different molecules. This model uses parameters related to the geometry (size and shape) of the molecules and binary interaction parameters between the components. These parameters can be obtained from experimental data or estimated using group contribution methods. The UNIQUAC model is widely used to predict liquid-liquid and liquid-vapor equilibria in binary and multicomponent systems, including polymer solutions. It is particularly useful in processes like pervaporation, where accurate prediction of sorption equilibria is crucial.

The Conventional Driving Force Model is used to describe membrane separation processes, such as pervaporation, as well as reverse osmosis and gas separation (Farhan et al., 2023). It refers to the use of a chemical potential gradient as the driving force for mass transport across the membrane. This model relies on the principle that the movement of components through the membrane is driven by differences in their chemical potential between the feed and permeate sides. The chemical potential is a measure of the thermodynamic potential of a substance to undergo a phase change or chemical reaction. In many practical applications, the chemical potential gradient is often approximated by the concentration gradient of the components. This simplification assumes that the concentration difference between the feed and permeate sides is the primary factor driving the transport. It helps to understand how different factors, such as feed concentration, temperature, and membrane properties, influence the separation performance.

10.2.4 Membrane distillation

Membrane distillation (MD) uses thermal energy to create a vapor pressure gradient to separate water vapor from a liquid feed by using a hydrophobic membrane (Teodoro et al., 2023; Tijing et al., 2015). The hydrophobic membrane repels liquid water while permitting water vapor to migrate across its pores. The temperature difference between the hot feed side and the cooler permeate side allows water to evaporate at the membrane interface and then condense on the permeate side, while nonvolatile components are retained in the feed solution. Typical operating feed temperatures for MD can vary between 40°C and 90°C, although specific operational temperatures are required depending on the application and the membrane configuration employed.

Membrane distillation (MD) can be implemented in several configurations, each designed to optimize specific performance parameters. In direct contact membrane distillation (DCMD), both the feed and permeate streams are in direct contact with the membrane surface, with water vapor condensing into a cooled permeate stream. In air gap membrane distillation (AGMD), an air layer is placed between the membrane and the condensation surface, reducing conductive heat loss and enabling vapor condensation after diffusion across the air gap. Vacuum membrane distillation (VMD) utilizes a vacuum on the permeate side to intensify the vapor pressure gradient and thereby enhance permeate flux. In sweeping gas membrane distillation (SGMD), an inert carrier gas flows along the permeate side to remove the water vapor, which is subsequently condensed in an external condenser.

MD is largely used for desalination of seawater, industrial wastewater treatment, and concentration of liquid food products (i.e., fruit juices and dairy streams) (Teodoro et al., 2023; Tijing et al., 2015). Nonvolatile solutes can be recovered almost 100% by using membrane distillation. Even though heat is used to create a water vapor pressure difference, MD operates at lower temperatures than conventional distillation. However, if the membrane surface is wetted even though its hydrophobic nature, the selectivity can decrease. The accumulation of contaminants, such as organic components, in feed effluent can cause fouling.

Typical materials employed for the fabrication of membranes to be used in MD processes are;

- Polytetrafluorethylene (PTFE): Highly hydrophobic and chemically resistant.
- Polyvinylidene Fluoride (PVDF): Good thermal and chemical stability.
- Polypropylene (PP): Cost-effective and hydrophobic.
- Polyethylene (PE): Hydrophobic and chemically resistant.

Models for MD can be categorized based on their focus on heat transfer, mass transfer, or overall system performance (Aytaç et al., 2024; Eleiwi et al., 2016; Hitsov et al., 2015; Rizki and Ottens, 2023). Heat Transfer Models, such as Nusselt Equation-Based Models, Computational Fluid Dynamics (CFD) Models, are used to predict heat transfer coefficients (Feed side, permeate side, and membrane heat transfer coefficients) in the water compartments and temperature profiles (bulk feed, membrane-feed interface, membrane permeate ınterface and bulk permeate temperatures). Nusselt Equation-Based Models are simple and have low computational burden, but predict a single, uniform interfacial temperature. CFD models have a high computational burden and often oversimplify the geometry or exclude heat transfer in spacer materials. Membrane thermal conductivity can be calculated using models like Isostrain, Isostress, or Maxwell type I. Sensible and latent heat fluxes, Reynolds number, Prandtl number, and Nusselt number are important parameters used in calculations.

Mass transfer models vary in complexity and computational requirements, offering different levels of insight into the mass transfer mechanisms in MD processes (Eleiwi et al., 2016; Hitsov et al., 2015). Sherwood equation-based models use semiempirical correlations to predict the mass transfer coefficient in the feed channels. The Sherwood relation is a dimensionless number indicating the ratio of convective to diffusive mass transfer. It is used to model the concentration polarization, which occurs when the concentration of solutes near the membrane surface increases through the rejection of salts, revealing a reduction in water activity and flux. They are simple and have low computational cost but predict a single, uniform interfacial concentration. The Dusty gas model (DGM) provides good predictions with low computational cost. It considers molecular and Knudsen diffusion besides viscous flow, but typically ignores surface diffusion due to its negligible effect in MD processes. It treats the pores of the membrane as stationary pseudo gas molecules with a large size (dust), allowing for the modeling of multi-component gas mixtures. The DGM assumes isothermal conditions, which may not always be the case in practical applications. Schofield's model is simpler but has difficulties in evaluating membrane parameters experimentally. Schofield's model provides a semiempirical approximation of the flux in the viscous-Knudsen transition region. It is designed to account for both Knudsen diffusion and viscous flow within the membrane pores. However, it is less favored compared to the Dusty Gas Model due to its limitations and experimental challenges. Pore size distribution models are more mechanistic and consider the effect of pore size distribution. Pore size distribution models in MD are designed to account for the variability in pore sizes within the membrane, which can affect the mass transfer process. These models aim to provide a more accurate representation of the mass transfer by accounting for the different transport mechanisms that may dominate in pores of different sizes. They are complex and suppose that the influence of pore size variance is small. However, Structural Network Models (SNM) simulate the membrane structure using a network of nodes and pores. They are mechanistic and need high computational studies due to their complexity. The ballistic transport model is used for predicting flux in extremely short or long pores, which is typically used to simulate the motion of electrons in the environment with a characteristic size lower than the mean free path of electrons. In membrane distillation, it is applied under low pressures where the mean free path of the water vapor molecules is remarkably greater than the pore size, resulting in Knudsen type of flow ($Kn > 1$). It is mechanistic but has high complexity and computational work.

Their overall system models include a variety of approaches that aim to simulate the entire MD process, incorporating both heat and mass transfer mechanisms (Eleiwi et al., 2016; Hitsov et al., 2015). Artificial Neural Network (ANN) models can mimic the behavior of real systems and have a low computational work. However, they are black-box models requiring a large amount of experimental data for calibration. They are unable to extrapolate beyond the calibration range. They can be used for optimization of processing parameters, control strategies, and imagination of the operational area. Design of Experiments (DoE)-Based Models: These statistical models use techniques like factorial design and response surface methodology (RSM) for system performance optimization. They are nonmechanistic and cannot extrapolate beyond the calibration range, but are simple and have low computational work. It is useful for revealing interactions between system parameters and for statistical assessments. Combined Nusselt and Sherwood equations with membrane mass transfer models are simple and have limited computational workload but are semiempirical and predict uniform temperature and concentration, making them unsuitable for suitable module design. Computational fluid dynamics system models provide the most mechanistic approach, but require high computational work. It is used for module design, investigating local temperature and concentration polarization, flux, and pressure drop-in lab-scale modules, and can facilitate the design of innovative spacers and module configurations. Additional efforts are required for developing CFD models that accurately simulate the complex interactions in MD modules, including the effects of concentration polarization and convective flow near the membrane. Validation of CFD models with experimental data is crucial to ensure their accuracy and reliability. Each type of model has advantages and limitations, and the selection of a model depends on the particular needs of the relevant system, such as the need for detailed local predictions, computational resources, and the level of mechanistic understanding required (Aytaç et al., 2024; Eleiwi et al., 2016; Hitsov et al., 2015; Rizki and Ottens, 2023).

10.2.5 Separation by liquid membranes

Liquid membranes are a type of membrane technology used for separation processes, where the membrane phase is a liquid rather than a solid. Liquid membranes have a liquid phase that acts as the selective barrier for separating components in a mixture based on their solubility and diffusivity in the liquid phase. They offer high selectivity and permeability due to their unique properties.

Basically, there are three major membrane types: Supported liquid membranes have a porous solid substrate where liquid fills the pores and creates a selective barrier. The solid porous support provides mechanical stability, and the liquid acts as a membrane. The commonly known usage is to remove ions and organic solvents from water. Emulsion membranes consist of two immiscible liquids to form an emulsion. The emulsion is required to be stabilized by surfactants to keep the stability of the emulsion. This kind of membrane is used in the extraction and separation of specific solutes from aqueous solutions. Bulk liquid membranes have a much simpler system. The liquid membrane is localized between two immiscible liquids. The target solute diffuses from one phase to another through the liquid membrane. This technique is mostly used to extract and concentrate a specific solute.

Liquid membranes selectively allow specific components to dissolve and diffuse through the membrane phase while rejecting others. Often, a carrier molecule is incorporated into the liquid membrane to facilitate the target solutes transportation. The carrier links to the solute at the feed interface, diffuses across the membrane, and subsequently releases the solute on the permeate side.

Although the composition of liquid membranes can be readily tailored to target a wide variety of solutes, these systems are prone to stability issues, such as leakage and degradation of the liquid phase over time. Furthermore, the liquid membrane systems are generally more complex in design and operation requirements than solid membranes due to the challenges for process scale-up and industrial application.

Various mathematical modeling approaches are investigated for separation processes in liquid membrane by liquid membrane, particularly in hollow fiber contactors and other configurations such as single function fiber contractor, tubes and shells (aqueous solution flows in tube side and organic phase in shell size) U-type bundels, Fiber-in-Fiber and Fiber-by-Fiber, etc (Bringas et al., 2009; Divakar et al., 2022; Kostanyan et al., 2023; Lantto, 2015). One of the modeling approaches is rigorous modeling, which utilizes mass conservation equations and relevant boundary conditions, and characterizes spatial variation in fluid properties using two-dimensional flow models. Simplified modeling uses an overall permeability coefficient related to average concentrations and assumes averaged conditions on either side of the membrane for quick calculations. Solution-diffusion and facilitated transport mechanisms can be used for modeling of liquid membranes (Bringas et al., 2009; Kostanyan et al., 2023). SD describes transfer without chemical reactions, and facilitated transport involves selective extractants reacting with target species. There are diffusional-kinetic and mixed-kinetic regime approaches, in which diffusional-kinetic regime focuses on the rate of extraction reaction and mixed-kinetic regime is used for selective separation. Ideal plug-flow modeling assumes ideal plug flow in both the aqueous feed solution and the organic phase (Bringas et al., 2009; Kostanyan et al., 2023). Although the membrane approach assumes all resistance lies in the membrane, with steady-state diffusion through the membrane, the integral approach takes all changes in all three phases into account, including concentration and volume changes, and mass balances. There would be steady-state and unsteady-state diffusion with chemical reactions during liquid membrane separation. The models are constructed for steady-state or unsteady-state diffusion combined with fast, slow, or intermediate chemical reactions. The Finite Discretization Method solves model equations using finite discretization, considering resistances in mass transfer at boundary layers. The complexity and variety of approaches in modeling liquid membrane separation processes emphasize the need for accurate and efficient models for industrial applications (Bringas et al., 2009; Divakar et al., 2022; Kostanyan et al., 2023; Lantto, 2015).

10.2.6 Fouling and blocking

Fouling is a common and important problem in membrane separation processes, referring to unwanted materials accumulation on the membrane surface and/or within its pores, which leads to a decrease in membrane performance. Fouling can be categorized into several types, including colloidal/particulate fouling, biofouling/organic fouling, inorganic fouling (scaling), and chemical fouling.

Colloidal/Particulate Fouling: Colloidal and particulate fouling arises from the accumulation of suspended solids, such as dust, dirt, and other particulates, onto the membrane surface. These particles can aggregate and form a cake layer, blocking membrane pores, reducing permeate flux, and increasing the transmembrane pressure drop. Water quality in feed, operational parameters, and surface properties of the membrane strongly influence the extent of particulate fouling. Pretreatment strategies, including coagulation, flocculation, and medium filtration, are typically employed to mitigate this type of fouling.

Biofouling/Organic Fouling: Biofouling involves the adhesion and subsequent proliferation of microorganisms (i.e., algae, bacteria, fungi, and protozoa) on the membrane surface. The process initiates with microbial attachment, followed by colonization and biofilm formation, which significantly impairs membrane flux and elevates energy consumption. Organic molecules, including proteins and polysaccharides, can also adsorb onto the membrane, promoting microbial proliferation. Factors like membrane surface charge, pore size, surface roughness, and hydrophilicity play critical roles in biofouling susceptibility. Biofouling typically manifests as a gel-like or cake layer that clogs membrane pores.

Inorganic Fouling: Inorganic fouling occurs because of the precipitation and deposition of salts and minerals, including calcium carbonate, calcium sulfate, and silica, onto the membrane surface or within its cavities. Scaling is promoted by the presence of ions like PO_4^{3-}, COO^-, CO_3^{2-}, OH^-, and SO_4^{2-}, as well as concentration polarization effects near the membrane interface. Chemical cleaning protocols and the application of antiscalants are commonly used to manage inorganic fouling.

Chemical Fouling: Chemical fouling occurs when chemical reactions—such as oxidation—alter either the membrane material or the foulants themselves, leading to the generation of insoluble compounds that accumulate on the membrane surface. It can be particularly challenging to reverse and may necessitate specialized cleaning strategies or membrane replacement.

The fouling process typically progresses through multiple stages, including initial adsorption, pore blocking, and subsequent formation of cake layers, gel layers, biofilms, precipitation deposits, or crystalline scaling, related to the feed stream composition (Table 10.2).

Fouling in membrane separation processes can be prevented or controlled by carefully managing characteristics of the feed solution, membrane properties, and operating conditions. The concentration and composition of foulants (i.e., pH, ionic strength, temperature, and the presence of proteins, minerals, and microorganisms) in the feed solution—significantly influence fouling propensity. Membrane properties (i.e., the membrane material, surface roughness, hydrophilicity or hydrophobicity, surface charge, pore size, and pore size distribution) also play a critical role in fouling behavior. In addition, operating conditions such as transmembrane pressure, crossflow velocity, feed flow rate, and operating temperature can either mitigate or exacerbate fouling by influencing the rate and extent of foulant deposition on the membrane surface.

Fouling formation prevention needs different strategies.

- Pretreatment of feed stream: Removing or reducing foulants in the feed solution through filtration, coagulation, flocculation, or other pretreatment methods, such as the use of chemicals for the inhibition of fouling.
- Modification of membrane: modification of the membrane surface may reduce fouling formation by changing surface nature, such as hydrophilicity, adding antifouling coatings, or incorporating antimicrobial agents.
- Optimization of operating conditions: Fine-tuning conditions of membrane operation, such as reducing transmembrane pressure and operation temperature, increasing crossflow velocity, and using pulse flow.

Table 10.2 Summary of fouling types and mechanisms.

Fouling type	Fouling mechanism
Adsorption	Solutes or particles initially adhere to the membrane surface through interactions like van der Waals forces, hydrogen bonding, and electrostatic attraction.
Pore blocking	Complete blocking: Particles or solutes entirely obstruct membrane pores, preventing any flow through the affected areas. Partial blocking: Solutes partially occlude pores, decreasing effective pore size and reducing permeability. Pore narrowing: Gradual accumulation of foulants within pores leads to a reduction in pore diameter, increasing hydraulic resistance.
Cake formation	Accumulated foulant layer forms on the membrane surface, known as a "cake layer." this layer imposes extra resistance to flow, further diminishing membrane permeability. The thickness and density of the cake layer increase with time, exacerbating performance decline.
Gel layer formation	Organic and biological foulants can aggregate to generate a gel-like layer on the membrane surface, which serves as an extra barrier to mass transfer.
Biofilm formation	Microorganisms adhering to the membrane surface can establish biofilms, consisting of microbial cells embedded in extracellular polymeric substances (EPS). Biofilm development not only increases resistance to water flow but also leads to significant flux reduction as the microbial matrix expands across the membrane surface.
Precipitation and crystallization	Inorganic salts and other compounds may precipitate from the feed solution and crystallize on the membrane surface.
Compaction and consolidation	Under filtration pressure, the cake layer can become compacted and consolidated, making it more resistant to removal and intensifying the decline in membrane performance.

- Cleaning membranes regularly: applying and improving cleaning protocols, including both physical and chemical cleaning.

10.3 Transport phenomena in membrane processes

As mentioned above, various membrane separation technologies are widely used for industrial applications today. To select or design the most appropriate membrane system for the targeted separation, the transport phenomena and relevant mechanisms through membranes should be well-understood. Transport phenomena in membrane processes are characterized by the particle movement through a membrane under the driving forces of mainly concentration, pressure, and electrical potential gradients. Characteristics of the membrane, operation conditions, feed content, and membrane and feed matrix interaction are critical aspects of membrane separation, contributing to the process efficiency in the target application. Basically, chemical and electrical potential differences are the main forces considered in mass, momentum, and heat transfer mechanisms, composing transport phenomena in membrane processes. In particular, membrane process, mathematical relations based on these mechanisms can be obtained from simplification of the Navier–Stokes flow equations (Nagy, 2018).

Transport in dense membranes is described through three major theories including the thermodynamics of irreversible process, the sorption capillary flow model, and the solution-diffusion model. Diffusion coefficient is expressed in these equations since transport occurs mainly through diffusion (Bitter, 2012; Van den Berg and Smolders, 1992). Transport rate, namely mass flux, J for i species, retention and other parameters can be derived from thermodynamic relations.

$$J_i = \sum_{k=1}^{n} L_{ik} X_k \tag{10.1}$$

where, L_{ik} is phenomenological coefficient as a function of concentration or composition, and X_k is the driving force as a chemical potential gradient.

In sorption-capillary flow, molecules are transported due to diffusion along the pore walls, resulting in component flux, N_{si};

$$N_{si} = -C' D_{si} \left(\frac{dC_{si}}{dp} \right) \nabla P \tag{10.2}$$

where, C' is the geometrical parameter, D_{si} is the diffusion coefficient, $\frac{dC_{si}}{dp}$ is the concentration gradient, and P is the pressure.

Theories based upon Nernst–Planck flux equations, principles of classical thermodynamics and irreversible thermodynamics, and rate processes enable descriptive understanding of transport phenomena in membranes (Lakshminarayanaiah, 1965). According to the Nernst–Planck equations, the diffusional flux of molecular or ionic species, i, caused by the chemical potential gradient $\left(\frac{d\mu_i}{dx}\right)$ and expressed in;

$$J_{i(d)} = -\frac{D_i}{RT} C_i \left(-\frac{d\mu_i}{dx} \right) \tag{10.3}$$

where, D_i is the diffusion coefficient, R is the gas constant, T is the absolute temperature. Diffusion can generate an electrical field, namely diffusion potential, which can also be considered as a driving force. The flux based on the electrical field gradient $\left(\frac{d\psi}{dx}\right)$ is represented by;

$$J_{i(e)} = -\upsilon_i \, z_i \, C_i \left(\frac{d\psi}{dx}\right) \tag{10.4}$$

where, υ_i is the electrochemical mobility, z_i is the valance, C_i is the concentration of the moving species.

Characteristics of the membranes, composition of the feed, and operational conditions determine the type of transport mechanism, such as diffusion and/or sorption. Permeation selectivity is described as a function of sorption ratio and/or diffusion coefficient (Lee et al., 1989). Several basic theoretical equations have been introduced for the determination of these parameters. The Flory-Huggins equation expresses the sorption of the permeants in a polymer matrix by considering a homogenous mixture of polymer and permeants within the swollen membrane network (Van den Berg and Smolders, 1992). Thus, the molar free energy, F_s of a solvent and the polymer is given by;

$$F_s = RT\left[ln\,\varnothing_s + (1 - 1/m)\varnothing_p + \chi\phi_p\right] \tag{10.5}$$

where, $\varnothing_s$ and $\varnothing_p$ are the volume fractions of the polymer and solvent, respectively, m is the molar volume ratio of polymer and solvent, and χ is the Huggins' parameter.

On the other hand, gas separation can be described by the dual sorption model combining Henry's law with a Langmuir isotherm (Koros et al., 1977). The amount of sorped gas per amount of polymer, C can be expresses by;

$$C = k_D p + \left(C'_H bp\right)/(1 + bp) \tag{10.6}$$

where, k_D is Henry's law solubility coefficient, p is the relative pressure, C'_H is the Langmuir capacity constant, and b is the affinity constant.

Maxwell–Stefan equation given below is used to describe the diffusive transport for binary systems (Van den Berg and Smolders, 1992).

$$\frac{\nabla\mu_i}{RT} = \sum_{\substack{J=1 \\ J\neq i}}^{n} \frac{x_i N_j - x_j N_i}{D_{ij}} \tag{10.7}$$

where, μ_i is the chemical potential, x_i and x_j are the mole fractions of i and j , N_i and N_j are the molar flux of i and j, and D_{ij} is the binary diffusivity.

Based on the transport mechanism, membranes allow the solvent flux and can selectively allow or reject solute molecules based on size and charge. The rejected solute molecules accumulate at the surface of the membrane and lead to the generation of a fouling cake resisting the flux. Principles of mass, heat, and momentum transport enable the quantification of the fluxes and resistance across the membranes for individual separation systems. Thus, the design and monitoring of an efficient membrane separation process is achieved.

10.3.1 Mass transport

Investigation regarding the mass transport across membranes started about the 1830s, with the osmotic pressure studies based on the semipermeable membrane concept by van't Hoff (Läuger, 1969). After the first and focused interest on semipermeable membranes, transport phenomena have been resolved and more efficient separation mechanisms have been introduced in time. Mass transport in membrane processes refers to the traveling of molecules or ions through a membrane, which is a selective barrier that allows certain particles to pass while blocking others. Movement of the solutes and gas molecules across the membrane is described by diffusion, convection, and electromigration mechanisms. Diffusion, the key mechanism in NF, RO, and PV processes, driven by concentration gradient, reveals movement of molecules from high to low concentration of solutes and is described by Fick's law. One-dimensional form of mass transfer rate, namely mass flux (J; kg/m^2s) is given by;

$$J = -D\frac{d\varphi}{dx} \tag{10.8}$$

where, D is the diffusivity, $\frac{d\varphi}{dx}$ is the concentration gradient. In case of two or more dimensions, mass flux is given by;

$$J = -D\nabla\varphi \tag{10.9}$$

Convective mass transport, considered mostly in MF and UF processes, driven by pressure gradient, involves the shipment of the solutes within the fluid through the membrane. Convective mass transport flux can be given by;

$$J = \frac{\varphi}{\rho}\left(J_i + J_j\right) \tag{10.10}$$

where, $\frac{\varphi}{\rho}$ is the concentration in weight fraction, J_i and J_j are the fluxes of the i and j components in binary mixtures.

Rejection of molecules in the feed solution due to several factors, for example, molecular weight exceeding the limits of the membrane and charge, can be described by film theory based on Fickian diffusion as well (Bird et al., 1960). According to film layer theory, a thin layer with a thickness of δ is formed in the membrane surface where the concentration declines from C on the surface to C_b in the bulk. Mass flux of permeate, J_v is obtained by integrating mass balance equation in the following equation (Todisco et al., 2002).

$$J_v = k.ln\frac{C - C_p}{C_b - C_p} \tag{10.11}$$

where, k is the mass transfer coefficient C_p and C_b are the permeate and bulk concentration of the solute, respectively. Mass transfer coefficient is a significant parameter required for membrane system design and estimation of membrane behavior against the permeate flux. Concentration polarization occurs due to the deposition of solids at the membrane surface. Rejection coefficient, R, representing the maximum concentration at the membrane surface through concentration polarization can be defined as;

$$R = 1 - \frac{C_p}{C} \tag{10.12}$$

Reverse solute diffusion, movement of solutes from the draw solution to a feed solution due to concentration difference in both sites, occurring in FO is a challenging phenomenon distinguished from classical pressure-driven membrane process (Phillip et al., 2010; Wu et al., 2023). The amount of reverse solute flux, J_s is given by;

$$J_s = \frac{J_w C_D}{1 - \left(1 + \frac{J_w}{B}\right)\exp\left(\frac{J_w S}{D}\right)} \tag{10.13}$$

where, J_w is water flux, C_D is the concentration of the draw solution, B is the solute permeability coefficient, S is the membrane structural parameter, D is the diffusivity coefficient of the draw solution. External and internal concentration polarization can occur on the surface and inner side of membranes during the FO process, revealing a reduction in the pressure gradient, lowering the water flux (Mehta and Loeb, 1978; Phillip et al., 2010; Wu et al., 2023). By considering this limitation based on external and internal concentration polarization, the reverse flux model is modified as follows (Suh and Lee, 2013).

$$J_s = B\left(\frac{C_{D,b} + C_{sw}}{\exp\left(J_s\frac{K}{J_{sw}}\right)\exp\left(\frac{J_s}{J_{sw}k_D}\right)} - (C_{F,b} + J_{sw})\exp\left(\frac{J_s}{J_{sw}k_D}\right)\right) \tag{10.14}$$

where, J_{sw} is the specific reverse solute flux, $\left(\frac{J_s}{J_w}\right)$, $C_{D,b}$ and $C_{F,b}$ are the concentrations of bulk draw and feed solutions, respectively, K represents the resistivity of the substrate.

10.3.2 Momentum transport

Momentum transport refers to the movement of fluids through the membrane via a potential difference and is described by fluid dynamics. The rate of solute mass transfer is influenced by the fluid velocity in pressure-driven membrane processes like MF and UF. Thus, the viscous flow behavior of feed solutions determines the effectiveness of the momentum transport in certain membrane processes. Besides, the tangential flow leading shear stress occurs through the membrane surface and prevents deposition of solute particles over membrane cavities, thus limiting fouling.

The pressure distribution along the porous membrane is described by the application of overall linear momentum balance, which includes the effects of inertial, viscous, gravitational, and pressure forces, besides external forces keeping the membrane stationary (Nagy, 2018). Momentum and continuity equations are used to determine velocity distribution and pressure fields of flow through the pores of membranes. Those equations have been solved by considering boundary conditions of the relevant membrane systems (Bird et al., 1960; Mondor and Moresoli, 1999). Radial convective velocity, υ in the membrane based on Darcy law is given below.

$$\upsilon = -\frac{K}{\mu}\frac{dp}{dr} \tag{10.15}$$

where, K is the Darcy permeability, μ is the fluid viscosity, and p is the transmembrane pressure. Membrane pores can be considered as hollow cylinders or capillaries perpendicular to the membrane surface in which convective transport occurs. The laminar fluid flow and momentum flux across the membrane pores are described by Newton's law of viscosity.

$$\tau_{ry} = -\mu\frac{d\vartheta_y}{dr} \tag{10.16}$$

where, τ_{ry} is the shear stress per unit area (N/m^2), μ is the viscosity $(Pa.s)$, ϑ_y is the fluid velocity (m/s), and r is the cylindrical coordinate (m) in membrane pores. If the pressure drops along with the capillary length (δ, m) is considered, shear stress can be expressed as;

$$\tau_{ry} = \frac{P - P_\delta}{2\delta}r \tag{10.17}$$

where, P is the pressure (Pa) at the feed side and P_δ is the pressure (Pa) at the permeate side.

10.3.3 Heat transport

Mass transfer is driven by temperature gradients in some membrane processes like membrane distillation (Field et al., 2013). During the heating of the feed solution, water evaporates, and the generated vapor diffuses across the membrane. Subsequent cooling at the other side of the membrane leads to condensation of the vapor, thus the water fraction in the feed is separated from the solute fraction. The flux, N is expressed in;

$$N = K\left(\Delta T_b - \Delta T_b^{BP}\right) \tag{10.18}$$

where, K is the mass transfer coefficient, ΔT_b is the local bulk temperature difference, ΔT_b^{BP} is corresponded to the boiling point rise because of the salinity.

In the membrane distillation process, concentration and temperature gradients are both considered. By considering vapor diffusion, heat transfer rate along with the membrane is expressed by Martínez-Diez and Vazquez-Gonzalez (2000);

$$q = h(T_1 - T_2) + J\left(C_p T_{ave} + \Delta H\right) \tag{10.19}$$

where, q is the heat flux (W/m^2), h is the heat transfer coefficient of the membrane (W/m^2K), T_1 and T_2 are the temperatures at the two sides of the membrane (K), J is the mass flux (kg/m^2s), C_p is the specific heat of vapor ($j/kg\ K$), T_{ave} is the average temperature in the membrane pores (K), and ΔH is the latent heat of the vapourization (j/kg).

Heat transfer coefficient of a membrane can be determined by;

$$h = \frac{\lambda}{\delta} \tag{10.20}$$

where, λ is the thermal conductivity and δ is the thickness of the membrane. In the case of hydrophobic membranes, the heat transfer coefficient can be determined by considering the thermal conductivities of both the membrane and the air kept in the membrane pores (λ_g) (Qtaishat et al., 2008).

$$h = \frac{\lambda_g \varepsilon + \lambda(1 - \varepsilon)}{\delta} \tag{10.21}$$

where, ε is the porosity of the membrane.

10.4 Modeling of mass transport in membrane processes

Modeling of mass transport in membrane processes is based on a mathematical description of the movement of solid particles or solvents, which are expressed as flux through the membrane. It aims to estimate the performance of the membrane process under various operation conditions. Concentration gradient, pressure, and solute-membrane interactions are critical factors to be considered in modeling of membrane processes (Ahmad et al., 2005; Bardhan et al., 2023; Lee and Clark, 1998). Several approaches exist to describe mass transport in membrane processes, depending on the type of membrane process, application, and specific transport mechanism involved (Akhtar et al., 2021; Banerjee and De, 2011; Bowen and Welfoot, 2002; Hubach et al., 2023; Ma et al., 2023). Modeling of mass transport in membrane processes facilitates optimizing design, operation, and performance prediction. Transport models vary from basic empirical relations to complex mechanistic equations based on membrane type, diffusion mechanism, and application system. Besides, fouling models are considered significant tools in determining the performance of the separation method and selected membranes. Key factors affecting mass transport models are given as follows (Teodoro et al., 2023; Wang and Lin, 2021).

1. Membrane characteristics include porosity, pore size, and thickness for porous membranes, permeability and selectivity for dense membranes, surface charge, hydrophobicity, and chemical composition.
2. Feed properties include solute concentration and molecular size, viscosity, pH, and ionic strength.
3. Operational conditions such as flow rate, temperature, and pressure.
4. Concentration polarizing affects the boundary layer and reduces driving forces, and membrane fouling reduces membrane performance and elevates resistance to mass transport.

The primary equations governing mass transport in membrane processes are based on diffusion-driven and pressure-driven transport (Lakshminarayanaiah, 1965). Diffusion-driven solute transport is described by Fick's Law with the flux corresponding to the amount of solute passing across the membrane per unit area and unit time.

$$J = -D\frac{dC}{dx} \tag{10.22}$$

where, J is the solute flux; D is the diffusion coefficient; C is the solute concentration and x is the position across the membrane. The pressure-driven solvent transport is described by Darcy Law with the flux corresponding to the solvent flow rate per unit membrane area.

$$J_v = \frac{\Delta P - \Delta\pi}{\mu R_m} \tag{10.23}$$

where, J_v is the volumetric flux of solvent; ΔP is the pressure difference through the membrane; $\Delta\pi$ is the osmotic pressure variation, μ is the solvent viscosity, R_m is the membrane resistance. The average convective velocity (ϑ) across the membrane pores via pressure drop is described by Hagen-Poiseuille equation.

$$\vartheta = \frac{\Delta P r^2}{8\mu\delta} \tag{10.24}$$

where, ΔP is the pressure variation between two sides of the membrane, r is the pore radius, μ is the fluid viscosity and δ is the membrane thickness. Transfer of partially rejected molecules through the membrane is represented by the sieving coefficient, S as follows.

$$S = \frac{C_p}{C_f} \tag{10.25}$$

where, C_p and C_b are concentration of solute molecules in the permeate and the feed solution. Besides, apparent rejection coefficient, R is given by;

$$R = 1 - \left(\frac{C_p}{C_f}\right) \tag{10.26}$$

Membrane diffusion-controlled transport rate of ions and solvent is calculated mainly through Nernst-Plank and Maxwell–Stephan models in ion exchange membranes (Kraaijeveld, 1994; Model, 1990). In fact, different modeling approaches, including mechanistic models, theoretical, predictive, or simulation models, are reported to define solute and solvent transport across the membranes, as well as rejection and fouling behavior of the membranes. Those are separately discussed below. Additionally, some recent reports on the mentioned models with different membrane separation applications are presented in Table 10.3.

10.4.1 Mechanistic models

Mechanistic models in mass transport focus on understanding the fundamental physical and chemical processes governing the movement of solutes and solvents through semipermeable or selectively permeable membranes. These models combine several transport mechanisms, including diffusion, convection, and permeation, to describe complex separation systems (Heiranian et al., 2023). The mechanistic models commonly used to analyze membrane processes in foods are given as follows.

Solution-Diffusion (SD) Model assumes that the pressure within the membrane is uniform, thus diffusion occurs across the membrane according to a concentration gradient. It is predominantly applied for nonporous, dense, and homogenous membranes, where differences in permeates' solubility and diffusivity govern the transport phenomena (Nagy, 2018). Permeability of the membrane and solubility of the solute in the membrane are important in the processes to be defined by this model, such as forward and reverse osmosis, pervaporation, and dialysis (Merdaw et al., 2011). The solute flux is expressed in (Sekino, 1993):

$$J_s = p\left(C_{feed} - C_{permeate}\right) \tag{10.27}$$

where, p is the solute permeability (g/m^2h) and C is the solute concentration (g/L). The solvent flux is defined by (Lin, 2016);

Table 10.3 Models representing mass transport in membrane separation.

Model		Type of membrane separation	Application	References
Mechanistic models (J: flux)				
Donnan-steric partitioning pore model (DSPM)	$J_i = -D_{i,p}\frac{dC_i}{dx} - \frac{Z_i C_i D_{i,p}}{RT}F\frac{d\psi}{dx} + K_{i,c}C_iJ_v$	NF (NF90 membrane)	Fractionation of aqueous saline solution	Saavedra et al. (2024)
Spiegler-kedem-katchalsky (SKK) model	$J_s = \nu_w C_v(1-\sigma) - P\frac{dC_v}{dx}$	RO (commercial membrane, SW30XLE)	Desalination	Du et al. (2023)
Solution-diffusion model	$J_w = L_p\left[\left(\pi_{s,d} - \pi_{s,f} - \sum_i^n \pi_f\right) - (P_d - P_f)\right]$ $J_s = B(C_{s,d} - C_{s,f})$	FO (aquaporin embedded hollow fiber membrane module)	Concentration of black tea extract	Bardhan et al. (2023)
Donnan-steric pore model	$J_i = -K_{i,d}D_{i,p}\frac{dC_i}{dx} + K_{i,c}C_iJ_v - \frac{Z_iC_iD_{i,p}}{RT}F\frac{d\psi}{dx}$	NF (NF270 and DK membrane)	Separation of succinic acid from fermentation broth	Mancini et al. (2022)
Molecular diffusion or Knudsen diffusion model	$J_w = \frac{P}{RT}\left(\frac{D_{w-air}}{\tau\delta}\right)\ln\left(\frac{P - a_w^B P_w^B}{P - P_w^B}\right)$ $J_{WF} = k_F(C_{f,m} - C_{f,b})$	OMD (hydrophobic porous hollow fiber membrane)	Concentration of apple juice	Ahmad et al. (2020)
Solution-diffusion model	$J_w = A(\Delta\pi - \Delta P)$ $J_s = B(\Delta C) = B(C_{s,d} - C_{s,f})$	FO (axial flow hallow fiber membrane)	Desalination	Munubarthi et al. (2020)
Pore flow and gel polarization model	$J_i = -D_{i,m}\frac{dC_i}{dx} + K_{i,c}C_iJ_v - \frac{Z_iC_iD_{i,m}}{RT}F\frac{d\psi}{dx}$	UF (lanthanum cobalt oxide nanoparticle doped hollow fiber membrane)	Salt removal from seawater	Mondal et al. (2020)

Continued

Table 10.3 Continued

Model		Type of membrane separation	Application	References
Mass balance and Fick's law of diffusion	$J_s = \frac{\Delta m}{A_m \Delta t}$	FO (cellulose triacetate (CTA) membrane)	Concentration of skim milk and whey	Chen et al. (2019)
Solution-diffusion model	$J_{w(x)} = A_w(\Delta P_{b(x)} - \Delta P_{total(x)})$ $J_{s(x)} = B_s(C_{v(x)} C_{p(av)})$	RO (MSCB 2521 RE99 spiral-wound membrane)	Concentration and clarification of apple juice	Al-Obaidi et al. (2017)
Nernst Planck model	$J_A = -\frac{D_A D_B (z_A^2 C_A + z_B^2 C_B)}{D_A z_A^2 C_A + D_B z_B^2 C_B} \frac{dC_A}{dx}$	ED (Nafion-117 ion-exchange membrane)	Desalination	Agarwal and Goswami (2016)
Nernst–Planck equation and concentration polarization	$J_{si} = -P_i \frac{dC_i}{dx} + J_v C_i (1 - \sigma_i)$	NF (HDS-12–2540 membrane)	Xylo-oligosaccharide syrup	Hua et al. (2010)
Predictive models and simulation				
CFD model	$J_w = D_e(P_{mf} - P_f)$	MD (PVDF and PTFE membranes)	Evaporation efficiency	Ali et al. (2023)
Regression model	$J_p = \frac{V_p}{A_m + t}$	NF: (Polypiperazine amide, fully aromatic polyamide (NF90), polyethersulfones (NP010 and NP030)	Purification of xylitol from fermentation broth	Cerceau Alves, Fernandes Antunes, Silverio da Silva and Forte (2021)

Multi-parametric numerical simulation	$\Delta v - 0.5Re(v \cdot \nabla)v = \nabla p$	A crossflow past a row of parallel hollow fiber membranes	Viscous incompressible fluid	Kirsch & Bazhenov (2020)
Box-Jenkins autoregressive integrated moving average (ARIMA) model	$J_p = \frac{\Delta M_p}{\Delta t \cdot A}$	UF (hollow fiber, tubular, and spiral-wound membrane modules)	Clarification of fruit (bergamot, kiwifruit, and pomegranate) juice	Ruby-Figueroa et al. (2017)
Molecular dynamics simulation	$J_w = \frac{\Delta M_w}{A \cdot \Delta t}$	FO (thin film composite polyamide membrane)	Concentration of fructose sugar	Madhumala et al. (2017)
Response surface methodology and Monte Carlo simulation models	$J_w = \frac{1}{t_n} \int_{t_1}^{t_n} J_w(t)dt$	FO (spiral wound membrane module, 2521 FO-CS)	Desalination	Khayet et al. (2016)
Fouling models				
Multilayer perceptron and long short-term memory ANN model	$J = \frac{dV}{A.dt}$	MF (polyvinylidene fluoride hollow-fiber membrane module)	Wastewater treatment through anaerobic membrane bioreactor	T. Wang and Li (2024)
Hermia's model	$J_v = -\frac{1}{A} \frac{dV_f}{dt}$	UF (ETNA01PP, a composite fluoropolymer with a 1000 Da molecular weight cut-off)	Lycopene enrichment in watermelon juice	Doan et al. (2023)

Continued

Table 10.3 Continued

Model		Type of membrane separation	Application	References
Resistance in a series model	$J_p = \frac{V_p}{t \cdot A_m}$	FO (flat-sheet cellulose triacetate–polyester membrane)	Concentration of orange juice	Y. Li et al. (2020)
Mass balance	$J_w = \frac{\Delta W}{\Delta t \cdot A \cdot \rho_w}$ $J_s = \frac{C_0 V_0 - C_t V_t}{\Delta t \cdot A}$	FO (thin film composite and cellulose triacetate membranes)	Polyphenol&pectin solution	Pei et al. (2021)
Hermia's model	$J_w = \frac{W}{A.t}$ $-\frac{dJ}{dt} = K_n A^{2-n} J^{3-n}$	MF (dual-layer ceramic hollow fiber membrane)	Clarification of genipap fruit juice extract	Terra et al. (2019)
Complete pore blocking, standard blocking, intermediate fouling and cake fouling models	$J = J_L\left[1 - \exp\left(\frac{-TMP_e}{TMP_c}\right)\right]$	UF (ceramic membranes)	Synthesis of galacto-oligosaccharides in a membrane bioreactor	Córdova et al. (2016)
Hermia's equations and resistance-in-series models	$-\frac{dJ}{dt} = K(J - J_{SS})J^{2-n}$ $J = \frac{\Delta P}{\mu \cdot (R_m + R_a + R_b)}$	UF (monotubular ZrO_2–TiO_2 membrane, flat-sheet polyethersulfone membrane, flat-sheet permanently hydrophilic membrane)	Clarification of whey model solutions	Corbatón-Báguena et al. (2015)
Resistance-in-series model	$J_p = \frac{\Delta P}{\mu \cdot \left(R_m + R_c + R_{frev} + R_{firr}\right)}$	UF (polyvinylidenefluoride tubular membrane module)	Clarification of kiwi fruit juice	Cassano et al. (2007)

Sorption-capillary flow model	$N_A = \frac{D_{AM}}{K\delta}(C_{A2} - C_{A3})$ $N_B = A(\Delta P - \Delta\pi)$	RO (MSCB 2521 R99 spiral-wound membrane)	Aroma compounds from apple juice	Álvarez et al. (2001)
Resistance-in-series model	$J = \frac{\Delta P}{\mu_p \cdot (R_m + R_a + R_p + R_c)}$	MF (ceramic tubular membrane)	Clarification of lactic acid fermentation broths	Carrère et al. (2001)

ANN: artificial neural networks, *CFD*: computational fluid dynamics, *ED*: electrodialysis, *FO*: forward osmosis, *MD*: membrane distillation, *MF*, microfiltration, *NF*: nanofiltration, *OMD*: osmotic membrane distillation, *RO*: reverse osmosis.

$$J_w = \mathcal{P}\ (\Delta\pi - \Delta \mathrm{P}) \tag{10.28}$$

where, $\mathcal{P}$ is the solvent permeability, $\Delta\pi$ is the osmotic pressure difference at the active layer of the membrane surface on the feed and draw solution sides (Pa) and ΔP is the pressure difference of solvent (Pa) between the feed and draw solution sides.

The SD model is most widely used to define solvent and solute flux along with the membrane layer in terms of pressure and osmotic pressure difference, film theory is used to define mass transport in external and internal concentration polarization layers (Bardhan et al., 2023; Munubarthi et al., 2020). It is also used to predict rejection of components in RO membranes and optimize operational parameters, including feed flow rate, temperature, concentration, permeate flux, and osmotic pressure (Al-Obaidi et al., 2017; Álvarez et al., 2001).

Diffusion/Convection Model involves convective diffusion due to both osmotic pressure variation and concentration gradient through the porous membrane (Kirsch et al., 2016).

Pore flow (PF) Model undertakes that the concentration of the solute and solvent is uniform across the membrane, and convective or diffusive flow forces the solute particles to move through membrane pores via a pressure gradient (Mondal et al., 2020; Mondal et al., 2019). The transport of solutes depends on the pore size distribution, pore morphology, and the hydrodynamic interactions between the solute and the pore walls. The permeate flux is expressed as:

$$J = k \cdot \frac{\Delta P}{L} \tag{10.29}$$

where, k is the permeability coefficient of membrane, ΔP is the transmembrane pressure difference and L is the thickness of the membrane.

Nernst–Planck (NP) Model is used to express the transport of charged species (ions) through a membrane under the impact of both concentration gradients and electric fields. It combines diffusion, electromigration (movement due to electric potential gradients), and convection (movement due to pressure gradients) to give a full description of ion transport (Biesheuvel et al., 2023; Hua et al., 2010; Mondal et al., 2020). The flux of ion i is expressed as:

$$J_i = -D_i\left(\frac{dC_i}{dx} + \frac{z_i F C_i}{RT}\frac{d\psi}{dx}\right) + C_i v_x \tag{10.30}$$

where, D_i is the diffusion coefficient of ion i; C_i is the concentration of ion i; z_i is the charge number of ion i; F is Faraday's constant; R is the gas constant; T is the temperature; ψ is the electric potential, v_x is the convective velocity.

Donnan-Equilibrium (DE) Model involves the transport of charged molecules in ion-exchange membrane systems, promoted by electrochemical gradient potential. DE is commonly applied to estimate the partition of the ions at the solution and membrane interfaces for only considering the ideal Donnan effect (Wang et al., 2023). This model accounts for the electrostatic interactions between ions and the fixed charges on the membrane (Mancini et al., 2022). The Donnan potential is the electrostatic potential that develops across the membrane due to unequal ion concentrations on either side. The Donna potential can be given as:

$$\Delta\varphi = \frac{RT}{zF} \, ln\left(\frac{a_+}{a_-}\right) \tag{10.31}$$

where, a_+ and a_- are the activities of positively and negatively charged ions in the membrane; z is the ionic charge; F is Faraday's constant; R is the gas constant; T is the temperature.

Donna Steric Pore Flow Model (DSPM) involves two groups, in which the first one PF model, is used for the prediction of ion rejection, and the second one PF model coupled with the concentration polarization, for the estimation of both ion rejection and permeate flux (Mondal et al., 2020). This model is based on an extended NP equation to describe fractionation of solutes at the membrane interface, including steric effects and Donnan electrostatic exclusion, and movement of solutes through membrane pores, including transport mechanisms of diffusion, convection, and electro-migration (Saavedra et al., 2024; Wang and Lin, 2021). The model is given by;

$$J_i = -D_{i,p}\frac{dC_i}{dx} - \frac{Z_i C_i D_{i,p}}{RT} F \frac{d\psi}{dx} + K_{i,c} C_i J_v \tag{10.32}$$

where, $K_{i,c}$ is the hindrance factor for convection and J_v is the permeate flow rate. Percentage rejection (R) is given by;

$$R_{dif,conv,elect/total} = \left(\frac{J_{i,dif,conv,elect}}{J_i} + \frac{J_{i,conv}}{J_i} + \frac{J_{i,elect}}{J_i}\right) x100 \tag{10.33}$$

Maxwell-Staphen (MS) Model is used for describing multicomponent mass transfer, particularly when there are interactions between different species. It is frequently used in gas separation processes and in liquid mixtures with high concentrations of solutes, where the interaction between different solutes affects their transport rates (Krishna and Wesselingh, 1997). The general equation of MS for component i in a multi-component system is:

$$\nabla X_i = \sum_j D_{ij}\left(\frac{X_j}{D_{ij}}\right)(J_j - J_i) \tag{10.34}$$

where, ∇X_i is the gradient of the mole fraction of component i; D_{ij} is the binary diffusion coefficient between components i and j; J_i and J_j are the fluxes; X_j is the mole fraction of component j.

Molecular Diffusion (MD) Model is used to define separating solids, liquids, and gases in various applications, including filtration, concentration, and dehydration based on concentration gradient. Basically, the MD model equation is based on Fick's first law, stating that the flux of molecules via the membrane is proportional to the concentration gradient.

$$J = -D\nabla C \tag{10.35}$$

where, J is the diffusive flux; D is the diffusion coefficient; ∇C is the concentration gradient through the membrane.

The molecular diffusion model is a significant model in numerous membrane-based food processes such as filtration, reverse osmosis, gas separation, and pervaporation (Ahmad et al., 2020; Khayet, 2011). Type of the membrane, pore size, and operational parameters are several factors to be considered in modeling of the molecular diffusion in the membrane process.

Knudsen Diffusion (KD) Model is applied to describe separating gases, water vapor, and volatile components from liquids and other gas components. In this mechanism, the mean free path of gas molecules is larger than the membrane pore size, meaning that intensive interaction of gas molecules with the pore walls than with each other (Gilron and Soffer, 2002). Molecular weight is another key parameter providing selective separation as the lighter molecules diffuse faster than the heavier ones. The flux of the transport mechanism depends on size, porosity, and pore structure of the membrane, weight of the gas molecules, and pressure gradient through the membrane. The Knudsen diffusion coefficient, $D_{K,i}$ and flux, J_i can be calculated as follows (Mason & AP, 1983; Nagy, 2018).

$$D_{K,i} = \frac{d_p}{3}\frac{\varepsilon}{\tau}\sqrt{\frac{8RT}{\pi M_i}} \tag{10.36}$$

$$J_i = -\frac{D_{K,i}}{RT}\frac{dp_i}{dy} \tag{10.37}$$

where, d_p is the pore size, ε is the porosity, τ tortuosity and L is the thickness of the membrane; R is the universal gas constant; T is the temperature; M is the molecular weight of the gas; and P_i is the pressure in the membrane.

The KD model is applicable to describe aroma and flavor preservation during fruit juice concentration, dehydration, and concentration during freeze-drying and vacuum-drying, controlling the passage of gas in modified atmosphere packaging, and separation of volatiles like ethanol and organic acids from beverages and dairy foods (Ahmad et al., 2020).

Spiegler-Kedem-Katchalsky (SKK) Model is used to define mass transport in membrane processes by combining both convection (pressure-driven flow) and diffusion (concentration gradient-driven flow) (Kedem and Katchalsky, 1958; Spiegler and Kedem, 1966). It is commonly applied in ultrafiltration and nanofiltration processes where solutes are partially rejected by the membrane (Casado-Coterillo et al., 2023).

$$J_s = \sigma \, J_v \, \left(C_{feed} - C_{permeate}\right) \tag{10.38}$$

where, σ is the reflection coefficient which indicates solute rejection efficiency; J_v is the solvent flux and J_s is the solute flux. According to Kedem-Katchalsky equation, the water and solute fluxes are given as follows.

$$J_v = L_p(\Delta P - \sigma \Delta \pi) \tag{10.39}$$

$$J_s = C_{ln}(1 - \sigma) \cdot J_v \frac{C_{ln}}{\Delta C} \tag{10.40}$$

In osmotic membrane distillation of fruit juices, water vapor migration along with the membrane boundary layer is described by the Fick's law of diffusion, in which the concentration gradient is the driving force of the transport mechanism between the liquid and the liquid-membrane interface as well (Ahmad et al., 2020).

10.4.2 Predictive permeation models and simulation

In comparison to experiment-based observational studies, simulation approaches usually serve easier and cost-efficient performances, providing spatially resolved data for all geometries and membrane modules with high accuracy (Marriott and Sørensen, 2003). Predictive models in membrane processing for food technology are quite helpful for optimizing the separation, filtration, and purification processes of various food products including dairy, fruit juices, wine, and edible oils. These models use experimental data and theoretical approaches, aiming to estimate performance, efficiency, and selectivity in membrane processes under different operating conditions by considering the complex nature of food components (Alves, Antunes, da Silva and Forte, 2021; Khayet et al., 2016). Predictive models based on regression analysis and response surface methodology (RSM) approaches are successfully applied to optimize water and solute permeate fluxes, and performance of membrane separation processes like FO and UF (Khayet et al., 2016; Ruby-Figueroa et al., 2017).

Numerical techniques such as finite element analysis (FEA) or computational fluid dynamics (CFD) are often used to solve complex mass transport models that cannot be solved analytically. Numerical simulations consider the geometry of the membrane module, porosity, flow conditions (e.g., rate and temperature), and material properties to estimate the performance of real systems, besides the design and optimization of the separation membranes (Ali et al., 2023; Carrère et al., 2001). For instance, a multiparametric simulation of solute transport and removal through hollow fiber membranes by considering dimensionless parameters of Reynolds number (Re), Peclet number (Pe), and Schmidt number (Sc) was employed to describe membrane efficiency by successful validation with the analytical asymptotic solutions for low and high Pe numbers (Kirsch & Bazhenov, 2020). Such findings are promising for the design and optimization of hollow membrane modules, sorbent fibrous filters, and heat exchangers.

CFD models are used to simulate the flow dynamics, mass and heat transfer, and pressure distribution in different membrane modules and contactors in 1D, 2D, and 3D geometries (Haddadi et al., 2018; Pellerin et al., 1995; X. Yang et al., 2012). By the numerical solution of the Navier–Stokes equations and mass transport equations with the relevant boundary conditions, CFD models can predict how flow patterns affect mass transport, heat transport, evaporation, and solute retention efficiency and fouling inside the membrane module (Ali et al., 2023; Kirsch et al., 2016; H. Yang et al., 2022). Thus, these models are applicable to optimize the construction and performance of membrane modules in food processing, such as optimizing the flow distribution in tubular or spiral-wound membranes for the filtration of dairy streams, fruit juices, and oils. CFD modeling is also used to examine the behavior of fluid flow in membrane testing cells by analyzing velocity and shear stress distributions (Ahmed et al., 2012).

Artificial Neural Networks (ANN), is a favorite machine learning method, including nonlinear regression and discriminative statistical methods capable of prediction with the computational systems aiming at imitation of the natural neural networks (Guiné, 2019). Due to several advantages such as high accuracy, easy adaptation in complex relationships, nonlinearity, and model independence, they are increasingly used as predictive models in food membrane processing (Yang et al., 2020). ANNs rely on experimental data and can predict permeate flux, rejection, and fouling due to variables including pressure, feed concentration, temperature, and flow rate in the MF and UF of milk, whey protein recovery, fruit juice concentration, and vegetable oil refining (Nourbakhsh et al., 2014). Generally, input variables of a model are membrane process parameters such as pressure, concentration, temperature, pH, and flow velocity, while the output variables are chosen as permeate flux and rejection coefficient (Niemi et al., 1995; Shetty et al., 2003). Key advantages of ANN models for membrane separation processes are easy to use, handling nonlinearity and noise in the data, accurate predictions based on a wide range of input variables, learn from complex datasets without requiring a mechanistic understanding (Jawad et al., 2021; Mahadeva et al., 2022).

10.4.3 Theoretical and predictive models for membrane fouling

Membrane separation systems are widely used for the concentration of liquid foods, thus operations with high rejection rates indicate deposition of functional nutrients in the retentate. On the other hand, pollutant rejection encountered in water/wastewater treatment, food waste stream treatment, such as protein recovery from whey, and biotechnological treatments are essential to analyze for the estimation of fouling behavior in membrane processes (Alves et al., 2021; Carrère et al., 2001; Cifuentes-Cabezas et al., 2023; Shetty et al., 2003). Organic and inorganic content of the liquid foods can cause to occurrence of the severe fouling issues during pressure-driven membrane processes like MF, UF, NF, and RO, limiting operation capacity and reducing membrane flux (Hu et al., 2013; Jiao et al., 2004; Merdaw et al., 2011). These compounds and suspended colloids caused the blockage of the membrane pores and surface, which can be taken away by physical, chemical, and hydrodynamic cleanup (Berg et al., 2014; Hou et al., 2020). Various mathematical approaches are used to modeling membrane separation performance and fouling profile.

Empirical models are commonly used in food processes like milk ultrafiltration, fruit juice clarification, and wine filtration to predict flux behavior and fouling tendencies. Hermia's theory is the most commonly utilized empirical description of the relation between permeate flux decline and filtration time under dead-end invariable pressure conditions (Kirschner et al., 2019). Four major fouling mechanisms are proposed: filter cake formation which is reversible and can be cleaned by backwashing and flushing; intermediate blocking, standard blocking, and complete blocking are irreversible and need chemical cleaning. The flux decay arisen from membrane fouling over time is estimated by Hermia's model as follows (Hermia, 1985);

$$\frac{d^2t}{dV^2} = k\left(\frac{dt}{dV}\right)^n \tag{10.41}$$

The ordinary differential equation is solved for discrete n values to describe four fouling mechanism (Pereira et al., 2023);

$$\text{in case of cake formation } (n=0): \ \frac{1}{J^2} = \left(\frac{1}{J_0^2}\right) - k.t \tag{10.42}$$

$$\text{in case of intermediate blocking } (n=1): \ \frac{1}{J} = \left(\frac{1}{J_0}\right) - k.t \tag{10.43}$$

$$\text{in case of standard blocking } (n=3/2): \ \frac{1}{\sqrt{J}} = \left(\frac{1}{\sqrt{J_0}}\right) - k.t \tag{10.44}$$

$$\text{in case of complete blocking } (n=2): \ ln\, J = ln\ J_0 - k.t \tag{10.45}$$

where, t is the time measured during the filtration, V is the cumulative permeate volume, k is the constant, n is an empirical fouling constant, J is the permeate flux at time t, J_0 is the permeate flux at time $t=0$.

In the estimation of permeate flux and rejection of valuable nutrients like aroma compounds in fruit juices, sorption-capillary flow model is successfully used (Alvarez et al., 1998; Álvarez et al., 2001; Álvarez et al., 2002). In this model, it is assumed that mass transport takes place along with the pores of the membrane on which water is absorbed by the pore walls and the solutes are rejected at the surface due to physicochemical reasons (Kimura & Sourirajan, 1967; Sourirajan and Agrawal, 1969). Equations describing the fluxes are given below.

$$N_A = \frac{D_{AB}}{K\delta}\ (C_{A2} - C_{A3}) \tag{10.46}$$

$$N_B = A\ (\Delta P - \Delta \Pi) \tag{10.47}$$

where, N_A and N_B are solute and solvent fluxes, respectively; $\frac{D_{AB}}{K\delta}$ is solute transport parameter, C_{A2} and C_{A3} are the solute concentrations at the feed and permeate sides on the membrane surface, respectively; A represents the solvent permeability of the membrane; ΔP is the pressure variation through the membrane; and $\Delta \Pi$ is the osmotic pressure variation at both surfaces of the membrane.

Resistance-in-series model is used to analyze the fouling contribution to the membrane filtration process and enables prediction of the relation between the membrane structure and the foulants (Taghavijeloudar et al., 2019). Deposition of the foulants and concentration polarization give rise to the increase in hydraulic resistance during the filtration. Darcy's Law describes the hydraulic permeability and states the controlling of flux by several hydraulic resistance-in-series expressed as (Li et al., 2022);

$$\Delta P = J \cdot \mu \,(R_m + R_c + R_b) \tag{10.48}$$

where, ΔP is the pressure decline through the filter cake and membrane (Pa), J is the permeate flux (m/s), μ is the dynamic fluid viscosity ($Pa.s$), R_m is the resistance arising from membrane thickness (m^{-1}), R_c is the resistance of the accumulated foulants (m^{-1}), and R_b is the boundary layer resistance (m^{-1}).

Hagen-Poiseuille theory-based model involves a basic hydrodynamic equation used to determine pressure decline across a hollow cylinder. To investigate the features of the fibers considered as porous cylinders composing filter membrane and axial pressure drop through the fibers, Hagen-Poiseuille equation was extended by the addition of the local flux distribution ($J_i(x)$) which is expressed as (Chang et al., 2002; Li et al., 2022);

$$J_i(x) = \frac{\lambda\, L\, J_m}{e^{\lambda L} - e^{-\lambda L}} \left(e^{\lambda x} + e^{-\lambda x}\right) \tag{10.49}$$

$$\lambda = 4R_i^{3/2} \cdot R_t^{-1/2} \tag{10.50}$$

where, J_m is the average permeate flux (m/s), L is the fiber (hollow cylinder) length (m), R_i is the inner layer resistance (m^{-1}) and R_t is the overall resistance (m^{-1}).

Convection and diffusion model is used to describe particle motion around the membrane surface, particularly in crossflow filtration. A two-dimensional transport model based on convection-diffusion equations provides a detailed analysis regarding the influence of membrane module geometry and transport characteristics on the solute transport and separation efficiency (Donato et al., 2017). For hollow fiber membranes, the local solute flux (J_i) based on convection and diffusion model is expressed by (Li et al., 2022);

$$J_i = \varepsilon_b \left(\nu C K_c - D_\infty K_D \frac{\partial C}{\partial z} \right) \tag{10.51}$$

where, ε_b is local porosity, C is the concentration of the solute (kg/m^3), D_∞ is the Brownian diffusion coefficient (m/s^2), K_c and K_D are the hindrance factors of convection and diffusion, respectively.

Concentration polarization model is commonly used as a successful model in predicting permeate flux for different particulate feed solutions. Film theory is used to predict concentration polarization, a common issue in food membrane processing where solutes accumulate near the membrane surface, lowering the effective driving force for filtration by creating an extra resistance to mass transfer (Álvarez et al., 2001). It assumes a stagnant boundary layer (film) near the membrane, where mass transfer occurs primarily by diffusion. The permeate flux (J) can be expressed by (Quezada et al., 2021);

$$J = \frac{D}{\delta}\left(\frac{C_m - C_p}{C_b - C_p}\right) \tag{10.52}$$

where, D is the diffusivity coefficient, δ is the thickness of the boundary layer, C_m is the solute concentration on the membrane surface, C_p is the solute concentration in the filtrate stream, and C_b is the solute concentration in the feed stream.

Predictive models and simulations successfully describe membrane fouling via feeding by polydispersed suspensions in crossflow membrane separation processes (Chellam, 2005). Current studies use statistical and machine learning methodologies to estimate membrane fouling with the help of rapid computational and algorithmic advances (Wang and Li, 2024). Within this scope, CFD and ANN models are considered fast, powerful, and competitive approaches for modeling of membrane fouling. The effect of simultaneous aeration on membrane fouling and solute withdrawal efficiency in a submerged membrane reactor, dynamic simulation-based hydraulic resistance, and permeate flux in protein crossflow UF were determined via ANN modeling (Mirbagheri et al., 2015; Razavi et al., 2004).

10.4.4 Model validation

Model validation is performed by comparing the model predictions with the experimental data. In membrane separation processes, model validation is required to ensure that the mechanistic, predictive, and simulated models precisely represent the real behavior of membrane separation technology. Relative errors between experimental data set and theoretical results of permeate flux, rejection rates, and fouling coefficient via different operating conditions are acceptable in the validated models (Al-Obaidi et al., 2017; Munubarthi et al., 2020).

Generally, the validation process involves the key steps, including data collection, model development, calibration, validation, sensitivity analyses, quantification of uncertainty, and application to real systems. A mathematical model based on empirical relations, mechanistic equations, and simulation is developed to represent the membrane process. Model parameters like concentration, flow rate, and pore size are adjusted to align model predictions with experimental data, resulting in the improvement of the model accuracy. Independent data sets are used for comparison of the model predictions and observed results for diffusivity, permeability, mass, and heat flux (S. Ahmad et al., 2020; Singh et al., 2023; Vinh-Thang and Kaliaguine, 2013). Subsequently, statistical measures are employed to assess the data fitting. Sensitivity analysis of the model outcomes to changes in model parameters provides information regarding significant effects on performance, to be considered in the following designs. Quantification of uncertainty in model predictions due to variability in the inputs and parameters can provide insights into the reliability of model outputs and thus guide decision-making (Bijak et al., 2022). The validated model can be applied to real processes, including process optimization, scale-up studies, or troubleshooting existing operations. Regular updates and retraining of the model with new empirical data can further enhance its reliability and predictive power over time (Al-Obaidi et al., 2017).

10.5 Conclusion

Membrane processes used for separation and concentration purposes in food systems are mainly classified as pressure-driven, ion-exchange, pervoration, and distillation processes. In all systems, movement of solute and solvent in the bulk solution through a well-characterized membrane is achieved by a chemical potential gradient based on concentration, pressure, and electrical potential. The performance of the membrane modules and efficiency of the process are directly influenced by membrane type and characteristics, feed content and properties, process conditions, and relevant parameters. Transport phenomena occurring throughout the membrane process consist of mass, momentum, and heat transfer mechanisms. Modeling of transport phenomena facilitates determining solute and solvent fluxes in feed and permeate streams, sieving, rejection, and fouling profiles in the membrane separation process. Mechanistic, theoretical, and predictive models have been developed for many years, but still, they are in use and provide valuable data for deep analyzing of the membrane processes. Recent models based on machine learning, such as ANN and CFD, have superiorities via easiness to use, handling of nonlinearity, efficient time, and dedication. All reported modeling approaches successfully target the design and optimization of membrane processes with different membrane modules and operational parameters for various feed suspensions.

References

Agarwal, C., Goswami, A., 2016. Nernst planck approach based on non-steady state flux for transport in a donnan dialysis process. J. Membr. Sci. 507, 119–125. https://doi.org/10.1016/j.memsci.2016.02.021.

Ahmad, A., Chong, M., Bhatia, S., 2005. Mathematical modeling and simulation of the multiple solutes system for nanofiltration process. J. Membr. Sci. 253 (1–2), 103–115.

Ahmad, S., Marson, G.V., Zeb, W., Rehman, W.U., Younas, M., Farrukh, S., Rezakazemi, M., 2020. Mass transfer modelling of hollow fiber membrane contactor for apple juice concentration using osmotic membrane distillation. Separ. Purif. Technol. 250, 117209.

Ahmed, I., Hussain, A., Hasani, S., Shakaib, M., Yunus, R.M., 2012. Computational modeling for visualization of flow patterns in a membrane testing device. Separ. Purif. Technol. 90, 1–9.

Akhtar, A., Singh, M., Subbiah, S., Mohanty, K., 2021. Modelling, experimental validation and process design of forward osmosis process for sugarcane juice concentration. Lebensm. Wiss. Technol. 141, 110852.

Al-Obaidi, M., Kara-Zaïtri, C., Mujtaba, I.M., 2017. Optimum design of a multi-stage reverse osmosis process for the production of highly concentrated apple juice. J. Food Eng. 214, 47–59.

Al Kholif, M., Hermana, J., Bilad, M.R., Chao, H.-P., 2024. Comprehensive updates on recent advances, fouling mechanisms, and future perspectives of nanofiltration. J. Water Proc. Eng. 63, 105565.

Ali, K., Arafat, H.A., Hassan Ali, M.I., 2023. Detailed numerical analysis of air gap membrane distillation performance using different membrane materials and porosity. Desalination 551, 116436. https://doi.org/10.1016/j.desal.2023.116436.

Alvarez, S., Riera, F., Alvarez, R., Coca, J., 1998. Permeation of apple aroma compounds in reverse osmosis. Separ. Purif. Technol. 14 (1–3), 209–220.

Álvarez, S., Riera, F.A., Álvarez, R., Coca, J., 2001. Prediction of flux and aroma compounds rejection in a reverse osmosis concentration of apple juice model solutions. Ind. Eng. Chem. Res. 40 (22), 4925–4934.

Álvarez, S., Riera, F.A., Álvarez, R., Coca, J., 2002. Concentration of apple juice by reverse osmosis at laboratory and pilot-plant scales. Ind. Eng. Chem. Res. 41 (24), 6156–6164.

Alves, Y.P.C., Antunes, F.A.F., da Silva, S.S., Forte, M.B.S., 2021. From by-to bioproducts: selection of a nanofiltration membrane for biotechnological xylitol purification and process optimization. Food Bioprod. Process. 125, 79–90.

Aytaç, E., Contreras-Martínez, J., Khayet, M., 2024. Mathematical and computational modeling of membrane distillation technology: a data-driven review. Int. J. Thermofluids 21, 100567.

Banerjee, P., De, S., 2011. Modeling of nanofiltration of dye using a coupled concentration polarization and pore flow model. Separ. Sci. Technol. 46 (4), 561–570.

Bardhan, A., Subbiah, S., Mohanty, K., 2023. Modeling and experimental validation for the preparation of concentrated tea extract using a forward osmosis process using a food-grade inorganic draw solute. Ind. Eng. Chem. Res. 62 (7), 3315–3326.

Bechhold, H., 1907. Kolloidstudien mit der filtrationsmethode. Z. Phys. Chem. 60 (1), 257–318.

Berg, T.H., Knudsen, J.C., Ipsen, R., van den Berg, F., Holst, H.H., Tolkach, A., 2014. Investigation of consecutive fouling and cleaning cycles of ultrafiltration membranes used for whey processing. Int. J. Food Eng. 10 (3), 367–381.

Bertan, A.S., Cremasco, M.A., 2024. Multi-steps microfiltration of micelles from fermentation of streptomyces tsukubaensis and its impact on proteins retention and tacrolimus yield. Food Bioprod. Process. 148.

Biesheuvel, P., Rutten, S., Ryzhkov, I., Porada, S., Elimelech, M., 2023. Theory for salt transport in charged reverse osmosis membranes: novel analytical equations for desalination performance and experimental validation. Desalination 557, 116580.

Bijak, J., Bijak, J., Hilton, J., 2022. Uncertainty quantification, model calibration and sensitivity. Towards Bayesian Model-Based Demography: Agency, Complexity and Uncertainty in Migration Studies 10, 71–92.

Bird, R., Stewart, W., Lightfoot, E., 1960. Transport Phenomena. John Wiley and Sons, New York.

Bird, R.B., Stewart, W.E., Lightfoot, E.N., 1960. Transport Phenomena. John Wiley & Sons, New York, p. 362.

Bitter, J.G., 2012. Transport Mechanisms in Membrane Separation Processes. Springer Science & Business Media.

Blais, H., Ho, Q.T., Murphy, E.G., Schroën, K., Tobin, J.T., 2021. A cascade microfiltration and reverse osmosis approach for energy efficient concentration of skim milk. J. Food Eng. 300, 110511.

Bowen, W.R., Welfoot, J.S., 2002. Modelling the performance of membrane nanofiltration—critical assessment and model development. Chem. Eng. Sci. 57 (7), 1121–1137.

Bringas, E., San Roman, M., Irabien, J., Ortiz, I., 2009. An overview of the mathematical modelling of liquid membrane separation processes in hollow fibre contactors. J. Chem. Technol. Biotechnol. 84 (11), 1583–1614.

Cadotte, J., King, R., Majerle, R., Petersen, R., 1981. Interfacial synthesis in the preparation of reverse osmosis membranes. J. Macromol. Sci.-Chem. 15 (5), 727–755.

Cadotte, J.E., 1981. Interfacially synthesized reverse osmosis membrane. In: Google Patents.

Carrère, H., Blaszkow, F., Balmann, H.R.d., 2001a. Modelling the clarification of lactic acid fermentation broths by cross-flow microfiltration. J. Membr. Sci. 186 (2), 219–230. https://doi.org/10.1016/S0376-7388(00)00677-3.

Carrère, H., Blaszkow, F., de Balmann, H.R., 2001b. Modelling the clarification of lactic acid fermentation broths by cross-flow microfiltration. J. Membr. Sci. 186 (2), 219–230.

Casado-Coterillo, C., Díaz-Guridi, P., Otero, J.A., Ibáñez, R., 2023. Modeling of lactic acid rejection from lactose in acidified cheese whey by nanofiltration. J. Dairy Sci. 106 (7), 4533–4544.

Cassano, A., Donato, L., Drioli, E., 2007. Ultrafiltration of kiwifruit juice: operating parameters, juice quality and membrane fouling. J. Food Eng. 79 (2), 613–621. https://doi.org/10.1016/j.jfoodeng.2006.02.020.

Castro-Muñoz, R., 2019. Pervaporation: the emerging technique for extracting aroma compounds from food systems. J. Food Eng. 253, 27–39.

Cerceau Alves, Y.P., Fernandes Antunes, F.A., Silverio da Silva, S., Forte, M.B.S., 2021. From by- to bioproducts: selection of a nanofiltration membrane for biotechnological xylitol purification and process optimization. Food Bioprod. Process. 125, 79–90. https://doi.org/10.1016/j.fbp.2020.10.005.

Chandrappa, L., Tabain, Z., Pastrana, E.F., Dons, T., Ahrné, L., 2024. Separation of oil from rapeseed protein rich extracts by microfiltration using hydrophilic ceramic membranes. Future Foods 10, 100453.

Chang, S., Fane, A., Vigneswaran, S., 2002. Modeling and optimizing submerged hollow fiber membrane modules. AIChE J. 48 (10), 2203–2212.

Chellam, S., 2005. Artificial neural network model for transient crossflow microfiltration of polydispersed suspensions. J. Membr. Sci. 258 (1–2), 35–42.

Chen, G.Q., Artemi, A., Lee, J., Gras, S.L., Kentish, S.E., 2019. A pilot scale study on the concentration of milk and whey by forward osmosis. Separ. Purif. Technol. 215, 652–659. https://doi.org/10.1016/j.seppur.2019.01.050.

Chen, Q., Zhao, L., Yao, L., Chen, Q., Ahmad, W., Li, Y., Qin, Z., 2018. The application of membrane separation technology in the dairy industry. Technological Approaches for Novel Applications in Dairy Processing 23.

Cifuentes-Cabezas, M., Bohórquez-Zurita, J.L., Gil-Herrero, S., Vincent-Vela, M.C., Mendoza-Roca, J.A., Álvarez-Blanco, S., 2023. Deep study on fouling modelling of ultrafiltration membranes used for OMW treatment: comparison between semi-empirical models, response surface, and artificial neural networks. Food Bioprocess Technol. 16 (10), 2126–2146.

Corbatón-Báguena, M.-J., Álvarez-Blanco, S., Vincent-Vela, M.-C., 2015. Fouling mechanisms of ultrafiltration membranes fouled with whey model solutions. Desalination 360, 87–96. https://doi.org/10.1016/j.desal.2015.01.019.

Córdova, A., Astudillo, C., Giorno, L., Guerrero, C., Conidi, C., Illanes, A., Cassano, A., 2016. Nanofiltration potential for the purification of highly concentrated enzymatically produced oligosaccharides. Food Bioprod. Process. 98, 50–61. https://doi.org/10.1016/j.fbp.2015.11.005.

Dammak, L., Fouilloux, J., Bdiri, M., Larchet, C., Renard, E., Baklouti, L., Pismenskaya, N., 2021. A review on ion-exchange membrane fouling during the electrodialysis process in the food industry, part 1: types, effects, characterization methods, fouling mechanisms and interactions. Membranes 11 (10), 789.

Daneluz, J., da Silva, G.F., Duarte, J., Turossi, T.C., dos Santos, V., Baldasso, C., Daneluz, A. C., 2023. Membrane separation process of microfiltration applied to the filtration of kombuchas. Food Chemistry Advances 3, 100451.

Divakar, S., Padaki, M., Balakrishna, R.G., 2022. Review on liquid–liquid separation by membrane filtration. ACS Omega 7 (49), 44495–44506.

Doan, N.T.T., Lai, Q.D., Nguyen, H.D., Nabetani, H., 2023. Enrichment of lycopene in watermelon juice by ultrafiltration: technical assessment and fouling analysis. Int. J. Food Sci. Technol. 58 (12), 6285–6295. https://doi.org/10.1111/ijfs.16732.

Donato, D., Boschetti-de-Fierro, A., Zweigart, C., Kolb, M., Eloot, S., Storr, M., Segers, P., 2017. Optimization of dialyzer design to maximize solute removal with a two-dimensional transport model. J. Membr. Sci. 541, 519–528. https://doi.org/10.1016/j.memsci.2017.07.018.

Du, Y., Wang, L., Belgada, A., Younssi, S.A., Gilron, J., Elimelech, M., 2023. A mechanistic model for salt and water transport in leaky membranes: Implications for low-salt-rejection reverse osmosis membranes. J. Membr. Sci. 678, 121642. https://doi.org/10.1016/j.memsci.2023.121642.

El-Okazy, M.A., Chen, G.Q., Hadde, E., Akbari, A., Kentish, S.E., 2024. The performance of a graphene oxide thin film composite membrane for sweet whey ultrafiltration. Separ. Purif. Technol. 10, 129868.

Eleiwi, F., Ghaffour, N., Alsaadi, A.S., Francis, L., Laleg-Kirati, T.M., 2016. Dynamic modeling and experimental validation for direct contact membrane distillation (DCMD) process. Desalination 384, 1–11.

Farhan, N.M., Ibrahim, S.S., Alsalhy, Q.F., 2023. Modeling and simulation of pervaporation (PV) separation for alcohol dehydration. Heliyon 9 (2).

Ferry, J.D., 1936. Ultrafilter membranes and ultrafiltration. Chem. Rev. 18 (3), 373–455.

Field, R.W., Wu, H.Y., Wu, J.J., 2013. Multiscale modeling of membrane distillation: some theoretical considerations. Ind. Eng. Chem. Res. 52 (26), 8822–8828.

Filippov, A., 2022. A Cell model of an ion-exchange membrane. capillary-osmosis and reverse-osmosis coefficients. Colloid J. 84 (3), 332–343.

Gilron, J., Soffer, A., 2002. Knudsen diffusion in microporous carbon membranes with molecular sieving character. J. Membr. Sci. 209 (2), 339–352.

Guiné, R., 2019. The use of artificial neural networks (ANN) in food process engineering. Int. J. Food Eng. 5 (1), 15–21.

Haddadi, B., Jordan, C., Miltner, M., Harasek, M., 2018. Membrane modeling using CFD: combined evaluation of mass transfer and geometrical influences in 1D and 3D. J. Membr. Sci. 563, 199–209.

Hairom, N.H.H., Mohammad, A.W., Kadhum, A.A.H., 2014. Nanofiltration of hazardous Congo red dye: performance and flux decline analysis. J. Water Proc. Eng. 4, 99–106.

Heiranian, M., Fan, H., Wang, L., Lu, X., Elimelech, M., 2023. Mechanisms and models for water transport in reverse osmosis membranes: history, critical assessment, and recent developments. Chem. Soc. Rev. 52.

Hermia, J., 1985. Blocking filtration. Application to non-Newtonian fluids. In: Mathematical Models and Design Methods in Solid-Liquid Separation. Springer, pp. 83–89.

Herrera-Valdez, M.A., 2012. Membranes with the same ion channel populations but different excitabilities. PLoS One 7 (4), e34636.

Hitsov, I., Maere, T., De Sitter, K., Dotremont, C., Nopens, I., 2015. Modelling approaches in membrane distillation: a critical review. Separ. Purif. Technol. 142, 48–64.

Hou, H., Duan, L., Zhou, B., Tian, Y., Wei, J., Qian, F., 2020. The performance and degradation mechanism of sulfamethazine from wastewater using IFAS-MBR. Chin. Chem. Lett. 31 (2), 543–546.

Hu, D., Xu, Z.L., Wei, Y.M., 2013. A high performance silica–fluoropolyamide nanofiltration membrane prepared by interfacial polymerization. Separ. Purif. Technol. 110, 31–38.

Hua, X., Zhao, H., Yang, R., Zhang, W., Zhao, W., 2010. Coupled model of extended Nernst–Planck equation and film theory in nanofiltration for xylo-oligosaccharide syrup. J. Food Eng. 100 (2), 302–309.

Hubach, T., Schlüter, S., Held, C., 2023. Model-based optimization of multi-stage nanofiltration using the solution-diffusion–electromigration model. Processes 11 (8), 2355.

Jawad, J., Hawari, A.H., Zaidi, S.J., 2021. Artificial neural network modeling of wastewater treatment and desalination using membrane processes: a review. Chem. Eng. J. 419, 129540.

Kadival, A., & Ghosh, S. (2018). Applications of membrane separation technology in food industry. In (pp. 43–55).

Jiao, B., Cassano, A., Drioli, E., 2004. Recent advances on membrane processes for the concentration of fruit juices: a review. Journal of food engineering 63 (3), 303–324.

Kapoor, A., Datta, S., Gupta, G., Vishwakarma, A., Singh, A., Chattopadhyay, S., Ambatipudi, K., 2024. One-step separation of fat globules based on size from bovine milk using a cross-flow microfiltration. Separ. Purif. Technol. 330, 125251.

Kedem, O., Katchalsky, A., 1958. Thermodynamic analysis of the permeability of biological membranes to non-electrolytes. Biochim. Biophys. Acta 27, 229–246.

Khayet, M., 2011. Membranes and theoretical modeling of membrane distillation: a review. Adv. Colloid Interface Sci. 164 (1–2), 56–88.

Khayet, M., Sanmartino, J., Essalhi, M., García-Payo, M., Hilal, N., 2016. Modeling and optimization of a solar forward osmosis pilot plant by response surface methodology. Sol. Energy 137, 290–302.

Kim, H.-Y., Park, H.-J., Ko, G., 2009. Hollow-fiber ultrafiltration for the concentration and simultaneous recovery of multiple pathogens in contaminated foods. J. Food Protect. 72 (12), 2547–2552.

Kim, Y.-B., Myagmarsukh, Z., Kim, Y.-D., 2024. Influence of microbubble two-phase feed flow on reverse osmosis desalination using different salt concentrations. Desalination 21, 118178.

Kimura, S., Sourirajan, S., 1967. Analysis of data in reverse osmosis with porous cellulose acetate membranes used. AIChE J. 13 (3), 497–503.

Kirsch, V., Bazhenov, S., 2020. Numerical simulation of solute removal from a cross-flow past a row of parallel hollow-fiber membranes. Separ. Purif. Technol. 242, 116834.

Kirsch, V., Roldugin, V., Bildyukevich, A., Volkov, V., 2016. Simulation of convective-diffusional processes in hollow fiber membrane contactors. Separ. Purif. Technol. 167, 63–69.

Kirschner, A.Y., Cheng, Y.-H., Paul, D.R., Field, R.W., Freeman, B.D., 2019. Fouling mechanisms in constant flux crossflow ultrafiltration. J. Membr. Sci. 574, 65–75.

Koros, W.J., Chan, A., Paul, D., 1977. Sorption and transport of various gases in polycarbonate. J. Membr. Sci. 2, 165–190.

Kostanyan, A.E., Voshkin, A.A., Belova, V.V., Zakhodyaeva, Y.A., 2023. Modelling and comparative analysis of different methods of liquid membrane separations. Membranes 13 (6), 554.

Kraaijeveld, G. (1994). The Maxwell-Stefan description of mass transfer in ion exchange and electrodialysis.

Krishna, R., Wesselingh, J.A., 1997. The Maxwell-Stefan approach to mass transfer. Chem. Eng. Sci. 52 (6), 861–911.

Lakshminarayanaiah, N., 1965. Transport phenomena in artificial membranes. Chem. Rev. 65 (5), 491–565.

Lantto, J. (2015). Analytical model of mass transfer through supported liquid membranes. In.

Läuger, P., 1969. Transport phenomena in membranes. Angew Chem. Int. Ed. Engl. 8 (1), 42–54.

Lee, Y., Clark, M.M., 1998. Modeling of flux decline during crossflow ultrafiltration of colloidal suspensions. J. Membr. Sci. 149 (2), 181–202.

Lee, Y.M., Bourgeois, D., Belfort, G., 1989. Sorption, diffusion, and pervaporation of organics in polymer membranes. J. Membr. Sci. 44 (2–3), 161–181.

Li, X., Younas, M., Rezakazemi, M., Ly, Q.V., Li, J., 2022. A review on hollow fiber membrane module towards high separation efficiency: process modeling in fouling perspective. Chin. Chem. Lett. 33 (8), 3594–3602.

Li, Y., Luo, J., Wan, Y., 2020. Biofouling in sugarcane juice refining by nanofiltration membrane: fouling mechanism and cleaning. J. Membr. Sci. 612, 118432. https://doi.org/10.1016/j.memsci.2020.118432.

Lin, S., 2016. Mass transfer in forward osmosis with hollow fiber membranes. J. Membr. Sci. 514, 176–185.

Liu, B., Wang, Z., Tan, X., Liu, S., 2024. Ultra-high flux ceramic hollow fiber membranes for nanofiltration. J. Membr. Sci. 709, 123104.

Ma, S., Wu, X., Fan, L., Wang, Q., Hu, Y., Xie, Z., 2023. Effect of different draw solutions on concentration polarization in a forward osmosis process: theoretical modeling and experimental validation. Ind. Eng. Chem. Res. 62 (8), 3672–3683.

Madhumala, M., Moulik, S., Sankarshana, T., Sridhar, S., 2017. Forward-osmosis-aided concentration of fructose sugar through hydrophilized polyamide membrane: molecular modeling and economic estimation. J. Appl. Polym. Sci. 134 (13). https://doi.org/10.1002/app.44649.

Mahadeva, R., Kumar, M., Patole, S.P., Manik, G., 2022. Employing artificial neural network for accurate modeling, simulation and performance analysis of an RO-based desalination process. Sustainable Comput. Inf. Syst. 35, 100735.

Mancini, E., Ramin, P., Styrbæck, P., Bjergholt, C., Mansouri, S.S., Gernaey, K.V., Pinelo, M., 2022. Separation of succinic acid from fermentation broth: dielectric exclusion, Donnan effect and diffusion as the most influential mass transfer mechanisms. Separ. Purif. Technol. 281, 119904.

Mareev, S., Gorobchenko, A., Ivanov, D., Anokhin, D., Nikonenko, V., 2022. Ion and water transport in ion-exchange membranes for power generation systems: guidelines for modeling. Int. J. Mol. Sci. 24 (1), 34.

Marriott, J., Sørensen, E., 2003. A general approach to modelling membrane modules. Chem. Eng. Sci. 58 (22), 4975–4990.

Martínez-Diez, L., Vazquez-Gonzalez, M., 2000. A method to evaluate coefficients affecting flux in membrane distillation. J. Membr. Sci. 173 (2), 225–234.

Marx, M., Bernauer, S., Kulozik, U., 2018. Manufacturing of reverse osmosis whey concentrates with extended shelf life and high protein nativity. Int. Dairy J. 86, 57–64.

Mason, E. A., & Ap, M. (1983). Gas transportin porous media: the dusty-gas model.

Mehta, G.D., Loeb, S., 1978. Internal polarization in the porous substructure of a semipermeable membrane under pressure-retarded osmosis. J. Membr. Sci. 4, 261–265.

Merdaw, A.A., Sharif, A.O., Derwish, G.A.W., 2011. Mass transfer in pressure-driven membrane separation processes, Part II. Chemical Engineering Journal 168 (1), 229–240.

Mirbagheri, S.A., Bagheri, M., Bagheri, Z., Kamarkhani, A.M., 2015. Evaluation and prediction of membrane fouling in a submerged membrane bioreactor with simultaneous upward and downward aeration using artificial neural network-genetic algorithm. Process Saf. Environ. Prot. 96, 111–124.

Model, I.M.M., 1990. Ion and solvent transport in ion-exchange membranes. J. Electrochem. Soc. 137 (3).

Mondal, M., Bhattacharjee, S., De, S., 2020. Prediction of long term filtration by coupled gel layer and pore transport model for salt removal using mixed matrix hollow fiber ultrafiltration membrane. Separ. Purif. Technol. 250, 117213.

Mondal, S., Griffiths, I.M., Ramon, G.Z., 2019. Forefronts in structure–performance models of separation membranes. J. Membr. Sci. 588, 117166.

Mondor, M., Moresoli, C., 1999. Theoretical analysis of the influence of the axial variation of the transmembrane pressure in cross-flow filtration of rigid spheres. J. Membr. Sci. 152 (1), 71–87.

Munubarthi, K.K., Gautam, D.K., Reddy, K.A., Subbiah, S., 2020. Distributed parameter system modeling approach for the characterization of a high flux hollow fiber forward osmosis (HFFO) membrane. Desalination 496, 114706.

Nagy, E., 2018. Basic Equations of Mass Transport through a Membrane Layer. Elsevier.

Nambi Krishnan, J., Venkatachalam, K.R., Ghosh, O., Jhaveri, K., Palakodeti, A., Nair, N., 2022. Review of thin film nanocomposite membranes and their applications in desalination. Front. Chem. 10, 781372.

Nath, K., Dave, H.K., Patel, T.M., 2018. Revisiting the recent applications of nanofiltration in food processing industries: progress and prognosis. Trends Food Sci. Technol. 73, 12–24.

Nath, S., Astapati, A.D., Naha, A., Sharma, I., Shah, M.P., 2024. Algal-based membrane bioreactors for effective removal of hazardous and toxic contaminants: a comprehensive review. Current Research in Biotechnology 7, 100201.

Niemi, H., Bulsari, A., Palosaari, S., 1995. Simulation of membrane separation by neural networks. J. Membr. Sci. 102, 185–191.

Nourbakhsh, H., Emam-Djomeh, Z., Omid, M., Mirsaeedghazi, H., Moini, S., 2014. Prediction of red plum juice permeate flux during membrane processing with ANN optimized using RSM. Comput. Electron. Agric. 102, 1–9.

Oren, Y.S., Nir, O., Freger, V., 2024. Analyzing ion uptake in ion-exchange membranes using ion association model. J. Membr. Sci. 690, 122202.

Pei, J., Wang, W., Wang, Y., Wang, H., Bucs, S.S., Vrouwenvelder, J.S., Li, Z., 2021. Fate of polyphenols in forward osmosis. J. Membr. Sci. 621, 118993.

Pellerin, E., Michelitsch, E., Darcovich, K., Lin, S., Tam, C., 1995. Turbulent transport in membrane modules by CFD simulation in two dimensions. J. Membr. Sci. 100 (2), 139–153.

Pereira, G.L.D., Cardozo-Filho, L., Jegatheesan, V., Guirardello, R., 2023. Generalization and expansion of the Hermia model for a better understanding of membrane fouling. Membranes 13 (3), 290.

Pervez, M.N., Mahboubi, A., Uwineza, C., Sapmaz, T., Zarra, T., Belgiorno, V., Taherzadeh, M.J., 2022. Feasibility of nanofiltration process for high efficient recovery and concentrations of food waste-derived volatile fatty acids. J. Water Proc. Eng. 48, 102933.

Phillip, W.A., Yong, J.S., Elimelech, M., 2010. Reverse draw solute permeation in forward osmosis: modeling and experiments. Environ. Sci. Technol. 44 (13), 5170–5176.

Porozhnyy, M., Sarapulova, V., Pismenskaya, N., Huguet, P., Deabate, S., Nikonenko, V., 2017. Mathematical modeling of concentration dependences of electric conductivity and diffusion permeability of anion-exchange membranes soaked in wine. Petrol. Chem. 57, 511–517.

Pouliot, Y., Conway, V., Leclerc, P.L., 2014. Separation and concentration technologies in food processing. Food Process: Principles and Applications 7, 33–60.

Purpura, G., Papiewska, E., Culcasi, A., Filingeri, A., Tamburini, A., Ferrari, M.C., Cipollina, A., 2024. Modelling of selective ion partitioning between ion-exchange membranes and highly concentrated multi-ionic brines. J. Membr. Sci. 700, 122659.

Qtaishat, M., Matsuura, T., Kruczek, B., Khayet, M., 2008. Heat and mass transfer analysis in direct contact membrane distillation. Desalination 219 (1–3), 272–292.

Quezada, C., Estay, H., Cassano, A., Troncoso, E., Ruby-Figueroa, R., 2021. Prediction of permeate flux in ultrafiltration processes: a review of modeling approaches. Membranes 11 (5), 368.

Rajendran, S.R., Mason, B., Doucette, A.A., 2021. Review of membrane separation models and technologies: processing complex food-based biomolecular fractions. Food Bioprocess Technol. 14, 415–428.

Razavi, M.A., Mortazavi, A., Mousavi, M., 2004. Application of neural networks for crossflow milk ultrafiltration simulation. Int. Dairy J. 14 (1), 69–80.

Reid, C., Breton, E., 1959. Water and ion flow across cellulosic membranes. J. Appl. Polym. Sci. 1 (2), 133–143.

Rizki, Z., Ottens, M., 2023. Model-based optimization approaches for pressure-driven membrane systems. Separ. Purif. Technol. 315, 123682.

Ruby-Figueroa, R., Saavedra, J., Bahamonde, N., Cassano, A., 2017. Permeate flux prediction in the ultrafiltration of fruit juices by ARIMA models. J. Membr. Sci. 524, 108–116. https://doi.org/10.1016/j.memsci.2016.11.034.

Saavedra, A., Valdés, H., Velásquez, J., Hernández, S., 2024. Comparative analysis of Donnan steric partitioning pore model and dielectric exclusion applied to the fractionation of aqueous saline solutions through nanofiltration. ChemEngineering 8 (2), 39.

Sadek, S.A., Al-Jubouri, S.M., 2023. Structure and performance of polyvinylchloride microfiltration membranes improved by green silicon oxide nanoparticles for oil-in-water emulsion separation. Mater. Today Sustain. 24, 100600.

Sekino, M., 1993. Precise analytical model of hollow fiber reverse osmosis modules. J. Membr. Sci. 85 (3), 241–252.

Shetty, G.R., Malki, H., Chellam, S., 2003. Predicting contaminant removal during municipal drinking water nanofiltration using artificial neural networks. J. Membr. Sci. 212 (1–2), 99–112.

Shilpashree, B., Arora, S., Sharma, V., Singh, A., 2015. Preparation of succinylated sodium caseinate–iron complex by adopting ultrafiltration technology: a novel food fortificant. Innov. Food Sci. Emerg. Technol. 32, 165–171.

Singh, S.K., Sharma, C., Maiti, A., 2023. Modeling and experimental validation of forward osmosis process: parameters selection, permeate flux prediction, and process optimization. J. Membr. Sci. 672, 121439.

Soni, V., Abildskov, J., Jonsson, G., Gani, R., 2009. A general model for membrane-based separation processes. Comput. Chem. Eng. 33 (3), 644–659.

Sourirajan, S., Agrawal, J., 1969. Reverse osmosis. Ind. Eng. Chem. 61 (11), 62–89.

Spiegler, K.S., Kedem, O., 1966. Thermodynamics of hyperfiltration (reverse osmosis): criteria for efficient membranes. Desalination 1 (4), 311–326.

Suh, C., Lee, S., 2013. Modeling reverse draw solute flux in forward osmosis with external concentration polarization in both sides of the draw and feed solution. J. Membr. Sci. 427, 365–374.

Taghavijeloudar, M., Park, J., Han, M., Taghavi, A., 2019. A new approach for modeling flux variation in membrane filtration and experimental verification. Water Res. 166, 115027.

Tanaka, Y. (2010). Ion exchange membrane electrodialysis: fundamentals. (Desalination, Separation).

Tanambell, H., Danielsen, M., Freund, S., Nissen, S.H., Møller, A.H., Corredig, M., Dalsgaard, T.K., 2024. Partitioning of proteins and small molecular weight compounds from alfalfa juice during ultrafiltration. Food Biosci. 62, 105044.

Teodoro, J.A., Arend, G.D., Proner, M.C., Verruck, S., Rezzadori, K., 2023. A review on membrane separation processes focusing on food industry environment-friendly processes. Crit. Rev. Food Sci. Nutr. 63 (32), 11275–11289.

Terra, N.M., Madrona, G.S., Ferreira, F.B., Cardoso, V.L., Reis, M.H.M., 2019. High performance of asymmetric alumina hollow fiber membranes for the clarification of genipap (genipa americana L.) fruit extract. Food Bioprocess Technol. 12 (1), 27–38. https://doi.org/10.1007/s11947-018-2185-3.

Tijing, L.D., Woo, Y.C., Choi, J.-S., Lee, S., Kim, S.-H., Shon, H.K., 2015. Fouling and its control in membrane distillation—a review. J. Membr. Sci. 475, 215–244. https://doi.org/10.1016/j.memsci.2014.09.042.

Todisco, S., Tallarico, P., Gupta, B., 2002. Mass transfer and polyphenols retention in the clarification of black tea with ceramic membranes. Innov. Food Sci. Emerg. Technol. 3 (3), 255–262.

Trudel-Ferland, M., Levasseur, M., Goulet-Beaulieu, V., Jubinville, E., Hamon, F., Jean, J., 2024. Concentration of foodborne viruses eluted from fresh and frozen produce: applicability of ultrafiltration. Int. J. Food Microbiol. 416, 110687.

Uzdenova, A., Urtenov, M., 2021. Mathematical modeling of the phenomenon of space-charge breakdown in the galvanostatic mode in the section of the electromembrane desalination channel. Membranes 11 (11), 873.

Van den Berg, G., Smolders, C., 1992. Diffusional phenomena in membrane separation processes. J. Membr. Sci. 73 (2–3), 103–118.
Vinh-Thang, H., Kaliaguine, S., 2013. Predictive models for mixed-matrix membrane performance: a review. Chem. Rev. 113 (7), 4980–5028.
Wang, R., Duddu, R., Lin, S., 2023. Extended donnan-Manning theory for selective ion partition and transport in ion exchange membrane. J. Membr. Sci. 681, 121782.
Wang, R., Lin, S., 2021. Pore model for nanofiltration: history, theoretical framework, key predictions, limitations, and prospects. J. Membr. Sci. 620, 118809.
Wang, T., Li, Y.-Y., 2024. Predictive modeling based on artificial neural networks for membrane fouling in a large pilot-scale anaerobic membrane bioreactor for treating real municipal wastewater. Sci. Total Environ. 912, 169164.
Wu, X., Zhang, X., Wang, H., Xie, Z., 2023. Smart utilisation of reverse solute diffusion in forward osmosis for water treatment: a mini review. Sci. Total Environ. 873, 162430.
Yadav, J.S.S., Yan, S., Ajila, C.M., Bezawada, J., Tyagi, R.D., Surampalli, R.Y., 2016. Food-grade single-cell protein production, characterization and ultrafiltration recovery of residual fermented whey proteins from whey. Food Bioprod. Process. 99, 156–165.
Yamamoto, K., Hiasa, M., Mahmood, T., Matsuo, T., 1988. Direct solid-liquid separation using hollow fiber membrane in an activated sludge aeration tank. In: Water Pollution Research and Control Brighton. Elsevier, pp. 43–54.
Yang, C., Peng, X., Zhao, Y., Wang, X., Fu, J., Liu, K., Li, P., 2020. Prediction model to analyze the performance of VMD desalination process. Comput. Chem. Eng. 132, 106619.
Yang, H., Yu, X., Liu, J., Tang, Z., Huang, T., Wang, Z., Wang, L., 2022. A concise review of theoretical models and numerical simulations of membrane fouling. Water 14 (21), 3537.
Yang, X., Yu, H., Wang, R., Fane, A.G., 2012. Optimization of microstructured hollow fiber design for membrane distillation applications using CFD modeling. J. Membr. Sci. 421, 258–270.
Younas, M., Rezakazemi, M., 2022. Membrane Contactor Technology: Water Treatment, Food Processing, Gas Separation, and Carbon Capture. John Wiley & Sons.
Yun, K.H., Sim, S., Hong, H.S., Jeong, K.J., Bae, T.-H., 2024. Optimizing PES microfiltration membranes for sterile filtration: a comparative study of polydopamine coating in acidic and basic conditions to minimize protein adsorption. J. Membr. Sci. 713, 123398.
Zin, M.M., Márki, E., Bánvölgyi, S., 2020. Evaluation of reverse osmosis membranes in concentration of beetroot peel extract. Periodica Polytech., Chem. Eng. 64 (3), 340–348.
Zsigmondy, R., Bachmann, W., 1918. On new filters. Zeitschrift Fur Anorganische und Allgemeine Chemie 103, 119.

Advances in modeling microbial inactivation in plant-based liquid products using emerging non-thermal technologies

11

Theocharia Tsagkaropoulou [1], *M. Teresa Fernández-Felipe* [2], *Arisa Thamsuaidee* [3, 4], *Poliana Mendes de Souza* [5], *Kimon Andreas G. Karatzas* [1], *Dolores Rodrigo* [2], *Claudia Siemer* [3] *and Arícia Possas* [6]

[1]Department of Food & Nutritional Sciences, University of Reading, Reading, United Kingdom; [2]Instituto de Agroquímica y Tecnología de Alimentos (IATA-CSIC), Paterna, Valencia, Spain; [3]ELEA Technology GmbH, Quakenbrueck, Germany; [4]Laboratory of Food Chemistry, Department of Chemistry, National and Kapodistrian University of Athens, Athens, Greece; [5]African Sustainable Agriculture Research Institute (ASARI), University Mohammed VI Polytechnic (UM6P), Laâyoune, Morocco; [6]Department of Food Science and Technology, UIC Zoonosis y Enfermedades Emergentes (ENZOEM), CeiA3, Universidad de Córdoba, Córdoba, Spain

11.1 Introduction

Plant-based beverages and fruit/vegetable juices are typically made by extracting, blending or processing plant-based raw materials. Widely available plant-based beverages include those obtained from soy, cereals and pseudocereals (e.g., rice and oat) and nuts (e.g., almond and cashew). On the other hand, fruit and vegetable juices are primarily obtained from mechanical extraction or pressing of fresh produce, including berries, citrus, and tropical fruits. Many factors drive the consumption of these products, including the growing awareness about animal welfare and sustainability (Kyrylenko et al., 2023), as well as the growing demand for healthier, minimally processed beverages. Health-conscious consumers and the growing vegan population often prefer plant-based beverages as dairy substitutes, as they have high nutritional value, low or zero fat, and zero lactose content (Canaviri Paz et al., 2020; Kyrylenko et al., 2023; Possas et al., 2018). Meanwhile, fruit and vegetable juices remain popular, attributable to their nutritional profile and perceived health-benefits, particularly in functional beverage markets. As a result, both sectors are experiencing significant market growth. The global plant-based beverages market is estimated to grow at a compound annual growth rate (CAGR) of 10.12% between 2023 and 2032 (Precedence Research, 2023), while the juices market is expected to grow annually by 4.24% (CAGR, 2025-29) (Statista, 2025). The development of novel plant-based products

Modeling of Unit Operations in the Food Industry. https://doi.org/10.1016/B978-0-12-822180-8.00005-5

and innovative juice formulations continues to stimulate market expansion (Precedence Research, 2023; Statista, 2025).

Plant-based beverages and fruit/vegetable juices can become contaminated with foodborne and spore-forming bacteria at various steps along the distribution chain. Contamination risks start with handling and sourcing raw materials, since bacteria from soil, water, and handling equipment can contaminate cereals, nuts, and fruits and vegetables (González-Tejedor et al., 2023). In addition, when equipment is not properly sanitized, cross-contamination events may take place in processing environments. Packaging in aseptic environments is critical, as contamination can occur if packaging materials are not sterile or if filling machinery is exposed to bacteria. Once contaminated, both plant-based beverages and juices support rapid growth of foodborne pathogens, including *Listeria monocytogenes* (Bartula et al., 2023, 2024), *Salmonella* (Bartula et al., 2023), and *Bacillus* spp. (Bartula et al., 2023). Food safety standards require pasteurization or an equivalent validation processing step for both categories. According to the U.S. Food and Drug Administration (FDA) regulation (21 CFR 120.24), Food Business Operators (FBOs) are required to implement the so-called Hazard Analysis and Critical Control Points (HACCP) plan, where a 5-log reduction of the pathogen of interest in juices (e.g., *Escherichia coli* 0157:H7, *L. monocytogenes*, *Salmonella* spp.) must be achieved during a validated control measure to ensure food safety (Usaga et al., 2021).

The introduction of the FDA regulation (21 CFR120) was driven by the occurrence of several illness outbreaks in the 1990s associated with raw juices (Krug et al., 2020). These include various *E. coli* O157:H7 outbreaks resulting from the consumption of contaminated unpasteurized apple juice (Krug et al., 2020). Furthermore, foodborne and spore-forming bacteria have been detected in plant-based product sources, such as *Bacillus cereus* from pea and oat samples (Kyrylenko et al., 2023). Numerous recalls of plant-based beverages have been issued due to their potential for microbial contamination, including *B. cereus* in oat milk (New Zealand Food Safety, 2024), *L. monocytogenes* in soy milk (CDC, 2014), and the opportunistic foodborne pathogen *Cronobacter sakazakii* in almond, coconut, and oat beverages (FDA, 2022). Additionally, an outbreak of *L. monocytogenes* infections linked to almond, coconut, oat, and cashew refrigerated beverages occurred in Canada in 2024, resulting in 20 illness cases and 3 deaths. The outbreak strain was found within the production environment (Public Health Agency of Canada, 2024).

As for dairy beverages, thermal heat treatments have usually been applied for microbial inactivation in both plant-based beverages and fruit/vegetable juices (Kyrylenko et al., 2023). However, the application of heat may cause undesirable changes to the nutritional features of these products, besides compromising their organoleptic characteristics, such as taste, color, and texture. In this sense, the use of nonthermal technologies as substitutes for thermal treatments has been intensively investigated, offering effective microbial inactivation while preserving nutritional and sensory attributes. Overall, nonthermal methods require less energy than thermal treatments, leading to lower energy consumption and reduced greenhouse gas emissions, aligning well with the environmental ethos of consumers, who demand sustainable, minimally processed plant-based and juice products.

Mathematical models are valuable tools for estimating microbial inactivation in plant-based beverages, helping to optimize processing parameters to ensure safety while maintaining quality (González-Tejedor et al., 2023; Possas et al., 2018). These models allow for precise control of key conditions of nonthermal methods such as temperature, pressure, UV dose, and treatment duration, which helps minimizing nutrient loss and preserves sensory qualities. Moreover, predictive models account for the natural variability in plant-based ingredients and fruit matrices, enabling tailored processing approaches for different formulations. By identifying the most effective and efficient inactivation parameters, these models reduce trial-and-error testing, lower energy costs, and support compliance with food safety criteria, contributing to both product quality and sustainability.

This chapter presents the most recent modeling approaches for bacterial inactivation induced by nonthermal technologies in plant-based and fruit/vegetable juices. Mechanisms of action, the most important factors affecting process efficacy, existing limitations, and prospective developments are also discussed to present an integrated view of the application of these technologies in ensuring microbial safety and product quality.

11.2 Emerging nonthermal technologies

Several nonthermal technologies are gaining attention for the decontamination of plant-based beverages and fruit/vegetable juices. Some of the promising technologies, which will be further discussed in this chapter, include high-pressure processing (HHP), ultraviolet-C (UV-C) radiation, pulsed electric fields (PEF), and cold plasma. These nonthermal methods can effectively decontaminate beverages while preserving their quality. The selection of an appropriate technology is influenced by factors like the target microorganisms and the type of beverage. A detailed description of these technologies follows below.

11.2.1 High-pressure processing

High-pressure processing (HPP), commonly referred to as high hydrostatic pressure, is a nonthermal food preservation technology aligned with consumer interest in minimally processed and healthier foods. HPP is generally performed on vacuum-packaged products enclosed in compressible materials ("in-pack" processing), which are arranged in baskets and introduced into a pressure vessel. The machine fills the vessel with water as the pressure transmitting fluid and hydrostatic pressures of 400–600 MPa (6,000 bar) are applied for several minutes (1.5–6 min; holding time). After pressure release (come-down time; CDT) the products are typically supplied to the market under refrigeration. Processing conditions, including pressure level, treatment duration, and temperature, are selected based on the intended application and the microorganisms of primary concern. Although most of HPP-treated foods are processed 'In-pack', latest advances in commercial equipment have made the 'In-bulk' process possible, where beverages are processed before bottling. This innovation simplifies the processing steps and increases productivity by reducing the costs and lowering energy consumption. In addition, the manufacturer can select any type of bottling container after HPP processing, with no material, design, or size limitations, providing a more sustainable solution.

Extension of product shelf life is achieved through inactivation of spoilage microorganisms and enzymes, while food safety is ensured by reduction of foodborne pathogens like *E. coli* and *L. monocytogenes*. Recontamination events taking place throughout the food supply chain can be prevented, since the product is treated in its final packaging. HPP may function either as the primary lethality step or as a post-lethality treatment (PLT). According to guidance from the USDA Food Safety and Inspection Service (FSIS), its use is intended to control or eliminate postprocessing contamination by *L. monocytogenes* in Ready-to-Eat (RTE) meat products (FSIS, 2014). The latter has been estimated to significantly reduce the risk of listeriosis cases/year compared with scenarios in which no postlethality intervention is applied. When implemented, PLTs need to be formally integrated into the HACCP plans of the food business operators, and their effectiveness must be appropriately validated. Currently, no process criteria for HPP-based PLTs are defined within EU regulation, and assessments are largely based on scientific opinions and recommendations issued by food safety organisations (EFSA, 2022).

Commercial applications of HPP are most frequently found in fruit and vegetable juices and RTE meat products, each accounting for approximately 30% of the current market (Aganovic et al., 2021). Despite its broad applicability, HPP presents several limitations, including reduced effectiveness in low moisture foods (<40%) and aerated products, as well as potential modifications to food structure and texture. In addition, spore-forming microorganisms are not efficiently inactivated by HPP, and their control generally requires the application of very high pressures (>600 MPa) in combination with high temperatures (>100°C). Nevertheless, the technology can be applied to a wide range of food products and has long been implemented by the industry not only for food production with superior quality but also for meat extraction from molluscs and product texturization. Currently, its potential for the production of plant-based foods is being investigated as a more sustainable alternative to conventional pasteurization.

The expanding plant-based food sector has created new opportunities for HPP as a promising approach to meet consumer demands while aligning with sustainability goals. Traditional plant-based foods such as fruit juices have long been a focus for HPP applications, while recent innovations in plant-based dairy (e.g., cheese, milk) and meat replacement products (e.g., burger patties, sliced deli meats) highlight the expanding potential of this technology.

11.2.1.1 Mechanisms of HPP-induced microbial inactivation

HPP operates according to fundamental physicochemical principles, including Le Chatelier's principle, isostatic pressure, and microscopic ordering (Aganovic et al., 2021; Georget et al., 2015). In particular, Le Chatelier's principle indicates that the application of pressure favors reactions and phase transitions associated with a decrease in system volume, while processes involving volume expansion are disfavored under high-pressure conditions. The isostatic principle refers to uniform pressure transmission regardless of sample size or geometry, while microscopic ordering reflects the increased molecular organization under pressure at constant temperature. HPP-induced phenomena include electrostriction, where increased pressure reduces system volume by ordering water molecules, causing pH reduction and ionic dissociation. Additionally, pressure influences phase transitions, such as raising lipid melting points (10°C/100 MPa) and lowering the melting temperature of water, driven by the respective volume differences between liquid and solid states of these substances. Electrostatic and hydrophobic interactions are pressure-sensitive, whereas covalent and hydrogen bonds remain largely unaffected.

HPP induces the disruption of cellular structures and functions. These effects have significant implications for biological functions, impacting key cellular components like membranes, structural and functional proteins, enzymatic systems, and ribosomes. Notably, the membrane of vegetative cells undergoes morphological changes and serves as a primary target for microbial inactivation. The level of damage depends on treatment intensity and can range from disruption of membrane permeability to formation of buds, wrinkles, and swelling, to eventual loss of integrity, leakage of intracellular components, and cell death (Ferreira et al., 2016; Huang et al., 2014). Diameter reduction, filamentation, cell elongation, and loss of motility (10 MPa) are usually observed upon HPP (Ritz et al., 2002) along with protein aggregation in the cytosol. Proteins undergo conformational changes, with quaternary and tertiary structures being HPP sensitive, while secondary structures are only affected by higher pressures (>700 MPa). Primary structures, on the other hand, remain unaffected. HPP may lead to reversible but also irreversible changes in protein structure, and as a result, physiological functions are also compromised. Membrane-bound proteins act as transporters to facilitate nutrient absorption or efflux of metabolites to the extracellular environment for maintenance of homeostasis. Thus, membrane damage and protein denaturation (e.g., F0F1-ATPases) by HPP result in disruption of metabolic pathways such as adenosine triphosphate (ATP) synthesis, intracellular accumulation of waste products, and pH imbalance. The activity and specificity of some enzymes, especially those involved in DNA replication, transcription, and protein translation is affected in pressures of 50–100 MPa, leading to growth inhibition (Gayán et al., 2017). Sublethal injury caused by HPP has also been associated with the generation of reactive oxygen species (ROS), contributing to increased microbial sensitivity (Aertsen et al., 2005).

While moderate pressures cause sublethal injuries, higher pressures (>200 MPa) lead to irreversible damage. However, some microorganisms may adapt to stress and induce repair mechanisms, presenting a potential hazard in food processing. Pagán and Mackey (2000) reported that stationary phase *E. coli* cells recovered upon HPP due to membrane repair and resealing. Some other strategies include the upregulation of Heat Shock proteins (HSP), mutations in the genes encoding for repressors (e.g., CtsR in *L. monocytogenes*) of such proteins, and upregulation of general stress response regulators like SigB in *L. monocytogenes* (Karatzas et al., 2005) and RpoS in *E. coli*, orchestrating the expression of multiple genes to increase survival and recovery.

As previously mentioned, increasing pressure causes progressive cell damage with pleiotropic effects, ultimately leading to cell death. However, besides pressure, other factors may also influence microbial inactivation by implication of different mechanisms. For instance, the incubation temperature during culturing of the microorganism and before the HPP treatment affects the subsequent inactivation (Bull et al., 2005). This is attributed to membrane fluidity alterations that render the microorganism more HPP resistant. Pagán and Mackey (2000) showed that prior heat shock increased the HPP resistance of *E. coli*, as also evidenced by Hayman et al. (2008) for *L. monocytogenes* due to de novo protein synthesis. Interestingly, prior exposure to low temperatures can confer cellular protection during subsequent HPP treatment, which has been attributed to an increased proportion of polyunsaturated fatty acids in the membrane (Rendueles et al., 2011). However, HPP resistance decreases when cells are subjected to low processing temperatures again due to structural membrane modifications such as crystallization of phospholipids. Moreover, pH reduction significantly enhances HPP inactivation (Koseki and Yamamoto, 2006). Cells already suspended in low pH might be further affected by pressure, but this also depends on the type of acid since higher dissociation of organic acids during HPP leads to reduced antimicrobial effect (Hayert et al., 1999). Tsagkaropoulou and Karatzas (2024) found that the growth of *L. monocytogenes* in two substrates (BHI, TSB-D) led to different inactivation profiles and changes in strain variability, potentially due to the different pH of the cultures before HPP treatment. Other factors in the suspending medium can also influence piezo tolerance, such as water activity, fat content, ionic solutes (e.g., NaCl; Linton, McClements and Patterson, 1999), and sugars (e.g., trehalose) due to increased uptake and/or synthesis of protective disaccharides and compatible solutes (betaine, L-carnitine). The presence of antimicrobials such as nisin or chitosan has a synergistic effect with HPP. Cell membrane is the common target, with nisin and chitosan enhancing this effect by inducing pore formation and membrane destabilization (Giannoulis and Karatzas, 2024).

11.2.1.2 Modeling approaches for HPP-induced microbial inactivation

Traditional HPP-treated plant-based foods, such as juices, beverages, and products thereof, hold more than 30% of the market share and have been available in supermarkets for years due to their superior quality (Aganovic et al., 2021). HPP has been widely implemented in juice manufacturing, with nearly 24% of installed HPP units located in juice production facilities and annual production volumes reaching approximately 540 billion liters (González-Angulo et al., 2021).

The optimization of a processing treatment relies on a detailed understanding of the inactivation kinetics of the target pathogen in the given food matrix that can be achieved through the development of predictive models (Zwietering et al., 2021). Mathematical models detailing HPP inactivation of pathogens across various plant-based beverages are summarized in Table 11.1. Multiple studies have reported that microbial inactivation follows first-order kinetics in apple (Ramaswamy et al., 2003), peach (Dogan and Erkmen, 2004), orange (Dogan and Erkmen, 2004), carrot (Van Opstal et al., 2005), mango (Bermúdez-Aguirre et al., 2011; Hiremath and Ramaswamy, 2012), and tomato (Xu et al., 2009) juices. Linear models express the logarithmic reduction of microorganisms as a function of treatment time (Eq. 11.1) with the assumption that all cells have the same sensitivity to static Pressure/Temperature conditions and decrease proportionally with time (Possas et al., 2017).

$$\log\left(N/N_0\right) = -kt = -t/D \qquad (11.1)$$

where N represents the surviving population following treatment; N_o corresponds to the initial cell concentration; t denotes the processing time; k is the first-order inactivation rate constant; and D refers to the time required to achieve a one-log reduction.

Van Opstal et al. (2005) recorded a linear *E. coli* MG1655 inactivation in carrot juice (pH 6.5 ± 0.3) during HPP at different pressure-temperature combinations (150–600 MPa and 5–45°C) under isobaric and isothermal conditions. The D-values ranged from 46.7 (±3.0) to 2.0 (±0.4) minutes when the temperature increased from 5 to 45°C, demonstrating the enhanced piezotolerance of *E. coli* at lower temperatures. The authors reported that a five-log reduction was achieved at 600 MPa applied for 12.5 min at 20°C, although equivalent results with lower pressures required higher temperatures. Interestingly, Zhu et al. (2017) studied the inactivation of *E. coli* ATCC 25922 in chilled carrot juice (4°C) at 300, 350, and 400 MPa at the same processing temperature (20°C) and found considerably lower D values. In the same study, HPP treatment of frozen carrot juice (−20°C) resulted in further *E. coli* reduction due to crystal formation and phase transitions in cells.

First-order inactivation of *E. coli* ATCC 29055 has also been observed in other juices, with the microorganism demonstrating increased sensitivity. No survivors were detected after HPP treatment at 400 MPa in apple juice (pH 3.5; Ramaswamy et al., 2003), 414 MPa for 1 min in fresh mango nectar, or 375 MPa for 4 min in sterile mango nectar (pH 3.56; Bermúdez-Aguirre et al., 2011). Hiremath and Ramaswamy (2012) observed higher D values (e.g., $D_{400} = 2.88$ min) for *E. coli* O157:H7 (ATCC 43894) in mango juice, although this could be due to the higher pH (4.5) and/or different strain used. This strain was found to be more HPP resistant than *L. monocytogenes*. In this case, a 5-D reduction at 400 MPa could be achieved in 4 min ($D_{400} = 0.8$ min) while similar inactivation under these HPP conditions required approximately 2.65 (pH~3.46; Xu et al., 2009) or 9 min (pH~3.55; Dogan and Erkmen, 2004) in orange juice, 16.95 min in peach juice (pH 5.21; Dogan and Erkmen, 2004) and 3.05 min in tomato juice (pH 4.11; Xu et al., 2009). In contrast, the application of HPP at 300 MPa for 2 min can result in more than five-log reductions of *Salmonella* Enteritidis in orange and tomato juices (pH~3.46; Xu et al., 2009).

Table 11.1 Summary of modeling approaches describing the microbial inactivation in beverages subjected to high-pressure processing.

Microorganism	Food matrix/ medium	Modeling approach	Independent variables (values or ranges)[a]	References
Escherichia coli	Wheatgrass juice	Log linear with tail	P (400–600 MPa), t (1–3 min)	Ali et al. (2020)
	Orange juice	Weibull	P (400 MPa), t (1–7 min)	Torres et al. (2016)
	Apple juice	First order kinetics	P (150–400 MPa), t (0–80 min)	Ramaswamy et al. (2003)
	Carrot juice	First order kinetics	P (150–600 MPa), t (0–60 min), T (5–45°C)	van Opstal et al. (2005)
	Mango nectar	First order kinetics	P (275–414 MPa), t (1, 2 and 4 min)	Bermúdez-Aguirre et al. (2011)
	Apple juice	Weibull and Fermi's equations	P (300–600 MPa), t (0.5–7 min)	Moody et al. (2014)
	Carrot juice	First order kinetics	P (200–400 MPa), t (0–10 min)	Zhu et al. (2017)
Escherichia coli O157:H7	Carrot juice	First order kinetics and Secondary linear models	P (200–400 MPa), t (5–40 min)	Pilavtepe-Çelik et al. (2009)
	Mango juice	First order kinetics	P (250–550 MPa), t (0–60 min), Temperature (20–25°C)	Hiremath and Ramaswamy (2012)
	Orange juice	Weibull	P (100–400 MPa), t (1–20 min)	Torres et al. (2016)
	Apple juice	Log–linear and Weibull	P (400, 450, 475 MPa), t (1–9 min)	Usaga et al. (2021)
Listeria innocua	Wheatgrass juice	Log linear with tail	P (400–600 MPa), t (1–3 min)	Ali et al. (2020)
	Pitaya juice	Polynomial	P (400–600 MPa), t (7.3–15 min)	Quiroz-González et al. (2018)

Listeria monocytogenes	Apple juice	Log–linear and Weibull	P (400, 450, 475 MPa), t (1–7 min)	Usaga et al. (2021)
	"Wild berries" smoothies	Polynomial	P (100–300 MPa), T (-5–45°C)	Scolari et al. (2015)
	Orange juice	First order kinetics	P (300–500 MPa), t (0–10 min)	Xu et al. (2009)
	Mango juice	First order kinetics	P (250–400 MPa), t (0–60 min), T (20–25°C)	Hiremath and Ramaswamy (2012)
	Tomato juice	First order kinetics	P (300–500 MPa), t (0–10 min)	Xu et al. (2009)
	Purple smoothie	Geeraerd	P (300–450 MPa), t (4–11 min)	González-Tejedor et al. (2023a)
	Peach juice	First order kinetics	P (300–700 MPa), t (0–24 min)	Dogan and Erkmen (2004)
	Orange juice	First order kinetics	P (300–700 MPa), t (0–24 min)	Dogan and Erkmen (2004)
Salmonella spp.	Tomato juice	First order kinetics	P (300–500 MPa), t (0–10 min)	Xu et al. (2009)
	Apple juice	Log–linear and Weibull	P (400, 450, 475 MPa), t (1–7 min)	Usaga et al. (2021)
	Orange Juice	Generalized linear model	P (35–380 MPa), t (1–10 min)	Allison et al. (2018)
	Orange juice	First order kinetics	P (300–500 MPa), t (0–10 min)	Xu et al. (2009)
	Wheatgrass juice	Log linear with tail	P (400–600 MPa), t (1–3 min)	Ali et al. (2020)
Staphylococcus aureus	Carrot juice	Weibull and Secondary linear	P (200–400 MPa), t (5–40 min)	Pilavtepe-Çelik et al. (2009)

[a]P = Pressure; t = Time; T = Temperature.

Despite its simplicity and widespread adoption by the food industry and regulatory bodies, the first-order kinetic model often fails to accurately describe experimental microbial inactivation curves. Several studies have demonstrated that HPP-induced inactivation frequently exhibits nonlinear patterns, including shoulders (downward concavity), tails (upward concavity), sigmoid shapes, and other deviations from linearity (Table 11.1). Some of the most frequently used nonlinear models to fit microbial inactivation curves in HPP-treated juices include the Weibull, Geeraerd, and log-logistic models, as well as polynomial approaches derived from Response Surface Methodology (RSM). Among these, the Weibull model (Eq. 11.2) is widely used because it accounts for heterogeneity in microbial resistance within a population, assuming the presence of subpopulations with different sensitivities. The model describes inactivation behavior based on the cumulative distribution of survival times, thereby capturing biological variability more effectively than linear approaches (Van Boekel, 2002).

$$\log\left(N/N_0\right) = -bt^n = -\left(t/\delta\right)^n \tag{11.2}$$

where N denotes the number of surviving cells after treatment; N_o represents the initial microbial population before treatments; b is the Weibull rate parameter associated with the speed of microbial inactivation and mathematically related to δ and n; δ is the scale parameter corresponding to the time required to achieve the first decimal reduction; n is the shape parameter describing the curvature of the curve ($n<1$indicates upward concavity; $n = 1$ linearity; and $n>1$ downward concavity); and t refers to the treatment duration.

The Weibull model (Eq. 11.2) was applied to describe tailed survival curves of *E. coli* O157:H7 and K12 in orange juice (pH 4.0) treated at 400 MPa and 25°C. Under these conditions, *E. coli* O157:H7 was more pressure-sensitive, reaching a 6.5-log reduction after 3 min of treatment (Torres et al., 2016). This model described well the *E. coli* ATCC 11775 inactivation in apple juice (pH 3.52), where a shoulder (b=0.4399 min^{-1}, n=1.3885) was found only at 300 MPa and tailing only at 400, 500, and 600 MPa (Moody et al., 2014). Usaga et al. (2021) used an *E. coli* multi-strain cocktail in the same type of juice (pH 4.51) with a considerably higher t_{5D} (9 ± 2 min) at 400 MPa, while that corresponded to 7.4 ±0.2 min and 7.7 ±0.7 min for *Salmonella* and *L. monocytogenes* strain cocktails, respectively.

Besides the primary nonlinear models, RSM (e.g., Central Composite Design) has been used to generate polynomial functions for fitting microbial inactivation data in juices and studying the interaction between influencing factors (Possas et al., 2017). However, these empirical models may include terms without biological meaning, and interpretation outside the given dataset for parameter estimation should be avoided. Scolari et al. (2015) found a dome-shaped *L. monocytogenes* inactivation curve in "Wild berries" smoothies at 300 MPa. Inactivation was reduced at 15°C, while higher reductions were observed at 45 or -5°C, reaching approximately 6.2 log and 5.6 log reduction, respectively. A quadratic polynomial equation was developed for *L.*

monocytogenes CECT 4032 inactivation data in mung bean sprouts juice (pH 5.68) and showed a maximum reduction of 1.8 logs at 400 MPa and 40°C (Muñoz et al., 2006).

Considering all the information outlined above, the inactivation pattern of pathogens in juices due to HPP, and consequently, the selection of an appropriate mathematical model to describe it, may be affected by multiple factors. They include the microbial strain (single strains or cocktails), prior cell history (e.g., growth phase and acid adaptation), HPP processing conditions (pressure, temperature, come-up time (CUT), and CDT), the laboratory medium used, food matrix composition (presence or absence of protective components), and the initial contamination level (Allison et al., 2018; Bermúdez-Aguirre et al., 2011; Dogan and Erkmen, 2004; Farakos and Zwietering, 2011; Ramaswamy et al., 2003; Tsagkaropoulou and Karatzas, 2024). In general, HPP efficacy is increased in highly acidic juices due to synergistic effects with low pH. Nevertheless, treatment at 600 MPa for up to 6 min has been shown to be sufficient to achieve 5-6 log reductions in juices and to assure compliance with target PC (Process Criteria) as recommended by food safety and regulatory authorities (EFSA, 2022).

11.2.1.3 Factors affecting HPP-induced microbial inactivation

Microbial characteristics

Resistance to HPP varies across different microorganisms, including differences between genera, species, and strains. Overall, eukaryotic microorganisms such as yeasts and molds exhibit higher sensitivity to HPP than prokaryotes (bacteria) due to higher structural complexity (Rendueles et al., 2011), although ascospores of some molds can be very resistant. Additionally, Gram-negative bacteria generally exhibit higher pressure sensitivity than Gram-positive bacteria, which is often attributed to differences in membrane and cell envelope structure (Smelt, 1998), and bacterial rods are more susceptible than cells of smaller size and coccoid shape. Inactivation of viruses is variable and mediated either by envelope damage and prevention of cell binding or by particle dissociation (Hogan et al., 2005). Moreover, there is variability in bacterial spore inactivation with *Clostridium* spp. producing more HPP-resistant spores than *Bacilllus* spp. (Smelt, 1998). However, in all cases, sufficient inactivation generally requires combining HPP with other treatments (e.g., heat). With regards to the inactivation of vegetative cells, resistance is typically highest for *S. aureus*, followed by *E. coli*, *L. monocytogenes*, and *Salmonella* (EFSA, 2022). Notably, exponential phase cells are more susceptible than stationary phase cells due to continuous cell division, lower stress response capacity, and incomplete formation of the cell membrane (Patterson, 2005). Stationary phase cells have higher levels of proteins and sigB (*L. monocytogenes*) or RpoS (*E. coli*) activity involved in general stress response and protection. Strain variability has a significant impact on HPP resistance among the foodborne pathogens most relevant to pressure-processed foods (Tsagkaropoulou and Karatzas, 2024), and it is affected by variations in pressure-time combinations as well as by the physiological state and prior history of the cells (e.g., origin, source of isolation), the genetic makeup, and the food matrix/medium, among other factors.

Process parameters

Microbial inactivation is enhanced with higher pressures or longer holding times, although the relative importance of the latter decreases for pressures above a certain level (500 MPa), probably due to the already high degree of inactivation. Equivalent levels of inactivation of a target pathogen within a given food matrix can be achieved using different pressure-time combinations, with commercial HPP applications typically operating between 400 and 600 MPa for treatment times up to 6 min. The CUT and CDT rates also influence inactivation kinetics. Some studies suggest that longer CUT may induce adaptation of the bacterial population to increasing levels of pressure, resulting in less efficient inactivation described by non-linear biphasic models and emergence of shoulders. Fast decompression may exert mechanical stress and membrane disruption, resulting in higher inactivation. However, the effects of CUT and CDT are complex and depend on the microorganism, equipment, and experimental setup (Ferreira et al., 2016).

Environmental conditions and food intrinsic factors

Temperature plays an important role in HPP efficacy. As previously mentioned, elevated incubation temperatures (43°C) during growth of *L. monocytogenes* can increase piezo tolerance. Processing temperatures deviating from ambient, higher or lower, have been shown to enhance inactivation, either due to higher protein damage and/or alterations in the fluidity and structure of the membrane (Rendueles et al., 2011). Thermal gradients are influenced by the initial temperature of the food, the thermophysical properties of the pressure-transmitting fluid, the insulating characteristics of the packaging material, as well as the design, insulation efficiency, and compression-unloading time of the HPP system (Balasubramaniam and Mart, 2015). During compression, adiabatic heating increases with higher target pressures and oily pressure-transmitting fluids or fatty matrices. Processing temperatures above 45°C can sensitize the cells through thermal inactivation effects but may also compromise food quality. To mitigate these effects, industrial HPP operations typically reduce the temperature of both the product and the pressure-transmitting fluid prior to processing. Microbial resistance to high-pressure processing is further modulated by factors including pH, water activity, and the presence of antimicrobial agents. Microbial inactivation is enhanced when HPP is combined with low pH due to inhibition of cell recovery (Ritz et al., 2006), as seen with fruit compared to reductions achieved in laboratory media (BHI) or milk. In addition, low water activity generally exerts a protective effect against HPP, but the extent of this effect depends on a lot of factors such as the process of water activity reduction (e.g., drying) and type of solutes (e.g., sucrose), and fats present in the matrix/medium (Georget et al., 2015). Moreover, foods are usually vacuum-packed in flexible pouches or bottles and subjected to HPP to avoid headspace air and prevent packaging collapse. However, studies have demonstrated that modified atmosphere packaging (MAP) can be combined with HPP to enhance microbial inactivation, an effect attributed to CO_2 dissociation and enhanced oxidative damage (Amanatidou et al., 2000). Finally, a combination of HPP with antimicrobials such as nisin (Black et al., 2008), chitosan (Giannoulis and Karatzas, 2024), and lysozyme (Nakimbugwe et al., 2006) as a hurdle approach has been found to induce inactivation of foodborne pathogens in several food products due to synergy, thereby offering an innovative solution to food preservation.

11.2.2 Pulsed electric fields

Pulsed electric fields (PEF) consist of applying short, high-voltage electric pulses, usually in the microsecond range, to a biological material placed between electrodes. PEF results in changes of the structure of cell membranes, with effects ranging from increased permeability to irreversible pore formation and cell death. These effects are often referred to as electropermeabilization or electroporation. Most research addressing PEF application in the food industry has concentrated on textural modification, improvement of mass transfer, preservation, and, most recently, functional modification. Application of PEF in plant-based foods could offer some quality advantages such as retention of fresh-like sensory attributes and nutritional value (Buckow et al., 2013; Timmermans et al., 2022; Toepfl, 2011; Tongdonyod et al., 2023; Walkling-Ribeiro et al., 2010).

Using PEF as a preservation technique often requires high electric field strengths (>10 kV/cm), defined as the ratio between the applied voltage and the distance between the electrodes. This implies the need for narrow equipment dimensions, making the process feasible only for liquid and pumpable products. In relation to plant-based matrices, numerous PEF decontamination studies have been conducted on fruit and vegetable juices, smoothies, and soups (Akin and Evrendilek, 2009; Buckow et al., 2013; Gurtler et al., 2011; Kayalvizhi et al., 2016; Mosqueda-Melgar et al., 2008a; Timmermans et al., 2014, 2016; Vega-Mercado et al., 1996; Walkling-Ribeiro et al., 2008). Recently, PEF has been investigated as a nonthermal pasteurization strategy of plant-based dairy analogs derived from legumes, nuts, and cereals such as soybean, almond, oat, and pea (Horlacher et al., 2024; Li and Zhang, 2006; Li et al., 2013; Manzoor et al., 2020; Thamsuaidee et al., 2024).

11.2.2.1 Mechanisms of PEF-induced microbial inactivation

The inactivation mechanism of PEF is believed to stem from its direct physical effects on the cell membrane. The first theory is based on the electromechanical effect, that is, the external electric field causes rearrangement of intracellular charged species, membrane compression, and eventual dielectric breakdown when a critical threshold transmembrane potential is exceeded (Crowley, 1973; Dimitrov, 1984). Another theory is based on the formation of metastable hydrophilic pores, whereby the application of an external electric field reduces the energy barrier for pore formation, leading to an increase in pore size and number with repeated pulses (Glaser et al., 1988; Powell and Weaver, 1986; Saulis and Wouters, 2007). Cell death is thought to occur as a consequence of increased permeability and structural damage leading to leakage of intracellular substances and/or osmotic lysis (Saulis and Wouters, 2007). Transmission electron microscopy (TEM) and scanning electron microscopy (SEM) imaging of microorganisms following PEF treatment revealed morphological changes, including increased roughness of the cell wall, cytoplasmic clumping, elongation of cell length, leakage, and rupture of the cell envelope (Calderón-Miranda et al., 1999; Harrison et al., 1997; Jayaram et al., 1992; Pothakamury et al., 1997).

After the exposure to PEF, membrane pores may reseal or remain permanently open, resulting in reversible or irreversible electroporation, respectively. Irreversible electroporation ultimately leads to cell death, while some reversibly electroporated cells are merely sublethally injured and could later recover into a state of viability when conditions are favorable (Schottroff et al., 2018). The occurrence of sublethal injury has not been consistently reported in all studies (Aronsson et al., 2004; Dutreux et al., 2000; García et al., 2005a; Russell et al., 2000; Simpson et al., 1999; Ulmer et al., 2002; Wuytack et al., 2003). Recovery may depend on the type of microorganism and the characteristics of the treatment medium. García et al. (2005a) detected sublethal injury in Gram-positive species following PEF treatment in a buffer at pH 7.0, but not at pH 4.0. Vice versa, the authors detected sublethal injury in Gram-negative species after treatment in a pH 4.0 buffer, but not at pH 7.0. In another study, *E. coli* O157:H7 in apple juice was initially resistant to PEF, showing less than a one-log reduction, but after 48 hours of refrigerated storage, up to five-log reductions could be observed in the same juice (García et al., 2005). These findings indicated that PEF inflicted sublethal injury and sensitized the cells to storage in apple juice. Similarly, Li et al. (2021) observed sublethally injured *Saccharomyces cerevisiae* in cantaloupe juice (pH 6.4) after PEF treatment, but the majority died off after 3 days of cold storage. In contrast, Horlacher et al. (2024) reported no recovery of either *E. coli* NZRM 916 or *Listeria innocua* NZRM 3024 following PEF treatment of an oat beverage enriched with pea protein (pH 7.0) after 48 hours of cold storage.

Evidence also suggests that PEF may induce indirect lethal effects through the formation of antimicrobial compounds in the treatment medium as a result of electrochemical reactions occurring at the electrodes, which are inherently complex processes. The extent and nature of these reactions depend on the composition of the treatment medium, the design of the equipment, and the PEF process parameters (Pataro and Ferrari, 2020). Reyns et al. (2004) reported that PEF treatment of media containing chloride ions resulted in the formation of chemical products with bactericidal activity, including chlorine, hypochlorite, and possibly other reactive species. The authors also observed that the concentrations of these compounds increased with higher electric field strengths and an increased number of pulses.

Although PEF is generally regarded as a nonthermal technology, the application of high-intensity PEF for the purpose of microbial inactivation may, in practice, lead to lethal effects arising from a combination of nonthermal and thermal effects (Schottroff et al., 2020a; Timmermans et al., 2019). The increase in sample temperature commonly observed after PEF is mainly attributed to Ohmic heating, that is, whereby part of the electrical energy is converted into heat during processing (Heinz et al., 2001). The lethal impact of temperature increase becomes particularly relevant at higher specific energy input levels (Buckow et al., 2010; Thamsuaidee et al., 2024; Timmermans et al., 2014, 2019). While an increase in temperature may enhance the decontamination efficiency of PEF, it may contribute to adverse effects on heat-sensitive food constituents (Schottroff et al., 2019, 2020a). Strategies such as submersion of the electrodes in a cooling medium, or modification of the treatment chamber and/or product flow characteristics may help to limit the extent of temperature increase and improve the uniformity of the treatment while preserving the product quality (Jaeger et al., 2009a; Meneses et al., 2011; Schottroff et al., 2020b).

11.2.2.2 Modeling approaches for PEF-induced microbial inactivation

The kinetic modeling approaches for PEF processing have been extensively reviewed by Huang et al. (2012), Delso et al. (2022), and most recently by Sastry (2023). Apart from microbial inactivation, the models were also used to describe electroporation phenomena at a molecular level (Sastry, 2023), as well as the change of enzyme activities, and degradation of health-related compounds associated with PEF processing (Huang et al., 2012). In plant-based foods, the approaches used to model microbial inactivation by PEF are, in general, similar to those applied in buffers, model food systems, and animal-based products. The mathematical models used to describe PEF inactivation of spoilage and pathogenic microorganisms in various plant-based beverages have been summarized in Table 11.2. The most used approach is based on the RSM, resulting in polynomial models with the aim of maximizing the reduction of the target microorganism(s) through optimization of multiple product- and process-related independent variables (Evrendilek, 2017; Marsellés-Fontanet et al., 2009; Mosqueda-Melgar et al., 2007, 2008a; Tongdonyod et al., 2023). For instance, Mosqueda-Melgar et al. (2007) applied RSM to describe the reduction of *S.* Enteritidis, *E. coli*, and *L. monocytogenes* populations using a quadratic function of pulse frequency and treatment time during PEF processing of melon and watermelon juices at 35 kV/cm, with 4 μs square-wave pulses in bipolar mode and a maximum outlet temperature of $39.1 \pm 0.1°C$. Recently, Tongdonyod et al. (2023) also used the RSM to describe PEF-induced inactivation of *E. coli* K12 and *L. innocua* in tender coconut water using a second-order polynomial function based on inlet temperature, electric field strength, and specific energy.

The Hülsheger, Peleg, Weibull, and first-order kinetic models are also frequently used to describe microbial survival curves under PEF treatment. In particular, the Hülsheger model expresses microbial inactivation as a function of electric field strength and/or treatment time (Hülsheger et al., 1981, 1983). It implies the existence of a critical treatment time and a critical electric field strength, both of which are related to microbial characteristics. For PEF inactivation of *E. coli* in carrot juice, Zhong et al. (2005) and Singh et al. (2017) found that the Hülsheger model showed the best performance compared to other predictive models. Also notable is the Weibull model (Eq. 11.2), which has often been cited as amongst the most accurate for predicting microbial survival subjected to PEF treatment in all types of matrices as a function of either treatment time or specific energy (Delso et al., 2022; Monfort et al., 2010; Rodrigo et al., 2001, 2003; Timmermans et al., 2014). As mentioned above, it is characterized by a scale parameter, which is considered a measure of microbial resistance related to electric field strength, and a shape parameter that describes the upward or downward concavity of the curve (Mafart et al., 2001). Timmermans et al. (2014)

Table 11.2 Summary of modeling studies on pulsed electric fields-induced microbial inactivation in plant-based liquid products.

Microorganism	Food matrix/medium	Modeling approach	Independent variables (ranges or values)[a]	References
Escherichia coli	Melon juice	Polynomial	EFS (35 kV/cm), t (500–2000 μs), SBPW (4 μs), F (100–250 Hz)	Mosqueda-Melgar et al. (2007)
	Watermelon juice	Polynomial	EFS (35 kV/cm), t (500–2000 μs), SBPW (4 μs), F (100–250 Hz)	Mosqueda-Melgar et al. (2007)
	Apple juice	Weibull	EFS (20 kV/cm), SE (0–100 kJ/kg), SMPW2 μs, F (120–964 Hz)	Timmermans et al. (2014)
	Orange juice	Weibull	EFS (20 kV/cm), SE (0–100 kJ/kg), SMPW2 μs, F (120–964 Hz)	Timmermans et al. (2014)
	Watermelon juice	Weibull	EFS (20 kV/cm), SE (0–100 kJ/kg), SMPW2 μs, F (120–964 Hz)	Timmermans et al. (2014)
	Orange juice	Gauss-Eyring	EFS (0.9–20 kV/cm), SE (0–225 kJ/kg), Monopolar (2 μs width) and bipolar (15–1000 μs width) square pulses, F (0–1700 Hz)	Timmermans et al. (2019)
	Watermelon juice	Gauss-Eyring	EFS (0.9–2.7 kV/cm), Bipolar square pulse 1000 μs width, F (0–238 Hz)	Timmermans et al. (2019)
	Coconut water	Gauss-Eyring	EFS (0.9–2.7 kV/cm), Bipolar square pulse 1000 μs width, F (0–100 Hz)	Timmermans et al. (2019)
	Orange-carrot juice	First order kinetic, Hülsheger, Weibull	EFS (25–40 kV/cm), t (40–340 μs), SBPW2.5 μs	Rodrigo et al. (2003)
	Orange juice-milk beverage	First order kinetic, Hülsheger, Weibull	EFS (15–40 kV/cm), t (0–700 μs), SBPW2.5 μs	Rivas et al. (2006)
	Carrot juice	Hülsheger, Peleg	EFS (5–25 kV/cm), t (207–1449 μs), exponential decay pulse width 1.5 μs, F 10 Hz	Zhong et al. (2005)

Escherichia coli	Coconut water	Polynomial	EFS (20–30 kV/cm), SE (80–120 kJ/L), SMPW (4 μs), F (130–350 Hz)	Tongdonyod et al. (2023)
	Carrot juice	First order kinetic, Hülsheger, Peleg, Weibull	EFS (9–21 kV/cm), t (100–500 μs), SMPW (1 μs), F (1 Hz)	Singh et al. (2017)
Escherichia coli O157:H7	Apple juice ± 50 ppm ethyl lauroyl arginate (LAE)	Polynomial	EFS (20–30 kV/cm), t (5–125 μs), F (1–52 Hz)	Saldaña et al. (2011)
	Apple juice ± citric acid and/or cinnamon bark oil	Polynomial	EFS (35 kV/cm), t (500–2000 μs), SBPW (4 μs), F (100–250 Hz)	Mosqueda-Melgar et al. (2008)
	Pear juice ± citric acid and/or cinnamon bark oil	Polynomial	EFS (35 kV/cm), t (500–2000 μs), SBPW (4 μs), F (100–250 Hz)	Mosqueda-Melgar et al. (2008)
	Orange juice ± citric acid and/or cinnamon bark oil	Polynomial	EFS (35 kV/cm), t (500–2000 μs), SBPW4 μs, F (100–250 Hz)	Mosqueda-Melgar et al. (2008)
	Strawberry juice ± citric acid and/or cinnamon bark oil	Polynomial	EFS (35 kV/cm), t (500–2000 μs), SBPW (4 μs), F (100–250 Hz)	Mosqueda-Melgar et al. (2008)
	Pomegranate juice	First order kinetic, Polynomial	EFS (0–30 kV/cm), SBPW (3 μs), F (500 Hz)	Evrendilek (2017)
	Apple cider (unfiltered apple juice)	Peleg-log logistic	EFS (20–35 kV/cm), SE (0–407.6 J/mL), SBPW (2 μs), F (500–1500 Hz)	Mendes-Oliveira et al. (2020)
Listeria innocua	Coconut water	Polynomial	EFS (20–30 kV/cm), SE (80–120 kJ/L), SMPW (4 μs), F (130–350 Hz)	Tongdonyod et al. (2023)

Continued

Table 11.2 Continued

Microorganism	Food matrix/medium	Modeling approach	Independent variables (ranges or values)[a]	References
Listeria monocytogenes	Melon juice	Polynomial	EFS (35 kV/cm), t (500–2000 μs), SBPW (4 μs), F (100–250 Hz)	Mosqueda-Melgar et al. (2007)
	Watermelon juice	Polynomial	EFS (35 kV/cm), t (500–2000 μs), SBPW (4 μs), F (100–250 Hz)	Mosqueda-Melgar et al. (2007)
	Orange juice	Gauss-Eyring	EFS (0.9–20 kV/cm), SE (0–225 kJ/kg), SMPW (2 μs), SBPW (15–1000 μs), F (0–1700 Hz)	Timmermans et al. (2019)
	Watermelon juice	Gauss-Eyring	EFS (0.9–2.7 kV/cm), SBPW (1000 μs), F (0–238 Hz)	Timmermans et al. (2019)
	Coconut water	Gauss-Eyring	EFS (0.9–2.7 kV/cm), SBPW (1000 μs), F (0–100 Hz)	Timmermans et al. (2019)
	Apple juice	Weibull	EFS (20 kV/cm), SE (0–100 kJ/kg), SMPW (2 μs), F (120–964 Hz)	Timmermans et al. (2014)
	Orange juice	Weibull	EFS (20 kV/cm), SE (0–100 kJ/kg), SMPW (2 μs), F (120–964 Hz)	Timmermans et al. (2014)
	Watermelon juice	Weibull	EFS (20 kV/cm), SE (0–100 kJ/kg), SMPW (2 μs), F (120–964 Hz)	Timmermans et al. (2014)
Salmonella enteritidis	Apple juice ± citric acid and/or cinnamon bark oil	Polynomial	EFS (35 kV/cm), t (500–2000 μs), SBPW (4 μs), F (100–250 Hz)	Mosqueda-Melgar et al. (2008)
	Pear juice ± citric acid and/or cinnamon bark oil	Polynomial	EFS (35 kV/cm), t (500–2000 μs), SBPW (4 μs), F (100–250 Hz)	Mosqueda-Melgar et al. (2008)
	Orange juice ± citric acid and/or cinnamon bark oil	Polynomial	EFS (35 kV/cm), t (500–2000 μs), SBPW (4 μs), F (100–250 Hz)	Mosqueda-Melgar et al. (2008)

	Strawberry juice ± citric acid and/or cinnamon bark oil	Polynomial	EFS (35 kV/cm), t (500–2000 μs), SBPW (4 μs), F (100–250 Hz)	Mosqueda-Melgar et al. (2008)
	Melon juice	Polynomial	EFS (35 kV/cm), t (500–2000 μs), SBPW (4 μs), F (100–250 Hz)	Mosqueda-Melgar et al. (2007)
	Watermelon juice	Polynomial	EFS (35 kV/cm), t (500–2000 μs), SBPW (4 μs), F (100–250 Hz)	Mosqueda-Melgar et al. (2007)
Salmonella Panama	Apple juice	Weibull	EFS (20 kV/cm), SE (0–100 kJ/kg), SMPW (2 μs), F (120–964 Hz)	Timmermans et al. (2014)
	Orange juice	Weibull	EFS (20 kV/cm), SE (0–100 kJ/kg), SMPW (2 μs), F (120–964 Hz)	Timmermans et al. (2014)
	Watermelon juice	Weibull	EFS (20 kV/cm), SE (0–100 kJ/kg), SMPW (2 μs), F (120–964 Hz)	Timmermans et al. (2014)
	Watermelon juice, acidified	Weibull	EFS (20 kV/cm), SE (0–100 kJ/kg), SMPW (2 μs), F (120–964 Hz)	Timmermans et al. (2014)
Salmonella senftenberg	Orange juice	Gauss-Eyring	EFS (0.9–20 kV/cm), SE (0–225 kJ/kg), SMPW (2 μs), SBPW (15–1000 μs), F (0–1700 Hz)	Timmermans et al. (2019)
Salmonella typhimurium	Apple cider (unfiltered apple juice)	Peleg-log logistic	EFS (20–35 kV/cm), SE (0–407.6 J/mL), SBPW (2 μs), F (500–1500 Hz)	Mendes-Oliveira et al. (2020)
Staphylococcus aureus	Pomegranate juice	First-order kinetic, Polynomial model	EFS (0–30 kV/cm), SBPW (3 μs), F (500 Hz)	Evrendilek (2017)

[a]EFS = Electric field strength; t = Time; SE = Specific Energy; SMPW = Square Monopolar Pulse Width; SBPW = Square Bipolar Pulse Width; F = Frequency.

also found the Weibull model to be the most suitable for describing PEF inactivation kinetic data of various spoilage and pathogenic microorganisms in apple, orange, and watermelon juices as a function of specific energy.

11.2.2.3 Factors influencing PEF-induced microbial inactivation

The factors that affect PEF-induced microbial inactivation can be divided into three categories of process-, microbial-, and matrix-related.

A typical PEF process is defined by several parameters including the electric field strength (kV/cm), treatment time (μs to ms), pulse number, pulse characteristics (shape, width, polarity), pulse repetition rate or frequency (Hz), temperature (°C), and specific energy input (kJ/kg). The PEF equipment could also vary in its design and operational mode, with some being batch equipment with static treatment chambers, while others are continuous, equipped with a fluid handling system capable of regulating both the flow and temperature. Furthermore, there are several possible arrangements of the electrodes (e.g., co-linear, co-axial, and parallel) that make up the treatment chamber (Toepfl et al., 2005; Witt et al., 2021). In principle, all components could be optimized for a specific food matrix and target microorganism. Irrespective of the factors chosen for consideration, a few key pieces of information should always be reported as part of any PEF decontamination study to facilitate comparison and experimental reproducibility. These recommendations have been published by Raso et al. (2016).

From a technical perspective, the design and configuration of the treatment chamber are critical aspects of PEF processing. The electric field is generated between two electrodes, and the configuration of these electrodes significantly influences the homogeneity of the field (Toepfl et al., 2005). In colinear treatment chambers, commonly employed in continuous processing due to their ease of integration into pipelines, pronounced inhomogeneities in the electric field are often observed (Jaeger et al., 2009a; Witt et al., 2021). These spatial nonuniformities may compromise microbial inactivation efficiency, since certain zones within the treatment chamber, particularly the central region, can be exposed to electric field strengths below the required threshold, resulting in reduced inactivation levels (Jaeger et al., 2009a; Schottroff et al., 2020b). Furthermore, elevated treatment intensities can lead to localized temperature increases, potentially introducing thermal effects that superimpose on electroporation mechanisms (Jaeger et al., 2010; Thamsuaidee et al., 2024). In contrast, parallel or coaxial treatment chambers provide improved electric field homogeneity. However, their implementation in continuous processing systems poses greater design challenges (Toepfl et al., 2005; Witt et al., 2021). Accurate modeling or simulation of the electric field and energy distribution within the chamber is essential for understanding and optimizing the effects of PEF. This approach facilitates the identification of key parameters influencing process efficiency and helps mitigate undesirable variations in treatment outcomes.

Among all the process parameters, the electric field strength is arguably the most important as it dictates the extent of electroporation. Higher field strengths are generally preferred for maximum lethality, but they are also technologically limited by the capacity of the equipment (Heinz et al., 2001). Longer treatment times have, as well, been associated with higher microbial inactivation. However, as the treatment duration tends to be very short, the specific energy is suggested as a more meaningful parameter for comparing the different conditions (Heinz et al., 2001; Raso et al., 2016). Higher specific energy, likewise, leads to increased inactivation, but it also raises the process temperatures (Heinz et al., 2001). For this reason, the outlet temperature should always be monitored and included during the interpretation of the results due to possible thermal effects. Meanwhile, the starting temperature prior to PEF treatment could alter the critical electric field strength required for electroporation, due to its effect on the fluidity of the membrane (Coster and Zimmermann, 1975; Kandušer et al., 2008). Higher initial temperatures are known to significantly enhance the microbial inactivation in plant-based beverages while lowering the PEF treatment intensity needed to achieve the target reduction (Buckow et al., 2013; Horlacher et al., 2024; Saldaña et al., 2011; Timmermans et al., 2014).

Given its direct effect on the membrane, PEF is known to be effective at inactivating vegetative microorganisms, but it has no effect on microbial spores under the conditions used for the inactivation of vegetative cells (Grahl and Märkl, 1996; Pagán et al., 1998). Differences in microbial sensitivity to PEF are strongly influenced by physiological and structural factors. For example, Gram-positive bacteria generally exhibit higher resistance than Gram-negative species, largely due to differences in cell envelope architecture (Hülsheger et al., 1983). Microbial physiological state also plays an important role, as cells in the exponential growth phase are typically more susceptible to PEF treatment than those in the stationary phase (Álvarez et al., 2002; Wouters et al., 1999). In addition, cultivation temperature has been shown to affect PEF sensitivity in microorganisms such as *L. monocytogenes*, *S. aureus*, and *E. coli* to PEF, with growth temperature influencing membrane fatty acid composition and, consequently, membrane stability (Álvarez et al., 2002; Ohshima et al., 2002; Wang et al., 2016). Cell size represents another relevant factor, and multiple studies, including those conducted in plant-based food matrices, have reported that yeast cells are generally more readily inactivated by PEF than bacterial cells (Marsellés-Fontanet et al., 2009; Timmermans et al., 2014, 2019). Both the transmembrane potential and induced electric field strength are higher in larger cells upon the application of PEF, making them more susceptible to the treatment (Heinz et al., 2001; Huang et al., 2014). With regards to the cell shape, nonspherical cells experience different induced transmembrane potentials depending on their spatial orientation in the electric field, making their resistance more variable in comparison to spherical cells (Kotnik et al., 2010; Valič et al., 2003). Regardless, a recent study indicates that the difference stemming from spatial orientation could be mitigated by applying short pulses in the microsecond range (Dermol-Černe et al., 2020). Strain-dependent differences in resistance to PEF have been reported within the same microbial species, including *E. coli*, *L. monocytogenes*, *L. plantarum*, and *S. cerevisiae* across buffers, model systems, and oat-based beverages (Lado and Yousef, 2003; Lytras et al., 2024b; Somolinos et al., 2008; Thamsuaidee et al., 2024). Although the mechanisms underlying this intraspecific variability remain incompletely understood, they are commonly attributed to strain-specific responses to PEF-induced abiotic stress, involving changes in gene expression and protein synthesis (Chueca et al., 2015; Liu et al., 2019; Lytras et al., 2024a; Rivas et al., 2013).

During microbial challenge tests, the matrix itself may influence inoculum viability due to the presence of certain substances (either naturally occurring or added during formulation). This potential effect should always be evaluated before initiating PEF decontamination experiments. Regarding matrix composition, pH can significantly influence species variability, as well as overall decontamination efficiency. According to García et al. (2005b), the Gram-positive *L. monocytogenes* was the most PEF-resistant microorganism in buffer at pH 7.0, but it became one of the most sensitive at pH 4.0. Conversely, the Gram-negative *E. coli* O157:H7 and *Salmonella* Senftenberg were the most PEF-resistant species at pH 4.0. Timmermans et al. (2014) reported that various microorganisms were more resistant to PEF in watermelon juice (pH 5.3), compared with orange (pH 3.7) and apple (pH 3.5) juices. When the same watermelon juice was acidified to pH 3.6, higher inactivation levels were achieved. In an oat-pea beverage (pH 7.0), *L. innocua* showed greater resistance to PEF than *E. coli* (Horlacher et al., 2024). In general, pH is thought to be associated with the maintenance of cellular homeostasis and the ability of cells to recover from sublethal injury after exposure to the electric field (García et al., 2005a; Timmermans et al., 2014).

With PEF being an electro-technology, the electrical conductivity of the matrix uniquely defines the range of food products that are suitable for treatment as well as dictates the design of the electrodes (Heinz et al., 2001; Toepfl et al., 2005). However, in comparison to the other intrinsic factors, its influence on microbial inactivation rates is not always direct (Delso et al., 2022; Lytras et al., 2024a; Toepfl et al., 2005). Due to different dielectric properties, PEF is not suitable for use in liquid matrices that contain air bubbles, as it could lead to dielectric breakdown and issues with treatment efficacy (Toepfl et al., 2005). Similarly, food components such as lipids, proteins, and sugars may enhance microbial resistance or promote recovery following PEF treatment (Eshtiaghi and Nakthong, 2021; Jaeger et al., 2009b; Mañas et al., 2001; Picart et al., 2002; Schottroff et al., 2017). However, the current findings are based on either animal-based matrices or buffers, and none have been consistent enough to draw overarching conclusions. On the other hand, the addition of compounds with antimicrobial properties is known to enhance the effects of PEF during the decontamination of fruit juices (Mosqueda-Melgar et al., 2008, 2008a, 2008b; Saldaña et al., 2011).

11.2.3 Ultraviolet radiation

Ultraviolet-C (UV-C) radiation, operating within wavelengths of 200–280 nm, is widely recognized as an effective nonthermal processing method for microbial inactivation in food matrices (Koutchma, 2009). The UV-C radiation equipment comprises several essential components, with the UV-C lamps, typically low-pressure mercury ones, serving as the core element. These lamps are housed within reactors and/or treatment chambers, where liquid and solid foods are exposed to the

UV-C light (Delorme et al., 2020; Tchonkouang et al., 2023). The chamber is designed to maximize exposure by incorporating reflective surfaces that direct the light to all parts of the product. To maintain optimal operating conditions, cooling systems are used to prevent overheating of the lamps and other components. Additionally, advanced control and monitoring systems track parameters such as UV light intensity, exposure time, and temperature to ensure consistent treatments. Safety features, such as shielding or protective covers, are integrated to prevent operator exposure to harmful UV-C radiation, with interlocks that prevent operation when the system is open or not properly sealed (Tchonkouang et al., 2023).

At the European level, the European Food Safety Authority has issued a scientific opinion on the UV-C treatment of pasteurized milk, concluding that the treatments result in a safer product besides increasing its vitamin D content (EFSA, 2016). On the other hand, the FDA has established specific criteria for UV-C radiation application in fruit juices to ensure microbial safety. UV-C treatments must ensure a minimum 5-log reduction of target pathogens, such as *E. coli* O157:H7, *Salmonella*, and *L. monocytogenes* (FDA, 2004). Juice processors are responsible for verifying the effectiveness of UV-C processing, demonstrating that the appropriate dose and exposure time are sufficient to inactivate these pathogens without compromising the juice's quality. UV-C equipment must meet FDA standards and ensure uniform exposure to the juice. Additionally, processors must implement monitoring systems to track treatment parameters and maintain records of UV-C dose, exposure time, and equipment performance. If UV-C radiation is used, the juice must be properly labeled to inform consumers about the treatment, providing transparency and ensuring regulatory compliance (FDA, 2003).

11.2.3.1 Mechanisms of UV-C radiation-induced inactivation

UV-C radiation primarily causes direct damage to microbial genetic material, that is, DNA and RNA, by forming covalent bonds between adjacent pyrimidine bases, leading to cyclobutane pyrimidine dimers (CPDs and 6-4 photoproducts. These lesions prevent proper DNA replication and transcription, resulting in the death of cells (Delorme et al., 2020). UV-C photons can cause breaks in DNA, which are lethal and prevent the proper functioning of genetic material. The radiation also inhibits the cell's DNA repair mechanisms, including the activity of photolyases and endonucleases, which further exacerbates the damage. While the primary target of UV-C is genetic material, oxidative damage to the cell membrane also occurs, resulting in increased permeability and the leakage of intracellular contents.

Proteins, including essential enzymes, undergo structural alterations due to UV-C-generated ROS and free radicals, which disrupt enzymatic activity and hinder vital metabolic processes. UV-C can also alter the fluidity of microbial membranes, making them more susceptible to damage and facilitating the uptake of ROS (Li and Yang, 2022). These combined effects compromise key cellular functions, such as nutrient uptake, waste removal, and ATP production, leading to cellular dysfunction and death. Furthermore, UV-C radiation induces stress response pathways within cells, but these responses are often insufficient to counteract the overwhelming damage caused by high doses of UV-C, resulting in irreversible cell damage and inactivation (Li and Yang, 2022).

11.2.3.2 Modeling approaches for UV-C radiation-induced inactivation

UV-C irradiation has demonstrated significant potential in ensuring the safety and quality of liquid plant-based foods. Its application in fruit and vegetable juices is particularly well-documented, with numerous modeling studies highlighting its ability to achieve substantial reductions in pathogenic microorganisms (Ochoa-Velasco et al., 2020). For instance, Fenoglio et al. (2020) reported that the Weibull model was the most versatile for describing the inactivation of *E. coli* in several fruit juices, treated at an UV-C dose of 390 mJ/cm^2 at 20°C and subsequently stored at 5°C. In their study, 4.8 to 5.1 log reductions of the microorganism were recorded in clear matrices, while for turbid juices, bacterial reductions were lower than 4 logs. Additionally, the Weibull and the biphasic models were suitable to describe the inactivation of *E. coli* and *Salmonella* in apple juices due to UV-C treatments at 1271 mJ/cm^2 (Casco et al., 2025).

Beyond juices, UV-C application has also been explored for microbial inactivation in plant-based milk alternatives, including almond, soy, and oat milk. These matrices present unique challenges due to their turbidity and the presence of insoluble particles, which scatter and absorb UV-C light, reducing penetration. Possas et al. (2018) reported that the Weibull model coupled with an empirical nonlinear model was suitable to describe the *Salmonella* Enteritidis inactivation in soymilk induced by UV-C when applying doses ranging from 0 to 10 J/cm^2 and temperatures ranging from 4 to 30°C. These authors computed more than 5-log reductions at temperatures ranging from 18 to 30°C when UV-C doses higher than 7.5 J/cm^2 were applied. Further research is needed to improve light penetration in alternative dairy-based beverages.

11.2.3.3 Factors affecting UV-C radiation-induced inactivation

The efficacy of UV-C irradiation for microbial inactivation is governed by a complex interplay of several factors. The most important one is the UV-C dose, typically expressed in millijoules per square centimeter (mJ/cm^2). Higher doses result in greater microbial inactivation due to increased energy absorption by the target cells (Possas et al., 2018). However, excessive UV-C exposure can compromise the quality of certain plant-based foods, leading to undesirable alterations in color, texture, or nutrient composition (Delorme et al., 2020). Exposure time also plays a crucial role, as prolonged treatment increases the cumulative dose and thereby enhances microbial inactivation (Ochoa-Velasco et al., 2020). An optimization of UV-C dose and exposure times is necessary to strike a balance between safety and quality, particularly in foods with delicate sensory profiles (Tchonkouang et al., 2023).

The properties of the food matrix significantly affect the penetration and effectiveness of UV-C application (Delorme et al., 2020). Factors such as turbidity, optical density, and the presence of suspended particles can impede the transmission of UV-C light, creating shadowed regions that protect microorganisms from exposure (Ochoa-Velasco et al., 2020). These challenges are particularly evident in plant-based liquids, such as fruit juices and alternative dairy products, where high turbidity and viscosity limit uniform irradiation. Moreover, food chemical composition, including pH and the presence of compounds with antioxidant properties, can interact with UV-C-induced ROS, potentially mitigating their antimicrobial effects (Koutchma, 2009).

Microbial characteristics, including species, morphology, and physiological state, further modulate susceptibility to UV-C (Li and Yang, 2022). Gram-positive bacteria, for instance, exhibit greater resistance than Gram-negative counterparts due to thicker peptidoglycan layers, which act as protective barriers (Li and Yang, 2022). Similarly, microorganisms in stationary-phase growth or biofilms demonstrate enhanced resilience compared to those in log-phase growth, necessitating higher UV-C doses for effective inactivation. Environmental factors, such as temperature and pH, also influence UV-C efficacy (Li and Yang, 2022). Acidic environments, for example, can amplify microbial sensitivity to UV-C by reducing the efficiency of DNA repair mechanisms, while suboptimal temperatures may inhibit microbial recovery posttreatment (Koutchma, 2009).

The susceptibility of microorganisms to UV-C treatment varies significantly depending on their structural and physiological characteristics (Li and Yang, 2022). For instance, vegetative cells are generally more susceptible than spore-forming bacteria due to the lack of protective structures such as spore coats or pigments that can absorb or deflect UV-C radiation (Koutchma, 2009). This variability necessitates a tailored approach when applying UV-C to plant-based foods, considering both the type of microorganisms likely to be present and the unique attributes of the food matrix (Koutchma, 2009).

11.2.4 Cold plasma

The term "plasma" was first introduced by Irving Langmuir in 1928 to designate this fourth state of matter, defined as a gas in a partially or fully ionized state (Thirumdas et al., 2015). Plasma can be generated through the application of different energy sources, such as thermal energy, electric or magnetic fields, as well as radio or microwave frequencies. These energy sources increase the kinetic energy of electrons, promoting more frequent collisions within the gas molecules and ultimately promoting plasma formation (Conrads and Schmidt, 2000; Thirumdas et al., 2015). Plasma consists of a complex mixture of highly reactive species, including radicals, positively and negatively charged ions, atoms and molecules, in addition to electrons and photons (Rauscher et al., 2010).

Thermal and nonthermal plasma are two significant categories of plasma that differ in their temperature ranges and characteristics. Thermal plasma is characterized by high temperatures, typically ranging from 2000 to 30,000 K, where the temperatures of electrons and heavy particles are approximately equal and are nearly fully ionized. In contrast, nonthermal plasma, also referred to as cold plasma, is only partially ionized and is characterized by a marked temperature difference between electrons and heavy particles, with electrons exhibiting much higher temperatures than the neutral gas molecules (Khan et al., 2022; Safa and Soucy, 2015). Thanks to its effective microbicidal properties and the fact that it is a nonthermal technology, cold plasma is a great option for preserving food products, keeping bioactive ingredients, and prolonging shelf life (Deshmukh et al., 2022).

Cold plasma can be produced over a wide range of pressures depending on the plasma source. In early studies, plasma was generated at low pressures. However, a vacuum system was required for plasma generation, which limited the applicability of low-pressure plasma to batch processing. For this reason, alternative plasma sources have been developed to produce cold plasma at atmospheric pressure (CAP) (Lee et al., 2017; Niemira, 2012). The main appeal of this plasma technology lies in its ability to achieve rapid decontamination under ambient temperature and pressure conditions, without significantly altering food quality or generating high costs, especially when using ambient air as the working gas (Misra et al., 2019).

Electric gas discharge is the primary method for producing cold plasma (Liao et al., 2017), and in food processing, the most widely used devices for generating it are plasma jets and dielectric barrier discharge (DBD) systems. Plasma jet devices feature two concentric electrodes, where the inner electrode is driven by a high-voltage power supply. This setup ionizes the working gas, which then exits through the nozzle, producing a plasma "jet". In contrast, DBD systems consist of two metallic electrodes separated by a gap, with at least one electrode covered by a dielectric barrier. This dielectric barrier acts as a stabilizing element, preventing arc formation and promoting the generation of multiple microdischarges, thereby contributing to more uniform treatment conditions (Nwabor et al., 2022; Pankaj et al., 2018). Among the different gas discharge techniques used to generate CAP, surface discharge plasma has recently gained attention as a promising method for food processing (Zhou et al., 2020). In this case, both electrodes are in contact with the dielectric material, and plasma is generated on the ground electrode surface (Maccaferri et al., 2023).

11.2.4.1 Mechanisms of cold plasma-induced microbial inactivation

Reactive species, UV photons, and charged particles generated in cold plasma treatments are responsible for microbial inactivation (Surowsky et al., 2015). The type of reactive species formed depends on the composition of the feed gas (Dharini et al., 2023), and air plasmas are excellent sources of reactive oxygen and nitrogen species (RONS like O, $^1O_2$, $\bullet O_2^-$, O_3, NO_2 NO_3^-, NO_2^-, and ONOOH). In addition, cold plasma generated at gas-liquid interface, in the context of plasma sterilization of liquids (plasma-activated water), produces secondary reactive species such as H_2O_2, $\bullet OH$, and HNO_3. RONS are able to cause cell wall or membrane disruption through lipid and protein oxidation, ultimately leading to intracellular protein alteration, DNA damage (Ding et al., 2023a; Laroussi and Leipold, 2004; Muhammad, 2021), and leakage of intracellular components, as well as acceleration of oxidative stress and disruption of intracellular redox homeostasis (Zhao et al., 2022). Regarding spores, reactive plasma species can contribute to the breakage of disulfide bonds in the protein coat (Valdez-Narváez et al., 2024) and diffuse into the spore, causing damage to membranes and microbial macromolecules. This disruption leads to the inactivation of spores, preventing them from germinating (Klämpfl et al., 2012).

On the other hand, the generation of UV photons at various wavelengths has been determined to play a role in the dimerization of thymine bases in DNA (Misra et al., 2011). Regarding charged particles generated during cold plasma treatments, they can also gather on the bacterial cell wall's surface, creating electrostatic forces and producing the wall rupture. In addition, the intense electric field produced by treatments causes perforations in the bacterial cell membrane when exposed, leading to the leakage of intracellular contents and ultimately resulting in bacterial death (Zhu et al., 2022). However, in the case of liquids treated by plasma, reactive species formed are the main cause of damage and microbial inactivation (Ma et al., 2015; Oehmigen et al., 2010).

11.2.4.2 Modeling approaches for cold plasma-induced microbial inactivation

Table 11.3 shows a compilation of various studies conducted to date on modeling microbial inactivation in plant-based beverages treated with plasma. The abovementioned Weibull model (Eq. 11.2), providing predictions for population parameters that are neither linear nor log-linear, has been used successfully to describe the microbial response to CAP technology.

For instance, Muhammad et al. (2019) studied the inactivation of *B. cereus* in tiger nut milk, treated by DBD-CAP with air. In this study, the Weibull model was used to fit the impact of input power (39, 43, and 46 W) and treatment time (0–270 s) on the bacterial kinetic inactivation in the beverage. The δ model parameter (Eq. 11.2), defined as the time for the first log reduction, decreased as the input power increased, indicating that the antimicrobial effectiveness of the CAP treatment improved with higher power. In fact, a maximum 5.28-log reduction was achieved at 46 W after 270 s of treatment. Moreover, the downward concave survival curves ($n > 1$) obtained might be explained by the accumulated damage to the remaining population, making the surviving cells progressively more susceptible to the lethal treatment (Pan et al., 2019). Surowsky et al. (2014) also observed downward concave survival curves for *Citrobacter freundii* in apple juice, treated with a jet-CAP and using pure argon mixed with (0.025%–0.1%) oxygen, during exposure times from 0 to 480 s. This trend was well described by the Weibull model. The increase in oxygen concentration in the process gas led to enhanced antimicrobial activity, resulting in decreasing δ values. *C. freundii* levels in apple juice were reduced by a maximum of five log cycles after 480 s of exposure plus 24 h of storage, using argon with 0.1% oxygen. Additionally, Liao et al. (2018) studied the inactivation of *E. coli* O157:H7 in apple juice by DBD-CAP with air at 30, 40, and 50 W of input power and 0–40 s of exposure time. Their findings showed that 40 s cold plasma treatment resulted in more than 4-log reductions at 50 W. The inactivation behavior of DBD-CAP on *E. coli* was well described by the Weibull model. Decreasing δ values were obtained as the input power of the

Table 11.3 Summary of modeling studies on cold plasma-induced microbial inactivation in beverages.

Microorganism	Food matrix/ medium	Modeling approach	Process variables (values or ranges)	References
Bacillus cereus	Tiger nut milk	Weibull	Input power (39–46 W), time (0–270 s), process gas: air	Muhammad et al. (2019)
Citrobacter freundii	Apple juice	Weibull	Voltage: 65 V, time (0–480 s), process gas: pure argon mixed with oxygen (0.025%–0.1%)	Surowsky et al. (2014)
Escherichia coli	Apple juice	Weibull	Input power (30–50 W), time (0–40 s), process gas: air	Liao et al. (2018)
Zygosaccharomyces rouxii	Apple juice	First-order kinetic/ Weibull	Voltage (11.3–21.3 kV), time (0–30 min), process gas: air, gas flow rate: 5 slm	Wang et al. (2018)
Alicyclobacillus acidoterrestris	Apple juice	Biphasic	Voltage (1.23–6.86 kV), time (0–720 s), process gas: air, gas flow rate: 130 mL/min	Ding et al. (2023)

treatment increased, and *n* values greater than one were observed, indicating downward concavity.

Wang et al. (2018) evaluated the inactivation of *Zygosaccharomyces rouxii* yeast (LB and 1130 strains) in apple juice, using gas-phase surface discharge CAP, 11.3–21.3 kV of voltage and exposure times from 0 to 30 min. Extended plasma exposure durations, combined with higher discharge voltages, led to increased inactivation rates. The cell count of *Z. rouxii* LB and 1130 in apple juice decreased by approximately 6.58 and 6.82 log units, respectively, at 21.3 kV of voltage and 30 min of exposure time. The authors compared the performance of the first-order kinetic and Weibull models in describing the survival curves of both yeast strains and found that the Weibull provided a better fit to the data. On the other hand, Ding et al. (2023) assessed the inactivation of *Alicyclobacillus acidoterrestris* in apple juice by surface discharge CAP at varying voltages. The biphasic model was applied, with parameters $K_{\max 1}$ and $K_{\max 2}$ representing the bactericidal rates of the first and second phases, respectively. Across different voltages, $K_{\max 1}$ was notably higher than $K_{\max 2}$ and increased significantly with voltage, indicating an enhanced antimicrobial effect. A 99% bacterial reduction was achieved within 120 s, with efficacy improved as voltage increased (1.32–6.86 kV).

11.2.4.3 Factors affecting cold plasma-induced microbial inactivation

The antimicrobial efficacy of cold plasma is influenced by various factors, and one of them is the microbial properties. For instance, bacteria, yeasts, and fungi have different susceptibility, and, in many cases, bacteria are more vulnerable to plasma (Jambrak et al., 2015). Previous research has shown that Gram-negative bacteria are generally less resistant to cold plasma compared to Gram-positive bacteria (Han et al., 2020). This difference in susceptibility is attributed to the thinner cell walls of Gram-negative bacteria, making them more vulnerable to the bactericidal agents produced by cold plasma (Wei et al., 2023). Additionally, plasma treatments are less effective on spores than on vegetative cells (Scholtz et al., 2010; Valdez-Narváez et al., 2024), primarily due to the protective layers of the spore coat and the spore encasement (Raguse et al., 2016). Earlier studies have also found that a lower initial inoculum load results in a stronger inactivation effect (Laroussi et al., 2011). Furthermore, the growth stage plays a significant role, as research on *E. coli* has demonstrated that cells in the 12-hour growth phase have a higher inactivation rate compared to those in the 24-hour and 48-hour growth phases (Deng et al., 2007).

On the other hand, the lethality of cold plasma treatments depends on various processing variables. Research has shown that the intensity of the treatment (input power/voltage), is a crucial factor in microbial inactivation, as it regulates the amount of the discharge responsible for creating reactive species. In addition to this, the duration of exposure to the treatment is also a critical process parameter, and it can be stated that inactivation generally increases with higher power or voltage levels and longer exposure times (Fernández-Felipe et al., 2024; Nasiru et al., 2021). Furthermore, the choice of process gas is critical. For instance, previous studies have demonstrated that cold plasma generated in air or oxygen is generally more effective at inactivating microbes than plasma produced in noble gases. This effectiveness is attributed to the specific plasma species formed (Pai et al., 2018).

Additionally, higher gas flow rates have been shown to enhance microbial inactivation. The configuration of the plasma system could also influence the results of decontamination, as plasma can be generated using different devices (Smet et al., 2016). Regarding the type of treatment, direct plasma treatments, where displacement and conduction currents flow through the sample, typically achieve greater inactivation compared to indirect plasma treatments. In indirect treatments, plasma generated between two electrodes is transported to the sample via diffusion and convection. The distance of exposure to the plasma also influences the outcome, with inactivation efficacy decreasing as the distance increases (Das et al., 2022).

Lastly, environmental factors like food matrix properties and pH can also impact the sterilization effectiveness. For instance, the plasma's efficacy is strongly affected by the chemical composition of the treated matrix, highlighting the need to carefully consider food compounds that can interfere with plasma treatments, such as vitamins, proteins, and fats (De Baerdemaeker et al., 2022). The pH is another relevant factor. An acidic environment is more conducive for reactive species to penetrate cell walls, decreasing bacterial resistance. H^+ ions enter the bacterial cell, disrupting the intracellular pH balance and ultimately leading to cell death. In addition, it is documented that plasma reactive species can cause acidification of liquid matrices (Muhammad, 2021; Ni et al., 2024; Oehmigen et al., 2010).

11.3 Innovations and future perspectives in nonthermal processing for plant-based liquid products

The application of nonthermal technologies for microbial inactivation in plant-based beverages continues to evolve, with future research focusing on enhancing process efficiency, predictive modeling, and industrial scalability. A key area of interest is the development of more sophisticated predictive models that integrate multiple variables, including process parameters, microbial variability, and food matrix characteristics. Current modeling approaches often overlook critical factors such as come-up time during HPP and temperature increases due to adiabatic heating during compression, despite evidence of their significant impact on microbial inactivation (Possas et al., 2017). Similarly, Cold Plasma inactivation kinetics are often described using primary models like the Weibull model, but future studies should focus on secondary models that better integrate the interaction of process variables, microbial variability, and food matrix effects. For PEF, multiphysics simulations have gained attention as a tool to better understand the complexity of continuous flow treatment systems (Buckow et al., 2010; Huang et al., 2013, 2014; Lombergar et al., 2024; Yan et al., 2025). Future research should focus on the collection of large-scale experimental data to enhance predictive modeling accuracy, particularly in plant-based food products with diverse compositions such as dairy analogs, where microbial inactivation kinetics remain underexplored (Horlacher et al., 2024).

Another major research direction is the combination of different non-thermal technologies to enhance microbial inactivation and product stability. Hybrid approaches, such as combining UV-C with PEF or Cold Plasma with HPP, have shown potential in addressing the limitations of individual treatments. These combinations promote complementary inactivation mechanisms, which can help overcome challenges like the limited penetration depth of UV-C or the inability of PEF to inactivate bacterial spores (Alkhafaji and Farid, 2012; Siemer et al., 2014). Additionally, combining low-intensity HPP with factors such as pH, temperature, and antimicrobials is being explored as a strategy to extend shelf life and reduce operational costs while maintaining sustainability (Farakos and Zwietering, 2011; Guillou and Membré, 2019).

For industrial adoption and regulatory approval, further efforts are required to improve equipment design, treatment uniformity, and process validation. Many of these technologies remain at the laboratory scale, with limited data on their efficacy under real processing conditions. The adoption of standardized validation protocols is crucial, particularly for foods undergoing combined preservation treatments (EFSA, 2022). Challenges such as adiabatic heating in HPP, uneven electric field distribution in PEF, and gas composition effects in Cold Plasma require targeted optimization for large-scale applications. Additionally, the reporting of all relevant process parameters is essential to ensure the reproducibility and applicability of predictive models across different studies and food matrices.

Lastly, sustainability and economic feasibility will play a crucial role in shaping the future of these technologies. Reducing energy consumption and operational costs while maintaining high efficiency is a challenge, especially for industrial applications. With the plant-based food market projected to grow significantly by 2030, optimizing nonthermal processing methods for cost-effective and scalable applications will be critical. Sustainability concerns, such as packaging compatibility with nonthermal treatments and the reduction of processing-induced waste, must also be addressed to align with consumer and industry expectations. As nonthermal technologies continue to advance, their integration into sustainable food processing frameworks will be essential for delivering safe, high-quality, and minimally processed plant-based beverages to the market.

References

Aertsen, A., De Spiegeleer, P., Vanoirbeek, K., Lavilla, M., Michiels, C.W., 2005. Induction of oxidative stress by high hydrostatic pressure in *Escherichia coli*. Appl. Environ. Microbiol. 71, 2226–2231. https://doi.org/10.1128/AEM.71.5.2226-2231.2005.

Aganovic, K., Hertel, C., Vogel, R.F., Johne, R., Schlüter, O., Schwarzenbolz, U., Jäger, H., Holzhauser, T., Bergmair, J., Roth, A., Sevenich, R., Bandick, N., Kulling, S.E., Knorr, D., Engel, K.H., Heinz, V., 2021. Aspects of high hydrostatic pressure food processing: perspectives on technology and food safety. Compr. Rev. Food Sci. Food Saf. 7. https://doi.org/10.1111/1541-4337.12763.

Akin, E., Evrendilek, G.A., 2009. Effect of pulsed electric fields on physical, chemical, and microbiological properties of formulated carrot juice. Food Sci. Technol. Int. 15, 275–282. https://doi.org/10.1177/1082013209341414.

Ali, N., Popović, V., Koutchma, T., Warriner, K., Zhu, Y., 2020. Effect of thermal, high hydrostatic pressure, and ultraviolet-C processing on the microbial inactivation, vitamins, chlorophyll, antioxidants, enzyme activity, and color of wheatgrass juice. J. Food Process. Eng. 43, e13036. https://doi.org/10.1111/JFPE.13036.

Alkhafaji, S.R., Farid, M.M., 2012. PEF assisted thermal sterilization (PEF-ATS) process-inactivation of *Geobacillus sterothermophilus* spores. J Food Sci Eng 2, 403–410. https://doi.org/10.17265/2159-5828/2012.07.007.

Allison, A., Daniels, E., Chowdhury, S., Fouladkhah, A., 2018. Effects of elevated hydrostatic pressure against mesophilic background microflora and habituated *Salmonella* serovars in orange juice. Microorganisms 9, 1–12. https://doi.org/10.3390/microorganisms6010023.

Álvarez, I., Pagán, R., Raso, J., Condón, S., 2002. Environmental factors influencing the inactivation of *Listeria monocytogenes* by pulsed electric fields. Lett. Appl. Microbiol. 35, 489–493. https://doi.org/10.1046/j.1472-765X.2002.01221.x.

Amanatidou, A., Schlüter, O., Lemkau, K., Gorris, L.G.M., Smid, E.J., Knorr, D., 2000. Effect of combined high pressure treatment and modified atmospheres on the shelf-life of fresh Atlantic salmon. Innov. Food Sci. Emerg. Technol. 1, 87–98. https://doi.org/10.1016/S1466-8564(00)00007-2.

Aronsson, K., Borch, E., Stenlöf, B., Rönner, U., 2004. Growth of pulsed electric field exposed *Escherichia coli* in relation to inactivation and environmental factors. Int. J. Food Microbiol. 93, 1–10. https://doi.org/10.1016/S0168-1605(03)00071-0.

Food Safety Inspection Service (FSIS), 2014. FSIS Compliance Guideline: Controlling LISTERIA Monocytogenes in Post-lethality Exposed Ready-to-Eat Meat and Poultry Products. United States Department of Agriculture (USDA), Washington (DC).

Balasubramaniam, V.M.B., Mart, S.I., 2015. Principles and application of high pressure – based technologies in the food industry. https://doi.org/10.1146/annurev-food-022814-015539.

Bartula, K., Begley, M., Latour, N., Callanan, M., 2023. Growth of food-borne pathogens *Listeria* and *Salmonella* and spore-forming *Paenibacillus* and *Bacillus* in commercial plant-based milk alternatives. Food Microbiol. 109. https://doi.org/10.1016/j.fm.2022.104143.

Bartula, K., Biagui, S., Begley, M., Callanan, M., 2024. Investigation of the growth of listeria in plant-based beverages. Food Microbiol. 121, 1–6. https://doi.org/10.1016/j.fm.2024.104530.

Bermúdez-Aguirre, D., Guerrero-Beltrán, J.Á., Barbosa-Cánovas, G.V., Welti-Chanes, J., 2011. Study of the inactivation of *Escherichia coli* and pectin methylesterase in mango nectar under selected high hydrostatic pressure treatments. Food Sci. Technol. Int. 17, 541–547. https://doi.org/10.1177/1082013211399681.

Buckow, R., Schroeder, S., Berres, P., Baumann, P., Knoerzer, K., 2010. Simulation and evaluation of pilot-scale pulsed electric field (PEF) processing. J. Food Eng. 101, 67–77. https://doi.org/10.1016/j.jfoodeng.2010.06.010.

Black, E.P., Linton, M., McCall, R.D., Curran, W., Fitzgerald, G.F., Kelly, A.L., Patterson, M. F., 2008. The combined effects of high pressure and nisin on germination and inactivation of *Bacillus* spores in milk. J. Appl. Microbiol. 105 (1), 78–87.

Buckow, R., Ng, S., Toepfl, S., 2013. Pulsed electric field processing of orange juice: a review on microbial, enzymatic, nutritional, and sensory quality and stability. Compr. Rev. Food Sci. Food Saf. 12, 455–467. https://doi.org/10.1111/1541-4337.12026.

Bull, M.K., Hayman, M.M., Stewart, C.M., Szabo, E.A., Knabel, S.J., 2005. Effect of prior growth temperature, type of enrichment medium, and temperature and time of storage on recovery of *Listeria monocytogenes* following high pressure processing of milk. Int. J. Food Microbiol. 101, 53–61. https://doi.org/10.1016/j.ijfoodmicro.2004.10.045.

Calderón-Miranda, M.L., Barbosa-Cánovas, G.V., Swanson, B.G., 1999. Transmission electron microscopy of *Listeria innocua* treated by pulsed electric fields and nisin in skimmed milk. Int. J. Food Microbiol. 51, 31–38. https://doi.org/10.1016/S0168-1605(99)00071-9.

Canaviri Paz, P., Janny, R.J., Håkansson, Å., 2020. Safeguarding of quinoa beverage production by fermentation with Lactobacillus plantarum DSM 9843. Int. J. Food Microbiol. 324, 108630. https://doi.org/10.1016/j.ijfoodmicro.2020.108630.

Casco, M. de los A., Schelegueda, L., Andreone, A., Schenk, M., Campos, C.A., Guerrero, S. N., 2025. A novel approach using UV-C light assisted by mild heat and natamycin as a hurdle strategy to preserve juice from non-commercial-grade apples. Food Control 171, 111147. https://doi.org/10.1016/J.FOODCONT.2025.111147.

CDC, 2014. 2014 Outbreak of Listeria Infections Linked to Wholesome Soy Products. Inc. Sprouts - Advice. CDC Archive. URL. https://archive.cdc.gov/#/details?archive_url=https://archive.cdc.gov/www_cdc_gov/listeria/outbreaks/bean-sprouts-11-14/advice-consumers.html.

Chueca, B., Pagán, R., García-Gonzalo, D., 2015. Transcriptomic analysis of *Escherichia coli* MG1655 cells exposed to pulsed electric fields. Innov. Food Sci. Emerg. Technol. 29, 78–86. https://doi.org/10.1016/j.ifset.2014.09.003.

Conrads, H., Schmidt, M., 2000. Plasma generation and plasma sources. Plasma Sources Sci. Technol. 9, 441. https://doi.org/10.1088/0963-0252/9/4/301.

Coster, H.G.L., Zimmermann, U., 1975. The mechanism of electrical breakdown in the membranes of Valonia utricularis. J. Membr. Biol. 22, 73–90. https://doi.org/10.1007/BF01868164.

Crowley, J.M., 1973. Electrical breakdown of bimolecular lipid membranes as an electromechanical instability. Biophys. J. 13, 711–724. https://doi.org/10.1016/S0006-3495(73)86017-5.

Das, S., Gajula, V.P., Mohapatra, S., Singh, G., Kar, S., 2022. Role of cold atmospheric plasma in microbial inactivation and the factors affecting its efficacy. Health Sciences Review 4, 100037. https://doi.org/10.1016/J.HSR.2022.100037.

De Baerdemaeker, K., Van der Linden, I., Nikiforov, A., Zuber, S., De Geyter, N., Devlieghere, F., 2022. Non-thermal plasma inactivation of *Salmonella* Typhimurium on different matrices and the effect of selected food components on its bactericidal efficacy. Food Res. Int. 151, 110866. https://doi.org/10.1016/J.FOODRES.2021.110866.

Delorme, M.M., Guimarães, J.T., Coutinho, N.M., Balthazar, C.F., Rocha, R.S., Silva, R., Margalho, L.P., Pimentel, T.C., Silva, M.C., Freitas, M.Q., Granato, D., Sant'Ana, A.S., Duart, M.C.K.H., Cruz, A.G., 2020. Ultraviolet radiation: an interesting technology to preserve quality and safety of milk and dairy foods. Trends Food Sci. Technol. 102. https://doi.org/10.1016/j.tifs.2020.06.001.

Delso, C., Martínez, J.M., Cebrián, G., Condón, S., Raso, J., Álvarez, I., 2022. Microbial inactivation by pulsed electric fields. In: Raso, J., Heinz, V., Alvarez, I., Toepfl, S. (Eds.), Pulsed Electric Fields Technology for the Food Industry: Fundamentals and Applications. Springer, Cham, Switzerland, pp. 169–207. https://doi.org/10.1007/978-3-030-70586-2_5.

Deng, S., Ruan, R., Mok, C.K., Huang, G., Lin, X., Chen, P., 2007. Inactivation of *Escherichia coli* on almonds using nonthermal plasma. J. Food Sci. 72, M62–M66. https://doi.org/10.1111/J.1750-3841.2007.00275.X.

Dermol-Černe, J., Batista Napotnik, T., Reberšek, M., Miklavčič, D., 2020. Short microsecond pulses achieve homogeneous electroporation of elongated biological cells irrespective of their orientation in electric field. Sci. Rep. 10, 1–17. https://doi.org/10.1038/s41598-020-65830-3.

Deshmukh, A.K., Talwar, G., Singla, M., 2022. Cold plasma technology – an overview of basics and principle. Environ. Conserv. J. 23, 87–101. https://doi.org/10.36953/ECJ.8722147.

Dharini, M., Jaspin, S., Mahendran, R., 2023. Cold plasma reactive species: generation, properties, and interaction with food biomolecules. Food Chem. 405, 134746. https://doi.org/10.1016/J.FOODCHEM.2022.134746.

Dimitrov, D.S., 1984. Electric field-induced breakdown of lipid bilayers and cell membranes: a thin viscoelastic film model. J. Membr. Biol. 78, 53–60. https://doi.org/10.1007/BF01872532.

Ding, H., Wang, T., Sun, Y., Zhang, Y., Wei, J., Cai, R., Guo, C., Yuan, Y., Yue, T., 2023. Role and mechanism of cold plasma in inactivating *alicyclobacillus acidoterrestris* in apple juice. Foods 12, 1531. https://doi.org/10.3390/FOODS12071531.

Dogan, C., Erkmen, O., 2004. High pressure inactivation kinetics of *Listeria monocytogenes* inactivation in broth, milk, and peach and orange juices 62, 47–52. https://doi.org/10.1016/S0260-8774(03)00170-5.

Dutreux, N., Notermans, S., Wijtzes, T., Góngora-Nieto, M.M., Barbosa-Cánovas, G.V., Swanson, B.G., 2000. Pulsed electric fields inactivation of attached and free-living *Escherichia coli* and *Listeria innocua* under several conditions. Int. J. Food Microbiol. 54, 91–98. https://doi.org/10.1016/S0168-1605(99)00175-0.

EFSA, 2016. Safety of UV-treated milk as a novel food pursuant to regulation (EC) No 258/97. EFSA J. 14. https://doi.org/10.2903/j.efsa.2016.4370.

EFSA, 2022. The efficacy and safety of high-pressure processing of food. EFSA J. 20. https://doi.org/10.2903/j.efsa.2022.7128.

Eshtiaghi, M.N., Nakthong, N., 2021. Application of pulsed electric field for inactivation of yeast s. cerevisiae in apple juice. In: Journal of Physics: Conference Series. IOP Publishing Ltd. https://doi.org/10.1088/1742-6596/1893/1/012008.

Evrendilek, G.A., 2017. Impacts of pulsed electric field and heat treatment on quality and sensory properties and microbial inactivation of pomegranate juice. Food Sci. Technol. Int. 23, 668–680. https://doi.org/10.1177/1082013217715369.

Farakos, S.M.S., Zwietering, M.H., 2011. Data analysis of the inactivation of foodborne microorganisms under high hydrostatic pressure to establish global kinetic parameters and influencing factors. J. Food Protect. 74, 2097–2106. https://doi.org/10.4315/0362-028X.JFP-11-162.

FDA, 2003. Small Entity Compliance Guide: Juice HACCP. Small Entity Compliance Guide. https://www.fda.gov/regulatory-information/search-fda-guidance-documents/small-entity-compliance-guide-juice-haccp.

FDA, 2004. Guidance for Industry: Juice Hazard Analysis Critical Control Point Hazards and Controls Guidance, First Edition. Guidance for Industry: Juice Hazard Analysis Critical Control Point Hazards and Controls Guidance. https://www.fda.gov/regulatory-information/search-fda-guidance-documents/guidance-industry-juice-hazard-analysis-critical-control-point-hazards-and-controls-guidance-first.

FDA, 2022. Lyons Magnus Voluntarily Recalls 53 Nutritional and Beverage Products Due to the Potential for Microbial Contamination. Recalls, Market, Withdrawals and Safety Alerts. https://www.fda.gov/safety/recalls-market-withdrawals-safety-alerts/lyons-magnus-voluntarily-recalls-53-nutritional-and-beverage-products-due-potential-microbial.

Fenoglio, D., Ferrario, M., Schenk, M., Guerrero, S., 2020. Effect of pilot-scale UV-C light treatment assisted by mild heat on *E. coli*, L. plantarum and *S. cerevisiae* inactivation in clear and turbid fruit juices. Storage study of surviving populations. Int. J. Food Microbiol. 332, 108767. https://doi.org/10.1016/j.ijfoodmicro.2020.108767.

Fernández-Felipe, M.T., Inés Valdez-Narváez, M., Martinez, A., Rodrigo, D., 2024. Oxygen and air cold plasma for the inactivation of *Bacillus cereus* in low-water activity soy powder. Food Res. Int. 193, 114861. https://doi.org/10.1016/J.FOODRES.2024.114861.

Ferreira, M., Almeida, A., Delgadillo, I., Saraiva, J., 2016. Susceptibility of *Listeria monocytogenes* to high pressure processing : a review. Food Rev. Int. 32, 377–399. https://doi.org/10.1080/87559129.2015.1094816.

García, D., Hassani, M., Mañas, P., Condón, S., Pagán, R., 2005. Inactivation of *Escherichia coli* O157:H7 during the storage under refrigeration of apple juice treated by pulsed electric fields. J. Food Saf. 25, 30–42. https://doi.org/10.1111/j.0149-6085.2005.25552.x.

García, D., Gómez, N., Mañas, P., Condón, S., Raso, J., Pagán, R., 2005a. Occurrence of sublethal injury after pulsed electric fields depending on the micro-organism, the treatment medium ph and the intensity of the treatment investigated. J. Appl. Microbiol. 99, 94–104. https://doi.org/10.1111/j.1365-2672.2005.02611.x.

García, D., Gómez, N., Raso, J., Pagán, R., 2005b. Bacterial resistance after pulsed electric fields depending on the treatment medium pH. Innov. Food Sci. Emerg. Technol. 6, 388–395. https://doi.org/10.1016/j.ifset.2005.04.003.

Gayán, E., Govers, S.K., Aertsen, A., 2017. Impact of high hydrostatic pressure on bacterial proteostasis. Biophys. Chem. 231, 3–9. https://doi.org/10.1016/J.BPC.2017.03.005.

Georget, E., Sevenich, R., Reineke, K., Mathys, A., Heinz, V., Callanan, M., Rauh, C., Knorr, D., 2015. Inactivation of microorganisms by high isostatic pressure processing in complex matrices: a review. Innov. Food Sci. Emerg. Technol. 27, 1–14. https://doi.org/10.1016/j.ifset.2014.10.015.

Giannoulis, N., Karatzas, K.A.G., 2024. The combined effect of chitosan and high hydrostatic pressure on *Listeria monocytogenes* and *Escherichia coli*. Innov. Food Sci. Emerg. Technol. 94, 103693. https://doi.org/10.1016/J.IFSET.2024.103693.

Glaser, R.W., Leikin, S.L., Chernomordik, L.V., Pastushenko, V.F., Sokirko, A.I., 1988. Reversible electrical breakdown of lipid bilayers: formation and evolution of pores. BBA - Biomembranes 940, 275–287. https://doi.org/10.1016/0005-2736(88)90202-7.

González-Angulo, M., Serment-Moreno, V., Clemente-García, L., Tonello, C., Jaime, I., Rovira, J., 2021. Assessing the pressure resistance of *Escherichia coli* O157:H7, *Listeria monocytogenes* and *Salmonella enterica* to high pressure processing (HPP) in citric acid model solutions for process validation. Food Res. Int. 140, 110091. https://doi.org/10.1016/J.FOODRES.2020.110091.

González-Tejedor, G., Garre, A., Iguaz, A., Wong-Zhang, R., Fernández, P.S., Possas, A., 2023. Dynamic thermal treatments in green coconut water induce dynamic stress adaptation of *Listeria innocua* that increases its thermal resistance. Foods 12, 4015. https://doi.org/10.3390/foods12214015.

González-Tejedor, G., Garre, A., Egea, J.A., Aznar, A., Artés-Hernández, F., Fernández, P.S., 2023a. Application of High Hydrostatic Pressure in fresh purple smoothie: microbial inactivation kinetic modelling and qualitative studies. Food Sci. Technol. Int. 29, 372–382. https://doi.org/10.1177/10820132221095607.

Grahl, T., Märkl, H., 1996. Killing of micro-organisms by pulsed electric fields. Applied Microbiology and Biotechnology. Appl. Microbiol. Biotechnol. 45, 148–157.

Guillou, S., Membré, J.M., 2019. Inactivation of *Listeria monocytogenes, Staphylococcus aureus*, and *Salmonella enterica* under High Hydrostatic Pressure: a quantitative analysis of existing literature data. J. Food Protect. 82, 1802–1814. https://doi.org/10.4315/0362-028X.JFP-19-132.

Gurtler, J.B., Bailey, R.B., Geveke, D.J., Zhang, H.Q., 2011. Pulsed electric field inactivation of *E. coli* O157:H7 and non-pathogenic surrogate *E. coli* in strawberry juice as influenced by sodium benzoate, potassium sorbate, and citric acid. Food Control 22, 1689–1694. https://doi.org/10.1016/j.foodcont.2011.03.029.

Han, J.Y., Song, W.J., Eom, S., Kim, S.B., Kang, D.H., 2020. Antimicrobial efficacy of cold plasma treatment against food-borne pathogens on various foods. J. Phys. D Appl. Phys. 53, 204003. https://doi.org/10.1088/1361-6463/AB761F.

Harrison, S.L., Barbosa-Cánovas, G.V., Swanson, B.G., 1997. *Saccharomyces cerevisiae* structural changes induced by pulsed electric field treatment. Lebensm. Wiss. Technol. 30, 236–240. https://doi.org/10.1006/fstl.1996.0171.

Hayert, M., Perrier-Cornet, J.M., Gervais, P., 1999. A simple method for measuring the pH of acid solutions under high pressure. J. Phys. Chem. A 103, 1785–1789. https://doi.org/10.1021/jp983204z.

Hayman, M.M., Anantheswaran, R.C., Knabel, S.J., 2008. Heat shock induces barotolerance in *Listeria monocytogenes*. J. Food Protect. 71, 426–430. https://doi.org/10.4315/0362-028X-71.2.426.

Heinz, V., Alvarez, I., Angersbach, A., Knorr, D., 2001. Preservation of liquid foods by high intensity pulsed electric fields - basic concepts for process design. Trends Food Sci. Technol. 12, 103–111. https://doi.org/10.1016/S0924-2244(01)00064-4.

Hiremath, N.D., Ramaswamy, H.S., 2012. High-pressure destruction kinetics of spoilage and pathogenic microorganisms in mango juice. J. Food Process. Preserv. 36, 113–125. https://doi.org/10.1111/j.1745-4549.2011.00559.x.

Hogan, E., Kelly, A.L., Sun, D.W., 2005. High pressure processing of foods: an overview. In: Sun, D.W. (Ed.), Emerging Technologies for Food Processing. Academic Press, pp. 3–32. https://doi.org/10.1016/B978-012676757-5/50003-7.

Horlacher, N., Oey, I., Ying, S., 2024. Effect of pulsed electric field processing on microbial and enzyme inactivation in blended plant-based milk alternatives: a case study on a microbial challenge test for a non-presterilized oat-based beverage enriched with pea protein. Innov. Food Sci. Emerg. Technol. 94, 103699. https://doi.org/10.1016/j.ifset.2024.103699.

Huang, K., Tian, H., Gai, L., Wang, J., 2012. A review of kinetic models for inactivating microorganisms and enzymes by pulsed electric field processing. J. Food Eng. 111, 191–207. https://doi.org/10.1016/j.jfoodeng.2012.02.007.

Huang, K., Yu, L., Gai, L., Wang, J., 2013. Coupled simulations in colinear and coaxial continuous pulsed electric field treatment chambers. Trans. ASABE 56, 1473–1484. https://doi.org/10.13031/trans.56.9167.

Huang, H.W., Lung, H.M., Yang, B.B., Wang, C.Y., 2014. Responses of microorganisms to high hydrostatic pressure processing. Food Control 40, 250–259. https://doi.org/10.1016/J.FOODCONT.2013.12.007.

Huang, K., Jiang, T., Wang, W., Gai, L., Wang, J., 2014. A comparison of pulsed electric field resistance for three microorganisms with different biological factors in grape juice via numerical simulation. Food Bioprocess Technol. 7, 1981–1995. https://doi.org/10.1007/s11947-014-1272-3.

Hülsheger, H., Potel, J., Niemann, E.-G., 1981. Killing of bacteria with electric pulses of high field strength. Radiat. Environ. Biophys. 20, 53–65. https://doi.org/10.1007/BF01323926.

Hülsheger, H., Potel, J., Niemann, E.G., 1983. Electric field effects on bacteria and yeast cells. Radiat. Environ. Biophys. 22, 149–162. https://doi.org/10.1007/BF01338893.

Jaeger, H., Meneses, N., Knorr, D., 2009a. Impact of PEF treatment inhomogeneity such as electric field distribution, flow characteristics and temperature effects on the inactivation of *E. coli* and milk alkaline phosphatase. Innov. Food Sci. Emerg. Technol. 10, 470–480. https://doi.org/10.1016/j.ifset.2009.03.001.

Jaeger, H., Schulz, A., Karapetkov, N., Knorr, D., 2009b. Protective effect of milk constituents and sublethal injuries limiting process effectiveness during PEF inactivation of *Lb. rhamnosus*. Int. J. Food Microbiol. 134, 154–161. https://doi.org/10.1016/j.ijfoodmicro.2009.06.007.

Jaeger, H., Meneses, N., Moritz, J., Knorr, D., 2010. Model for the differentiation of temperature and electric field effects during thermal assisted PEF processing. J. Food Eng. 100, 109–118. https://doi.org/10.1016/j.jfoodeng.2010.03.034.

Jambrak, A.R., Vukušić, T., Stulić, V., Mrvčić, J., Milošević, S., Šimunek, M., Herceg, Z., 2015. The effect of high power ultrasound and cold gas-phase plasma treatments on selected yeast in pure culture. Food Bioprocess Technol. 8, 791–800. https://doi.org/10.1007/S11947-014-1442-3/FIGURES/3.

Jayaram, S., Castle, G.S.P., Margaritis, A., 1992. Kinetics of sterilization of *Lactobacillus brevis* cells by the application of high voltage pulses. Biotechnol. Bioeng. 40, 1412–1420. https://doi.org/10.1002/bit.260401116.

Kanduser, M., Šentjurc, M., Miklavčič, D., 2008. The temperature effect during pulse application on cell membrane fluidity and permeabilization. Bioelectrochemistry 74, 52–57. https://doi.org/10.1016/j.bioelechem.2008.04.012.

Karatzas, K.A.G., Valdramidis, V.P., Wells-Bennik, M.H.J., 2005. Contingency locus in ctsR of *Listeria monocytogenes* Scott A: a strategy for occurrence of abundant piezotolerant isolates within clonal populations. Appl. Environ. Microbiol. 71, 8390–8396. https://doi.org/10.1128/AEM.71.12.8390-8396.2005.

Kayalvizhi, V., Pushpa, A.J.S., Sangeetha, G., Antony, U., 2016. Effect of pulsed electric field (PEF) treatment on sugarcane juice. J. Food Sci. Technol. 53, 1371–1379. https://doi.org/10.1007/s13197-016-2172-5.

Khan, S.U.D., Khan, R., Hussain, S., 2022. Suitability of thermal plasma for solid waste treatment and non-thermal plasma for nano-scale high-tech plasmonic materials: a concise review. Appl. Nanosci. 12, 3111–3126. https://doi.org/10.1007/S13204-022-02342-8/FIGURES/8.

Klämpfl, T.G., Isbary, G., Shimizu, T., Li, Y.F., Zimmermann, J.L., Stolz, W., Schlegel, J., Morfill, G.E., Schmidt, H.U., 2012. Cold atmospheric air plasma sterilization against spores and other microorganisms of clinical interest. Appl. Environ. Microbiol. 78, 5077–5082. https://doi.org/10.1128/AEM.00583-12.

Koseki, S., Yamamoto, K., 2006. pH and solute concentration of suspension media affect the outcome of high hydrostatic pressure treatment of *Listeria monocytogenes*. Int. J. Food Microbiol. 111, 175–179. https://doi.org/10.1016/j.ijfoodmicro.2006.05.008.

Kotnik, T., Pucihar, G., Miklavčič, D., 2010. Induced transmembrane voltage and its correlation with electroporation- mediated molecular transport. J. Membr. Biol. 236, 3–13. https://doi.org/10.1007/s00232-010-9279-9.

Koutchma, T., 2009. Advances in ultraviolet light technology for non-thermal processing of liquid foods. Food Bioprocess Technol. 2, 138–155. https://doi.org/10.1007/s11947-008-0178-3.

Krug, M., Chapin, T., Danyluk, M., Goodrich-Schneider, R., Schneider, K., Harris, L., Worobo, R., 2020. Outbreaks of foodborne disease associated with fruit and vegetable juices. Edis 2020, 1922–2019. https://doi.org/10.32473/edis-fs188-2020.

Kyrylenko, A., Eijlander, R.T., Alliney, G., de Bos, E.L. van, Wells-Bennik, M.H.J., 2023. Levels and types of microbial contaminants in different plant-based ingredients used in dairy alternatives. Int. J. Food Microbiol. 407, 110392. https://doi.org/10.1016/j.ijfoodmicro.2023.110392.

Lado, B.H., Yousef, A.E., 2003. Selection and identification of a *Listeria monocytogenes* target strain for pulsed electric field process optimization. Appl. Environ. Microbiol. 69, 2223–2229. https://doi.org/10.1128/AEM.69.4.2223-2229.2003.

Laroussi, M., Leipold, F., 2004. Evaluation of the roles of reactive species, heat, and UV radiation in the inactivation of bacterial cells by air plasmas at atmospheric pressure. Int. J. Mass Spectrom. 233, 81–86. https://doi.org/10.1016/J.IJMS.2003.11.016.

Laroussi, M., Karakas, E., Hynes, W., 2011. Influence of cell type, initial concentration, and medium on the inactivation efficiency of low-temperature plasma. IEEE Trans. Plasma Sci. 39, 2960–2961. https://doi.org/10.1109/TPS.2011.2143731.

Lee, J., Lee, C.W., Yong, H.I., Lee, H.J., Jo, C., Jung, S., 2017. Use of atmospheric pressure cold plasma for meat industry. Korean J Food Sci Anim Resour 37, 477–485. https://doi.org/10.5851/KOSFA.2017.37.4.477.

Li, X., Yang, H., 2022. Response of foodborne pathogens to ultraviolet light. In: Stress Responses of Foodborne Pathogens. Springer International Publishing, pp. 315–346. https://doi.org/10.1007/978-3-030-90578-1_11.

Li, S.-Q., Zhang, Q.H., 2006. Inactivation of *E. coli* 8739 in enriched soymilk using pulsed electric fields. J. Food Sci. 69, 169–174. https://doi.org/10.1111/j.1365-2621.2004.tb13616.x.

Li, Y.Q., Tian, W.L., Mo, H.Z., Zhang, Y.L., Zhao, X.Z., 2013. Effects of pulsed electric field processing on quality characteristics and microbial inactivation of soymilk. Food Bioprocess Technol. 6, 1907–1916. https://doi.org/10.1007/s11947-012-0868-8.

Li, L., Yang, R., Zhao, W., 2021. The effect of pulsed electric fields (PEF) combined with temperature and natural preservatives on the quality and microbiological shelf-life of cantaloupe juice. Foods 10, 2606. https://doi.org/10.3390/foods10112606.

Liao, X., Liu, D., Xiang, Q., Ahn, J., Chen, S., Ye, X., Ding, T., 2017. Inactivation mechanisms of non-thermal plasma on microbes: a review. Food Control 75, 83–91. https://doi.org/10.1016/J.FOODCONT.2016.12.021.

Liao, X., Li, J., Muhammad, A.I., Suo, Y., Chen, S., Ye, X., Liu, D., Ding, T., 2018. Application of a dielectric barrier discharge atmospheric cold plasma (Dbd-Acp) for *Escherichia coli* Inactivation in Apple juice. J. Food Sci. 83, 401–408. https://doi.org/10.1111/1750-3841.14045.

Linton, M., McClements, J.M.J., Patterson, M.F., 1999. Inactivation of *Escherichia coli* O157:H7 in orange juice using a combination of high pressure and mild heat. J. Food Protect. 62, 277–279. https://doi.org/10.4315/0362-028X-62.3.277.

Liu, Z., Zhao, L., Zhang, Q., Huo, N., Shi, X., Li, L., Jia, L., Lu, Y., Peng, Y., Song, Y., 2019. Proteomics-based mechanistic investigation of *Escherichia coli* inactivation by pulsed electric field. Front. Microbiol. 10, 1–16. https://doi.org/10.3389/fmicb.2019.02644.

Lombergar, P., Flisar, K., Miklavčič, D., Mahnič-Kalamiza, S., 2024. Time-dependent model of temperature distribution in continuous flow pulsed electric field treatment chambers. Innov. Food Sci. Emerg. Technol. 93, 103628. https://doi.org/10.1016/j.ifset.2024.103628.

Lytras, F., Psakis, G., Gatt, R., Cebrián, G., Raso, J., Valdramidis, V., 2024a. Exploring the efficacy of pulsed electric fields (PEF) in microbial inactivation during food processing: a deep dive into the microbial cellular and molecular mechanisms. Innov. Food Sci. Emerg. Technol. 95, 103732. https://doi.org/10.1016/j.ifset.2024.103732.

Lytras, F., Psakis, G., Gatt, R., Hummerjohann, J., Raso, J., Valdramidis, V., 2024b. Evaluation of strain variability of food microorganisms in response to decontamination by pulsed electric fields and thermal treatments. Innov. Food Sci. Emerg. Technol. 95, 103731. https://doi.org/10.1016/j.ifset.2024.103731.

Ma, R., Wang, G., Tian, Y., Wang, K., Zhang, J., Fang, J., 2015. Non-thermal plasma-activated water inactivation of food-borne pathogen on fresh produce. J. Hazard Mater. 300, 643–651. https://doi.org/10.1016/J.JHAZMAT.2015.07.061.

Maccaferri, C., Sainz-García, A., Capelli, F., Gherardi, M., Alba-Elías, F., Laurita, R., 2023. Evaluation of the antimicrobial efficacy of a large-area surface dielectric barrier discharge on food contact surfaces. Plasma Chem. Plasma Process. 43, 1773–1790. https://doi.org/10.1007/S11090-023-10410-2/FIGURES/11.

Mafart, P., Couvert, O., Gaillard, S., Leguerinel, I., 2001. On calculating sterility in thermal preservation methods: application of the weibull frequency distribution model. Acta Hortic. 566, 107–114. https://doi.org/10.17660/ActaHortic.2001.566.11.

Mañas, P., Barsotti, L., Cheftel, J.C., 2001. Microbial inactivation by pulsed electric fields in a batch treatment chamber: effects of some electrical parameters and food constituents. Innov. Food Sci. Emerg. Technol. 2, 239–249. https://doi.org/10.1016/S1466-8564(01)00041-8.

Manzoor, M.F., Zeng, X.A., Ahmad, N., Ahmed, Z., Rehman, A., Aadil, R.M., Roobab, U., Siddique, R., Rahaman, A., 2020. Effect of pulsed electric field and thermal treatments on the bioactive compounds, enzymes, microbial, and physical stability of almond milk during storage. J. Food Process. Preserv. 44. https://doi.org/10.1111/jfpp.14541.

Marsellés-Fontanet, À.R., Puig, A., Olmos, P., Mínguez-Sanz, S., Martín-Belloso, O., 2009. Optimising the inactivation of grape juice spoilage organisms by pulse electric fields. Int. J. Food Microbiol. 130, 159–165. https://doi.org/10.1016/j.ijfoodmicro.2008.12.034.

Mendes-Oliveira, G., Jin, Z.T., Campanella, O.H., 2020. Modeling the inactivation of *Escherichia coli* O157:H7 and *Salmonella* Typhimurium in juices by pulsed electric fields: the role of the energy density. J. Food Eng. 282, 110001. https://doi.org/10.1016/j.jfoodeng.2020.110001.

Meneses, N., Jaeger, H., Knorr, D., 2011. Minimization of thermal impact by application of electrode cooling in a co-linear pef treatment chamber. J. Food Sci. 76, E536–E543. https://doi.org/10.1111/j.1750-3841.2011.02368.x.

Misra, N.N., Tiwari, B.K., Raghavarao, K.S.M.S., Cullen, P.J., 2011. Nonthermal plasma inactivation of food-borne pathogens. Food Eng. Rev. 3, 159–170. https://doi.org/10.1007/S12393-011-9041-9/FIGURES/1.

Misra, N.N., Yadav, B., Roopesh, M.S., Jo, C., 2019. Cold plasma for effective fungal and mycotoxin control in foods: mechanisms, inactivation effects, and applications. Compr. Rev. Food Sci. Food Saf. 18, 106–120. https://doi.org/10.1111/1541-4337.12398.

Monfort, S., Gayán, E., Raso, J., Condón, S., Álvarez, I., 2010. Evaluation of pulsed electric fields technology for liquid whole egg pasteurization. Food Microbiol. 27, 845–852. https://doi.org/10.1016/j.fm.2010.05.011.

Moody, A., Marx, G., Swanson, B.G., Bermúdez-Aguirre, D., 2014. A comprehensive study on the inactivation of *Escherichia coli* under nonthermal technologies: high hydrostatic pressure, pulsed electric fields and ultrasound. Food Control 37, 305–314. https://doi.org/10.1016/J.FOODCONT.2013.09.052.

Mosqueda-Melgar, J., Raybaudi-Massilia, R.M., Martín-Belloso, O., 2007. Influence of treatment time and pulse frequency on *Salmonella* Enteritidis, *Escherichia coli* and *Listeria monocytogenes* populations inoculated in melon and watermelon juices treated by pulsed electric fields. Int. J. Food Microbiol. 117, 192–200. https://doi.org/10.1016/j.ijfoodmicro.2007.04.009.

Mosqueda-Melgar, J., Raybaudi-Massilia, R.M., Martín-Belloso, O., 2008. Inactivation of *Salmonella enterica* ser. enteritidis in tomato juice by combining of high-intensity pulsed electric fields with natural antimicrobials. J. Food Sci. 73, 47–53. https://doi.org/10.1111/j.1750-3841.2007.00646.x.

Mosqueda-Melgar, J., Raybaudi-Massilia, R.M., Martín-Belloso, O., 2008a. Non-thermal pasteurization of fruit juices by combining high-intensity pulsed electric fields with natural antimicrobials. Innov. Food Sci. Emerg. Technol. 9, 328–340. https://doi.org/10.1016/j.ifset.2007.09.003.

Mosqueda-Melgar, J., Raybaudi-Massilia, R.M., Martín-Belloso, O., 2008b. Combination of high-intensity pulsed electric fields with natural antimicrobials to inactivate pathogenic microorganisms and extend the shelf-life of melon and watermelon juices. Food Microbiol. 25, 479–491. https://doi.org/10.1016/j.fm.2008.01.002.

Muhammad, A.I., 2021. Application of cold plasma in liquid food products. applications of cold plasma in food safety 269–291. https://doi.org/10.1007/978-981-16-1827-7_11/TABLES/2.

Muhammad, A.I., Lv, R., Liao, X., Chen, W., Liu, D., Ye, X., Chen, S., Ding, T., 2019. Modeling the inactivation of *Bacillus cereus* in tiger nut milk treated with cold atmospheric pressure plasma. J. Food Protect. 82, 1828–1836. https://doi.org/10.4315/0362-028X.JFP-18-586.

Muñoz, M., De Ancos, B., Sánchez-Moreno, C., Cano, M.P., 2006. Evaluation of chemical and physical (high-pressure and temperature) treatments to improve the safety of minimally processed mung bean sprouts during refrigerated storage. J. Food Protect. 69, 2395–2402. https://doi.org/10.4315/0362-028X-69.10.2395.

Nakimbugwe, D., Masschalck, B., Anim, G., Michiels, C.W., 2006. Inactivation of gram-negative bacteria in milk and banana juice by hen egg white and lambda lysozyme under high hydrostatic pressure. Int. J. Food Microbiol. 112, 19–25. https://doi.org/10.1016/j.ijfoodmicro.2006.06.010.

Nasiru, M.M., Frimpong, E.B., Muhammad, U., Qian, J., Mustapha, A.T., Yan, W., Zhuang, H., Zhang, J., 2021. Dielectric barrier discharge cold atmospheric plasma: influence of processing parameters on microbial inactivation in meat and meat products. Compr. Rev. Food Sci. Food Saf. 20, 2626–2659. https://doi.org/10.1111/1541-4337.12740.

New Zealand Food Safety, 2024. All Good Bananas Ltd Is Recalling a Specific Batch of its All Good Brand Barista Oat Milk Due to the Possible Presence of Bacillus Cereus. Food Recalls and Complaints. https://www.mpi.govt.nz/food-safety-home/food-recalls-and-complaints/recalled-food-products/all-good-brand-barista-oat-milk/.

Ni, J.B., Ding, C.J., Zhang, J.S., Fang, X.M., Xiao, H.W., 2024. Insight into the surface discharge cold plasma efficient inactivation of Pseudomonas fluorescens in water based on exogenous reactive oxygen and nitrogen species: synergistic mechanism and energy benefits. J. Hazard Mater. 476, 134984. https://doi.org/10.1016/J.JHAZMAT.2024.134984.

Niemira, B.A., 2012. Cold plasma decontamination of foods. Annu. Rev. Food Sci. Technol. 3, 125–142. https://doi.org/10.1146/ANNUREV-FOOD-022811-101132/CITE/REFWORKS.

Nwabor, O.F., Onyeaka, H., Miri, T., Obileke, K., Anumudu, C., Hart, A., 2022. A cold plasma technology for ensuring the microbiological safety and quality of foods. Food Eng. Rev. 14, 535–554. https://doi.org/10.1007/S12393-022-09316-0.

Ochoa-Velasco, C.E., Ávila-Sosa, R., Hernández-Carranza, P., Ruíz-Espinosa, H., Ruiz-López, I.I., Guerrero-Beltrán, J.Á., 2020. Mathematical modeling used to evaluate the effect of uv-c light treatment on microorganisms in liquid foods. Food Eng. Rev. 12, 290–308. https://doi.org/10.1007/s12393-020-09219-y.

Oehmigen, K., Hähnel, M., Brandenburg, R., Wilke, C., Weltmann, K.D., Von Woedtke, T., 2010. The role of acidification for antimicrobial activity of atmospheric pressure plasma in liquids. Plasma Process. Polym. 7, 250–257. https://doi.org/10.1002/PPAP.200900077.

Ohshima, T., Okuyama, K., Sato, M., 2002. Effect of culture temperature on high-voltage pulse sterilization of *Escherichia coli*. J. Electrost. 55, 227–235. https://doi.org/10.1016/S0304-3886(01)00206-6.

Pagán, R., Mackey, B., 2000. Relationship between membrane damage and cell death in pressure-treated *Escherichia coli* cells: differences between exponential- and stationary-phase cells and variation among strains. Appl. Environ. Microbiol. 66, 2829–2834. https://doi.org/10.1128/AEM.66.7.2829-2834.2000.

Pagán, R., Esplugas, S., Góngora-Nieto, M.M., Barbosa-Cánovas, G.V., Swanson, B.G., 1998. Inactivation of *Bacillus subtilis* spores using high intensity pulsed electric fields in combination with other food conservation technologies. Food Sci. Technol. Int. 4, 33–44. https://doi.org/10.1177/108201329800400105.

Pai, K., Timmons, C., Roehm, K.D., Ngo, A., Narayanan, S.S., Ramachandran, A., Jacob, J.D., Ma, L.M., Madihally, S.V., 2018. Investigation of the roles of plasma species generated by surface dielectric barrier discharge. Sci. Rep. 8, 1–13. https://doi.org/10.1038/s41598-018-35166-0.

Pan, Y., Cheng, J.H., Lv, X., Sun, D.W., 2019. Assessing the inactivation efficiency of Ar/O2 plasma treatment against *Listeria monocytogenes* cells: sublethal injury and inactivation kinetics. Lebensm. Wiss. Technol. 111, 318–327. https://doi.org/10.1016/J.LWT.2019.05.041.

Pankaj, S.K., Wan, Z., Keener, K.M., 2018. Effects of cold plasma on food quality: a review. Foods 7, 4. https://doi.org/10.3390/FOODS7010004.

Pataro, G., Ferrari, G., 2020. Limitations of Pulsed Electric Field Utilization in Food Industry, Pulsed Electric Fields to Obtain Healthier and Sustainable Food for Tomorrow. INC. https://doi.org/10.1016/B978-0-12-816402-0.00013-6.

Patterson, M.F., 2005. Microbiology of pressure-treated foods. J. Appl. Microbiol. 98, 1400–1409. https://doi.org/10.1111/j.1365-2672.2005.02564.x.

Picart, L., Dumay, E., Cheftel, J.C., 2002. Inactivation of *Listeria innocua* in dairy fluids by pulsed electric fields: influence of electric parameters and food composition. Innov. Food Sci. Emerg. Technol. 3, 357–369. https://doi.org/10.1016/S1466-8564(02)00055-3.

Pilavtepe-Çelik, M., Buzrul, S., Alpas, H., Bozoğlu, F., 2009. Development of a new mathematical model for inactivation of *Escherichia coli* O157:H7 and *Staphylococcus aureus* by high hydrostatic pressure in carrot juice and peptone water. J. Food Eng. 90, 388–394. https://doi.org/10.1016/J.JFOODENG.2008.06.043.

Possas, A., Pérez-Rodríguez, F., Valero, A., García-Gimeno, R.M., 2017. Modelling the inactivation of *Listeria monocytogenes* by high hydrostatic pressure processing in foods: a review. Trends Food Sci. Technol. 70, 45–55. https://doi.org/10.1016/j.tifs.2017.10.006.

Possas, A., Valero, A., García-Gimeno, R.M., Pérez-Rodríguez, F., de Souza, P.M., 2018. Influence of temperature on the inactivation kinetics of *Salmonella* Enteritidis by the application of UV-C technology in soymilk. Food Control 94, 132–139. https://doi.org/10.1016/j.foodcont.2018.06.033.

Pothakamury, U.R., Barbosa-Cánovas, G.V., Swanson, B.G., Spence, K.D., 1997. Ultrastructural changes in *Staphylococcus aureus* treated with pulsed electric fields. Food Sci. Technol. Int. 3, 113–121. https://doi.org/10.1177/108201329700300206.

Powell, K.T., Weaver, J.C., 1986. Transient aqueous pores in bilayer membranes: a statistical theory. Bioelectrochem. Bioenerg. 15, 211–227. https://doi.org/10.1016/0302-4598(86)80029-0.

Precedence Research, 2023. Plant-based Beverages Market Size, Share, and Trends 2024 to 2034. Plant-based Beverages Market. https://www.precedenceresearch.com/plant-based-beverages-market#:~:text=The global plant-based beverages,known as plant-based beverages.

Public Health Agency of Canada, 2024. Public Health Notice: Outbreak of Listeria Infections Linked to Recalled Plant-Based Refrigerated Beverages. Public Health Notices. https://www.canada.ca/en/public-health/services/public-health-notices/2024/outbreak-listeria-infections-recalled-refrigerated-plant-based-beverages.html.

Quiroz-González, B., Rodríguez-Martínez, V., García-Mateos, M.R., Torres, J.A., Welti-Chanes, J., 2018. High hydrostatic pressure inactivation and recovery study of *Listeria innocua* and *Saccharomyces cerevisiae* in pitaya (*Stenocereus pruinosus*) juice. Innov. Food Sci. Emerg. Technol. 50, 169–173. https://doi.org/10.1016/j.ifset.2018.10.003.

Raguse, M., Fiebrandt, M., Denis, B., Stapelmann, K., Eichenberger, P., Driks, A., Eaton, P., Awakowicz, P., Moeller, R., 2016. Understanding of the importance of the spore coat structure and pigmentation in the *Bacillus subtilis* spore resistance to low-pressure plasma sterilization. J. Phys. D Appl. Phys. 49, 285401. https://doi.org/10.1088/0022-3727/49/28/285401.

Ramaswamy, H.S., Riahi, E., Idziak, E., 2003. High-pressure destruction kinetics of *E. coli* (29055) in apple juice. J. Food Sci. 68, 1750–1756. https://doi.org/10.1111/j.1365-2621.2003.tb12323.x.

Raso, J., Frey, W., Ferrari, G., Pataro, G., Knorr, D., Teissie, J., Miklavčič, D., 2016. Recommendations guidelines on the key information to be reported in studies of application of PEF technology in food and biotechnological processes. Innov. Food Sci. Emerg. Technol. 37, 312–321. https://doi.org/10.1016/j.ifset.2016.08.003.

Rauscher, H., Perucca, M., Buyle, G., 2010. Plasma Technology for Hyperfunctional Surfaces: Food, Biomedical, and Textile Applications. Plasma Technology for Hyperfunctional Surfaces: Food, Biomedical, and Textile Applications. https://doi.org/10.1002/9783527630455.

Rendueles, E., Omer, M.K., Alvseike, O., Alonso-calleja, C., Capita, R., Prieto, M., 2011. Microbiological food safety assessment of high hydrostatic pressure processing: a review. Lebensm. Wiss. Technol. 44, 1251–1260. https://doi.org/10.1016/j.lwt.2010.11.001.

Reyns, K.M.F.A., Diels, A.M.J., Michiels, C.W., 2004. Generation of bactericidal and mutagenic components by pulsed electric field treatment. Int. J. Food Microbiol. 93, 165–173. https://doi.org/10.1016/j.ijfoodmicro.2003.10.014.

Ritz, M., Tholozan, J.L., Federighi, M., Pilet, M.F., 2002. Physiological damages of *Listeria monocytogenes* treated by high hydrostatic pressure. Int. J. Food Microbiol. 79, 47–53. https://doi.org/10.1016/S0168-1605(02)00178-2.

Ritz, M., Pilet, M.F., Jugiau, F., Rama, F., Federighi, M., 2006. Inactivation of *Salmonella* Typhimurium and *Listeria monocytogenes* using high-pressure treatments: destruction or sublethal stress? Lett. Appl. Microbiol. 42, 357–362. https://doi.org/10.1111/j.1472-765X.2006.01868.x.

Rivas, A., Pina-Pérez, M.C., Rodriguez-Vargas, S., Zuñiga, M., Martinez, A., Rodrigo, D., 2013. Sublethally damaged cells of *Escherichia coli* by pulsed electric fields: the chance of transformation and proteomic assays. Food Res. Int. 54, 1120–1127. https://doi.org/10.1016/j.foodres.2013.01.014.

Rodrigo, D., Martínez, A., Martínez, M., Harte, F., Barbosa-Cánovas, G.V., Cánovas, C., Rodrigo, A.M., 2001. Study of inactivation of *Lactobacillus plantarum* in orange-carrot juice by means of pulsed electric fields: comparison of inactivation kinetics models. J. Food Protect. 64, 259–263. https://doi.org/10.4315/0362-028X-64.2.259.

Rivas, A., Sampedro, F., Rodrigo, D., Martínez, A., Rodrigo, M., 2006. Nature of the inactivation of *Escherichia coli* suspended in an orange juice and milk beverage. Eur. Food Res. Technol. 223, 541–545. https://doi.org/10.1007/s00217-005-0234-4.

Rodrigo, D., Barbosa-Cánovas, G.V., Cánovas, C., Martínez, A., Martínez, M., Rodrigo, A.M., 2003. Weibull distribution function based on an empirical mathematical model for inactivation of *Escherichia coli* by pulsed electric fields. J. Food Protect. 6, 1007–1012. https://doi.org/10.4315/0362-028X-66.6.1007.

Russell, N.J., Colley, M., Simpson, R.K., Trivett, A.J., Evans, R.I., 2000. Mechanism of action of pulsed high electric field (PHEF) on the membranes of food-poisoning bacteria is an "all-or-nothing" effect. Int. J. Food Microbiol. 55, 133–136. https://doi.org/10.1016/S0168-1605(00)00169-0.

Safa, S., Soucy, G., 2015. Application of a novel CO2 dc thermal plasma torch submerged in aqueous solution for treatment of dissolved carboxylic acid. Plasma Chem. Plasma Process. 35, 21–43. https://doi.org/10.1007/S11090-014-9577-6/FIGURES/11.

Saldaña, G., Puértolas, E., Monfort, S., Raso, J., Álvarez, I., 2011. Defining treatment conditions for pulsed electric field pasteurization of apple juice. Int. J. Food Microbiol. 151, 29–35. https://doi.org/10.1016/j.ijfoodmicro.2011.07.033.

Sastry, S.K., 2023. Modeling of pulsed electric field processing. Food Innovation and Advances 2, 171–183. https://doi.org/10.48130/fia-2023-0012.

Saulis, G., Wouters, P., 2007. Probable mechanisms of microorganism inactivation by pulsed electric fields. Food Preservation by Pulsed Electric Fields 10, 138–155. https://doi.org/10.1201/9781439823866.ch9.

Scholtz, V., Julák, J., Kříha, V., 2010. The microbicidal effect of low-temperature plasma generated by corona discharge: comparison of various microorganisms on an agar surface or in aqueous suspension. Plasma Process. Polym. 7, 237–243. https://doi.org/10.1002/PPAP.200900072.

Schottroff, F., Krottenthaler, A., Jäger, H., 2017. Stress induction and response, inactivation, and recovery of vegetative microorganisms by pulsed electric fields. In: Handbook of Electroporation. Springer International Publishing, pp. 2539–2557. https://doi.org/10.1007/978-3-319-32886-7_183.

Schottroff, F., Fröhling, A., Zunabovic-Pichler, M., Krottenthaler, A., Schlüter, O., Jäger, H., 2018. Sublethal injury and Viable but Non-culturable (VBNC) state in microorganisms during preservation of food and biological materials by non-thermal processes. Front. Microbiol. 9, 2773. https://doi.org/10.3389/fmicb.2018.02773.

Schottroff, F., Gratz, M., Krottenthaler, A., Johnson, N.B., Bédard, M.F., Jaeger, H., 2019. Pulsed electric field preservation of liquid whey protein formulations – influence of process parameters, pH, and protein content on the inactivation of *Listeria innocua* and the retention of bioactive ingredients. J. Food Eng. 243, 142–152. https://doi.org/10.1016/j.jfoodeng.2018.09.003.

Schottroff, F., Johnson, K., Johnson, N.B., Bédard, M.F., Jaeger, H., 2020a. Challenges and limitations for the decontamination of high solids protein solutions at neutral pH using pulsed electric fields. J. Food Eng. 268, 109737. https://doi.org/10.1016/j.jfoodeng.2019.109737.

Schottroff, F., Knappert, J., Eppmann, P., Krottenthaler, A., Horneber, T., McHardy, C., Rauh, C., Jaeger, H., 2020b. Development of a Continuous Pulsed Electric Field (PEF) vortex-flow chamber for improved treatment homogeneity based on hydrodynamic optimization. Front. Bioeng. Biotechnol. 8, 340. https://doi.org/10.3389/fbioe.2020.00340.

Scolari, G., Zacconi, C., Busconi, M., Lambri, M., 2015. Effect of the combined treatments of high hydrostatic pressure and temperature on *Zygosaccharomyces bailii* and *Listeria monocytogenes* in smoothies. Food Control 47, 166–174. https://doi.org/10.1016/J.FOODCONT.2014.07.006.

Siemer, C., Toepfl, S., Heinz, V., 2014. Inactivation of Bacillus subtilis spores by pulsed electric fields (PEF) in combination with thermal energy II. Modeling thermal inactivation of *B. subtilis* spores during PEF processing in combination with thermal energy. Food Control 39, 244–250. https://doi.org/10.1016/j.foodcont.2013.09.067.

Simpson, R.K., Whittington, R., Earnshaw, R.G., Russell, N.J., 1999. Pulsed high electric field causes "all or nothing" membrane damage in *Listeria monocytogenes* and *Salmonella* typhimurium, but membrane H+-ATPase is not a primary target. Int. J. Food Microbiol. 48, 1–10. https://doi.org/10.1016/S0168-1605(99)00022-7.

Singh, J., Singh, M., Singh, B., Nayak, M., Ghanshyam, C., 2017. Comparative analyses of prediction models for inactivation of *Escherichia coli* in carrot juice by means of pulsed electric fields. J. Food Sci. Technol. 54, 1538–1544. https://doi.org/10.1007/s13197-017-2585-9.

Smelt, J.P.P.M., 1998. Recent advances in the microbiology of high pressure processing i, 152–158.

Smet, C., Noriega, E., Rosier, F., Walsh, J.L., Valdramidis, V.P., Van Impe, J.F., 2016. Influence of food intrinsic factors on the inactivation efficacy of cold atmospheric plasma: impact of osmotic stress, suboptimal pH and food structure. Innov. Food Sci. Emerg. Technol. 38, 393–406. https://doi.org/10.1016/J.IFSET.2016.09.028.

Somolinos, M., García, D., Mañas, P., Condón, S., Pagán, R., 2008. Effect of environmental factors and cell physiological state on Pulsed Electric Fields resistance and repair capacity of various strains of *Escherichia coli*. Int. J. Food Microbiol. 124, 260–267. https://doi.org/10.1016/j.ijfoodmicro.2008.03.021.

Statista, 2025. Juices - worldwide market outlook. URL Juices - Worldwide market outlook. (accessed 1.20.25).

Surowsky, B., Fröhling, A., Gottschalk, N., Schlüter, O., Knorr, D., 2014. Impact of cold plasma on Citrobacter freundii in apple juice: inactivation kinetics and mechanisms. Int. J. Food Microbiol. 174, 63–71. https://doi.org/10.1016/J.IJFOODMICRO.2013.12.031.

Surowsky, B., Schlüter, O., Knorr, D., 2015. Interactions of non-thermal atmospheric pressure plasma with solid and liquid food systems: a review. Food Eng. Rev. 7, 82–108. https://doi.org/10.1007/S12393-014-9088-5/TABLES/9.

Tchonkouang, R.D., Lima, A.R., Quintino, A.C., Cristofoli, N.L., Vieira, M.C., 2023. UV-C light: a promising preservation technology for vegetable-based nonsolid food products. Foods 27, 3227. https://doi.org/10.3390/foods12173227.

Thamsuaidee, A., Schaefer, E., Schneider, D., Siemer, C., Valdramidis, P., 2024. Disentangling the effects of electroporation and heat during pulsed electric field (PEF) processing of oat-based milk alternative: a case study on *Lactiplantibacillus plantarum* inactivation. Innov. Food Sci. Emerg. Technol. 94, 103691. https://doi.org/10.1016/j.ifset.2024.103691.

Thirumdas, R., Sarangapani, C., Annapure, U.S., 2015. Cold plasma: a novel non-thermal technology for food processing. Food Biophys. 10, 1–11. https://doi.org/10.1007/S11483-014-9382-Z/FIGURES/3.

Timmermans, R.A.H., Nierop Groot, M.N., Nederhoff, A.L., van Boekel, M.A.J.S., Matser, A. M., Mastwijk, H.C., 2014. Pulsed electric field processing of different fruit juices: impact of pH and temperature on inactivation of spoilage and pathogenic micro-organisms. Int. J. Food Microbiol. 173, 105–111. https://doi.org/10.1016/j.ijfoodmicro.2013.12.022.

Timmermans, R.A.H., Nederhoff, A.L., Nierop Groot, M.N., van Boekel, M.A.J.S., Mastwijk, H.C., 2016. Effect of electrical field strength applied by PEF processing and storage temperature on the outgrowth of yeasts and moulds naturally present in a fresh fruit smoothie. Int. J. Food Microbiol. 230, 21–30. https://doi.org/10.1016/j.ijfoodmicro.2016.04.014.

Timmermans, R.A.H., Mastwijk, H.C., Berendsen, L.B.J.M., Nederhoff, A.L., Matser, A.M., Van Boekel, M.A.J.S., Nierop Groot, M.N., 2019. Moderate intensity pulsed electric fields (PEF) as alternative mild preservation technology for fruit juice. Int. J. Food Microbiol. 298, 63–73. https://doi.org/10.1016/j.ijfoodmicro.2019.02.015.

Timmermans, R.A.H., Roland, W.S.U., van Kekem, K., Matser, A.M., van Boekel, M.A.J.S., 2022. Effect of pasteurization by moderate intensity Pulsed Electric Fields (PEF) treatment compared to thermal treatment on quality attributes of fresh orange juice. Foods 11. https://doi.org/10.3390/foods11213360.

Toepfl, S., 2011. Pulsed electric field food treatment - scale up from lab to industrial scale. Procedia Food Sci. 1, 776–779. https://doi.org/10.1016/j.profoo.2011.09.117.

Toepfl, S., Heinz, V., Knorr, D., 2005. Overview of pulsed electric field processing for food. In: Emerging Technologies for Food Processing. Elsevier, pp. 69–97. https://doi.org/10.1016/B978-012676757-5/50006-2.

Tongdonyod, S., Thikham, S., Kittiwachana, S., Wichaphon, J., Klangpetch, W., 2023. Optimization of pulsed electric fields combined with mild heat treatment on microbial inactivation of tender coconut water and evaluation of quality attributes during storage. Innov. Food Sci. Emerg. Technol. 90, 103507. https://doi.org/10.1016/j.ifset.2023.103507.

Torres, E., González-M, G., Klotz, B., Rodrigo, D., 2016. Effects of high hydrostatic pressure and temperature increase on *Escherichia coli* spp. and pectin methyl esterase inactivation in orange juice. Food Sci. Technol. Int. 22, 173–180. https://doi.org/10.1177/1082013215582107.

Tsagkaropoulou, T., Karatzas, K.A.G., 2024. Microbial species and strain heterogeneity affect resistance to high pressure processing. Innov. Food Sci. Emerg. Technol. 94, 103645. https://doi.org/10.1016/j.ifset.2024.103645.

Ulmer, H.M., Heinz, V., Gänzle, M.G., Knorr, D., Vogel, R.F., 2002. Effects of pulsed electric fields on inactivation and metabolic activity of *Lactobacillus plantarum* in model beer. J. Appl. Microbiol. 93, 326–335. https://doi.org/10.1046/j.1365-2672.2002.01699.x.

Usaga, J., Acosta, Ó., Churey, J.J., Padilla-Zakour, O.I., Worobo, R.W., 2021. Evaluation of high pressure processing (HPP) inactivation of *Escherichia coli* O157:H7, *Salmonella enterica*, and *Listeria monocytogenes* in acid and acidified juices and beverages. Int. J. Food Microbiol. 339, 109034. https://doi.org/10.1016/J.IJFOODMICRO.2020.109034.

Valdez-Narváez, M.I., Fernández-Felipe, M.T., Martinez, A., Rodrigo, D., 2024. Inactivation of *Bacillus cereus* spores and vegetative cells in inert matrix and rice grains using low-pressure cold plasma. Foods 13, 2223. https://doi.org/10.3390/FOODS13142223.

Valič, B., Golzio, M., Pavlin, M., Schatz, A., Faurie, C., Gabriel, B., Teissié, J., Rols, M.P., Miklavčič, D., 2003. Effect of electric field induced transmembrane potential on spheroidal cells: theory and experiment. Eur. Biophys. J. 32, 519–528. https://doi.org/10.1007/s00249-003-0296-9.

Van Boekel, M.A.J.S., 2002. On the use of the Weibull model to describe thermal inactivation of microbial vegetative cells. Int. J. Food Microbiol. 74, 139–159. https://doi.org/10.1016/S0168-1605(01)00742-5.

Van Opstal, I., Vanmuysen, S.C.M., Wuytack, E.Y., Masschalck, B., Michiels, C.W., 2005. Inactivation of *Escherichia coli* by high hydrostatic pressure at different temperatures in buffer and carrot juice. Int. J. Food Microbiol. 98, 179–191. https://doi.org/10.1016/j.ijfoodmicro.2004.05.022.

Vega-Mercado, H., Martín-Belloso, O., Chang, F.-J., Barbosa-Cánovas, G.V., Swanson, B.G., 1996. Inactivation of *Escherichia coli* and *Bacillus subtilis* suspended in pea soup using pulsed electric fields. J. Food Process. Preserv. 20, 501–510. https://doi.org/10.1111/j.1745-4549.1996.tb00762.x.

Walkling-Ribeiro, M., Noci, F., Cronin, D.A., Lyng, J.G., Morgan, D.J., 2008. Inactivation of *Escherichia coli* in a tropical fruit smoothie by a combination of heat and pulsed electric fields. J. Food Sci. 73, M395–M399. https://doi.org/10.1111/j.1750-3841.2008.00927.x.

Walkling-Ribeiro, M., Noci, F., Cronin, D.A., Lyng, J.G., Morgan, D.J., 2010. Shelf life and sensory attributes of a fruit smoothie-type beverage processed with moderate heat and pulsed electric fields. Lebensm. Wiss. Technol. 43, 1067–1073. https://doi.org/10.1016/j.lwt.2010.02.010.

Wang, L.H., Wang, M.S., Zeng, X.A., Liu, Z.W., 2016. Temperature-mediated variations in cellular membrane fatty acid composition of *Staphylococcus aureus* in resistance to pulsed electric fields. Biochim. Biophys. Acta Biomembr. 1858, 1791–1800. https://doi.org/10.1016/j.bbamem.2016.05.003.

Wang, Y., Wang, T., Yuan, Y., Fan, Y., Guo, K., Yue, T., 2018. Inactivation of yeast in apple juice using gas-phase surface discharge plasma treatment with a spray reactor. Lebensm. Wiss. Technol. 97, 530–536. https://doi.org/10.1016/J.LWT.2018.07.049.

Wei, W., Yang, S., Yang, F., Hu, X., Wang, Y., Guo, W., Yang, B., Xiao, X., Zhu, L., 2023. Cold plasma controls nitrite hazards by modulating microbial communities in pickled radish. Foods 12, 2550. https://doi.org/10.3390/FOODS12132550.

Witt, J., Siemer, C., Bostelmann, A., Toepfl, S., 2021. Juice preservation by pulsed electric fields. In: Innovative Food Processing Technologies. Elsevier, pp. 305–317. https://doi.org/10.1016/b978-0-12-815781-7.00026-3.

Wouters, P.C., Dutreux, N., Smelt, J.P.P.M., Lelieveld, H.L.M., 1999. Effects of pulsed electric fields on inactivation kinetics of *Listeria innocua*. Appl. Environ. Microbiol. 65, 5364–5371. https://doi.org/10.1128/aem.65.12.5364-5371.1999.

Wuytack, E.Y., Phuong, L.D.T., Aertsen, A., Reyns, K.M.F., Marquenie, D., De Ketelaere, B., Masschalck, B., Van Opstal, I., Diels, A.M.J., Michiels, C.W., 2003. Comparison of sublethal injury induced in *Salmonella enterica* serovar Typhimurium by heat and by different nonthermal treatments. J. Food Protect. 66, 31–37. https://doi.org/10.4315/0362-028X-66.1.31.

Xu, H., Lee, H.-Y., Ahn, J., 2009. High pressure inactivation kinetics of *Salmonella enterica* and *Listeria monocytogenes* in milk, orange juice, and tomato juice. Food Sci. Biotechnol. 18, 861–866.

Yan, B., Nakamura, T., Katsuki, S., Masuda, N., Shimizu, Y., Sasahara, R., Kurihara, J., 2025. Optimization of co-linear-type pulsed electric field treatment chamber for milk pasteurization. J. Food Eng. 388. https://doi.org/10.1016/j.jfoodeng.2024.112373.

Zhao, Y., Shao, L., Jia, L., Meng, Z., Liu, Y., Wang, Y., Zou, B., Dai, R., Li, X., Jia, F., 2022. Subcellular inactivation mechanisms of *Pseudomonas aeruginosa* treated by cold atmospheric plasma and application on chicken breasts. Food Res. Int. 160, 111720. https://doi.org/10.1016/J.FOODRES.2022.111720.

Zhong, K., Chen, F., Wang, Z., Wu, J., Liao, X., Hu, X., 2005. Inactivation and kinetic model for the *Escherichia coli* treated by a co-axial pulsed electric field. Eur. Food Res. Technol. 221, 752–758. https://doi.org/10.1007/s00217-005-0015-0.

Zhou, R., Zhou, R., Mai-Prochnow, A., Zhang, X., Xian, Y., Cullen, P.J., Ostrikov, K., 2020. Surface plasma discharges for the preservation of fresh-cut apples: microbial inactivation and quality attributes. J. Phys. D Appl. Phys. 53, 174003. https://doi.org/10.1088/1361-6463/AB6F8F.

Zhu, S., Wang, C., Ramaswamy, H.S., Yu, Y., 2017. Phase transitions during high pressure treatment of frozen carrot juice and influence on *Escherichia coli* inactivation. Lebensm. Wiss. Technol. 79, 119–125. https://doi.org/10.1016/j.lwt.2017.01.022.

Zhu, Z., Bassey, A.P., Huang, T., Zhang, Y., Ali Khan, I., Huang, M., 2022. The formation, germination, and cold plasma inactivation of bacterial spore. Food Chemistry Advances 1, 100056. https://doi.org/10.1016/J.FOCHA.2022.100056.

Zwietering, M.H., Garre, A., den Besten, H.M.W., 2021. Incorporating strain variability in the design of heat treatments: a stochastic approach and a kinetic approach. Food Res. Int. 139, 109973. https://doi.org/10.1016/j.foodres.2020.109973.

Modeling of mechanical food processes

12

Reza Abedi-Firoozjah[1,2], *Hananeh Yazdanbakhsh*[2], *Farinaz Esmi*[3,4], *Mohammad Yousefi*[5] *and Seid Mahdi Jafari*[6]

[1]Student Research Committee, Kermanshah University of Medical Sciences, Kermanshah, Iran; [2]Department of Food Science and Technology, School of Nutrition Sciences and Food Technology, Kermanshah University of Medical Sciences, Kermanshah, Iran; [3]Student Research Committee, Tabriz University of Medical Sciences, Tabriz, Iran; [4]Department of Food Science and Technology, Faculty of Nutrition and Food Sciences, Tabriz University of Medical Sciences, Tabriz, Iran; [5]Food and Beverages Safety Research Center, Urmia University of Medical Sciences, Urmia, Iran; [6]Department of Food Materials and Process Design Engineering, Faculty of Food Science and Technology, Gorgan University of Agricultural Sciences and Natural Resources, Gorgan, Iran

12.1 Introduction

Mechanical food processing constitutes a cornerstone of modern food manufacturing, enabling the efficient transformation of raw agricultural materials into safe, stable, and consumer-acceptable food products through the application of mechanized equipment and engineered unit operations (Khan et al., 2022). Processes such as mixing, grinding, cutting, extrusion, and drying induce controlled modifications in the physical and structural characteristics of food matrices, thereby determining key quality attributes including texture, stability, moisture distribution, and sensory performance (Bidyalakshmi et al., 2025). Compared to traditional manual practices, mechanical processing offers superior consistency, scalability, and process control, which are essential for meeting the demands of contemporary food supply chains (Szpicer et al., 2025). Driven by increasing global consumption and heightened expectations regarding quality, efficiency, and sustainability, the food industry has progressively adopted advanced mechanical technologies to enhance productivity while optimizing energy and resource utilization. Modern mechanical processing systems not only reduce labor intensity and processing time but also contribute to improved food safety, preservation of nutritional value, and reliable large-scale production (Ghiasi and Golmakani 2022; Khan et al., 2022). Equipment such as size-reduction mills, industrial mixers, continuous extruders, and automated sorting systems is specifically designed to handle complex material behaviors and operational challenges inherent to industrial food processing environments. In parallel with these technological advances, computational modeling and simulation tools, most notably Finite Element Analysis (FEA) and Computational Fluid Dynamics (CFD) have gained increasing importance in the analysis and optimization of mechanical food processing

Modeling of Unit Operations in the Food Industry. https://doi.org/10.1016/B978-0-12-822180-8.00012-2

operations (Marchand et al., 2024). These numerical approaches enable detailed investigation of flow dynamics, heat and mass transfer, stress distribution, and material deformation under realistic processing conditions (Hu et al., 2023). By facilitating virtual experimentation, computational modeling reduces reliance on costly trial-and-error approaches and supports rational process design, scale-up, and performance optimization. Importantly, such tools provide mechanistic insight into the relationships between process parameters, product quality attributes, and energy efficiency (Said and Sarbon 2022; Wang et al., 2022). More recently, the integration of computational modeling with data-driven methodologies, including machine learning algorithms and digital twin frameworks, has opened new avenues for predictive process control, real-time monitoring, and adaptive optimization in automated food manufacturing systems. These developments align closely with Industry 4.0 paradigms, promoting smart, flexible, and sustainable processing strategies capable of responding to variability in raw materials and evolving market requirements (Delfiya et al., 2022; Nadimi et al., 2023). Nevertheless, significant challenges persist, particularly in the accurate representation of complex food materials that often exhibit nonNewtonian, multiphase, or viscoelastic behavior, as well as in the experimental validation of simulation results. Overcoming these challenges is critical for developing robust, reliable, and cost-effective modeling frameworks that can fully support innovation in mechanical food engineering. The aim of this study is to critically review recent applications of computational modeling, particularly CFD, in mechanical food processing (Khashkhashi-Moghadam et al., 2022). Figure 12.1 provides a visual summary of the process discussed in this chapter.

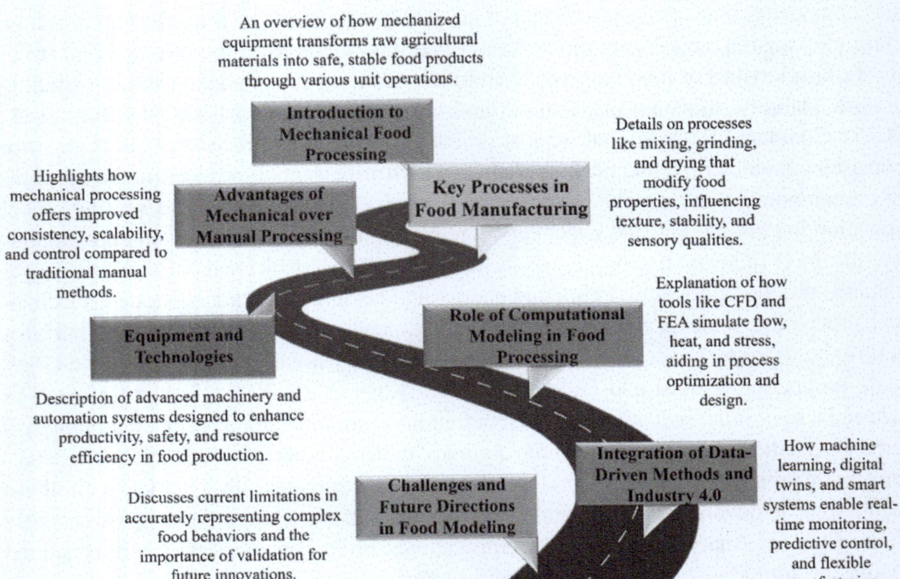

Figure 12.1 Overview of how mechanical processing and advanced modeling techniques enhance efficiency, quality, and innovation in modern food manufacturing.

This Chapter focuses on key unit operations such as mixing, extrusion, and drying, highlighting current modeling approaches, major achievements, existing challenges, and emerging trends to support process optimization and innovation in food manufacturing.

12.2 Fundamental concepts and governing equations

Mechanical food processing operations, including mixing, extrusion, drying, and grinding, are essential for transforming raw agricultural materials into safe, shelf-stable, and high-quality products with desirable sensory and nutritional characteristics (Iftekar et al., 2023). These processes inherently involve complex interactions among fluid flow, heat and mass transfer, mechanical stresses, and material deformation, all of which must be accurately understood to optimize processing performance and ensure product consistency (Mancuso et al., 2023). The underlying physics of these operations can be rigorously described using the principles of continuum mechanics, thermodynamics, and rheology, which together provide the foundation for predictive computational modeling. Among various modeling approaches, CFD has emerged as a powerful and versatile tool in food engineering, allowing detailed simulation of flow patterns, temperature fields, moisture transport, and stress distributions under a wide range of processing conditions (Taneja et al., 2023; Gupta et al., 2025). At the core of CFD are the fundamental conservation equations: the continuity equation, which ensures mass conservation by accounting for spatial and temporal variations in fluid density and velocity; the Navier–Stokes equations, which describe momentum conservation by capturing the influence of pressure gradients, viscous stresses, and external forces; and the energy equation, which governs heat transfer through conduction, convection, and internal energy generation (Marano et al., 2022). Accurate representation of food materials requires incorporation of constitutive relationships that account for nonNewtonian, viscoelastic, and multiphase behaviors commonly observed in doughs, pastes, emulsions, and suspensions (Mora and Toldrá 2023). Models such as the Power-law, Herschel–Bulkley, and Maxwell-based viscoelastic formulations are frequently employed to relate shear stress and strain rate, allowing realistic simulation of deformation and flow under varying shear conditions. Advanced CFD frameworks often integrate additional submodels for turbulence, multiphase flow, and species transport, enabling capture of complex phenomena such as droplet dispersion, bubble formation, phase transitions, and moisture migration (Abideen et al., 2021; Marano et al., 2022). These numerical simulations provide insights into the effects of process parameters, including impeller geometry, screw speed, temperature gradients, and residence time distributions on product quality attributes such as texture, porosity, moisture uniformity, and structural integrity (Ng et al., 2024). Despite the considerable capabilities of CFD and related modeling techniques, several challenges remain: accurate characterization of thermophysical and rheological properties, validation against experimental observations, computational cost, and the treatment of highly deformable or multiphase food systems (Yazdani et al., 2022). Addressing these challenges requires careful integration of experimental data, constitutive modeling, and advanced numerical schemes. Nevertheless, by providing detailed, predictive, and mechanistic understanding of mechanical food processing, CFD and related computational approaches have become indispensable tools for process optimization, equipment design, scale-up, and the development of innovative food manufacturing technologies, ultimately contributing to improved efficiency, sustainability, and product quality in modern food engineering (Chang et al., 2022). A thorough understanding and accurate mathematical representation of mass, momentum, energy, and rheological behavior are essential for reliable CFD modeling and optimization of mechanical food processing operations Fig. 12.2.

12.2.1 Classification of modeling approaches

Modeling approaches in mechanical food processing encompass a wide spectrum of methodologies, ranging from purely empirical correlations to sophisticated computational simulations, each with distinct theoretical foundations, levels of complexity, and intended applications (Beć et al., 2022). Empirical models are derived from experimental observations and statistical analyses, providing simplified mathematical relationships between process variables such as temperature, shear rate, or residence time and key product quality attributes, including texture, moisture content, and structural integrity (Eslami et al., 2023). These models are advantageous due to their simplicity and low computational cost, making them particularly useful in preliminary design and rapid process assessments (Yazdani et al., 2022; Islam et al., 2024). However, their applicability is inherently limited to the conditions under which they were developed, and they generally lack a mechanistic understanding of the underlying physical phenomena, which restricts their predictive capabilities for novel formulations, nonstandard equipment geometries, or scaled-up industrial processes (Barbaros et al., 2022). Mechanistic models, on the other hand, are grounded in first principles, including conservation of mass, momentum, and energy, complemented by constitutive equations that describe the rheological and thermal behavior of food materials (Pathare and Al-Dairi 2022). By representing the actual physics of fluid flow, heat and mass transfer, and material deformation, mechanistic models enable detailed analysis of process dynamics and facilitate optimization beyond the experimental domain. Within mechanistic modeling, continuum-based approaches such as CFD and FEA have gained widespread adoption (Emin et al., 2021; Oyinloye and Yoon 2021). CFD is particularly suitable for simulating complex fluid flow, shear stress distribution, and multiphase interactions in operations like mixing, extrusion, and drying, whereas FEA is commonly used to investigate solid deformation, structural integrity, and stress–strain relationships in viscoelastic or deformable food matrices (Pandiselvam et al., 2021). The predictive power of these models allows engineers to visualize flow patterns, temperature gradients, moisture distribution, and mechanical stresses, providing crucial insights for equipment design, process scale-up, and quality control (Narsaiah et al., 2021). In recent years, hybrid modeling approaches that integrate mechanistic simulations with empirical data or data-driven techniques, including machine learning algorithms, artificial neural networks, and digital twin frameworks, have emerged as a promising direction. These hybrid models not only enhance predictive accuracy but also reduce computational

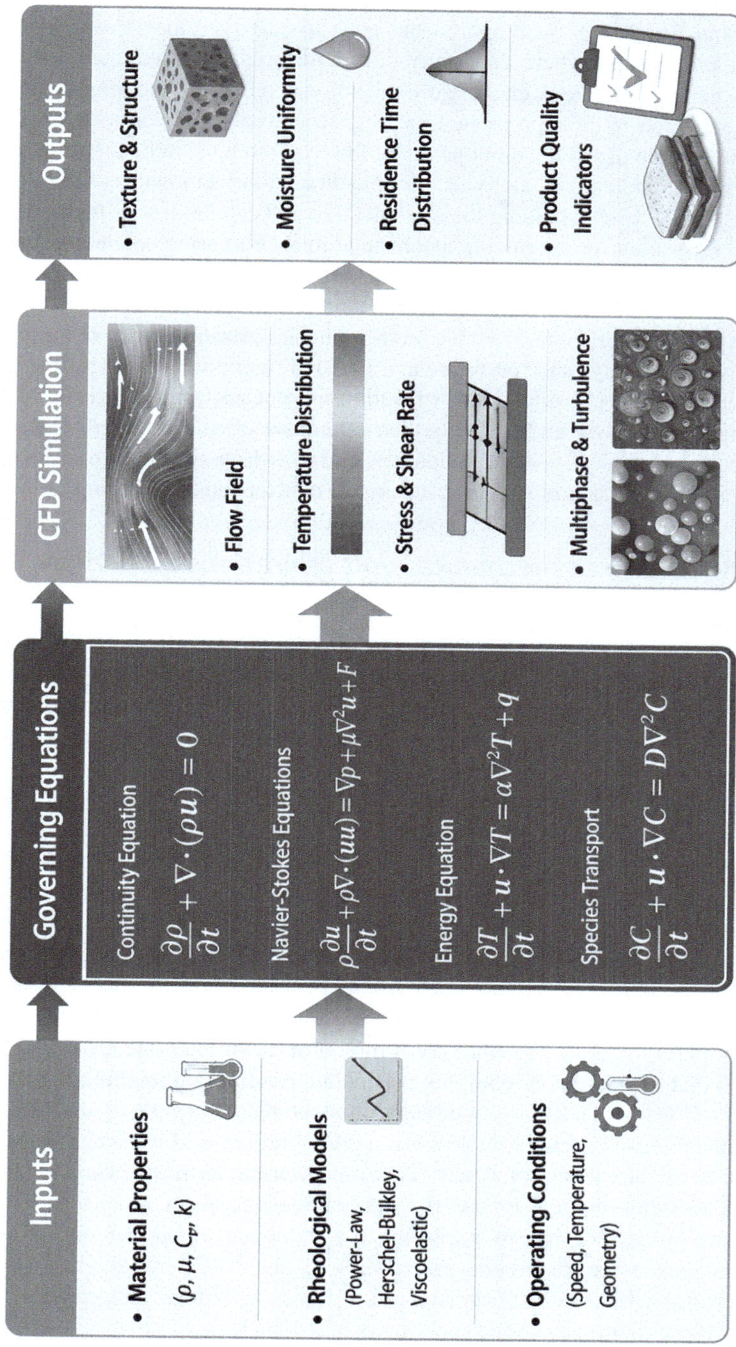

Figure 12.2 Overview of fundamental governing equations and rheological models applied in CFD simulations of mechanical food processing.

requirements, support real-time process monitoring, and enable adaptive control strategies (Chen and Wu 2021). Such approaches are particularly valuable in modern food processing environments where variability in raw materials, complex geometries, and dynamic operating conditions challenge conventional modeling techniques (Rahman 2024). The selection of an appropriate modeling strategy depends on multiple factors, including the complexity of the unit operation, the required level of detail, the availability of experimental data for model validation, and the computational resources at hand (Kong et al., 2022). Understanding the classification and capabilities of these modeling approaches is essential for optimizing mechanical food processing operations, ensuring product quality, and driving innovation (Yin et al., 2022). By systematically choosing and applying the appropriate modeling framework, researchers and industry practitioners can achieve efficient, reliable, and scalable process solutions, ultimately advancing sustainability and performance in food manufacturing (Kamat et al., 2023). In summary, the classification of modeling approaches from empirical and mechanistic to hybrid frameworks highlights the importance of selecting appropriate tools based on process complexity, data availability, and modeling objectives to achieve reliable prediction, optimization, and scale-up in mechanical food processing.

12.2.2 Modeling of mechanical unit operations involving particulates

Modeling mechanical unit operations that handle particulates is essential for improving process efficiency, especially in manufacturing and pharmaceutical industries. It uses computational methods to study how particulate materials behave during processing (Garofalo and Sansalone 2020). The key elements of this modeling approach are discussed in this section.

Mechanical processes with particulates, like grinding and conveying, are complex due to interactions between particles and equipment. Modeling these systems involves factors like particle size, shape, density, and mechanical traits. Further, discrete element models (DEM) is used to simulate individual particle mechanics, while continuum methods address the overall bulk behavior (Balliu et al., 2002; Garofalo and Sansalone 2020). CFD simulates fluid and particle dynamics during events like rainfall-runoff. It helps understand how particles separate and wash out. This supports better environmental and process management (Garofalo and Sansalone 2020). The dynamic modeling system described by Balliu et al. is an integrated, flexible framework developed for analyzing complex particulate processes like granulation (Balliu et al., 2002). It features a library of detailed models that incorporate particle size distribution, mechanisms such as nucleation, agglomeration, and breakage, and allows simulation at various levels of detail. This object-oriented, hierarchical system enables users to tailor models to specific applications, perform sensitivity analyses, and optimize process design by exploring the influence of operational parameters and mechanisms on system behavior (Balliu et al., 2002; Ingram and Cameron 2010). In a study, Balliu (2005) introduced an object-oriented, integrated modeling approach for granulation systems, enabling detailed analysis of processes like drying, size reduction, and crystallization. The Particulate Processing Library allows simulation at various levels of detail, from overall plant behavior to individual mechanisms, with models validated against real plant data. It includes regime-specific granulation models and detailed mechanisms such as nucleation, influenced by parameters like moisture and binder flow (Rogers et al., 2013). A notable contribution is the reaction-crystallization model for fertilizer granulation, which accurately predicts liquid phase changes and moisture effects, aiding process control and optimization. The approach promotes better understanding, control, and efficiency in particulate manufacturing industries (Balliu 2005). Multiscale models integrate population balance models (PBM) and DEM techniques to analyze powder behavior in solid handling. They link upstream processes, like purification, with downstream operations, such as tablet making, showing how physical properties affect quality (Sen 2015; Grimard et al., 2016). Mechanistic models analyze particle interactions in granulation, showing how operating parameters affect granule properties. They improve understanding of inhomogeneities, leading to better process control and product quality (Chaudhury 2015). Although these models offer valuable insights, challenges persist in applying them universally due to the complex nature of particulate behavior, requiring continuous research for refinement.

12.2.3 Modeling of mechanical processes for fluids and semi-solids

Modeling fluids and semi-solids involves complex phase interactions, requiring advanced computational methods. Recent research emphasizes diverse approaches to accurately simulate their behavior under various conditions. The upcoming sections highlight these key modeling strategies (Sun et al., 2017). Finite Element Methods include a level-set-based model that simulates semisolid behavior, highlighting the impact of solid fraction and microstructure on hot tearing. Accordingly, Sun et al. (2017) used a level-set finite element approach to simulate the mechanical behavior of semisolid materials with varying microstructures during small deformations (Sun et al., 2017). The model incorporated Voronoï diagrams to define phases, with the solid phase following viscoplastic laws and the liquid as a Newtonian fluid. Findings revealed that microstructure and solid fraction significantly influence the material's mechanical response and hot tearing formation, offering insights for semisolid material processing. Another approach, the reactive two-phase model, considers phase interactions in semisolid slurries, accounting for relative motion, with successful flow simulation outcomes (Pineau 2004). A study models the rheology of semisolid alloys using a two-phase flow approach, treating both phases as interpenetrating Newtonian fluids. Simulations reveal that the flow behavior depends on interactions and momentum exchange between solid and liquid phases. The model predicts nonNewtonian behavior arising from phase interactions despite both phases being Newtonian (Ludwig et al., 2000). A study by Kincl et al. (2023) presents a novel Smoothed Particle Hydrodynamics (SPH)-based approach that successfully discretizes the unified Symmetric Hyperbolic Thermodynamically Compatible equations (SHTC) equations for both fluids and solids, enabling accurate simulation of complex behaviors like fluid-solid coexistence (Kincl et al., 2023). The model effectively captures key mechanisms such as elasticity, viscoplasticity, and phase interactions, with optimal conditions including energy conservation and tensile instability treatment. In the study by Peshkova et al. (2020) a unified hyperbolic model effectively describes nonNewtonian viscoplastic flows, capturing solid-fluid transformations driven by stress (Peshkov et al., 2020). Unlike traditional models relying on nonlinear viscosity, this approach treats solids as deformable and fluids as "melted" solids through stress relaxation, enabling accurate simulation of yield stress behavior. The semiimplicit scheme used is especially suited for low-Mach number flows, with benchmarks like Couette and lid-driven cavity demonstrating its robustness. The model provides a thermodynamically consistent framework for predicting complex flow transitions, extending applicability to industrial and environmental processes involving solid-fluid interactions. Bigot and coworkers developed a combined experimental and numerical indentation method to characterize semisolid materials, particularly alloys like Sn–15% Pb, under high strains and temperatures (Bigot et al., 2005). The study highlights microstructure effects, showing that globular morphologies require lower stresses for deformation compared to dendritic structures, which favor higher load levels. The microstructure, strain rate, and temperature significantly influence the material's rheological response, with globular microstructures enhancing thixotropic properties. The methodology effectively links microstructure to flow behavior, but is limited to low strain rates and local deformations, prompting the development of tests for larger strain rate regimes. Early one-phase thermomechanical models struggled with solid-liquid transitions, but recent improvements have increased accuracy (Koeune and Ponthot 2008). Studies also developed thermo-mechanical models accounting for porosity and the interaction between solid deformation and liquid flow, improving understanding of semisolid alloys (Nguyen 1991).

12.3 Applications in major food processes: Mechanical modeling perspectives

This section covers modeling and simulation in food science, emphasizing model selection based on system knowledge and complexity. It discusses key food processing aspects and introduces various balanced, knowledge-based, and hybrid modeling approaches.

Various models assess how processing parameters influence food quality, with frying conditions critically affecting moisture loss and oil absorption. A two-dimensional transport model studied how frying parameters impact potato chip quality. The results showed that frying oil temperature significantly affects the outcome. This emphasizes the importance of controlling process conditions for optimal quality. The successful model demonstrated the significant effect of the frying oil temperature, the mass flow rate of raw potato slices, and the movement speed of material during frying on the final moisture content and oil content in the potato slice

(Wu et al., 2013). Predictive insights assist in optimizing deep fat frying parameters. Regression models help identify the most influential factors on responses. This approach improves process efficiency and product quality. Frying temperature and time are crucial parameters in deep-fat frying. They greatly affect the oil absorption of the final product. Proper control of these factors is essential for desired oil levels (Maneerote et al., 2009; Deng et al., 2019). Ingredient percentages and pressure or vacuum levels are additional important parameters. Regression analysis helps clarify their interactions. For example, their model explains how these factors influence outcomes. Maneerote et al. (2009) found that frying temperature, time, and fish powder ratio had minimal impact on moisture and oil content, with no significant interactions. However, increasing these factors slightly reduced moisture (Maneerote et al., 2009). In a study, Deng et al. (2019) used response surface methodology to optimize vacuum frying of shiitake slices, focusing on temperature, pressure, and time. Results showed these variables significantly affected quality, with lower oil content and better sensory attributes at specific conditions (Deng et al., 2019). Optimal settings were 105°C, −0.085 MPa, and 27.3 min, reducing oil uptake by over 33% while improving acceptability. The models accurately predicted outcomes, confirming the benefits of vacuum frying over traditional methods. The findings suggest vacuum frying can produce healthier, high-quality mushroom snacks. Future work may explore scaling up and refining process parameters for industrial applications.

Coating raw potatoes, like with alginate, can affect heat and mass transfer during frying. A 3D CFD model was used to analyze how this impacts moisture loss and oil absorption. The model provided insights into heat distribution across the batch. Pretreating potato strips with hydrocolloid coatings affects mass transfer properties. Coatings with 2% Arabic gum or carrageenan significantly increase moisture diffusivity. In contrast, 2% sodium alginate does not notably change moisture diffusion. These differences influence frying behavior and product quality (Naghavi et al., 2018a; 2018b). Increasing sodium alginate concentration from 1% to 3% influenced the evaporation rate but did not significantly affect moisture diffusivity. However, higher alginate levels effectively reduced oil absorption in fried products. CFD modeling helped understand how pretreatment can lower final oil content (Naghavi et al., 2018a; 2018b). This study developed a comprehensive 3D simulation model to analyze heat, momentum, and mass transfer during deep-fat frying of potato strips coated with various alginate concentrations. The model evaluated key variables such as water content, oil uptake, and temperature profiles at different fryer locations, and validated predictions with experimental data. Results showed that coating with alginate reduces moisture loss and oil absorption, with minimal effects from potato position within the fryer. The simulation accurately predicted temperature, moisture, and oil dynamics, aiding process optimization. Findings highlighted the importance of coating in controlling transfer phenomena and improving fried product quality.

CFD models analyze how different heating methods, such as conduction, convection, radiation, microwave, infrared, ultrasound, and pulsed electric fields, affect temperature and moisture loss. Various drying techniques, such as vacuum, freeze, spray, and impingement drying, are also studied to identify suitable methods. CFD simulations effectively model heat, mass transfer, and flow patterns in both traditional and innovative drying methods (Malekjani and Jafari 2018). Besides continuum and CFD models, ANN models excel at analyzing complex process parameter effects due to their ability to mimic the human brain, making them effective for handling nonlinearity and uncertainty. ANN models overcome parameter limitations of traditional observation-based approaches. In thermal processing, factors like temperature influence lethality, with pasteurization and sterilization showing three stages: heating, holding, and cooling. While transport and CFD models simulate product temperature profiles, handling irregular bell-shaped curves is better achieved with ANN models, which more effectively capture such complex patterns. Guldas et al. (2017) proposed a nonlinear autoregressive ANN with external inputs to model temperature profiles during sterilization, aiding in F-value estimation (Guldas et al., 2017). However, further validation of the model in actual food containers is still needed. Ghiasi and Golmakani (2022) also explored the development of starch and ethylcellulose-based bigels as healthier fat replacers in food, particularly low-fat burgers (Ghiasi and Golmakani 2022). These biphasic systems exhibit favorable mechanical, thermal, and rheological properties, remaining stable and viscoelastic at elevated temperatures. Incorporating bigels improves cooking quality and sensory attributes while reducing fat content. However, increased lipid oxidation was observed, indicating a need for antioxidants. The study highlights bigels' potential for creating healthier meat products and other foods, emphasizing their structural stability and functional versatility. Future research should focus on optimizing stability and expanding applications across various food types.

Modeling, through continuum and CFD methods, helps optimize food processing conditions. During frying, parameters influence moisture loss and oil absorption in food. A 2D transport model studied factors affecting fried potato chip quality in an industrial-scale system. The process used foul gas as the energy source for continuous production. Transient equations simulated the evolving conditions in continuous frying, including oil, potato size, and oil absorption. The energy balance between hot oil and potato slices was computed within the control volume of the frying system. The energy balance also modeled heat transfer between oil and air. The potato's temperature profile was predicted in 2D using this balance, accounting for vapourization-induced temperature changes (Wu et al., 2013). Water movement was described with a volume fraction continuity equation, while oil absorption was modeled through a diffusion equation with temperature-dependent diffusivity following Arrhenius law. A numerical approach in Visual C++ was employed to solve the complex equations. This enabled the optimization of processing conditions to achieve target moisture and oil levels, typically favoring low values for both.

A time-dependent kinetic model based on reaction mechanisms effectively predicts quality changes during processing and storage. The rate constant helps design process parameters to prevent quality deterioration, inhibit microbial growth, and improve efficiency. During storage, the kinetic rate constant indicates reactions that determine shelf life, affected by packaging and storage conditions. Using this model, it's possible to estimate product shelf life and optimal best-before dates. Soto et al. (2020) studied carotenoid breakdown in vacuum-fried papaya chips during storage, modeling the process kinetically (Fig. 12.3). Packing conditions significantly influenced the structure of the kinetic model (Soto et al., 2020). Carotenoids in

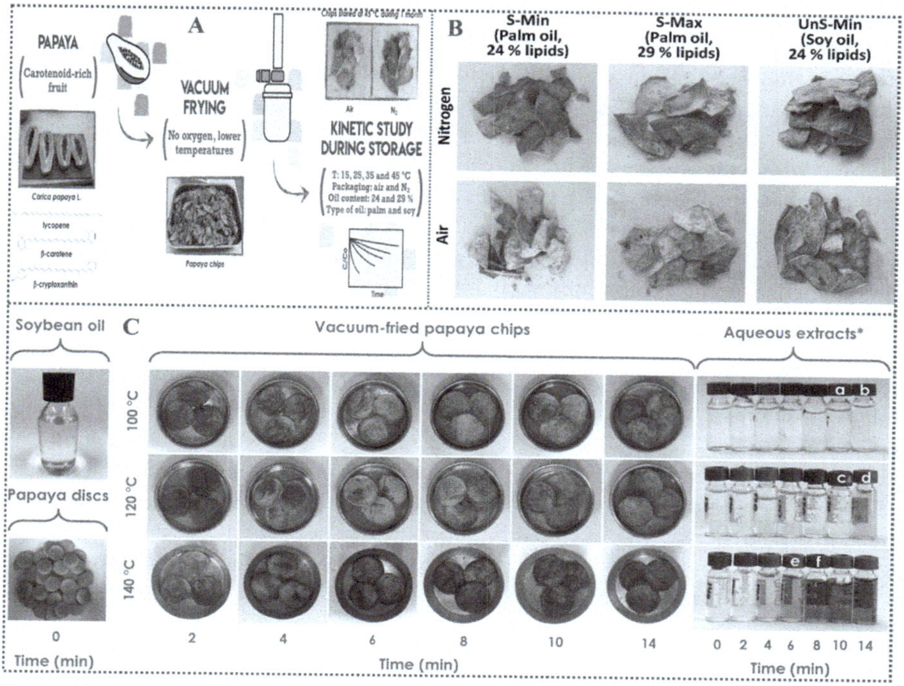

Figure 12.3 A) Schematic diagram of carotenoid degradation during storage of papaya chips, B) Papaya chips packaged under air or nitrogen conditions after 30 days of storage at 45°C (Soto et al., 2020), and C) Papaya discs, soybean oil, and papaya chips and their aqueous extracts obtained during vacuum frying at different temperatures (100–140°C) (Soto et al., 2021).
Reproduced with permission.

nitrogen-packed papaya chips remained stable over time, making kinetic modeling unsuitable. Conversely, carotenoid loss was rapid in air-packed chips and accurately modeled with a second-order kinetic equation. Storage temperature between 15 and 45°C followed the Arrhenius equation for carotenoid degradation, with higher temperatures shortening shelf life (Fig. 12.3). Additionally, the type of frying oil influenced the carotenoid's half-life. Frying with soy oil extended the half-life of certain carotenoids compared to palm oil. A kinetic model can guide the selection of frying oil, packing conditions, and storage temperature to enhance the shelf life of papaya chips. Soto and co-workers (2021) discuss the monitoring and modeling of physicochemical changes in papaya chips during vacuum frying, focusing on attributes like moisture, browning, sugar reactions, and carotenoid stability (Fig. 12.3) (Soto et al., 2021). It highlights how temperature and frying time influence quality parameters such as color, nutrient retention, and water activity. The study demonstrates that kinetic models can optimize frying conditions to produce healthy, visually appealing snacks with preserved nutritional value, especially carotenoids like β-carotene, lycopene, and β-cryptoxanthin. It emphasizes the importance of controlling moisture and temperature to minimize browning and nutrient degradation. In sterilization, actual temperature data are used to determine lethality and F-values, but predicted temperatures aid in designing retort conditions for new products.

A CFD model with FIDAP software simulated a continuous HTST sterilization process for research purposes. The model assumed laminar flow of a uniform fluid, but milk, with its low viscosity, typically exhibits turbulent flow in continuous sterilization. Enhancing radial mixing can help achieve better uniformity in turbulent milk processing. Jung and Fryer (1999) studied adding a static mixer to improve flow and temperature uniformity, but integrating it was complex due to cleaning requirements. CFD modeling was used to optimize the design process (Jung and Fryer 1999).

Many food products display complex flow behaviors like shear-thinning or viscoelasticity, deviating from Newtonian assumptions. Precise modeling of these rheological properties is crucial for realistic CFD simulations of processes like mixing and shaping. Accurately integrating nonNewtonian rheology in CFD models is vital for predicting pressure, shear, residence time, energy, and temperature accurately. Ignoring these behaviors can result in flawed process design and compromised product quality. Food products often exhibit nonNewtonian traits: shear-thinning in sauces and dairy emulsions, where viscosity drops with shear, and shear-thickening in concentrated starch slurries, where viscosity rises with shear. Yield stress fluids like ketchup and chocolate need a minimum force to flow, while thixotropic materials decrease in viscosity over time under shear. Viscoelastic foods, such as gels and gluten, show both elastic and viscous behaviors (Hicks et al., 2012; Martínez-Padilla 2024). Mechanical modeling improves food processing, but challenges exist in accuracy, scale-up, and consistency. Future work should aim to combine these models with real-time data for better predictions and control.

12.4 Emerging trends and future directions

This section discusses how these trends can enhance predictive capabilities, enable smarter process control, and facilitate the design of novel processing equipment (Fig. 12.4). Recent advances in computing power and data analytics are transforming mechanical process modeling. The integration of CFD, DEM, and finite element methods allows for multiphase and multiscale modeling of complex systems. Data-driven methods like machine learning and digital twins enhance real-time monitoring and optimization alongside traditional models. Future trends focus on better material data, hybrid physics-data models, and increased simulation for sustainable processes. Sensitivity analysis and parameter estimation help identify key variables and calibrate models using experimental data. Combining these with CFD improves accuracy, reduces uncertainty, and enhances process optimization and control. These methods are crucial in food systems with variable, hard-to-measure properties like viscosity and heat transfer. They help characterize complex, coupled phenomena such as phase changes and biochemical reactions (Kudryashova et al., 2023). Integrating these techniques improves model accuracy and process control. Despite their benefits, sensitivity analysis and parameter estimation in CFD face challenges like high

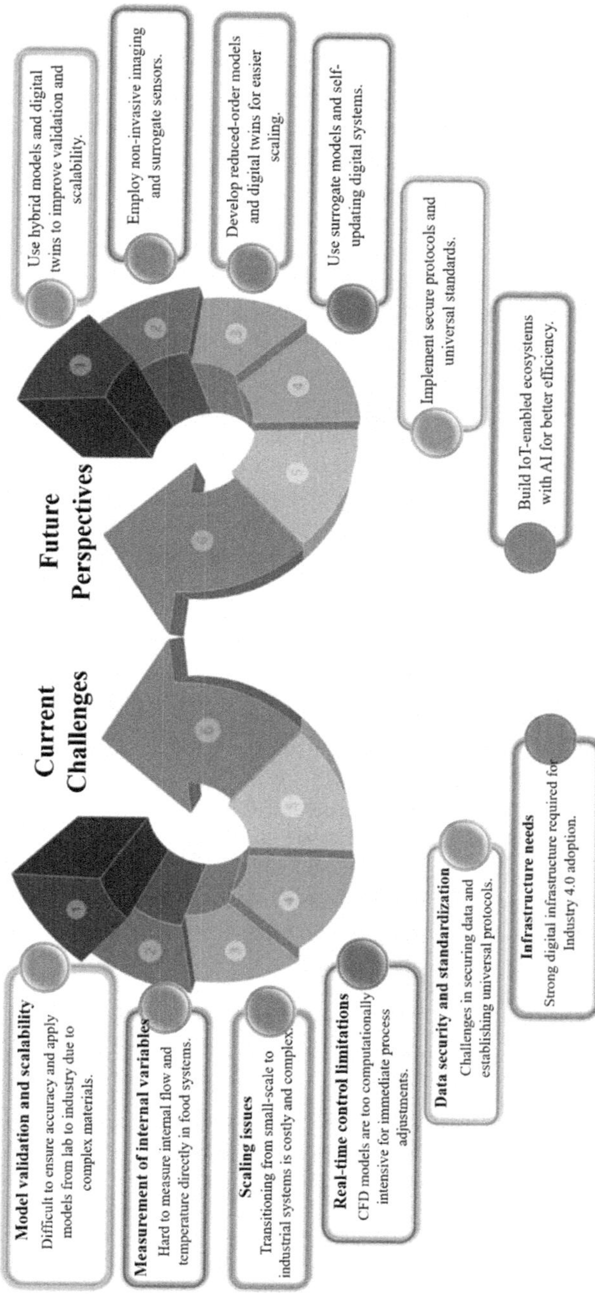

Figure 12.4 Challenges and future trends in mechanical process modeling.

computational demands, which can be mitigated with surrogate models. They also encounter identifiability issues, where different parameters produce similar results, needing regularization. Further, noisy or uncertain experimental data make accurate parameter estimation difficult (Ramirez-Asis et al., 2022; Luo and Koksel 2025).

Effective methods involve verifying mesh and time-step independence, limiting parameters with existing data, and using cross-validation to ensure model robustness. Future efforts focus on real-time parameter estimation with sensor data, integrating uncertainty analysis, and using machine learning for faster surrogate modeling. Open-source tools combining CFD solvers with libraries like SALib or UQLab will enhance accessibility. These advances aim to improve the accuracy and efficiency of data-driven CFD models in food engineering (Norton et al., 2013; Szpicer et al., 2025).

To enhance validation and scalability, hybrid models combining CFD with experiments, correlations, or machine learning are used. Testing subsystems separately improves accuracy, while sensitivity analysis and Bayesian methods help assess uncertainties and model robustness. Design of Experiments (DoE) offers a structured way to validate models across various conditions, while imaging techniques like MRI, PIV, and X-ray CT enable detailed measurements in research. Emerging digital twins allow real-time data to update CFD models, supporting industrial-scale validation. Future improvements include standardized property databases, automated meshing, and cloud GPU simulations. Physics-informed machine learning could enable real-time model updates and adaptive control. Addressing validation and implementing these systems needs a strong digital infrastructure, including IoT devices, edge computing, cloud storage, and communication protocols like OPC UA or MQTT. These enable seamless data exchange between simulations and control systems like SCADA or MES (Kannapinn et al., 2022; Abdurrahman and Ferrari 2025).

Addressing scaling issues is crucial for wider CFD adoption. Fully integrated models support predictive quality, scalable processes, and sustainability. In food processing, digital twins monitor spray drying, extrusion, mixing, and heating with real-time sensors, allowing operators to analyze, predict, and optimize in real time. Industrial gains include better product consistency, lower energy use, early fault detection, adaptability to raw material changes, and faster scale-up with virtual prototypes (Verboven et al., 2020; Krupitzer et al., 2022). These align with Industry 4.0 principles like interoperability, transparency, and cyber-physical integration. CFD models are too intensive for real-time, so surrogate approaches are used, requiring ongoing recalibration. Standardized integration protocols are under development, with cybersecurity being essential. Future trends include hybrid models combining physics and AI, self-updating digital twins, and sustainability metrics like CO_2 and energy use. Integrating CFD with real-time control will drive smarter, more adaptable, and eco-friendly food manufacturing processes (Szpicer et al., 2025).

12.5 Challenges and opportunities

Mechanical process modeling faces issues like high computational demands, scarce material data, and validation challenges, especially due to the variable behavior of complex food materials. These challenges open opportunities for innovation through advanced sensors, techniques, and numerical methods, with collaboration across disciplines essential for progress.

The application of CFD in food engineering faces limitations due to a lack of precise thermophysical data. Many food materials have properties such as thermal conductivity and diffusivity that vary with moisture, temperature, and structure. These composition-dependent variations complicate accurate modeling (Srikiatden and Roberts 2007; Carson 2015). Addressing this data gap is essential for improving CFD predictions in complex food systems. Many variables are poorly understood, hindering reliable modeling. While energy and momentum equations are established in processes like extrusion and drying, water transfer and mass transfer are less well characterized. Advanced CFD models need to include moisture sorption, complex diffusion, and evaporation kinetics, which are often lacking for real foods (Takhar 2008; Renzetti et al., 2012; Szpicer et al., 2025). Phase transitions like glass transition (Tg) also impact mass transfer and product properties, but are rarely integrated into CFD, despite their importance in predicting texture and structural changes during drying. In spray drying, surpassing Tg causes wall buildup, while in extrusion, Tg-related transitions affect melt viscosity and texture (Renzetti et al., 2012). CFD also struggles to directly predict sensory qualities like crispness or chewiness due to the absence of detailed texture-structure-process models, relying instead on indirect correlations such as porosity and crunchiness. Overcoming these challenges needs enhanced food property databases, improved CFD-experimental data integration (e.g., NIR, MRI), and combined multiphysics models for chemical, mechanical, and sensory aspects (Chanvrier et al., 2014; Joardder et al., 2018).

Multiphysics and CFD modeling in the food industry face significant hurdles such as high computational demands, limited experimental data, and difficulties in model validation. Addressing multiphase and thermal issues, especially around complex regions like screws and dies, requires fine meshes that increase computation time. Compatibility between different software platforms and access to internal process data also pose challenges. Additionally, accurately estimating parameters like activation energy and viscoelastic moduli needs advanced experimental measurements, which are often limited or costly (Szpicer et al., 2025). CFD in food engineering faces challenges in validation, generalization, and scaling. Reliable models need accurate data, realistic boundaries, and validation across scales. Bridging lab models to industrial use remains a key goal for researchers and industry (Rauh and Delgado 2015; Szpicer et al., 2025). Validating models in food systems is challenging because many materials are opaque, sticky, or heat-sensitive, preventing direct measurement of internal variables like velocity, temperature, or concentration. Model validation in food systems is challenging due to material complexity and limited internal measurements. Validation often relies on indirect metrics like temperature and quality attributes. Spatial validation, especially for flow fields, is mostly feasible at small scales (Kannapinn et al., 2022; Chai et al., 2024). This limits the accuracy of models for full-scale industrial applications. Scaling models from lab to industry is difficult due to increased geometry, changing flow regimes, and complex dynamics. Large equipment introduces control issues, wear, and operator effects, making high-resolution simulations costly and less practical for real-time use. Additionally, standard scaling laws may not hold in nonNewtonian or multiphase systems (Xia and Sun 2002; Norton and Sun 2006).

12.6 Conclusion

Modeling serves as a crucial tool for linking process variables with outcomes, enabling enhancements in efficiency and product quality based on specific objectives. This chapter underscores the pivotal role of mechanical process modeling in advancing food engineering, emphasizing its capacity to optimize processing operations, improve product quality, and foster sustainability. The integration of sophisticated computational tools such as CFD, DEM, FEA, and hybrid models with machine learning and digital twin technologies has opened new frontiers in real-time monitoring, predictive maintenance, and adaptive control. These innovations facilitate detailed insights into complex phenomena like fluid flow, heat and mass transfer, and material deformation, which are crucial for processes like extrusion, drying, and frying. Nonetheless, challenges persist, including high computational costs, limited material property data, and validation difficulties due to the complex, variable nature of food systems. Addressing these hurdles requires enhanced experimental techniques, comprehensive food property databases, and advances in multiphysics modeling. Future perspectives point toward the development of more accurate, scalable, and user-friendly models, coupled with real-time sensor data and AI-driven analytics. The emergence of digital twins and cyber-physical systems will enable smarter, more flexible, and sustainable food manufacturing processes aligned with Industry 4.0 principles. Continued interdisciplinary collaboration, standardization, and technological innovation are essential to fully harness the potential of mechanical process modeling, ultimately leading to improved efficiency, product consistency, and environmental sustainability in the food industry. Modeling and optimization can significantly decrease costs and time by minimizing reliance on trial-and-error testing. However, to ensure their reliability and broad applicability, these predictive models must be validated using independent data not involved in their development, thus confirming their accuracy and optimal performance.

References

Abdurrahman, E.E.M., Ferrari, G., 2025. Digital Twin applications in the food industry: A review. Front. Sustain. Food Syst. 9, 1538375.

Abideen, A.Z., Sundram, V.P.K., Pyeman, J., Othman, A.K., Sorooshian, S., 2021. Food supply chain transformation through technology and future research directions—a systematic review. Logistics 5 (4), 83.

Balliu, N., Cameron, I., Newell, R., 2002. An integrated dynamic modelling and simulation system for analysis of particulate processes. Comput. Aided Chem. Eng. 10, 427–432.

Balliu, N.E., 2005. An object oriented approach to the modelling and dynamics of granulation circuits.
Barbaros, I., Yang, Y., Safaei, B., Yang, Z., Qin, Z., Asmael, M., 2022. State-of-the-art review of fabrication, application, and mechanical properties of functionally graded porous nanocomposite materials. Nanotechnol. Rev. 11 (1), 321–371.
Beć, K.B., Grabska, J., Huck, C.W., 2022. Miniaturized NIR spectroscopy in food analysis and quality control: Promises, challenges, and perspectives. Foods 11 (10), 1465.
Bidyalakshmi, T., Jyoti, B., Mansuri, S.M., Srivastava, A., Mohapatra, D., Kalnar, Y.B., Narsaiah, K., Indore, N., 2025. Application of artificial intelligence in food processing: Current status and future prospects. Food Eng. Rev. 17 (1), 27–54.
Bigot, R., Favier, V., Rouff, C., 2005. Characterisation of semi-solid material mechanical behaviour by indentation test. J. Mater. Process. Technol. 160 (1), 43–53.
Carson, J.K., 2015. Thermal conductivity measurement and prediction of particulate foods. Int. J. Food Prop. 18 (12), 2840–2849.
Chai, J., Su, H., Li, X., Zhang, Y., Xiao, D., Kang, J., Dong, A., 2024. Computational fluid dynamics (CFD) modeling and application for cleaning of food-contact surfaces: A review. J. Food Process. Eng. 47 (4), e14605.
Chang, H., Xu, J., Macqueen, L.A., Aytac, Z., Peters, M.M., Zimmerman, J.F., Xu, T., Demokritou, P., Parker, K.K., 2022. High-throughput coating with biodegradable antimicrobial pullulan fibres extends shelf life and reduces weight loss in an avocado model. Nat. Food 3 (6), 428–436.
Chanvrier, H., Jakubczyk, E., Gondek, E., Gumy, J.-C., 2014. Insights into the texture of extruded cereals: Structure and acoustic properties. Innov. Food Sci. Emerg. Technol. 24, 61–68.
Chaudhury, A., 2015. Mechanistic modeling, simulation and optimization of wet granulation processes. Rutgers The State University of New Jersey, School of Graduate Studies.
Chen, L., Wu, B., 2021. Research progress in computational fluid dynamics simulations of membrane distillation processes: A review. Membranes 11 (7), 513.
Delfiya, D.A., Prashob, K., Murali, S., Alfiya, P., Samuel, M.P., Pandiselvam, R., 2022. Drying kinetics of food materials in infrared radiation drying: A review. J. Food Process. Eng. 45 (6), e13810.
Deng, K., Chen, J., Tian, Y., Miao, S., Zheng, B., 2019. Optimization of process variables on physical and sensory attributes of shiitake (Lentinula edodes) slices during vacuum frying. Innov. Food Sci. Emerg. Technol. 54, 162–171.
Emin, M., Wittek, P., Schwegler, Y., 2021. Numerical analysis of thermal and mechanical stress profile during the extrusion processing of plasticized starch by non-isothermal flow simulation. J. Food Eng. 294, 110407.
Eslami, Z., Elkoun, S., Robert, M., Adjallé, K., 2023. A review of the effect of plasticizers on the physical and mechanical properties of alginate-based films. Molecules 28 (18), 6637.
Garofalo, G., Sansalone, J., 2020. Modeling annual particulate matter separation and washout by unit operations with CFD. J. Environ. Eng. 146 (1), 04019101.
Ghiasi, F., Golmakani, M.-T., 2022. Fabrication and characterization of a novel biphasic system based on starch and ethylcellulose as an alternative fat replacer in a model food system. Innov. Food Sci. Emerg. Technol. 78, 103028.
Grimard, J., Dewasme, L., Vande Wouwer, A., 2016. A review of dynamic models of hot-melt extrusion. Processes 4 (2), 19.
Guldas, M., Kurtulmus, F., Gurbuz, O., 2017. Prediction of Lethality by nonlinear artificial neural network modeling. J. Food Process. Eng. 40 (3), e12457.

Gupta, D., Priyadarshi, R., Tammina, S.K., Rhim, J.-W., Agrawal, G., 2025. Fruit processing wastes as sustainable sources to produce multifunctional carbon quantum dots for application in active food packaging. Food Bioprocess Technol. 18 (3), 2145–2169.

Hicks, C.I., See, H., Fletcher, D.F., Ekwebelam, C., 2012. The shear rheology of bread dough: Analysis of local flow behaviour using CFD. Food Bioprod. Process. 90 (3), 361–369.

Hu, X., Zeng, Z., Zhang, J., Wu, D., Li, H., Geng, F., 2023. Molecular dynamics simulation of the interaction of food proteins with small molecules. Food Chem. 405, 134824.

Iftekar, S.F., Aabid, A., Amir, A., Baig, M., 2023. Advancements and limitations in 3D printing materials and technologies: A critical review. Polymers 15 (11), 2519.

Ingram, G.D., Cameron, I.T., 2010. Dynamic multiscale modeling–an application to granulation processes. Process Syst. Eng.: Vol. 7: Dyn. Process Model. 7, 35–65.

Islam, R., Shahriar, S., Jiang, L., Peng, X., Park, J., 2024. Effect of process parameters on mechanical properties of the 3D printed silk-PLA specimens fabricated via fused deposition modeling. In: ASME International Mechanical Engineering Congress and Exposition. American Society of Mechanical Engineers.

Joardder, M.U., Kumar, C., Karim, M., 2018. Prediction of porosity of food materials during drying: Current challenges and directions. Crit. Rev. Food Sci. Nutr. 58 (17), 2896–2907.

Jung, A., Fryer, P., 1999. Optimising the quality of safe food: Computational modelling of a continuous sterilisation process. Chem. Eng. Sci. 54 (6), 717–730.

Kamat, S., Cai, W., Rinker, T.J., Bracey, J., Xi, L., Tan, W., 2023. A novel integrated process-performance model for laser welding of multi-layer battery foils and tabs. J. Mater. Process. Technol. 320, 118121.

Kannapinn, M., Pham, M.K., Schäfer, M., 2022. Physics-based digital twins for autonomous thermal food processing: Efficient, non-intrusive reduced-order modeling. Innov. Food Sci. Emerg. Technol. 81, 103143.

Khan, M.I.H., Sablani, S.S., Nayak, R., Gu, Y., 2022. Machine learning-based modeling in food processing applications: State of the art. Compr. Rev. Food Sci. Food Saf. 21 (2), 1409–1438.

Khashkhashi-Moghadam, S., Ezazi-Toroghi, S., Kamkar-Vatanparast, M., Jouyaeian, P., Mokaberi, P., Yazdyani, H., Amiri-Tehranizadeh, Z., Saberi, M.R., Chamani, J., 2022. Novel perspective into the interaction behavior study of the cyanidin with human serum albumin-holo transferrin complex: Spectroscopic, calorimetric and molecular modeling approaches. J. Mol. Liq. 356, 119042.

Kincl, O., Peshkov, I., Pavelka, M., Klika, V., 2023. Unified description of fluids and solids in smoothed particle Hydrodynamics. Appl. Math. Comput. 439, 127579.

Koeune, R., Ponthot, J.P., 2008. A one phase thermomechanical model for semi-SOLID thixoforming. Solid State Phenom. 141, 629–635.

Kong, X., Liu, J., Yang, T., Su, Y., Geng, J., Niu, Z., 2022. Numerical simulation of feed pellet breakage in pneumatic conveying. Biosyst. Eng. 218, 31–42.

Krupitzer, C., Noack, T., Borsum, C., 2022. Digital food twins combining data science and food science: System model, applications, and challenges. Processes 10 (9), 1781.

Kudryashova, O., Toropkov, N., Lerner, M., Promakhov, V., Vorozhtsov, A., Mironov, E., 2023. Mathematical modeling of high-energy materials rheological behavior in 3D printing technology. Heliyon 9 (1).

Ludwig, A., Wu, M., Hofmeister, T., 2000. Simulation of the Rheology of semisolid alloys by using a two-phase flow approach, 652. MRS Online Proceedings Library (OPL). Y5. 6.

Luo, S., Koksel, F., 2025. Real-time thermal imaging of expansion dynamics during extrusion of protein-fortified snacks: Effects of nitrogen gas and protein concentration. Food Res. Int. 199, 115349.

Malekjani, N., Jafari, S.M., 2018. Simulation of food drying processes by Computational Fluid Dynamics (CFD); Recent advances and approaches. Trends Food Sci. Technol. 78, 206–223.
Mancuso, I., Petruzzelli, A.M., Panniello, U., 2023. Innovating agri-food business models after the Covid-19 pandemic: The impact of digital technologies on the value creation and value capture mechanisms. Technol. Forecast. Soc. Change 190, 122404.
Maneerote, J., Noomhorm, A., Takhar, P.S., 2009. Optimization of processing conditions to reduce oil uptake and enhance physico-chemical properties of deep fried rice crackers. LWT-Food Sci. Technol. 42 (4), 805–812.
Marano, S., Laudadio, E., Minnelli, C., Stipa, P., 2022. Tailoring the barrier properties of PLA: A state-of-the-art review for food packaging applications. Polymers 14 (8), 1626.
Marchand, T., Pey, B., Pautot, C., Lecerf, A., 2024. Mechanical traits as drivers of trophic interaction between macrodetritivores and leaf litter. Oecologia 204 (3), 641–651.
Martínez-Padilla, L.P., 2024. Rheology of liquid foods under shear flow conditions: Recently used models. J. Texture Stud. 55 (1), e12802.
Mora, L., Toldrá, F., 2023. Advanced enzymatic hydrolysis of food proteins for the production of bioactive peptides. Curr. Opin. Food Sci. 49, 100973.
Nadimi, M., Divyanth, L., Chaudhry, M.M.A., Singh, T., Loewen, G., Paliwal, J., 2023. Assessment of mechanical damage and germinability in flaxseeds using hyperspectral imaging. Foods 13 (1), 120.
Naghavi, E.-A., Dehghannya, J., Ghanbarzadeh, B., 2018a. 3D computational simulation for the prediction of coupled momentum, heat and mass transfer during deep-fat frying of potato strips coated with different concentrations of alginate. J. Food Eng. 235, 64–78.
Naghavi, E.-A., Dehghannya, J., Ghanbarzadeh, B., 2018b. Effect of hydrocolloid type on transfer phenomena during deep-fat frying of coated potato strips: Numerical modeling and experimental analysis. Comput. Electron. Agric. 154, 382–399.
Narsaiah, K., Bedi, V., Ghodki, B.M., Goswami, T.K., 2021. Heat transfer modeling of shrimp in tunnel type individual quick freezing system. J. Food Process. Eng. 44 (11), e13838.
Ng, W.L., Goh, G.L., Goh, G.D., Ten, J.S.J., Yeong, W.Y., 2024. Progress and opportunities for machine learning in materials and processes of additive manufacturing. Adv. Mater. 36 (34), 2310006.
Nguyen, T.G., 1991. Modélisation du comportement rhéologique d'alliages à l'état semi-solide, Grenoble INPG.
Norton, T., Sun, D.-W., 2006. Computational fluid dynamics (CFD)–an effective and efficient design and analysis tool for the food industry: A review. Trends Food Sci. Technol. 17 (11), 600–620.
Norton, T., Tiwari, B., Sun, D.-W., 2013. Computational fluid dynamics in the design and analysis of thermal processes: A review of recent advances. Crit. Rev. Food Sci. Nutr. 53 (3), 251–275.
Oyinloye, T.M., Yoon, W.B., 2021. Application of computational fluid dynamics (CFD) simulation for the effective design of food 3D printing (A review). Processes 9 (11), 1867.
Pandiselvam, R., Prithviraj, V., Kothakota, A., Raj, A.S., Pulivarthi, M.K., Siliveru, K., 2021. Numerical methods and modeling techniques in food processing. In: Handbook of Research on Food Processing and Preservation Technologies. Apple Academic Press, pp. 221–255.
Pathare, P.B., Al-Dairi, M., 2022. Effect of mechanical damage on the quality characteristics of banana fruits during short-term storage. Discov. Food 2 (1), 4.
Peshkov, I., Dumbser, M., Boscheri, W., Romenski, E., Chiocchetti, S., Ioriatti, M., 2020. Modeling solid-fluid transformations in non-Newtonian viscoplastic flows with a unified

flow theory. Comput. Fluids. 224, 104963. https://doi.org/10.1016/j.compfluid.2021.104963.
Pineau, F., 2004. Reactive two-phase flow model for materials in the semi-solid state. WIT Press.
Rahman, M.S., 2024. Computational fluid dynamics for predicting and controlling fluid flow in industrial equipment. Eur. J. Adva. Eng. Technol. 11 (9), 1–9.
Ramirez-Asis, E., Vilchez-Carcamo, J., Thakar, C.M., Phasinam, K., Kassanuk, T., Naved, M., 2022. A review on role of artificial intelligence in food processing and manufacturing industry. Mater. Today Proc. 51, 2462–2465.
Rauh, C., Delgado, A., 2015. Different modelling and simulation approaches for food processing operations. In: Modeling food processing operations. Elsevier, pp. 3–33.
Renzetti, S., Voogt, J., Oliver, L., Meinders, M., 2012. Water migration mechanisms in amorphous powder material and related agglomeration propensity. J. Food Eng. 110 (2), 160–168.
Rogers, A.J., Hashemi, A., Ierapetritou, M.G., 2013. Modeling of particulate processes for the continuous manufacture of solid-based pharmaceutical dosage forms. Processes 1 (2), 67–127.
Said, N.S., Sarbon, N.M., 2022. Physical and mechanical characteristics of gelatin-based films as a potential food packaging material: A review. Membranes 12 (5), 442.
Sen, M., 2015. Multiscale modeling and validation of particulate processes. Rutgers The State University of New Jersey, School of Graduate Studies.
Soto, M., Dhuique-Mayer, C., Servent, A., Jiménez, N., Vaillant, F., Achir, N., 2020. A kinetic study of carotenoid degradation during storage of papaya chips obtained by vacuum frying with saturated and unsaturated oils. Food Res. Int. 128, 108737.
Soto, M., Pérez, A.M., Servent, A., Vaillant, F., Achir, N., 2021. Monitoring and modelling of physicochemical properties of papaya chips during vacuum frying to control their sensory attributes and nutritional value. J. Food Eng. 299, 110514.
Srikiatden, J., Roberts, J.S., 2007. Moisture transfer in solid food materials: A review of mechanisms, models, and measurements. Int. J. Food Prop. 10 (4), 739–777.
Sun, Z., Bernacki, M., Logé, R., Gu, G., 2017. Numerical simulation of mechanical deformation of semi-solid material using a level-set based finite element method. Model. Simulat. Mater. Sci. Eng. 25 (6), 065020.
Szpicer, A., Bińkowska, W., Stelmasiak, A., Wojtasik-Kalinowska, I., Czajkowska, A., Mierzejewska, S., Domiszewski, Z., Rydzkowski, T., Piepiórka-Stepuk, J., Półtorak, A., 2025. Advances in computational fluid dynamics of mechanical processes in food engineering: Mixing, extrusion, drying, and process optimization. Appl. Sci. 15 (15), 8752.
Takhar, P.S., 2008. Role of glass-transition on fluid transport in porous food materials. Int. J. Food Eng. 4 (7).
Taneja, A., Nair, G., Joshi, M., Sharma, S., Sharma, S., Jambrak, A.R., Roselló-Soto, E., Barba, F.J., Castagnini, J.M., Leksawasdi, N., 2023. Artificial intelligence: Implications for the agri-food sector. Agronomy 13 (5), 1397.
Verboven, P., Defraeye, T., Datta, A.K., Nicolai, B., 2020. Digital twins of food process operations: the next step for food process models? Curr. Opin. Food Sci. 35, 79–87.
Wang, Z., Hirai, S., Kawamura, S., 2022. Challenges and opportunities in robotic food handling: A review. Front. Robot. AI 8, 789107.
Wu, H., Karayiannis, T.G., Tassou, S.A., 2013. A two-dimensional frying model for the investigation and optimisation of continuous industrial frying systems. Appl. Therm. Eng. 51 (1–2), 926–936.

Xia, B., Sun, D.-W., 2002. Applications of computational fluid dynamics (CFD) in the food industry: A review. Comput. Electron. Agric. 34 (1–3), 5–24.
Yazdani, M., Pamucar, D., Chatterjee, P., Torkayesh, A.E., 2022. A multi-tier sustainable food supplier selection model under uncertainty. Oper. Manag. Res. 15 (1), 116–145.
Yin, J., Guo, M., Liu, G., Ma, Y., Chen, S., Jia, L., Liu, M., 2022. Research progress in simultaneous heat and mass transfer of fruits and vegetables during precooling. Food Eng. Rev. 14 (2), 307–327.

Index

'*Note*: Page numbers followed by "f" indicate figures and "t" indicate tables.'